$$V^2 = \omega^2 R^2$$

pressure

Rotation _____ Bernoulli

$$\frac{P_1}{\gamma} + z_1 - \frac{V^2}{2g}$$

$$u = \mu \frac{y}{a}^{1/2} = V$$

$$G = \int u\, da = \int u\, width\, dy$$

$$\tan \alpha = \frac{a_x}{g}$$

★

$$\tan \alpha = \frac{a_x}{g}$$

$$\begin{array}{r} 18.5 \\ \underline{18.0} \\ 50 \\ 40 \end{array}$$

92.5

$$V = \omega^2 d^2 \qquad 2\omega$$

$D = 9''$
$r_{cph} = 460$

$\Sigma F_y = 0$

$T = 101.7 + 10.61 + 8.29$

$T = 101.7 - 10.6 - 8.21$

$r = 75$

$8''$

$r = 96$

$10''$

$$\boxed{\gamma = \rho g}$$

$$\omega = \forall r = \frac{4}{3}\pi (4.5)^2 \, 460 = 101.7$$
$$\frac{1}{2} 96 = 10.61$$
$$\frac{1}{2} 75 = 8.29$$

$$\rho = \frac{P}{RT}$$

$$\gamma = \rho g$$

$$W = \gamma \forall$$

$$\tau = \text{Sheer stress} = \mu \frac{dVel}{dy} \qquad \mu = \text{dynamic viscosity}$$

$$E_V = \frac{-dP}{d\forall/\forall} = \frac{dP}{dP/\rho}$$

sp. gr. merc = 13.6 $\rightarrow \gamma_{merc} = 13.6 \, \gamma_{H_2O}$

sp. vol = $v = \frac{1}{\rho}$

$$0 = \dot{m}\left(\frac{1}{2}V_1^2 + gz_1 + \frac{P_1}{\rho} - \frac{1}{2}V_2^2 - gz_2 - \frac{P_2}{\rho} - gh_a\right) + \dot{W}_s$$

$$\frac{dm}{dt} = \dot{m}_1 - \dot{m}_2 = -\rho A \frac{dh}{dt}$$

$$0 = \dot{m}(\bar{v}_1 - \bar{v}_2) - W\hat{k} - (P_1 - P_{atm})(A_1 \hat{m}_1) - (P_2 - P_{atm})(A_2 \hat{m}_2) + \bar{F}_s$$

Along streamline $\rightarrow \frac{1}{2}V_1^2 + \frac{P_1}{\rho} + gz_1 = \frac{1}{2}V_2^2 + \frac{P_2}{\rho} + gz_2$

Along $\perp$ to str. line $\rightarrow \frac{P_1}{\rho} + gz_1 = \frac{P_2}{\rho} + gz_2$.

$$Q = VA$$

$$\dot{m} = \rho Q$$

ENGINEERING
FLUID MECHANICS

$P = \gamma h$

$P = F/A$

$F = P A$

ENGINEERING FLUID MECHANICS

SECOND EDITION

John A. Roberson and Clayton T. Crowe

Washington State University, Pullman

Houghton Mifflin Company   Boston
Dallas   Geneva, Illinois   Hopewell, New Jersey
Palo Alto   London

CREDITS

Cover photo by Theodor Schwenk from his book *The Sensitive Chaos* (copyright © 1965, Rudolf Steiner Press, London, and Verlag Freies Geistesleben, Stuttgart)

Photographs

Printed in the U.S.A.
Library of Congress Catalog Card Number: 79-87855
ISBN: 0-395-28357-4

CONTENTS

PREFACE   ix

CHAPTER ONE   INTRODUCTION   2

1-1   Fluids   4
1-2   Flow Classification   5
1-3   Historical Note   6
1-4   Significance of Fluid Mechanics   7

CHAPTER TWO   FLUID PROPERTIES   10

2-1   Basic Units   12
2-2   System, Extensive and Intensive Properties   13
2-3   Properties Involving the Mass or Weight of the Fluid   14
2-4   Properties Involving the Flow of Heat   16
2-5   Viscosity   16
2-6   Elasticity   20
2-7   Surface Tension   21
2-8   Vapor Pressure   23

CHAPTER THREE   FLUID STATICS   28

3-1   Pressure   30
3-2   Pressure Variation with Elevation   34
3-3   Pressure Measurements   41
3-4   Hydrostatic Forces on Plane Surfaces   46
3-5   Hydrostatic Force on Curved Surfaces   52
3-6   Buoyancy   55
3-7   Stability of Immersed and Floating Bodies   58

CHAPTER FOUR   FLUIDS IN MOTION   84

4-1   Velocity and Flow Visualization   86
4-2   Rate of Flow   96
4-3   Acceleration   100

4-4   Basic Control-Volume Approach   103
4-5   Continuity Equation   110
4-6   Rotation and Vorticity   121
4-7   Separation   124

CHAPTER FIVE   PRESSURE VARIATION IN
               FLOWING FLUIDS   140

5-1   Basic Causes of Pressure Variation in a Flowing Fluid   142
5-2   Examples of Pressure Variation Resulting from Acceleration   144
5-3   The Bernoulli Equation   149
5-4   Application of the Bernoulli Equation   151
5-5   Separation and its Effect on Pressure Variation   161
5-6   Cavitation   163

CHAPTER SIX   MOMENTUM PRINCIPLE   180

6-1   The Momentum Equation   182
6-2   Applications of the Momentum Equation   188
6-3   Moment-of-Momentum Equation   205

CHAPTER SEVEN   ENERGY PRINCIPLE   232

7-1   Derivation of the Energy Equation   234
7-2   Discussion of the Energy Equation   238
7-3   Simplified Forms of the Energy Equation   239
7-4   Application of the Energy, Momentum, and Continuity Equations
        in Combination   248
7-5   Concept of the Hydraulic and Energy Grade Lines   251

CHAPTER EIGHT   DIMENSIONAL ANALYSIS
                AND SIMILITUDE   272

8-1   The Need for Dimensional Analysis   274
8-2   Dimensions and Equations   277
8-3   The Buckingham II Theorem   277
8-4   Dimensional Analysis   278
8-5   Common Dimensionless Numbers   283
8-6   Similitude   287
8-7   Model Studies for Flows Without Free-Surface Effects   291
8-8   Significance of the Pressure Coefficient   294
8-9   Approximate Similitude at High Reynolds Numbers   296
8-10 Free-Surface Model Studies   299
8-11 Closure   303

CHAPTER NINE   SURFACE RESISTANCE 312

9-1   Introduction 314
9-2   Surface Resistance with Uniform Laminar Flow 314
9-3   Qualitative Description of the Boundary Layer 321
9-4   Quantitative Relations for the Laminar Boundary Layer 323
9-5   Quantitative Relations for the Turbulent Boundary Layer 328

CHAPTER TEN   FLOW IN CONDUITS 358

10-1   Shear Stress Distribution Across a Pipe Section 360
10-2   Laminar Flow in Pipes 362
10-3   Criterion for Laminar or Turbulent Flow in a Pipe 365
10-4   Turbulent Flow in Pipes 368
10-5   Flow at Pipe Inlets and Losses from Fittings 380
10-6   Pipe Systems 386
10-7   Turbulent Flow in Noncircular Conduits 390

CHAPTER ELEVEN   DRAG AND LIFT 414

11-1   Basic Considerations 416
11-2   Drag of Two-Dimensional Bodies 419
11-3   Vortex Shedding from Cylindrical Bodies 426
11-4   Effect of Streamlining 428
11-5   Drag of Axisymmetric and Three-Dimensional Bodies 429
11-6   Terminal Velocity 432
11-7   Effect of Compressibility on Drag 435
11-8   Lift 437

CHAPTER TWELVE   COMPRESSIBLE FLOW 460

12-1   Wave Propagation in Compressible Fluids 462
12-2   Mach-Number Relationships 469
12-3   Normal Shock Waves 475
12-4   Isentropic Compressible Flow Through a Duct with Varying Area 481
12-5   Compressible Flow in a Pipe with Friction 494

CHAPTER THIRTEEN   FLOW MEASUREMENTS 518

13-1   Instruments for the Measurement of Velocity and Pressure 520
13-2   Instruments and Procedures for Measurement of Flow Rate 528
13-3   Measurement in Compressible Flow 545

CHAPTER FOURTEEN   TURBOMACHINERY 562

14-1   Propeller Theory 564
14-2   Axial-Flow Pumps 571

14-3   Radial- and Mixed-Flow Machines   575
14-4   Specific Speed   580
14-5   Suction Limitations of Pumps   581
14-6   Turbines   583
14-7   Viscous Effects   593

CHAPTER FIFTEEN   VARIED FLOW IN
                  OPEN CHANNELS   600

15-1   Energy Relations in Open Channels   602
15-2   The Hydraulic Jump   613
15-3   Surge or Tidal Bore   617
15-4   Gradually Varied Flow in Open Channels   618

APPENDIX   631

ANSWERS TO SELECTED PROBLEMS   645

INDEX   655

PREFACE

It is essential for students to get a "feel" for flow patterns, pressure variation, and shear stress in fluid flow if they are to have a clear understanding of fluid mechanics. Our aim is to present these concepts in a clear and elementary way so that students will develop confidence in the field. In this text it is done by

1. Covering flow patterns and associated topics on velocity and acceleration thoroughly in an early chapter on kinematics
2. Introducing a separate chapter on pressure variation in a flowing fluid immediately after the chapter on kinematics
3. Focusing attention on shear stress distribution in a separate chapter on surface resistance

Thus, we introduce the easier topics first and then move on to consider the more complex topics so that students will develop confidence at the outset and be able to carry through to the end without difficulty. After students have mastered flow patterns, pressure variation, and shear stress, they should be able to handle the material in the chapters on drag and lift, flow meters, compressible flow, pipe flow, and open-channel flow.

Because the integrated forms of the continuity, energy, and momentum equations are so important, more material on the development and application of these tools has been included than in most texts. The basic control-volume equation is derived in a simple and straightforward manner so that students may appreciate the physical principles without mathematical complications. Directly after its derivation, the control-volume equation is applied to the continuity principle. Separate chapters on the energy and momentum principles are then introduced, with heavy reliance on the basic control-volume approach. Because the control-volume equation is the key to most of the derivations, sufficient time and emphasis should be given by the instructor to its derivation for complete understanding of the basic principle involved.

Thus, we focus approximately one-half of this text on basic principles, and the other half on the applications of these principles to engineering problems.

A special note should be made with respect to the chapter on dimensional analysis and similitude. We have used a simple but rigorous approach to dimensional analysis without resort to the π theorem, adapted from Ipsen. With this simplified approach, students learn the why and how of dimensional analysis instead of just the how of the π theorem.

In this edition there are 858 problems, a 60% increase over the first edition. Also, the answers to even-numbered problems are provided. As in the first edition, problems and examples are presented in both the SI and traditional system of units.

Numerous minor changes have been made. For example, the Bernoulli equation has been derived in a more rigorous manner. Also, the section on compressible flow in pipes has been expanded to include isothermal flow.

This text may be used by students of all engineering disciplines for the first course in fluid mechanics. Any student who has had basic engineering mechanics should be able to read and understand it.

Special recognition is extended to Charles L. Barker, Professor Emeritus, who initially introduced the first author to the field of fluid mechanics and motivated him to write a text on the subject. The first author is also indebted to Hunter Rouse, who inspired him to further studies in the field. Indeed, students of Dr. Rouse will sense numerous similarities in approach between his texts and this one.

We wish to thank the many colleagues both at Washington State University and at other institutions who have made valuable suggestions for improving the text. Also, we wish to give special recognition to the reviewers of this second edition: Professors Philip M. Gerhart, University of Akron; Jerry Lee Hall and Theodore H. Okiishi of Iowa State University; Robert H. Kirchhoff, University of Massachusetts; Brian Launder, University of California, Davis; John S. Walker of the University of Illinois; and David J. Wilson, University of Alberta. Their in-depth review and constructive criticism is greatly appreciated.

Special thanks are extended to Amy Roberson who did all of the typing and manuscript preparation. We also acknowledge the encouragement and understanding of our families during the writing and editing of the text.

J.A.R.
C.T.C.

To

Amy and Linda

Fluid mechanics is related to our daily lives in many ways. In this photo we immediately sense the fluid-dynamic force on the sails if not the drag of the water on the hull. The waves generated by the wind involve an even more complex fluid mechanics phenomenon.

1 INTRODUCTION

1-1 FLUIDS

Physical characteristics of fluids

Fluid mechanics is the science dealing with the action of forces on fluids. In contrast to a solid, a fluid is a substance the particles of which easily move and change their relative position. More specifically, a fluid is defined as a substance that will continuously deform, that is, flow under the action of a shear stress no matter how small that shear stress may be. A solid, on the other hand, can resist a shear stress, assuming, of course, that the shear stress does not exceed the elastic limit of the material. The rate of deformation of the fluid is related to the applied shear stress by *viscosity*, a property of the fluid. Thus very viscous fluids, such as honey or cold oils, flow very slowly for a given shear stress. An example of this effect is observed if one pours a cup of oil or honey on an inclined surface. It takes a considerable length of time for these very viscous fluids to flow down the incline, whereas a cup of water would flow down and off very rapidly. At first glance, certain substances may appear to be solids but are in fact fluids. For example, one can break off a piece of tar from a large chunk of cold tar. Inspection by touch and sight will lead one to think of the tar as a solid. If the tar were placed in a container, for example, it would not immediately conform to the walls of the vessel. However, if one were to set the tar on an inclined plane, after a few hours or days it would be seen to deform under the action of gravitational shear forces, and it would eventually flow down the incline, thus indicating that it should be classified as a fluid rather than a solid.

Distinction between solids, liquids, and gases

A fluid can be either a gas or a liquid. The molecular structure of liquids is such that the spacing between molecules is essentially constant (only slight change with temperature and pressure), so that a given mass of liquid will occupy a definite volume of space. Therefore, when one pours a liquid into a container, it assumes the shape of the container for the volume it occupies. The molecules of solids also have definite spacing. However, the solid's molecules are arranged in a specific lattice formation and their movement is restricted, whereas liquid molecules can move with respect to each other when a shearing force is applied. The spacing of the molecules of gases is much larger than either that of solids or liquids, and it is also variable. Thus a gas completely fills the container in which it is placed, and the gas molecules travel in straight lines through space until they either bounce off the walls of the container or are deflected by interaction with other gas molecules.

Fluid as a continuum

In considering the action of forces on fluids, one can either account for the behavior of each and every molecule of fluid in a given field of flow or simplify the problem by considering the average effects of the molecules in a given volume. In most problems in fluid dynamics the latter approach is possible, which means that the fluid can be regarded as a *continuum* — that is, a hypothetically continuous substance.

The justification for treating a fluid as a continuum depends on the physical dimensions of the body immersed in the fluid and the number of molecules in a given volume. Let us say that we are studying the flow of air past a sphere with a diameter of 1 cm. A continuum is said to prevail if the number of molecules in a volume much smaller than the sphere's is sufficiently great so that the average effects (pressure, density, and so on) within the volume are constant or change smoothly with time. The number of molecules in a cubic meter of air at room temperature and sea-level pressure is about 10^{25}. Thus the number of molecules in a volume of 10^{-19} m^3 (about the size of a dust particle, which is very much smaller than the sphere) would be 10^6. This number of molecules is so large that the average effects within the microvolume will indeed be virtually constant. On the other hand, if the 1-cm sphere were at an altitude of 305 km, there would be only one chance out of 10^8 in finding a molecule in the microvolume, and the concept of an average condition would be meaningless. In this case, the continuum assumption would not be valid. It may thus be concluded that the assumption of a continuum is valid for fluid flow except in the rarest conditions, as encountered in outer space.

1-2 FLOW CLASSIFICATION

The subject of fluid mechanics can be subdivided into two broad categories: hydrodynamics and gas dynamics. Hydrodynamics deals primarily with the flow of fluids for which there is virtually no density change, such as liquid flow or the flow of gas at low speeds. Hydraulics, for example, the study of liquid flows in pipes or open channels, falls within this category. The study of fluid forces on bodies immersed in flowing liquids or in low-speed gas flows can also be classified as hydrodynamics.

Gas dynamics, on the other hand, deals with fluids that undergo significant density changes. High-speed gas flows through a nozzle or over a body, the flow of chemically reacting gases, or the movement of a body through the low-density air of the upper atmosphere fall within the general category of gas dynamics.

An area of fluid mechanics not classified as either hydrodynamics or gas dynamics is aerodynamics, which deals with the flow of air past aircraft or rockets, whether it be low-speed incompressible flow or high-speed compressible flow.

1-3   HISTORICAL NOTE

The science of fluid mechanics began with the need to control water for irrigation purposes in ancient Egypt, Mesopotamia, and India. Although these civilizations understood the nature of channel flow, there is no evidence that any quantitative relationships had been developed to guide them in their work. It was not until 250 B.C. that Archimedes discovered and recorded the principles of hydrostatics and flotation. Although the empirical understanding of hydrodynamics continued to improve with the development of fluid machinery, better sailing vessels, and more intricate canal systems, the fundamental principles of classical hydrodynamics were not founded until the seventeenth and eighteenth centuries. Newton (6), Daniel Bernoulli (1), and Euler (11) made the greatest contributions to the founding of these principles.

Classical hydrodynamics, though a fascinating subject that appealed to mathematicians, was not applicable to many practical problems because the theory was based on inviscid fluids. The practicing engineers at that time needed design procedures that involved the flow of viscous fluids; consequently, they developed empirical equations that were usable but narrow in scope. Thus, on the one hand, the mathematicians and physicists developed theories that in many cases could not be used by the engineers, and on the other, engineers used empirical equations that could not be used outside the limited range of application from which they were derived.

Near the beginning of the twentieth century, however, it was necessary to merge the general approach of the physicists and mathematicians with the experimental approach of the engineer to bring about significant advances in the understanding of flow processes. Osborne Reynolds' (9) paper in 1883 on turbulence and later papers on the basic equations of motion contributed immeasurably to the development of fluid mechanics. After the turn of the century, Ludwig Prandtl (8) proposed the concept of the boundary layer. This concept not only paved the way to sophisticated analyses needed in the development of the airplane, but also resolved many of the paradoxes involved with the flow of a low-viscosity fluid.

Gas dynamics is a relatively new field in that the earliest works did not appear until the nineteenth century. Riemann (10) published his paper on compression (shock) waves in 1860, and 20 years later Mach (5) observed such waves on supersonic projectiles. Once again it was Prandtl who organized the systematic study of gas dynamics around the turn of the century (8). Interest in gas dynamics increased tremendously after World War I, which led to the development of supersonic wind tunnels before World War II and supersonic flight soon after. The advent of space flight led to still another area of gas dynamics, called rarefied flow, in which the density is so low that the motion and impact of individual molecules must be considered. This is in contrast to the treatment of fluid as a continuum, which we do in most of our earthbound problems.

Modern developments in fluid mechanics, as in all fields, involve the use of high-speed computers in the solution of problems. Remarkable progress has been made in this area, and there is an increasing use of the computer in fluid dynamic design. Even though we presently have high-speed computers at our disposal, the solutions are only as valid as the equations we use to describe the basic flow phenomena. In fact, there is currently no general analytic model that describes the complete nature of turbulence. Hence, in fluid mechanics today, experimental tests continue to go hand in hand with theoretical analyses to yield answers to engineering problems.

1-4 SIGNIFICANCE OF FLUID MECHANICS

The significance of fluid mechanics becomes apparent when we consider the vital role it plays in our everyday lives. When we as a community turn on our kitchen faucets, we activate a complex hydraulic network of pipes, valves, and pumps. When we flick on a light switch, we are drawing energy either from a hydroelectric source that operates by the flow of water through turbines or from a thermal power source derived from the flow of steam past turbine blades. When we drive our automobiles, suspension is provided by pneumatic tires, road shocks are reduced by hydraulic shock absorbers, gasoline is pumped through tubes and later atomized, and air resistance creates a drag on the auto as a whole; and when we stop, we are confident of the operation of the hydraulic brakes. Very complex fluid processes are also involved in the manufacture of the paper on which this book is printed. Finally, the very essence of our lives is dependent on a very important fluid mechanic process—the flow of blood through our veins and arteries.

REFERENCES

1. Bernoulli, Daniel, *Hydrodynamics,* and Bernoulli, Johann, *Hydraulics.* Trans. Thomas Carmody and Helmut Kobus. Dover Publications, Inc., New York, 1968.

2. Durand, W. F. (ed.) *Aerodynamic Theory*, vol. 1. Dover Publications, Inc., New York, 1963.

3. Goldstein, S. (ed.) *Modern Developments in Fluid Dynamics*. Dover Publications, Inc., New York, 1965.

4. Karman, Theodore von. *Aerodynamics, Selected Topics in the Light of Their Historical Development*. Cornell University Press, Ithaca, New York, 1954.

5. Mach, E., and P. Salcher. "Photographische Fixierung der durch Projektile in der Luft eingeleiteten Vorgänge." *S. B. Akad. Wiss. Wien.*, 95, (1887), 764–780.

6. Newton, Issac S. *Philosophie Naturalis Principia Mathematica*. Trans. Florian Cajori. University of California Press, Berkeley, 1934.

7. Oswatitsch, K. *Gas Dynamics*. Trans. G. Kuerti. Academic Press, Inc., New York, 1956.

8. Prandtl, L. "Über Flussigkeitsbewegung bei sehr kleiner Reibung." *Verhandlungen des III. Internationalen Mathematiker Kongresses*, Leipzig, 1905.

9. Reynolds, O. "An Experimental Investigation of the Circumstances Which Determine Whether the Motion of Water Shall Be Direct or Sinuous, and of the Law of Resistance in Parallel Channels." *Phil. Trans. Roy. Soc. London*, 174 (1883).

10. Riemann, B. "Über die Fortplanzung ebener Luftwellen von endlicher Schwingungsweite," in *Gesammelte Werke*. Leipzig, 1876.

11. Rouse, H., and Ince, S. *History of Hydraulics*. Iowa Institute of Hydraulic Research, State University of Iowa, 1957.

Two opposing jets of water produce this sheet of liquid, which is concentric with the jets and which, through the action of surface tension, breaks up into droplets. Research involving this phenomenon was undertaken by J. H. Lienhard and C. P. Huang at Washington State University, Pullman, Washington. Photo by College of Engineering Photo Laboratory, Washington State University.

2 FLUID PROPERTIES

2-1 BASIC UNITS

Introduction

Every fluid has certain characteristics by which its physical condition may be described. We call such characteristics *properties* of the fluid; these properties are expressed in terms of a limited number of basic dimensions (length, mass or force, time, and temperature), which in turn are quantified by basic units. The traditional system of units in the United States has been the foot-pound-second system. However, because all the engineering societies are urging the adoption of the SI (Système Internationale) system, this text uses both SI and traditional units. Approximately half of the problems and the majority of the examples are given in the SI system, with the remainder in the foot-pound-second system.

SI system of units

The basic unit of temperature in the SI system, the degree Kelvin (K), is defined as zero at absolute zero and 273.16 K at the freezing point of water. The Celsius scale (°C) is defined as zero at the freezing point of water. Therefore, the conversion formula is

$$K = 273° + °C \qquad (2\text{-}1)$$

The basic units of mass, length, and time in the SI system are the kilogram (kg), meter (m), and second (s). The corresponding unit of force is derived from Newton's second law: The force required to accelerate a kilogram at one meter per second per second is defined as the *newton* (N). The acceleration due to gravity at the earth's surface is 9.81 m/s². Thus the weight of a kilogram at the earth's surface is

$$F = Mg \qquad (2\text{-}2)$$
$$= (1)(9.81) \text{ kg} \cdot \text{m/s}^2$$
$$= 9.81 \text{ N}$$

From Eq. (2-2) we can also determine the units for kilograms in terms of the meter-newton-second units. From Eq. (2-2),

$$M = \frac{F}{g} \frac{\text{N}}{\text{m/s}^2}$$

Thus the mass of a body in kilograms is given by its weight in newtons at the earth's surface divided by 9.81 m/s² (the acceleration due to gravity at the earth's surface). As shown, the units of mass may also be expressed in equivalent meter-newton-second units as $N \cdot s^2/m$.

The unit of work and energy in the SI system is the *joule* (J), which is a newton meter ($N \cdot m$). The unit of power is the *watt* (W), which is a joule per second.

The prefixes used in the SI system to indicate multiples and submultiples are

$$G \text{ (giga)} = 10^9$$
$$M \text{ (mega)} = 10^6$$
$$k \text{ (kilo)} = 10^3$$
$$c \text{ (centi)} = 10^{-2}$$
$$m \text{ (milli)} = 10^{-3}$$
$$\mu \text{ (micro)} = 10^{-6}$$

For example, km stands for kilometer or 1,000 meters, and mm signifies millimeter or one-thousandth of a meter.

Traditional units

The system of units that preceded SI units in several countries is the so-called English system. The fundamental units of length and mass are the foot (ft), equal to 30.48 cm, and the slug, equal to the 14.59 kg. The time unit of second is common to both systems. The force required to accelerate a mass of one slug at one foot per second per second is one pound force (lbf). The mass unit more common to mechanical engineers in the traditional system is the pound mass (lbm), which is a factor of 32.2. (g_c) smaller than the slug.

The traditional unit for temperature is the degree Fahrenheit (°F), which is $\frac{5}{9}$ of the Celsius degree. The corresponding absolute temperature scale is in degrees Rankine (°R). The Fahrenheit temperature at the freezing point of water is 32°F.

2-2 SYSTEM, EXTENSIVE AND INTENSIVE PROPERTIES

From Chapter 4 through the remainder of the text, wide use is made of the control-volume approach. In its application we use the concept of a system of particles and the intensive and extensive properties related to this concept. A *system* is defined as a given quantity of matter. To illustrate, if at an instant a quantity of matter were designated as a system and dyed red to distinguish it from the remaining matter, this red matter would always constitute the system even though it moved through space and might change in shape and volume. Because a system always consists of the same matter, the mass of a given system is constant. Properties related to the total mass of the system are called *extensive* properties and are usually represented by uppercase letters, for example, mass M and weight W. Properties that are independent of the amount of fluid are called *intensive* properties and are often designated by lowercase letters, such as pressure p (force per unit area) and *mass density* ρ (mass per unit volume). The distinction between extensive and intensive properties is very important in the derivation and application of the control-volume approach introduced in Chapter 4.

2-3 PROPERTIES INVOLVING THE MASS OR WEIGHT OF THE FLUID

Specific weight, γ

The gravitational force per unit volume of fluid or simply weight per unit volume is defined as *specific weight* and it is given the symbol γ (gamma). Water at 20°C has a specific weight of 9.79 kN/m³. In contrast, the specific weight of air at the same temperature and at standard atmospheric pressure is 11.9 N/m³. Specific weights of common liquids are given in Table A-4.

Mass density, ρ

The mass per unit volume is *mass density*; hence it has units of kilograms per cubic meter or $N \cdot s^2/m^4$. Because the specific weight is weight per unit of volume, the mass density will be given by the specific weight at the earth's surface divided by g. The mass density of water at 10°C is $9,810/g = 1,000$ kg/m³; and for air at 20°C and at standard pressure, the mass density is 1.20 kg/m³. Mass density, often simply called *density,* is given the Greek symbol ρ (rho). The density of common fluids is given in Tables A-2 to A-5.

Density variation

The density of some fluids is more easily changed than others. For example, air can easily be compressed with a consequent density change, whereas a very large pressure is needed to effect a relatively small density change in water. For most applications, water can be considered incompressible and, in turn, it would be assumed to have constant density. Air, on the other hand, is a relatively compressible fluid with variable density. However, there are even some air-flow problems for which the air density changes only slightly; one case is wind flow past buildings.

Incompressibility does not always imply constant density. For example, a mixture of salt in water will change the density without changing its volume. Therefore, there are some flows, such as in estuaries, where density variations may occur within the flow field even though the fluid is essentially incompressible. Such a fluid is termed *nonhomogeneous*. This text emphasizes the flow of *homogeneous* fluids, so the term *incompressible* used throughout the text implies constant density.

Specific gravity (sp. gr.)

The ratio of the specific weight of a given liquid to the specific weight of water at a standard reference temperature is defined as *specific gravity*. The standard reference temperature for water is often taken as 4°C, where the specific weight of water at atmospheric pressure is 9,810 N/m³. With this reference, the specific gravity of mercury at 20°C is

$$\text{sp. gr.}_{\text{Hg}} = \frac{133 \text{ kN/m}^3}{9.81 \text{ kN/m}^3} = 13.6$$

Because specific gravity is a ratio of specific weights, it is dimensionless and, of course, independent of system of units used.

Equation of state and density of gases

The equation of state for an ideal gas is given as follows:

$$\frac{p}{\rho} = RT \qquad (2\text{-}3)$$

where
p = absolute pressure[1]
ρ = mass density
T = absolute temperature
R = gas constant

Actually no gas is ideal; however, a gas far removed from the liquid phase, which is generally the case in gas-flow problems, behaves like an ideal gas. Values of R for a number of gases are given in Table A-2. To determine the mass density of a gas we simply solve Eq. (2-3) for ρ:

$$\rho = \frac{p}{RT} \qquad (2\text{-}4)$$

EXAMPLE 2-1   Air at standard sea-level pressure ($p = 101 \text{ kN/m}^2$) has a temperature of 4°C. What is the density of the air?

---

[1] We discuss pressure in detail in Chapter 3.

Solution We apply Eq. (2-4) to solve for ρ:

$$\rho = \frac{p}{RT}$$

$$= \frac{101 \times 10^3 \,\mathrm{N/m^2}}{287 \,\mathrm{J/kg\ K} \times (273 + 4)\mathrm{K}}$$

$$= 1.27 \,\mathrm{kg/m^3}$$

2-4 PROPERTIES INVOLVING THE FLOW OF HEAT

Specific heat, c

The property that describes the capacity of a substance to store thermal energy is called *specific heat*. By definition it is the amount of heat that must be transferred to a unit mass of substance to raise its temperature by one degree. The specific heat of gases depends on the process accompanying the change in temperature. If the specific volume v of the gas ($v = 1/\rho$) remains constant while the temperature changes, then the specific heat is identified as c_v; however, if the pressure is held constant during the change in state, then the specific heat is identified as c_p. The ratio c_p/c_v is given the symbol k. Data for c_p and k for various gases are given in Table A-2.

Internal energy, u

The energy that a substance possesses because of the state of the molecular activity in the substance is termed *internal energy*. Internal energy is given in joules per kilogram in the SI system. The internal energy is generally a function of temperature and pressure. However, for an ideal gas, it is a function of temperature only.

Enthalpy, h

The combination $u + p/\rho$ occurs frequently in thermodynamic and compressible flow calculations; it has been given the name *enthalpy*. In ideal gases, u and p/ρ are functions of temperature only; consequently, their sum, enthalpy, is also only a function of temperature for an ideal gas.

2-5 VISCOSITY

The most important distinction between a solid such as steel and a viscous fluid such as water or air is that shear stress in a solid material is generally proportional to shear strain, and the material will cease to deform when equilibrium is reached, whereas the shear stress in a viscous fluid is proportional to the *time rate* of strain. The proportionality factor for the solid is the shear modulus; the proportionality factor for the viscous fluid is the *dynamic*, or *absolute, viscosity*. For example, the shear stress of a fluid near a wall is given by

$$\tau = \mu \frac{dV}{dy} \tag{2-5}$$

where τ (tau) is the shear stress, μ (mu) is the dynamic viscosity, and dV/dy is the time rate of strain, which is also called the velocity gradient. Thus the definition of the viscosity μ is the ratio of the shear stress to the velocity gradient, $\mu = \tau/(dV/dy)$. Consider the flow shown in Fig. 2-1. This velocity distribution is typical of that for laminar (nonturbulent) flow next to a solid boundary. Several observations relating to this figure will help one appreciate the interaction between viscosity and velocity distribution. First, the velocity gradient at the boundary is finite. The curve of velocity variation cannot be tangent to the boundary because this would imply an infinite velocity gradient, and in turn, an infinite shear stress, which is impossible. Second, a velocity gradient that becomes less steep (dV/dy becomes smaller) with distance from the boundary has a maximum shear stress at the boundary, and the shear stress decreases with distance from the boundary. Also note that the velocity of the fluid is zero at the stationary boundary. This is characteristic of all flows dealt with in this basic text; that is, at the boundary surface the fluid will have the velocity of the boundary—no slip occurs.

Many of the equations of fluid mechanics include the combination μ/ρ in them. Because this occurs so frequently, the combination has been

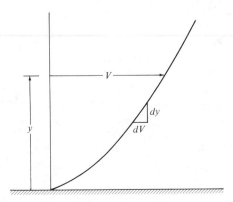

FIGURE 2-1   Velocity distribution next to a boundary.

given the special name *kinematic viscosity* (so called because the force dimension cancels in the combination μ/ρ). The symbol used to identify kinematic viscosity is ν (nu).

Whenever shear stress is applied to fluids, motion will occur. This is the basic difference between fluids and solids. Solids can resist shear stress in a static condition, but fluids deform continuously under the action of a shear stress. Another important characteristic of fluids is that the viscous resistance is independent of the normal force (pressure) acting within the fluid. In contrast, for two solids sliding relative to each other the shearing resistance is totally dependent on the normal force between the two.

The manner in which viscous forces are produced can be seen in the conveyer-belt analogy. Consider a type of transit system in which people are carried from one part of a city to another on conveyer belts (Fig. 2-2a). People ride the fast-moving belt from left to right—an equilibrium condition exists. Next visualize the action when the people step off the fast belt onto a slower-moving belt (Fig. 2-2b). Before they step off the fast-moving belt they will each possess a certain amount of momentum in the x direction, but as soon as they acquire the speed of the slower belt, their momentum in the x direction will have been reduced by a significant amount. It is known from basic mechanics that a change in momentum of a body results from an external force acting on that body; in our example,

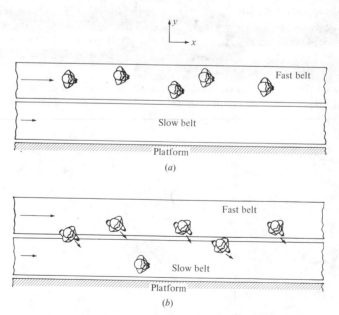

FIGURE 2-2 Conveyer-belt transportation system.

it is the slower belt that exerts a force in the negative x direction as each person steps on the belt. Conversely, as each person steps on the slower belt, a force is exerted on the belt in the positive x direction. Now, if the people step from the faster belt to the slower belt at a rather steady rate, then a rather continuous force is exerted on the slower belt. In effect, by the action of the people moving in the negative y direction, they produce a force on the slow belt in the positive x direction. One may think of this as a shear force in the x direction. In a similar manner, it can be visualized that if people stepped from the slow-moving belt to the faster one, a "shear force" in the negative x direction would be imposed on the faster belt.

If the people were continuously going both ways (back and forth) from one belt to the other, there would be, in effect, a continuous augmenting force (force in the direction of motion) on the slow belt and a like retarding force on the fast belt. Furthermore, it should be appreciated that as the relative speed between the belts is changed (analogous to a change in velocity gradient), the shear force will be increased or decreased in direct proportion to the increase or decrease in relative velocity. Thus if both belts were made to have the same speed, the shear force would be zero.

In fluid flow we can think of streams of fluid traveling in a given general direction, such as in a pipe with the fluid nearer the pipe center traveling faster (analogous to the faster belt) while the stream nearer the wall is traveling slower. The interaction between streams, as in the case of gas flow, occurs when the molecules of gas travel back and forth between adjacent streams, thus creating a shear stress in the fluid. Because the rate of activity (back-and-forth motion) of the gas molecules increases with an increase in temperature, it follows that the viscosity of gases should increase with the temperature of the gas. Such is indeed the case, as can be seen in Fig. 2-3.

For liquids, the shear stress is involved with the cohesive forces between molecules. These forces decrease with temperature, which results in a decrease in viscosity with an increase in temperature (see Fig. 2-3). The variation of viscosity (dynamic and kinematic) for other fluids is given in Figs. A-2 and A-3.

Units of viscosity

From Eq. (2-5) it can be seen that the units of μ are $N \cdot s/m^2$.

$$\mu = \frac{\tau}{dV/dy} = \frac{N/m^2}{(m/s)/m} = N \cdot s/m^2$$

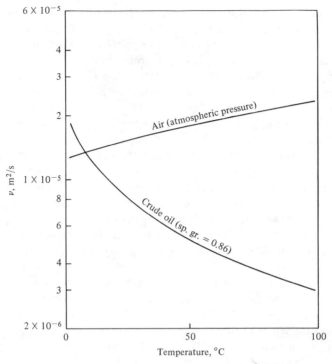

FIGURE 2-3 Kinematic viscosity for air and crude oil.

The units of kinematic viscosity ν are m²/s

$$\nu = \frac{\mu}{\rho} = \frac{N \cdot s/m^2}{N \cdot s^2/m^4} = m^2/s$$

Newtonian versus non-Newtonian fluid

Fluids for which the shear stress is directly proportional to the rate of strain are called Newtonian fluids. For some fluids, however, the shear stress may not be directly proportional to the rate of strain. These fluids are classified as non-Newtonian, examples of which are blood, certain plastics, and clay-water mixtures. This book will be limited to theory and applications involving Newtonian fluids only.

2-6 ELASTICITY

When pressure is applied to a fluid, it contracts; when the pressure is released, it expands. The elasticity of a fluid is related to the amount of deformation (expansion or contraction) for a given pressure change. Quantitatively the degree of elasticity is given by E_v, the definition of which is

$$dp = -E_v \frac{d\forall}{\forall} \tag{2-6}$$

or

$$E_v = -\frac{dp}{d\forall/\forall} \tag{2-7}$$

where

E_v = bulk modulus of elasticity

dp = incremental pressure change

$d\forall$ = incremental volume change

$\forall$ = original volume of fluid

Because $d\forall/\forall$ is negative for a positive dp, a negative sign is used in the definition to yield a positve E_v. An alternative form of Eq. (2-7) is

$$E_v = \frac{dp}{d\rho/\rho} \tag{2-8}$$

By comparing Eqs. (2-7) and (2-8), it can be seen that $d\rho/\rho = -d\forall/\forall$. We can verify this equality by considering a given mass of fluid M, where

$$M = \rho\forall \tag{2-9}$$

If we differentiate both sides of Eq. (2-9) we have

$$dM = \rho\, d\forall + \forall\, d\rho \tag{2-10}$$

But $dM = 0$ because the mass is constant; hence we find that

$$\forall\, d\rho = -\rho\, d\forall \qquad \text{or} \qquad \frac{d\rho}{\rho} = -\frac{d\forall}{\forall}$$

The bulk modulus of elasticity of water is approximately 2.2 GN/m^2, which corresponds to 0.05% change in volume for a 1 MN/m^2 change in pressure. Obviously, the term "incompressible" as often applied to water is well founded.

For gases it can be shown that $E_v = kp$ in an isentropic process and $E_v = p$ for an isothermal process, where k is c_p/c_v.

2-7 SURFACE TENSION

According to the theory of molecular attraction, molecules of liquid considerably below the surface act on each other by forces that are equal in all directions. However, molecules near the surface have a greater attraction for each other. This produces a surface on the liquid that acts like a stretched membrane. Because of this membrane effect, the liquid surface

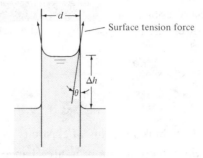

FIGURE 2-4   Capillary action in a tube.

exerts "tension" upon adjacent portions of the surface or upon other objects that are in contact with the liquid. This tension acts in the plane of the surface, and its magnitude per unit of length is defined as *surface tension, σ* (sigma). Surface tension for a water-air surface is 0.073 N/m at room temperature. The effect of surface tension is illustrated for the case of capillary action in a small tube (Fig. 2-4). Here the end of a small-diameter tube is put into the water and the characteristic curved water surface occurs within the tube. The relatively great attraction of the water molecules to the glass causes the water surface to curve upward in the region of the glass wall. Then the surface tension force acts around the circumference of the tube and in a direction as indicated. It may be assumed that θ (theta) is equal to 0° for water against glass. This produces a net upward force on the water that causes the water in the tube to rise above the water surface in the reservoir. An example problem will give a quantitative illustration of the principle.

EXAMPLE 2-2   To what height above the reservoir level will water (20°C) rise in a glass tube, such as that shown in Fig. 2-4, if the inside diameter of the tube is 1.6 mm?

Solution   By taking summation of forces in the vertical direction of the water in the tube that has risen above the reservoir level, we have

$$\sigma \pi d \cos \theta - \gamma(\Delta h)(\pi d^2/4) = 0$$

However, θ for water against glass is small; therefore, $\cos \theta \approx 1$. Then

$$\sigma \pi d - \gamma(\Delta h)\left(\frac{\pi d^2}{4}\right) = 0$$

or
$$\Delta h = \frac{4\sigma}{\gamma d}$$

$$= \frac{4 \times 0.073 \; N/m}{9{,}790 \; N/m^3 \times 1.6 \times 10^{-3} \; m}$$

or $\qquad\qquad \Delta h = 18.6 \; mm$ ◀

Other manifestations of surface tension include the excess pressure (over and above atmospheric pressure) created inside droplets and bubbles, the transformation of a liquid jet into droplets, and the binding together of wetted granular material, such as fine, sandy soil.

2-8 VAPOR PRESSURE

For every liquid, the internal molecular activity is such that molecules escape from the surface until the pressure within the space next to the surface reaches such a value that the net exchange of molecules between the liquid and vapor is zero. This pressure is called the *saturated vapor pressure* or, simply, *vapor pressure* p_v. Because the molecular activity depends upon temperature, the vapor pressure in turn is a function of the temperature of the liquid; therefore boiling can be brought about either by increasing the temperature or by reducing pressure.

PROBLEMS

2-1 Determine the density of air, helium, and carbon dioxide at an absolute pressure of 140 kN/m² (20 psia) and a temperature of 30°C (86°F).

2-2 Calculate the density and specific weight of carbon dioxide at 300 kN/m² absolute and 60°C.

2-3 Calculate the density and specific weight of helium at 300 kN/m² absolute and 60°C.

2-4 Meteorologists often refer to air masses in forecasting the weather. Estimate the mass of 1 mile³ of air in slugs and kilograms. Make your own reasonable assumptions with respect to the conditions of the atmosphere.

2-5 At a temperature of 10°C (50°F) and an absolute pressure of 103 kN/m² (15 psia), what is the ratio of the density of water to the density of air, ρ_w/ρ_a?

2-6 What is the weight of a 3-ft³ (0.0849 m³) tank of air if the air is under pressure of 1,000 psia (6.89 MN/m²) and the tank itself weighs 80 lbf (356 N)? Temperature = 70°F (21°C).

2-7 What is the specific weight and density of air at an absolute pressure of 345 kPa (50 psia) and a temperature of 38°C (100°F)?

2-8 Classify the following according to whether they are extensive or intensive properties: specific weight γ, density ρ, mass M, surface tension σ, vapor pressure p_v, weight W, velocity V, and acceleration a.

2-9 Two plates are spaced $\frac{1}{4}$ in. (6.35 mm) apart. The lower plate is stationary and the upper plate moves at a velocity of 10 ft/sec (3.05 m/s). Oil (SAE 10W-30, 100°F or 38°C) fills the space between the plates and has the same velocity as the plates at the surface of contact. If the velocity variation in the oil is linear, what is the shear stress in the oil?

2-10 What is the change in water viscosity between 10°C and 60°C? What is the change in air viscosity between 10°C and 60°C?

2-11 What is the change in kinematic air viscosity between 10°C and 60°C? Assume standard atmospheric pressure.

2-12 The velocity distribution for flow of kerosene at 100°F ($\mu = 3 \times 10^{-5}$ lb-sec/ft²) between two walls is given by $u = 100y\,(0.1 - y)$ ft/sec, where y is measured in feet and the spacing between the walls is 0.1 ft. Plot the velocity distribution and determine the shear stress at the walls.

2-13 The velocity distribution for flow of kerosene at 20°C ($\mu = 4 \times 10^{-3}$ N · s/m²) between two walls is given by $u = 1,000y\,(0.01 - y)$ m/s, where y is measured in meters and the spacing between the walls is 1 cm. Plot the velocity distribution and determine the shear stress at the walls.

2-14 The velocity distribution for viscous flow between stationary plates is given as follows:

$$V = \frac{1}{2\mu}\frac{dp}{dx}(By - y^2)$$

If glycerin is flowing ($T = 20°C$) and the pressure gradient dp/dx is 1.6 kN/m³, what is the velocity and shear stress at a distance of 12 mm from the wall if the spacing B is 5.0 cm? What are the shear stress and velocity at the wall?

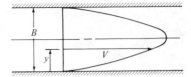

PROBLEMS 2-12, 2-13, 2-14

2-15 Find the dynamic and kinematic viscosity of kerosene, SAE 10W motor oil, and water at a temperature of 38°C (100°F).

2-16 What is the ratio of the dynamic viscosity of air to that of water at standard pressure and $T = 20°C$? What is the ratio of the kinematic viscosity of air to water for the same conditions?

2-17 The device shown consists of a disk that is rotated by a shaft. The disk is positioned very close to a solid boundary. Between the disk and boundary is viscous oil.

   (a) If the disk is rotated at a rate of 1 rad/s, what will be the ratio of the shear stress in the oil at $r = 2$ cm to the shear stress at $r = 3$ cm?

   (b) If the rate of rotation is 2 rad/s, what is the speed of oil in contact with the disk at $r = 3$ cm?

   • (c) If the oil viscosity is 0.01 N · s/m² and the spacing y is 2 mm, what is the shear stress for the conditions noted in (b)?

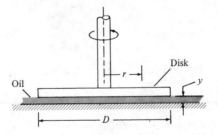

PROBLEMS 2-17, 2-18

2-18 The device shown consists of a disk that is rotated by a shaft. The disk is positioned very close to a solid boundary. Between the disk and boundary is viscous oil.

   (a) If the disk is rotated at a rate of 1 rad/sec, what will be the ratio of the shear stress in the oil at $r = 2$ in. to the shear stress at $r = 3$ in.?

   (b) If the rate of rotation is 2 rad/sec, what is the speed of oil in contact with the disk at $r = 3$ in.?

(c) If the oil viscosity is 0.01 lb-sec/ft² and the spacing y is 0.001 ft, what is the shear stress for the conditions noted in (b)?

2-19 Find the kinematic and dynamic viscosity of air and water at a temperature of 10°C (50°F) and absolute pressure of 103 kPa (15psia).

2-20 A pressure of 2×10^6N/m² is applied to a mass of water that initially filled a 1,000-cm³ volume. Estimate its volume after the pressure is applied.

2-21 What pressure increase must be applied to water to reduce its volume by 1%?

2-22 A spherical soap bubble has an inside radius R, a film thickness t, and a surface tension σ. Derive a formula for the pressure within the bubble relative to the outside atmospheric pressure. What is the pressure difference for a bubble with a 2-mm radius? Assume σ the same as for pure water.

2-23 A water column in a glass tube is used to measure the pressure in a pipe. If the tube is $\frac{1}{4}$ in. (6.35 mm) in diameter, how much of the water column is due to surface-tension effects? What would be the surface-tension effects if the tube were $\frac{1}{8}$ in. (3.2 mm) and $\frac{1}{32}$ in. (0.8 mm) in diameter?

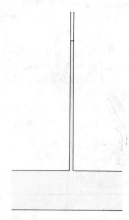

2-24 Calculate the maximum capillary rise of water (10°C) between two vertical glass plates spaced 1 mm apart.

2-25 What is the pressure within a 1-mm spherical droplet of water relative to the atmospheric pressure outside?

2-26 It can be shown for an ideal gas that the difference between the specific heat at constant pressure and that at constant volume is the gas constant R, that is, $c_p - c_v = R$. Derive expressions for c_p and c_v in terms of k and R.

2-27 The vapor pressure of water at 100°C is 101 kN/m² since water boils under these conditions. The vapor pressure of water decreases approximately linearly with decreasing temperature at a rate of 3.1 kN/m² °C. Calculate the boiling temperature of water at 3,000-m altitude, where the atmospheric pressure is 69 kN/m² abs.

REFERENCES

1. Bolz, R. E., and Tuve, G. L. *Handbook of Tables for Applied Engineering Science.* Chemical Rubber Company, 1973.
2. Metzner, A. B. "Flow of Non-Newtonian Fluids," in V. L. Streeter (ed.), *Handbook of Fluid Dynamics.* McGraw-Hill Book Company, New York, 1961, sec. 7.
3. "Physical Properties of Liquids and Gases for Plant and Process Design." *Proc. Symp. Collab. Scottish Br. Inst. Chem. Eng.,* Glasgow (1968).
4. Washburn, E. W. (ed.) *International Critical Tables of Numerical Data.* McGraw-Hill Book Company, New York, 1929.
5. Weast, Robert C. (ed.) *Handbook of Chemistry and Physics,* 53rd ed. Chemical Rubber Company, 1972.

FILMS

6. Lumley, John L. *Deformation of Continuous Media.* National Committee for Fluid Mechanics Films, distributed by Encyclopaedia Britannica Educational Corporation.
7. Markovitz, Hershel. *Rheological Behavior of Fluids.* National Committee for Fluid Mechanics Films, distributed by Encyclopaedia Britannica Educational Corporation.
8. Trefethen, Lloyd. *Surface Tension of Fluids.* National Committee for Fluid Mechanics Films, distributed by Encyclopaedia Britannica Educational Corporation.

Shasta Dam on the Sacramento River in northern California is over 1000 m long, 180 m high, and impounds 5 billion m³ of water upstream. This dam is the keystone to the Bureau of Reclamation's Central Valley Project. Photo courtesy of the U.S. Bureau of Reclamation.

3 FLUID STATICS

I N GENERAL, FLUIDS EXERT BOTH NORMAL AND SHEARING FORCES on surfaces that are in contact with them. However, only fluids with velocity gradients produce shearing forces; hence, for fluids at rest only normal forces exist. These normal forces in fluids are called *pressure forces*.

3-1   PRESSURE

Definition of pressure

At every point in a static fluid a certain pressure intensity exists. Specifically, the pressure intensity, usually simply called pressure, is defined as follows:

$$p = \lim_{\Delta A \to 0} \frac{\Delta F}{\Delta A} = \frac{dF}{dA} \tag{3-1}$$

where F is the normal force acting over the area A. Pressure intensity is a scalar quantity, that is, it has magnitude only and acts equally in all directions. This is easily demonstrated by considering the wedge-shaped element of fluid in equilibrium in Fig. 3-1. The forces that act on the element are the surface forces and the weight force.

By writing the equation of equilibrium for the x direction, we obtain

$$(p_n \, \Delta y \, \Delta l) \sin \alpha - p_x(\Delta y \, \Delta l \sin \alpha) = 0$$

or
$$p_n = p_x \tag{3-2}$$

For the z direction, we obtain

$$-(p_n \, \Delta y \, \Delta l) \cos \alpha + p_z(\Delta y \, \Delta l \cos \alpha) - \tfrac{1}{2}\gamma \, \Delta l \cos \alpha \, \Delta l \sin \alpha \, \Delta y = 0$$

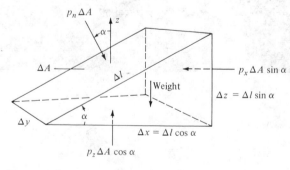

FIGURE 3-1   Pressure forces on a fluid element.

Now, when we let the element shrink to a point (Δy, $\Delta l \to 0$), the last term disappears because it is an infinitesimal of higher order than the other terms; therefore, in the limit,

$$p_n = p_z \tag{3-3}$$

Combining Eqs. (3-2) and (3-3), we finally arrive at the result

$$p_n = p_x = p_z \tag{3-4}$$

Since the angle α (alpha) is arbitrary and p_n is independent of α, we conclude that the pressure at a point in a static fluid acts with the same magnitude in all directions:

$$p_n = p_x = p_y = p_z$$

Pressure transmission

In a closed system a pressure change produced at one point in the system will be transmitted throughout the entire system. The principle is known as Pascal's law after Blaise Pascal, the French scientist, who first stated it in 1653. This phenomenon of pressure transmission plus the ease with which we can move fluids has led to the widespread development of hydraulic controls for operating equipment such as aircraft-control surfaces, heavy earthmoving equipment, and hydraulic presses. Figure 3-2 is an illustration of the application of this principle in the form of a hydraulic lift used in service stations. Here air pressure from a compressor establishes the pressure in the oil system, which in turn acts against the piston in the lift. It can be seen that if a pressure of 600 kN/m², for example, acts on

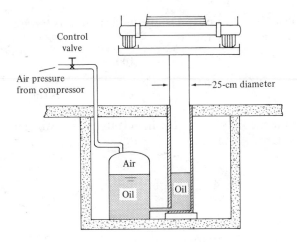

FIGURE 3-2 Hydraulic hoist.

the 25-cm piston, then a force equal to pA or 29.45 kN will be exerted on the piston. To handle larger or smaller loads it is necessary only to increase or decrease the pressure.

EXAMPLE 3-1 A hydraulic jack has the dimensions shown in the accompanying figure. If one exerts a force F of 100 N on the handle of the jack, what load, F_2, can be supported by the jack?

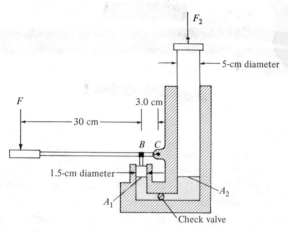

Solution The force F_1 exerted on the small piston will be obtained by taking moments about C. Therefore,

$$(0.33 \text{ m}) \times (100 \text{ N}) - (0.03 \text{ m})F_1 = 0$$

$$F_1 = \frac{0.33 \text{ m} \times 100 \text{ N}}{0.03 \text{ m}} = 1{,}100 \text{ N}$$

Then, because the small piston is in equilibrium, this force is equal to the pressure force on the piston, or

$$p_1 A_1 = 1{,}100 \text{ N}$$

Hence, $$p_1 = \frac{1{,}100}{A_1} = \frac{1{,}100}{\pi d^2/4} = 6.22 \times 10^6 \text{ N/m}^2$$

Now we know the pressure in the liquid; therefore, we can solve for the force on the large piston. Since $p_1 = p_2$,

$$F_2 = p_1 A_2$$

where A_2 = area of large piston

Finally,

$$F_2 = 6.22 \times 10^6 \frac{\text{N}}{\text{m}^2} \times \frac{\pi}{4} \times (0.05 \text{ m})^2 = 12.22 \text{ kN} \quad \blacktriangleleft$$

Absolute pressure, gage pressure, and vacuum

In a region such as outer space, which is virtually void of gases, the pressure is essentially zero. Such a condition can be approached very nearly in the laboratory when a vacuum pump is used to evacuate a bottle. The pressure in a vacuum is called *absolute zero*, and all pressures referenced with respect to this zero pressure are termed *absolute pressures*. Therefore, atmospheric pressure at sea level on a particular day might be given as 101 kN/m², which is equivalent to 760 mm of deflection on a mercury barometer.

Many pressure-measuring devices measure not absolute pressure but only differences in pressure. For example, a common Bourdon-tube pressure gage (see Sec. 3-3) indicates only the difference between the pressure in the fluid to which it is tapped and the pressure in the atmosphere. In this case then, the reference pressure is actually the atmospheric pressure at the gage. This type of pressure reading is called gage pressure.

The fundamental unit of pressure in the SI system is the pascal (Pa), which is one newton per square meter (N/m²). Gage and absolute pressures are usually identified after the unit.[1] For example, if a pressure of 50 kPa were measured with a gage referenced to the atmosphere and the

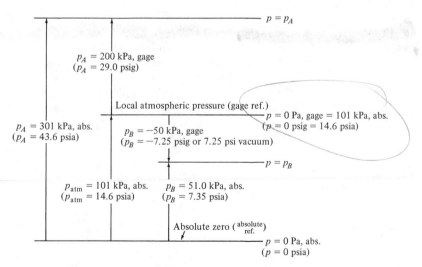

FIGURE 3-3   Example of pressure relations.

---

[1] In the traditional ft-lb-sec system of units the gage or absolute designations are usually included as part of the abbreviated unit. For example, a gage pressure of 10 pounds per square foot is designated as psfg. Other combinations are psfa, psig, psia. The latter two designations are for pounds per square-inch gage and pounds per square-inch absolute.

absolute atmospheric pressure were 100 kPa, then the pressure could be expressed

$$p = 50 \text{ kPa gage}$$

or $$\quad p = 150 \text{ kPa absolute}$$

Whenever atmospheric pressure is used as a reference (or, in other words, when gage pressure is being measured), the possibility exists that the pressure thus measured can be either positive or negative. Negative gage pressures are also termed *vacuum* pressures. Hence, if a gage tapped into a tank indicates a vacuum pressure of 31.0 kPa, this can also be stated as 70.0 kPa absolute, or −31.0 kPa gage pressure, assuming that the atmospheric pressure is 101 kPa absolute. An example of this reference system is depicted in Fig. 3-3 for arbitrary pressures of $p_A = 200$ kPa gage and $p_B = 51$ kPa absolute with an atmospheric pressure of 101 kPa absolute.

3-2   PRESSURE VARIATION WITH ELEVATION

Basic differential equation

For a static fluid, pressure varies only with change in elevation in the fluid. This may be shown by isolating a cylindrical element of fluid and by applying the equation of equilibrium to the element. Consider the element shown in Fig. 3-4. Here the element is oriented so that its longitudinal axis is parallel to an arbitrary ℓ direction. The element is $\Delta\ell$ long, ΔA in cross-sectional area, and inclined at an angle α with the horizontal. The

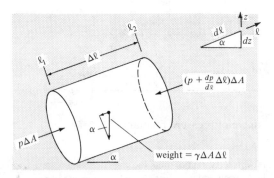

FIGURE 3-4   Pressure variation with elevation.

equation of equilibrium for the ℓ direction considering the pressure forces and gravitational force acting on the element in this direction is

$$\sum F_\ell = 0$$

$$p \, \Delta A - \left(p + \frac{dp}{d\ell} \Delta\ell\right) \Delta A - \gamma \, \Delta A \, \Delta\ell \sin \alpha = 0$$

Upon simplifying, this reduces to the following[1]:

$$\frac{dp}{d\ell} = -\gamma \sin \alpha$$

However, one notes that $\sin \alpha = dz/d\ell$; therefore,

$$\frac{dp}{d\ell} = -\gamma \frac{dz}{d\ell} \tag{3-5}$$

This can also be written as

$$\frac{dp}{dz} = -\gamma \tag{3-6}$$

which is the basic equation for hydrostatic pressure variation with elevation. Equation (3-5) states that for static fluids a change of pressure in the ℓ direction, $dp/d\ell$, occurs only when there is a change of elevation in the ℓ direction, $dz/d\ell$. In other words, if one considers a path through the fluid that lies in a horizontal plane, the pressure everywhere along this path will be constant. On the other hand, the greatest possible change in hydrostatic pressure will occur along a vertical path through the fluid. Furthermore, Eqs. (3-5) and (3-6) state that the pressure changes inversely with elevation. If one travels upward in the fluid (positive z direction), the pressure decreases; and if one goes downward (negative z), the pressure increases. Of course, a pressure increase is exactly what a diver experiences when descending in a lake or pool.

EXAMPLE 3-2   Compare the rate of change of pressure with elevation for air at sea level, 101.3 kPa absolute, at a temperature of 15.5°C, and for fresh water at the same pressure and temperature. Assuming constant specific weights for air and water, determine also the total pressure change that occurs with a 4-m decrease in elevation.

[1] The foregoing derivation includes only the first-order change of pressure over the incremental distance, $\Delta\ell$; however, it can be shown that even if higher-order terms are included, the identical differential equation for pressure variation results as $\Delta\ell$ approaches zero.

Solution Determine specific weights of water and air:

$$\rho_{air} = \frac{p}{RT} = \frac{101.3 \times 10^3 \text{ N/m}^2}{287 \text{ J/kg K} \times (15.5 + 273)\text{K}}$$

Then

$$\rho_{air} = 1.22 \text{ kg/m}^3$$

$$\gamma_{air} = \rho g = 1.22 \text{ kg/m}^3 \times 9.81 \text{ m/s}^2$$

$$= 11.97 \text{ kg/m}^2\text{s}^2$$

$$= 11.97 \text{ N/m}^3$$

and

$$\gamma_{water} = 9{,}799 \text{ N/m}^3 \text{ (interpolated from Table A-3)}$$

$$\frac{dp}{dz} = -\gamma$$

Then

$$\left(\frac{dp}{dz}\right)_{air} = -11.97 \text{ N/m}^3$$

$$\left(\frac{dp}{dz}\right)_{water} = -9{,}799 \text{ N/m}^3$$

Total pressure change for air $= (-11.97 \text{ N/m}^3) \times (-4 \text{ m})$

$$= 47.9 \text{ N/m}^2$$

$$= 47.9 \text{ Pa} \qquad \blacktriangleleft$$

Total pressure change for water $= (-9{,}799 \text{ N/m}^3) \times (-4 \text{ m})$

$$= 39.2 \text{ kN/m}^2$$

$$= 39.2 \text{ kPa} \qquad \blacktriangleleft$$

Pressure variation for a uniform-density fluid

Equations (3-5) and (3-6) are completely general in the sense that they describe the rate of change of pressure for all fluids in static equilibrium; however, much simplification accrues in practical application of the equations if it can be assumed that the density, thus the specific weight, of the fluid is the same throughout. Then γ is a constant in Eqs. (3-5) and (3-6). The reason for the simplification is that the integration of Eq. (3-5) or (3-6) becomes easier and the resulting equation is simpler than if γ were a function of z. With constant specific weight the following equation results upon integration of Eq. (3-6):

$$p = -\gamma z + \text{constant} \tag{3-7}$$

or

$$\left(\frac{p}{\gamma} + z\right) = \text{constant} \tag{3-8}$$

The sum of the terms p/γ and z on the left-hand side of Eq. (3-8) (p. 36) is called the *piezometric head*, and as shown by the equation, this is constant throughout an incompressible static fluid. Therefore, one can refer the pressure and elevation at one point to the pressure and elevation at another point in the fluid in the following manner:

$$\frac{p_1}{\gamma} + z_1 = \frac{p_2}{\gamma} + z_2 \tag{3-9}$$

or
$$\Delta p = -\gamma\,\Delta z \tag{3-10}$$

Note, however, that Eqs. (3-7) through (3-10) are applicable only in fluids with constant specific weights. In other words, Eqs. (3-9) and (3-10) can be applied between two points in a given fluid but not across an interface between two fluids having different specific weights.

EXAMPLE 3-3 What is the water pressure at a depth of 35 ft (10.67 m) in the tank shown?

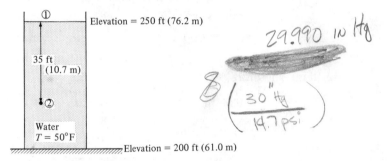

Solution   At an elevation of 250 ft the gage pressure is zero, so

$$\frac{p_1}{\gamma} + z_1 = \frac{p_2}{\gamma} + z_2$$

$$0 + 250 = \frac{p_2}{\gamma} + 215$$

$$\frac{p_2}{\gamma} = 35 \text{ ft}$$

$$p_2 = 35 \times 62.4 = 2{,}180 \text{ psfg}$$

$$= 15.2 \text{ psig} \qquad \blacktriangleleft$$

SI units $p_2 = 104.8$ kPa gage $\blacktriangleleft$

EXAMPLE 3-4 Oil with a specific gravity of 0.80 is 0.90 m deep in an open tank that is otherwise filled with water. If the tank is 3.0 m deep, what is the gage pressure at the bottom of the tank?

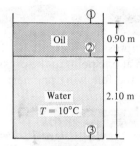

Solution   First determine the pressure at the oil-water interface staying within the oil and then calculate the pressure at the bottom.

$$\frac{p_1}{\gamma} + z_1 = \frac{p_2}{\gamma} + z_2$$

where p_1 = pressure at free surface of oil

z_1 = elevation of free surface of oil

p_2 = pressure at interface between oil and water

z_2 = elevation at interface between oil and water

For this example, $p_1 = 0$, $\gamma = 0.80 \times 9{,}810$ N/m³, $z_1 = 3$ m, and $z_2 = 2.10$ m. Therefore,

$$p_2 = 0.90 \text{ m} \times 0.80 \times 9{,}810 \text{ N/m}^3 = 7.06 \text{ kPa gage}$$

Now obtain p_3 from

$$\frac{p_2}{\gamma} + z_2 = \frac{p_3}{\gamma} + z_3$$

where p_2 has already been calculated and $\gamma = 9{,}810$ N/m³.

$$p_3 = 9{,}810 \left(\frac{7{,}060}{9{,}810} + 2.10\right)$$

$$= 27.7 \text{ kPa gage} \qquad \blacktriangleleft$$

Pressure variation for compressible fluids

The previous section dealt with pressure variation in fluids for which the specific weight is constant. However, when the specific weight varies significantly, it must be expressed in such a form that Eq. (3-6) can be integrated. For the case of an ideal gas, this is accomplished through the equation of state, Eq. (2-4), which relates the density of the gas to pressure and temperature:

$$\frac{p}{\rho} = RT$$

or
$$\rho = \frac{p}{RT} \tag{3-11}$$

which can be expressed as follows when both sides of Eq. (3-11) are multiplied by g:

$$\gamma = \frac{pg}{RT} \tag{3-12}$$

where
R = gas constant, 287 J/kg · K, for dry air

T = absolute temperature K

p = absolute pressure Pa

Because Eq. (3-12) introduces another variable, temperature, it is now necessary to have additional data relating temperature and elevation. If one is interested in the pressure variation in the atmosphere, and if temperature-versus-elevation data for a local area at a given time are available, then one can quite accurately compute pressure versus elevation. Lacking such data, one can resort to the so-called *U.S. standard atmosphere* (1). This is a set of data compiled by the U.S. Weather Bureau that represents average conditions over the United States at 40°N latitude. At sea level the standard atmospheric pressure is 101.3 kPa and the temperature is 288 K. Also, the atmosphere is divided into two layers, the *troposphere* and *stratosphere*. In the troposphere, defined as the layer between sea level and 10,769 m, the temperature decreases linearly with increasing elevation at a *lapse rate* of 6.50 K/km. The stratosphere begins at the top of the troposphere and extends to 32.3 km elevation. In the stratosphere the temperature is constant at −55°C.

We have sufficient information, then, to calculate the pressure and density at any elevation. Let us first consider the troposphere.

PRESSURE VARIATION IN THE TROPOSPHERE   Let the temperature T be given by

$$T = T_0 - \alpha(z - z_0) \tag{3-13}$$

In this equation T_0 is the temperature at a reference level where the pressure is known and α is the lapse rate. If we use the specific weight of a gas from Eq. (3-12) in the basic hydrostatic equation we obtain

$$\frac{dp}{dz} = -\frac{pg}{RT} \tag{3-14}$$

Upon substituting Eq. (3-13) for T, we get

$$\frac{dp}{dz} = -\frac{pg}{R[T_0 - \alpha(z - z_0)]}$$

Now we must separate the variables and integrate to obtain

$$\frac{p}{p_0} = \left[\frac{T_0 - \alpha(z - z_0)}{T_0}\right]^{g/\alpha R}$$

$$p = p_0 \left[\frac{T_0 - \alpha(z - z_0)}{T_0}\right]^{g/\alpha R} \tag{3-15}$$

EXAMPLE 3-5 If at sea level the absolute pressure and temperature are 101.3 kPa and 15°C, respectively, what is the pressure at 2,000-m elevation, assuming that standard atmospheric conditions prevail?

Solution Use Eq. (3-15):

$$p = p_0 \left[\frac{T_0 - \alpha(z - z_0)}{T_0}\right]^{g/\alpha R}$$

where
$$p_0 = 101,300 \text{ N/m}^2$$
$$T_0 = 273 + 15 = 288 \text{ K}$$
$$\alpha = 6.50 \times 10^{-3} \text{ K/m}$$
$$z - z_0 = 2,000 \text{ m}$$
$$g/\alpha R = 5.259$$

then
$$p = p_0 \left(\frac{288 - 6.50 \times 10^{-3} \times 2,000}{288}\right)^{5.259}$$

$$= 79.5 \text{ kPa} \qquad \blacktriangleleft$$

PRESSURE VARIATION IN THE STRATOSPHERE In the stratosphere the temperature is assumed to be constant; therefore, when Eq. (3-14) is integrated, we obtain

$$\ln p = -\frac{zg}{RT} + C$$

At $z = z_0$, $p = p_0$; therefore, the foregoing equation reduces to

$$\frac{p}{p_0} = e^{-(z-z_0)g/RT}$$

or
$$p = p_0 e^{-(z-z_0)g/RT} \tag{3-16}$$

EXAMPLE 3-6 If the pressure and temperature are 3.28 psia ($p = 22.6$ kPa absolute) and $-67°F$ ($-55°C$) at an elevation of 36,000 ft

(10,973 m), what is the pressure at 56,000 ft (17,069 m), assuming iso-thermal conditions over this range of elevation?

Solution   For isothermal conditions

$$T = -67 + 460 = 393°R$$

$$p = p_0 e^{-(z-z_0)g/RT}$$

or

$$p = 3.28 e^{-(20,000)(32.2)/(1,716 \times 393)}$$

$$= 3.28 e^{-0.955}$$

Therefore, the pressure at 56,000 ft is

$$p = 1.26 \text{ psia}$$ ◄

SI units $$p = 8.69 \text{ kPa absolute}$$ ◄

3-3   PRESSURE MEASUREMENTS

Introduction

Numerous instruments have been devised to indicate the magnitude of pressure intensity, and most of these operate on either the principle of manometry or the principle of flexing of an elastic member, the deflection of which is directly proportional to the applied pressure. These principles and representative pressure gages are described in the following sections.

Manometry

Basically, this method utilizes the change in pressure with elevation to evaluate pressure. By measuring the height of liquid in a single tube or by measuring the deflection of a liquid in a U-tube (a *differential manometer*), the pressure at the point where the tubes are connected can be evaluated. First consider the *piezometer*, or simple manometer, attached to the pipe as in Fig. 3-5. It is easy to compute the gage pressure at the center of the pipe; here the pressure is simply $p = \gamma h$, which follows directly from Eq. (3-10). This type of pressure-indicating device is accurate and simple; however, the student may visualize how impractical it might become for measuring high pressures, and of course it is useless in its present form for pressure measurement in gases. For both these cases, a U-tube such as is shown in Fig. 3-6 can be employed. In this case, a knowledge of the specific weights of fluids involved and of the linear measurements ℓ and Δh is needed to calculate the pressure in the pipe. Here the procedure is to calculate the pressure changes, step by step, from one level to the next in each fluid and to apply these changes finally to

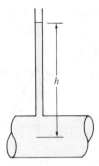

evaluate the unknown pressure. The following example will illustrate the procedure for the case shown in Fig. 3-6.

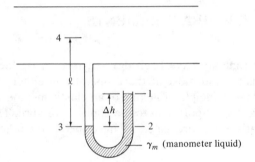

FIGURE 3-6 U-tube manometer.

EXAMPLE 3-7 Water is the liquid in the pipe of Fig. 3-6 and mercury is the manometer fluid. If the deflection Δh is 60 cm and ℓ is 180 cm, what is the gage pressure at the center of the pipe?

Solution Since the manometer is open to the atmosphere, we know that the gage pressure at point 1, the mercury surface, is zero. Then the pressure at 2 will be

$$p_2 = p_1 + \text{change in pressure between 1 and 2}$$
$$= 0 - \gamma_m \, \Delta z$$
$$= 0 - \gamma_m(-0.60) \quad \text{where } \gamma_m = 133 \text{ kN/m}^3$$
$$= 79.8 \text{ kPa}$$

Point 3 is at the same elevation as point 2 and in the same fluid; therefore, $p_3 = p_2$. The next step is to evaluate Δp from 3 to 4 and to apply this to the pressure at 3:

$$\Delta p_{3\rightarrow 4} = \gamma \times 1.80 \qquad \text{where } \gamma = 9,810 \text{ N/m}^3$$
$$= -17.66 \text{ kPa}$$

Then
$$p_4 = p_3 - 17.66 \text{ kPa}$$
$$= 62.1 \text{ kPa gage} \qquad \blacktriangleleft$$

Once the student becomes familiar with the basic principle of manometry, it should be easy to write one single equation rather than separate equations for each step in Example 3-7. The single equation for evaluation of the pressure in the pipe of Fig. 3-6 is

$$0 + \gamma_m \, \Delta h - \gamma \ell = p_p$$

One can read the equation in this way: zero pressure at the open end plus the change in pressure from 1 to 2 minus the change in pressure from 3 to 4 equals the pressure in the pipe. The main point that the student must remember in this process is that when one travels downward in the fluid, the pressure increases; and when one travels upward, the pressure decreases.

Up to now we have considered only liquid-filled manometers. Let us consider Fig. 3-6 again with a gas-filled pipe. This is illustrated in the next example.

EXAMPLE 3-8   Air at 20°C is the fluid in the pipe of Fig. 3-6 and water is the manometer fluid. If the deflection Δh is 70 cm and ℓ is 140 cm, what is the gage pressure in the pipe? Also compute this pressure by neglecting the pressure change due to the 140 cm column of air. Assume standard atmospheric pressure.

Solution   The specific weight of air is found from Eq. (3-12), which requires that the air pressure be known; therefore, this is first calculated. The air pressure at the bottom of the 70 cm column is given as

$$p_{\text{air}} = 9,790 \text{ N/m}^3 \times 0.70 \text{ m}$$
$$= 6,853 \text{ Pa gage}$$

Then the absolute air pressure is given as

$$p_{\text{air}} = 6,853 \text{ Pa} + 101,300 \text{ Pa} = 108.15 \text{ kPa}$$

Then
$$\rho_{\text{air}} = \frac{p}{RT} = \frac{108,150 \text{ N/m}^2}{287 \text{ J/kg K} \times (20 + 273)\text{K}}$$
$$= 1.286 \text{ kg/m}^3$$

or
$$\gamma_{\text{air}} = 1.286 \text{ kg/m}^3 \times 9.81 \text{ m/s}^2 = 12.62 \text{ N/m}^3$$

Now compute the gage pressure in the pipe:

$$p_{pipe} = 6{,}853 \text{ Pa} - 1.4 \text{ m} \times 12.62 \text{ N/m}^3$$
$$= 6{,}835 \text{ Pa} \qquad \blacktriangleleft$$

If we neglect the effect of the air column, the gage pressure in the pipe is

$$p_{pipe} = 9{,}790 \text{ N/m}^3 \times 0.70 \text{ m} = 6{,}853 \text{ Pa} \qquad \blacktriangleleft$$

Results of the foregoing example show that when liquids and gases are both involved in a manometer problem, it is well within engineering accuracy to neglect the pressure changes due to the columns of gas.

EXAMPLE 3-9   What is the pressure of the air in the tank shown in the accompanying figure if $\ell_1 = 40$ cm (1.31 ft), $\ell_2 = 100$ cm (3.28 ft), and $\ell_3 = 80$ cm (2.62 ft)?

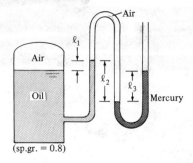

Solution

SI units $0 + 0.80 \text{ m} \times 133{,}000 \text{ N/m}^3 + 0.4 \text{ m} \times 9{,}810 \text{ N/m}^3$
$$\times 0.8 = p_{air}$$

$$p_{air} = 109.5 \text{ kPa gage} \qquad \blacktriangleleft$$

Traditional units

$$0 + 2.62 \text{ ft} \times 846 \text{ lbf/ft}^3 + 1.31 \text{ ft} \times 62.4 \text{ lbf/ft}^3 \times 0.8 = p_{air}$$
$$p_{air} = 2{,}282 \text{ psfg}$$
$$= 15.85 \text{ psig} \qquad \blacktriangleleft$$

DIFFERENTIAL MANOMETERS   It is often desirable to measure the difference in pressure between two points in a pipe, and for this application a manometer is connected to the two points between which the pressure difference is to be measured. Such a setup is shown in Fig. 3-7. In this case, a gas is flowing and the pressure difference between points 1 and 2 is given by $\Delta p = \gamma_m \, \Delta h$, where γ_m is the specific weight of the manometer liquid and Δh is the deflection of this liquid.

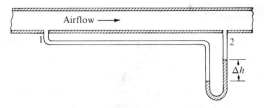

FIGURE 3-7   Differential manometer.

BOURDON-TUBE GAGE   This type of gage consists of a tube having an elliptical cross section bent into a circular arc as shown in Fig. 3-8*b*. When atmospheric pressure (zero gage pressure) prevails in the gage, the tube is undeflected, and for this condition the gage pointer is calibrated to read zero pressure. When pressure is applied to the gage, the curved tube tends to straighten, much like the type of party favors that straighten out when one blows into them, thereby actuating the pointer to read correspondingly higher pressure. The Bourdon-tube gage is a very common type of gage, which is reliable if not subjected to excessive pressure pulsations or undue external shock. However, because both these conditions sometimes prevail in engineering applications, it is desirable that pulsation dampers be installed in the line leading to such gages and that the gages be periodically calibrated to check their accuracy.

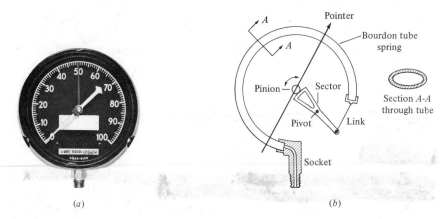

FIGURE 3-8   Bourdon-tube gage. (*a*) View of typical gage. (*b*) Internal mechanism (schematic).

Pressure transducers

Modern factories and systems that involve flow processes are controlled automatically, and much of their operation involves sensing of pressure at critical points of the system. Therefore, pressure-sensing devices, such as pressure transducers, are designed to produce electronic signals that can be transmitted to oscillographs or digital devices for record-keeping and/or to control other devices for process operation. Basically, most transducers are tapped into the system with one side of a small diaphragm exposed to the active pressure of the system. When the pressure changes, the diaphragm flexes and a sensing element connected to the other side of the diaphragm produces a signal that is usually linear with the change in pressure in the system. There are many types of sensing elements; however, one common type is the resistance-wire strain gage attached to a flexible diaphragm. As the diaphragm flexes, the wires of the strain gage change length, thereby changing the resistance of the wire. It is this resistance change that is utilized electronically to produce a voltage change that can then be used in various ways. Figure 3-9 shows a schematic arrangement for a pressure transducer and associated electronic equipment used to record pressure data.

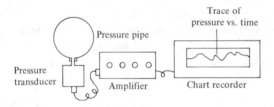

FIGURE 3-9   Schematic of pressure transducer and associated equipment.

3-4   HYDROSTATIC FORCES ON PLANE SURFACES

Introduction

Surfaces that are horizontal or are subjected to gas pressure have essentially constant pressure over their entire surface; therefore, the total force resulting from the pressure is equal to the product of the pressure and the area of the surface. For this case the resultant force acts at the centroid of the area and its line of action is normal to the area.

If a plane surface is not horizontal and if it is acted upon by a hydrostatic force such as that produced by static liquids, then the pressure is linearly distributed over the surface and a more general type of analysis must be made to evaluate the magnitude of the resultant force and the location of its line of action. The following derivations assume atmospheric pressure at the liquid surface.

Magnitude of resultant hydrostatic force

Consider the top side of the plane surface A-B in Fig. 3-10. Here line A-B is the end view of a surface entirely submerged in the liquid. The plane of this surface intersects the horizontal liquid surface at axis 0-0 with an angle α. The distance from the axis 0-0 to the horizontal axis through the centroid of the area is given by $\bar{y}$. The distance from 0-0 to the differential area dA is y. The pressure on the differential area can be computed if the y

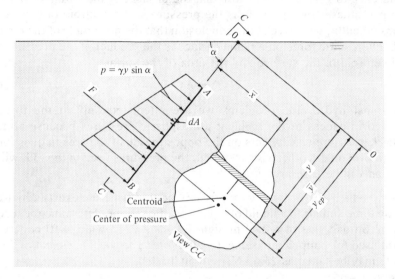

FIGURE 3-10   Hydrostatic pressure distribution on a plane surface.

distance to the point is known; that is, $p = \gamma y \sin \alpha$. Then it follows that the differential force on the differential area is

$$dF = p \, dA$$

or
$$dF = \gamma y \sin \alpha \, dA$$

The total force on the area will be obtained by integrating the differential force over the entire area:

$$F = \int_A p \, dA$$

or
$$F = \int_A \gamma y \sin \alpha \, dA \qquad (3\text{-}17)$$

In Eq. (3-17), γ and $\sin \alpha$ are constants; therefore, we obtain

$$F = \gamma \sin \alpha \int_A y \, dA \qquad (3\text{-}18)$$

Now, the integral in Eq. (3-18) is the first moment of the area; consequently, this is replaced by its equivalent, $\bar{y}A$. Therefore, we obtain

$$F = \gamma \, \bar{y}A \sin \alpha$$

which can be rewritten in the following form:

$$F = (\gamma\bar{y} \sin \alpha)A \qquad (3\text{-}19)$$

Reference to Fig. 3-10 will show that the product of the variables within the parentheses of Eq. (3-19) is the pressure at the centroid of the area. Consequently, we arrive at the conclusion that the magnitude of the resultant hydrostatic force on a plane surface is the product of the pressure at the centroid of the surface and the area of the surface:

$$F = \bar{p}A \qquad (3\text{-}20)$$

For most hydrostatic problems we are interested only in the forces created in excess of the ambient atmospheric pressures because atmospheric pressure usually acts on the opposite side of the area in question. Therefore, unless otherwise specified, the pressures used in the following section will be gage pressure.

EXAMPLE 3-10 Assuming that freshly poured concrete exerts a hydrostatic force similar to a liquid of equal specific weight, determine the force acting on one side of a 2.44-m-high (8-ft) × 1.22-m-wide (4-ft) concrete form used for pouring a basement wall. *Note:* The specific weight of concrete may be taken as 23.6 kN/m³ (150 lbf/ft³).

Solution

$$F = \bar{p}A$$
$$\bar{p} = 1.22 \text{ m} \times 23.6 \times 10^3 \text{ N/m}^3 = 28.79 \text{ kPa}$$
$$A = 1.22 \times 2.44 = 2.98 \text{ m}^2$$

Then $\qquad F = 28.79 \times 10^3 \text{ N/m}^2 \times 2.98 \text{ m}^2 = 85.8 \text{ kN}$ ◄

Vertical location of line of action of resultant hydrostatic force

In general, the location of the line of action of the resultant hydrostatic force lies below the centroid because pressure increases with depth. We can derive an equation for this location by taking moments of the pressure forces about the horizontal axis 0-0. We call the point where the resultant force intersects the surface the *center of pressure* and identify the slant distance from 0-0 to this point by y_{cp} (Fig. 3-10). Then, by definition of the location of a resultant force, the following moment equation can be written:

$$y_{cp}F = \int y \, dF$$

But dF is given by $dF = p \, dA$; therefore,

$$y_{cp}F = \int_A yp \, dA$$

Also,
$$p = \gamma y \sin \alpha$$

so
$$y_{cp}F = \int_A \gamma y^2 \sin \alpha \, dA \qquad (3\text{-}21)$$

Again, as in Eq. (3-17), γ and $\sin \alpha$ are constants, so we obtain

$$y_{cp}F = \gamma \sin \alpha \int_A y^2 \, dA \qquad (3\text{-}22)$$

Here we have the second moment of the area (often called area moment of inertia) at the right; therefore, this shall be identified as I_0. However, for engineering applications it is convenient to express the second moment with respect to the horizontal centroidal axis of the area; hence by the transfer equation we have

$$I_0 = \bar{I} + \bar{y}^2 A \qquad (3\text{-}23)$$

When this is substituted into Eq. (3-22) we obtain

$$y_{cp}F = \gamma \sin \alpha \, (\bar{I} + \bar{y}^2 A)$$

However, from Eq. (3-19), $F = \gamma \bar{y} \sin \alpha \, A$; therefore,

$$y_{cp}(\gamma \bar{y} \sin \alpha \, A) = \gamma \sin \alpha \, (\bar{I} + \bar{y}^2 A)$$

This then reduces to

$$y_{cp} = \bar{y} + \frac{\bar{I}}{\bar{y}A} \qquad (3\text{-}24)$$

or
$$y_{cp} - \bar{y} = \frac{\bar{I}}{\bar{y}A} \qquad (3\text{-}25)$$

It can be seen from Eq. (3-25) that for a given area the center of pressure comes closer to the centroid as the area is lowered deeper into the liquid. Equation (3-25) is valid only when one liquid is involved. In addition, it is restricted to the case where $p = 0$ gage at the liquid surface. If the pressure is not zero at the surface, then an equivalent problem can be found that satisfies the restriction. That is, y must be measured from an equivalent free surface located above the centroid of the area a distance $\bar{p}/\gamma$.

EXAMPLE 3-11   An elliptical gate covers the end of a 4-m-diameter pipe. If the gate is hinged at the top, what normal force F is required to open the gate when water is 8 m deep above the top of the pipe and the pipe is open to the atmosphere on the other side? Neglect the weight of the gate.

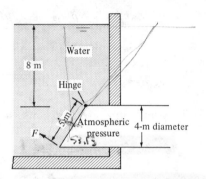

Solution   First evaluate the magnitude of the hydrostatic force:

$$F = \bar{p}A$$

The area in question is an ellipse with major and minor axes of 5 m and 4 m. The area is given by the formula $A = \pi ab$ (from Fig. A-1)

Then $F = 10\ \text{m} \times 9{,}810\ \text{N/m}^3 \times \pi \times 2\ \text{m} \times 2.5\ \text{m}$

$= 1.541\ \text{MN}$

Now calculate the slant distance between the centroid of the elliptical area and the center of pressure:

$$y_{cp} - \bar{y} = \frac{\bar{I}}{\bar{y}A} = \frac{\frac{1}{4}\pi a^3 b}{\bar{y}\pi ab} = \frac{\frac{1}{4}a^2}{\bar{y}}$$

Here $\bar{y} = 12.5$ m (slant distance from the water surface to the centroid). Thus

$$y_{cp} - \bar{y} = \frac{1}{4} \times \frac{6.25\ \text{m}^2}{12.5\ \text{m}} = 0.125\ \text{m}$$

Now take moments about the hinge at the top to obtain F:

$$\sum M_{\text{hinge}} = 0$$

$$1.541 \times 10^6\ \text{N} \times 2.625\ \text{m} - F \times 5\ \text{m} = 0$$

$$F = 809\ \text{kN}$$ ◀

Note: Students are sometimes confused about the axis from which to take the moment of inertia when computing the distance to the center of pressure. A check of the derivation will reveal that the area moment of inertia as used in Eq. (3-24) or Eq. (3-25) is always taken about the *horizontal-centroidal axis*.

Lateral location of line of action of resultant hydrostatic force

The same principles used for the vertical location of the line of action may
be used for the lateral location—that is, by taking moments about a line
normal to line 0-0 in Fig. 3-10. Areas that are symmetrical about an axis
normal to 0-0 will always yield a position for the center of pressure that is
along the axis of symmetry and below the centroid; however, for asym-
metrical areas it will be necessary to carry out the analysis to evaluate the
location.

EXAMPLE 3-12   Determine the magnitude of the hydrostatic force acting
on one side of this submerged vertical plate and determine the location of
the center of pressure (see the figure).

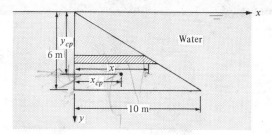

Solution   The centroid of the plate is at 4-m depth; therefore, $F =
4\ m \times 9{,}810\ N/m^3 \times \frac{1}{2} \times 60\ m^2 = 1.177\ MN$. The vertical location of
the center of pressure is obtained from Eq. (3-25):

$$y_{cp} - \bar{y} = \frac{\bar{I}}{\bar{y}A} = \frac{bh^3/36}{\bar{y}\frac{1}{2}bh} = \frac{h^2}{18\bar{y}} = \frac{36}{72}$$

$$y_{cp} = 4 + \tfrac{1}{2} = 4.50\ m$$

Obtain the lateral location of the center of pressure by summing moments
of forces acting on the elemental strips and then divide by F. Moments are
taken about the vertical edge:

$$dM = \tfrac{1}{2}x\ dF$$

$$= \tfrac{1}{2}x\gamma y x\ dy$$

but

$$x = \frac{10}{6}\ y$$

so

$$M = \frac{50}{36}\gamma \int_0^6 y^3\ dy$$

Then $$M = \frac{50}{36}\,9{,}810 \text{ N/m}^3 \left.\frac{y^4}{4}\right|_0^6 = 4.414 \text{ MN} \cdot \text{m}$$

But $$Fx_{cp} = M$$

Therefore $$x_{cp} = \frac{M}{F} = \frac{4.414 \text{ N} \cdot \text{m}}{1.177 \text{ N}} = 3.75 \text{ m} \qquad \blacktriangleleft$$

3-5   HYDROSTATIC FORCE ON CURVED SURFACES

This type of problem is analyzed by constructing horizontal and vertical projections of the surface (Fig. 3-11). Then the equations of equilibrium are applied to the fluid enclosed by these projections and by the surface in question. Consider the two-dimensional curved surface AB shown in Fig. 3-11. Assume that this surface has unit length normal to the paper. If we consider the mass of fluid OAB as a freebody and analyze the forces

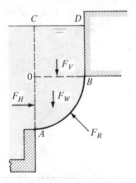

FIGURE 3-11   Analysis of hydrostatic force on a curved surface.

that act on it, we can identify the components that we are looking for. Because this is a hydrostatic condition, only normal forces act on the hypothetical surfaces OA and OB, and these forces are identified as F_H and F_V, respectively. The only other forces to act on this body of fluid are then the weight of the fluid itself, F_W, and the reaction from the curved surface, F_R. Therefore, all the forces acting on the body have been identified and the freebody is depicted as in Fig. 3-12. We now apply the equations of equilibrium to the freebody. In the horizontal direction $\Sigma F_x = 0$. Then $F_H - F_{Rx} = 0$, from which we get $F_{Rx} = F_H$. In the vertical direction $\Sigma F_y = 0$. Hence, we have

$$-F_V - F_W + F_{Ry} = 0$$
$$F_{Ry} = F_V + F_W$$

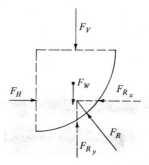

FIGURE 3-12   Freebody diagram.

In the analysis above we have shown that the magnitude of the horizontal reaction of the curved surface is equal to the hydrostatic force that acts on a vertical projection of the curved surface, and the magnitude of the vertical reaction is equal to the sum of the vertical forces acting above the curved surface—in this case, the weight of the fluid above.

Since we have found which forces make up the reaction of the curved surface, we need only reverse the signs on the components of the reaction to arrive at the components of the resultant hydrostatic force acting on the surface. By vectorially adding the component forces, we can determine both the magnitude and line of action of the resultant.

In summary, the horizontal component of force acting on a curved surface is equal to the force acting on a vertical projection of that surface—which includes both magnitude and line of action. The magnitude of the vertical force acting on a curved surface is equal to the sum of all vertical forces acting on it. The lines of action of these vertical components are used to determine the line of action of the resultant vertical force.

EXAMPLE 3-13   In Fig. 3-11 the surface AB is a circular arc with a radius of 2 m. The distance DB is 4 m. If water is the liquid supported by the surface and if atmospheric pressure prevails on the other side of AB, determine the magnitude and line of action of the resultant hydrostatic force on AB per unit length.

Solution   The vertical component is equal to the weight of water in volume $AOCDB$:

$$W_{OCDB} = \gamma\forall_{OCDB} = 9{,}810 \times 4 \times 2 \times 1 = 78.5 \text{ kN}$$
$$W_{AOB} = \gamma\forall_{AOB} = \gamma\tfrac{1}{4}\pi r^2 \times 1 = 30.8 \text{ kN}$$

Therefore, the vertical component is $F_{Ry} = 109.3$ kN. The line of action of the vertical force component acts through the centroid of the volume of water considered above, and this is calculated by taking moments about a horizontal axis through D and normal to plane ODB.

$$\bar{x}F_{Ry} = 1 \times 78.5 + r\left(1 - \frac{4}{3\pi}\right)30.8 = 114.0 \text{ kN}$$

Note: The $4/3\pi$ quantity is the distance to the centroid of the quadrant and is obtained from Fig. A-1 in the Appendix.

$$\bar{x} = \frac{114.0}{109.3} = 1.04 \text{ m}$$

The magnitude of the horizontal component is given by the force on OA:

$$F_H = \bar{p}A$$
$$= 5 \times 9,810 \times 2 \times 1 = 98.1 \text{ kN}$$

Location of the line of action of the horizontal component is

$$y_{cp} - \bar{y} = \frac{\bar{I}}{\bar{y}A}$$

$$= \frac{1 \times 2^3/12}{5 \times 2} = 0.0667 \text{ m}$$

$$y_{cp} = 5.067 \text{ m}$$

The resultant force is then as shown in the figure.

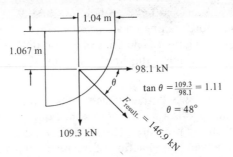

To this point we have considered a curved surface for which the external vertical forces act downward on the surface; however, it is easy to visualize the reverse sense for the hydrostatic forces if, for instance, the liquid and atmosphere are reversed from what is shown in Fig. 3-11. Such is the case in Fig. 3-13. If we consider the pressure point by point on the surface, we will find that it is exactly the same as when the liquid is to the left. However, the pressure forces on the curved surface in the second case act in the opposite sense to the forces in the first. Therefore, the

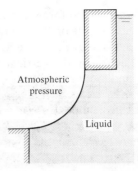

FIGURE 3-13   Upward hydrostatic force on a curved surface.

hydrostatic force acting on the curved surface in Fig. 3-13 will be reversed from that for Fig. 3-11 but will have the same line of action.

3-6 BUOYANCY

Basic development of the principle of buoyancy

The basic principles of buoyancy are readily grasped by referring to the principles of forces on curved surfaces that were introduced in Sec. 3-5. Let us consider a submerged body *ABCDEF* as shown in Fig. 3-14. We first examine the horizontal forces acting on the body in the y direction.

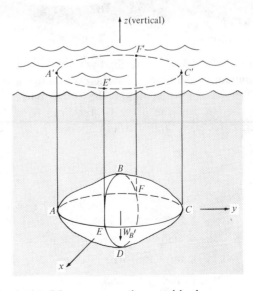

FIGURE 3-14   Analysis of forces on a submerged body.

We may think of these forces as the force that acts on the left end of the body *AEBFD* and the force that acts on the right end *CEBFD*. But to evaluate the force on the left end is to evaluate the force on the curved surface *AEBFD*. This is the force acting on the vertical projection of the surface, the plane surface *EBFD*, and the force will act in the positive *y* direction. Similarly, the horizontal force acting on the right end will be the force that acts on the same vertical projection *EBFD*; however, the sense of the right-end force will be reversed from the left-end force. Therefore, in the *y* direction we have two forces of equal and opposite magnitude acting on the body; consequently, they cancel one another. The same result would occur if we carried out a similar analysis for horizontal forces acting in the *x* direction. Thus it is seen that the net horizontal force acting on a submerged body in a static fluid is zero.

Consider next the vertical forces acting on the submerged body. We must take into account the vertical force acting on the top part of the body and then on the bottom part of the body. On the top the vertical force acts downward. Let us call this force F_{BV}, which will be equal to the weight of the fluid above the top surface of the body. To visualize the force that acts on the bottom, we should think about the equilibrium condition if the body were replaced by an equal volume of fluid having the same properties as the surrounding fluid. One should sense that equilibrium would prevail. Therefore, the force acting on the bottom surface is equal to F_{BV} plus the weight of fluid having a volume of the submerged body. The difference in forces on the bottom and top surfaces of the body, the buoyant force, is thus *equal to the weight of the displaced fluid and acts vertically upward*. The line of action of the buoyant force acts through the centroid of the displaced volume.

EXAMPLE 3-14 A 50-gal oil drum filled with air is to be used to assist a diver raise an ancient ship anchor from the bottom of the ocean. If the anchor weighs 400 lbf in sea water, and if the empty barrel weighs 50 lbf in air, how much weight will the diver be required to lift when the submerged air-filled barrel is attached to the anchor?

Solution

$$\sum F_y = 0$$

$$F_D - W_B + F_B - W_A = 0$$

where
$\quad F_D$ = force applied by diver

$\quad W_B$ = weight of barrel = 50 lbf

$\quad F_B$ = buoyant force of barrel

$\quad W_A$ = weight of anchor

Here $\qquad F_B = \dfrac{50 \text{ gal}}{7.48 \text{ gal/ft}^3} \times 64 \text{ lbf/ft}^3 = 427.8 \text{ lbf}$

Therefore, $\qquad F_D = W_B + W_A - F_B$

$$= 50 + 400 - 427.8 = 22.2 \text{ lbf} \qquad \blacktriangleleft$$

The preceding discussion pertains to totally submerged bodies; however, the buoyant force on floating bodies is also equal to the weight of the displaced liquid, and the force also acts vertically upward through the centroid of the volume displaced. This buoyant force just balances the weight of the body itself for equilibrium to prevail, which is why the gross tonnage of a ship is often referred to as the "displacement" of the ship.

Hydrometry

Precise measurement of the specific weight of a liquid is done by utilizing the principle of buoyancy. The device used for this, the *hydrometer*, is a glass bulb weighted on one end to make the hydrometer float in a vertical position and a stem of constant diameter extending from the other end (Fig. 3-15). The hydrometer is so designed that only the stem end extends above the liquid surface; therefore, appreciable vertical movement of the hydrometer is required to change the buoyant force or displaced volume of the device. Because the buoyant force (equal to the weight of the hydrometer) must be constant, the hydrometer will float deeper or shallower depending upon the specific weight of the liquid. Consequently,

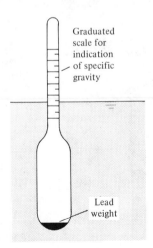

Graduated scale for indication of specific gravity

Lead weight

FIGURE 3-15   Hydrometer.

graduations on the stem corresponding to different depths of submergence of the hydrometer can be made to indicate directly the specific
weight or specific gravity of the liquid being measured.

3-7   STABILITY OF IMMERSED AND FLOATING BODIES

Immersed bodies

The stability of an immersed body depends upon the relative positions of
the *center of gravity* of the body and the centroid of the displaced volume
of fluid, called the *center of buoyancy*. If the center of buoyancy is above
the center of gravity, such as in Fig. 3-16a, any tipping of the body will
produce a righting couple; consequently, the body will be stable. However, if the situation is reversed and the center of gravity is above the
center of buoyancy, any tipping will produce an increasing overturning
moment, thus causing the body to turn through 180°. This is the condition
shown in Fig. 3-16c. Finally, if the center of buoyancy and center of gravity are coincident, the body is neutrally stable, that is, has the tendency
for neither righting nor overturning.

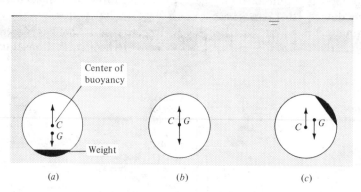

FIGURE 3-16   Conditions of stability for immersed bodies. (a) Stable. (b) Neutral.
(c) Unstable.

Floating bodies

The question of stability is more involved for floating bodies than for immersed bodies because the center of buoyancy may take different positions with respect to the center of gravity depending upon the shape of the
body and the position in which it is floating. For example, consider the

ship cross section shown in Fig. 3-17*a*. Here the center of gravity *G* is above the center of buoyancy *C*; therefore, at first glance it would appear that the ship is unstable and could flip over. However, if we observe the position of *C* and *G* after the ship has taken a small angle of heel, as shown in Fig. 3-17*b*, we see that the center of gravity is in the same position but the center of buoyancy has moved outward of the center of gravity, thus producing a righting moment. A ship having such characteristics is stable.

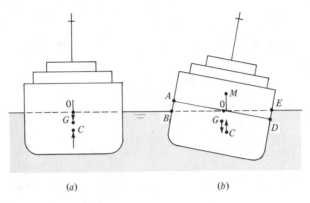

FIGURE 3-17   Ship stability relations.

The reason for the change in the center of buoyancy for the ship is that part of the original buoyant volume, as shown by the wedge shape *AOB*, is transferred to a new buoyant volume *EOD*. Because the buoyant center is at the centroid of the displaced volume, it follows that the buoyant center must move laterally to the right for this case. The point of intersection of the lines of action of the buoyant force before and after heel is called the *metacenter M*, and the distance *GM* is called the *metacentric height*. If *GM* is positive, that is, if *M* is above *G*, the ship is stable; however, if *GM* is negative, the ship is unstable. Quantitative relations involving these basic principles of stability are presented in the next paragraph.

Consider the ship in Fig. 3-18, which has taken a small angle of heel α. First we evaluate the lateral displacement of the center of buoyancy, *CC'*; then it will be easy by simple trigonometry to solve for the metacentric height *GM* or to evaluate the righting moment. Recall that the center of buoyancy is at the centroid of the displaced volume; therefore, we must resort to the basic fundamentals of centroids to evaluate the displacement *CC'*. From the basic definition of the centroid of a volume we can write the following equation:

$$\bar{x} \forall = \sum x_i \, \Delta \forall_i \tag{3-26}$$

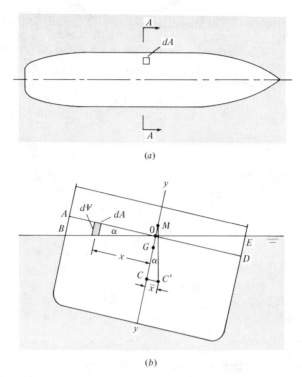

(a)

(b)

FIGURE 3-18   (a) Plan view of ship at waterline. (b) Section A-A of ship.

where $\bar{x} = CC'$ = distance from plane about which moments are taken to centroid of $\forall$

$\forall$ = total volume displaced

$\Delta\forall_i$ = volume increment

x_i = moment arm of increment of volume

Here we take moments about the plane of symmetry of the ship. It should be recalled from mechanics that in applying this equation, volumes to the left produce negative moments while volumes to the right produce positive moments. For the right side of Eq. (3-26) we write terms for the moment of the submerged volume about the plane of symmetry; however, the convenient way to do this is to consider the moment of the volume before heal, subtract the moment of the volume represented by the wedge *AOB*, and then add the moment represented by the wedge *EOD*. In a general way this is given by the following equation:

$\bar{x}\forall$ = moment of $\forall$ before heel

$$- \text{ moment of } \forall_{AOB} + \text{ moment of } \forall_{EOD} \quad (3\text{-}27)$$

Because the original buoyant volume is symmetrical with y-y, the moment for the first term on the right is zero. Also, the sign of the moment of $\forall_{AOB}$ is negative; therefore, when this negative moment is subtracted from the right-hand side of Eq. (3-27), we arrive at the following equation:

$$\bar{x}\forall = \sum x_i \, \Delta\forall_{i_{AOB}} + \sum x_i \, \Delta\forall_{i_{EOD}} \qquad (3\text{-}28)$$

Now expressing Eq. (3-28) in integral form,

$$\bar{x}\forall = \int_{AOB} x \, d\forall + \int_{EOD} x \, d\forall \qquad (3\text{-}29)$$

But it may be seen from Fig. 3-18b that $d\forall$ can be given as the product of the length of the differential volume $x \tan \alpha$ and the differential area dA. Consequently, Eq. (3-29) can be written as

$$\bar{x}\forall = \int_{AOB} x^2 \tan \alpha \, dA + \int_{EOD} x^2 \tan \alpha \, dA$$

Here $\tan \alpha$ is a constant with respect to the integration; and since the two terms on the right-hand side are identical except for the area over which integration is to be performed, we combine them as follows:

$$\bar{x}\forall = \tan \alpha \int_{A_{\text{waterline}}} x^2 \, dA \qquad (3\text{-}30)$$

The second moment or moment of inertia of the area defined by the waterline will be given the symbol I_{00}; therefore, the following is obtained:

$$\bar{x}\forall = I_{00} \tan \alpha$$

Next, replacing $\bar{x}$ by CC' and solving for CC',

$$CC' = \frac{I_{00} \tan \alpha}{\forall}$$

From Fig. 3-18, $\qquad\qquad CC' = CM \tan \alpha$

Thus eliminating CC' and $\tan \alpha$ yields

$$CM = \frac{I_{00}}{\forall}$$

However, $\qquad\qquad GM = CM - CG$

Therefore, $\qquad\qquad GM = \frac{I_{00}}{\forall} - CG \qquad (3\text{-}31)$

Equation (3-31) is used to determine the stability of floating bodies. As already noted, if GM is positive the body is stable, and if GM is negative it is unstable.

EXAMPLE 3-15   A block of wood 30 cm square in cross section and 60 cm long weighs 318 N. Will the block float with sides vertical?

Solution   First determine the depth of submergence of the block. This is calculated by applying the equation of equilibrium in the vertical direction.

$$\sum F_y = 0$$

$$-\text{weight} + \text{buoyant force} = 0$$

$$-318 \text{ N} + 9{,}810 \text{ N/m}^3 \times 0.30 \text{ m} \times 0.60 \text{ m} \times d = 0$$

$$d = 0.18 \text{ m} = 18 \text{ cm}$$

Determine whether the block is stable about the longitudinal axis:

$$GM = \frac{I_{00}}{\forall} - CG$$

$$= \frac{\frac{1}{12} \times 60 \times 30^3}{18 \times 60 \times 30} - (15 - 9)$$

$$= 4.167 - 6$$

$$= -1.833 \text{ cm}$$

Because the metacentric height is negative, the block is not stable about the longitudinal axis; thus a slight disturbance will make it tip. Now check to see if the block is stable about the transverse axis:

$$GM = \frac{\frac{1}{12} \times 30 \times 60^3}{18 \times 30 \times 60} - 6$$

$$= 10.67 \text{ cm}$$

The block is stable about the transverse axis and will float with the short sides vertical.

Note that for small angles of α the righting moment or overturning moment is given as follows:

$$\text{R.M.} = \gamma \forall GM\alpha \qquad (3\text{-}32)$$

However, for large angles of heel, direct methods of calculation based upon these same principles would have to be employed to evaluate the righting or overturning moment.

PROBLEMS

3-1   Some skin divers go as deep as 50 m. What is the gage pressure at this depth in fresh water and the ratio of the absolute pressure at this depth to normal atmospheric pressure? Assume $T = 20°C$.

3-2   A mercury barometer reads 750 mm of mercury. What is the atmospheric pressure in pascals? Assume the vapor pressure of mercury is nil.

3-3   The Crosby gage tester in the figure is used to calibrate or to test pressure gages. When the weights and the piston together weigh 20 lbf (88.0 N) the gage being tested indicates 26.0 psi (179 kPa). If the piston diameter is 1 in., what percent error exists in the gage?

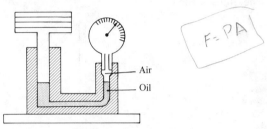

Air

Oil

3-4   Two hemispherical shells are perfectly sealed together, and the internal pressure is reduced to 4.0 psia. The inner radius is 7.5 in. and the outer radius is 8.0 in. If the atmospheric pressure is 14.0 psia, what force is required to pull the shells apart?

3-5   Two hemispheric shells are perfectly sealed together and the internal pressure is reduced to 20 kPa. The inner radius is 15 cm and the outer radius is 15.5 cm. If the atmospheric pressure is 100 kPa, what force is required to pull the shells apart?

3-6   Calculate the specific weight of air in pounds per cubic foot and in newtons per cubic meter in a tank that is at 300 K and 300 kPa gage, assuming an atmospheric pressure of 100.0 kPa.

3-7   If the cabin of a jet liner is pressurized to 100 kPa absolute and the outside pressure is 20 kPa absolute when the plane is cruising at 12 km, what net force is exerted by the air on the cabin door, which is 2.2 m × 1.0 m in size?

3-8   If exactly twenty 2-cm-diameter bolts are needed to hold the air chamber

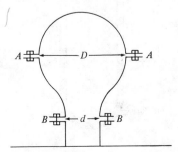

together at *A-A* as a result of the high pressure within, how many bolts will be needed at section *B-B*? Here $D = 50$ cm and $d = 25$ cm.

3-9   What is the pressure in an open tank of crude oil at a depth of 10 m?

3-10   The gage pressure at a depth of 5 m in an open tank of liquid is 73.6 kPa. What are the specific weight and specific gravity of the liquid?

3-11   Determine the pressure at the bottom of a container 4 m deep filled with a liquid the specific weight of which varies with depth according to the equation $\gamma = (10,000 + 100\ d)$ N/m³. Here *d* is in meters and is measured downward from the surface.

3-12   Determine the gage pressure at the center of pipe *A* if $h = 7$ m and the temperature is 20°C.

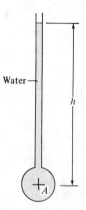

PROBLEMS 3-12, 3-13

3-13   Determine the gage pressure at the center of pipe *A* in pounds per square inch gage and pounds per square foot gage if $h = 22$ ft and the temperature is 60°F.

3-14   Determine the gage pressure at the center of pipe *A* in pounds per square inch if the temperature is 70°F.

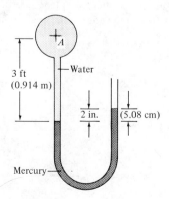

3-15   How many meters of water are equivalent to 120 kPa?

3-16   Water occupies the bottom 60 cm of depth of a cylindrical tank. On top of the water is 1.0 m of kerosene, which is open to the atmosphere. If the temperature is 20°C, what is the gage pressure at the bottom of the tank?

3-17   If the atmospheric pressure is 98.0 kPa, what will be the reading on a mercury barometer at that location?

3-18   What is the maximum gage pressure in the ~~oil~~ tank shown in the figure? Where will the maximum pressure occur? What is the hydrostatic force acting on the top (C-D) of the last chamber on the right-hand side of the tank? Assume $T = 10°C$.

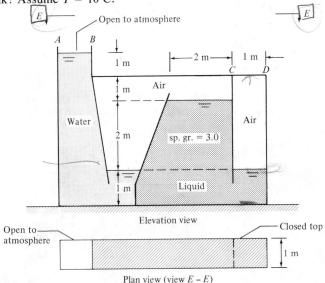

Elevation view

Plan view (view E - E)

3-19   Usually water is assumed to be incompressible for hydrostatic computations; however, extreme pressures may cause significant changes in density. Estimate the percentage change in density of sea water between the surface and at a point 6 km deep. Assume a constant temperature of 20°C and bulk modulus of elasticity of 2.2 GPa.

3-20   Determine the gage pressure in pipe A.

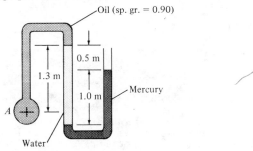

3-21 Considering the effects of surface tension estimate the true value of gage pressure at the center of pipe A.

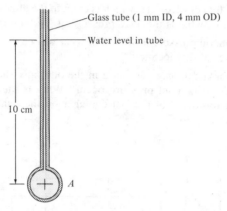

Glass tube (1 mm ID, 4 mm OD)

Water level in tube

10 cm

A

3-22 Given an atmospheric pressure of 100 kPa absolute, at what depth in a lake is the absolute pressure double that at the lake's surface?

3-23 What is the pressure at the center of pipe B?

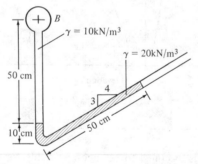

B

$\gamma = 10\text{kN/m}^3$

$\gamma = 20\text{kN/m}^3$

50 cm

4

3

50 cm

10 cm

3-24 The cistern diameter-to-tube diameter ratio is 10. When the air in the tank is at atmospheric pressure the free surface in the tube is at position 1. Then the cistern is pressurized, causing the liquid in the tube to move 50 cm up the tube from equilibrium position 1 to position 2. What is the cistern pressure that caused this deflection? The liquid density is 800 kg/m³.

$P = h\gamma$

$h\left(\frac{\pi}{4}D^2\right) = .5\ \frac{\pi}{4}d^2$

$h = .005$

$P = h\gamma$

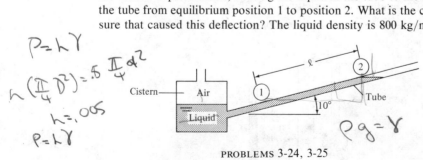

Cistern — Air

ℓ

2

1

Tube

10°

$\rho g = \gamma$

Liquid

PROBLEMS 3-24, 3-25

3-25 The cistern diameter-to-tube diameter ratio is 10. When the air in the tank is at atmospheric pressure the free surface in the tube is at position 1. Then the cistern is pressurized, causing the liquid in the tube to move 2 ft up the tube from position 1 to position 2. What will be the cistern pressure that caused this deflection? The specific weight of the liquid is 50 lbf/ft³.

3-26 The inclined manometer is filled with oil that has a specific gravity of 0.85. What angle α will yield a deflection 20 cm in the inclined tube when the air pressure in the cistern is increased 600 Pa?

$h = .20 \sin\alpha$

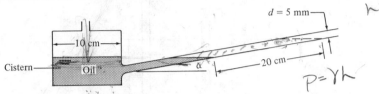

$d = 5$ mm

10 cm

Cistern — Oil

α 20 cm

$P = \gamma h$

3-27 Determine the gage pressure at the center of pipe A in pounds per square inch gage and kilopascals.

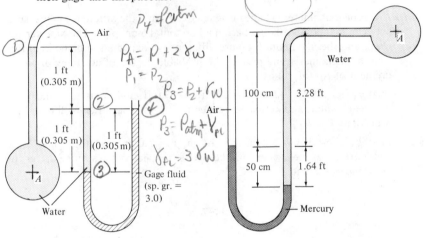

$P_4 \neq P_{atm}$

$P_A = P_1 + 2\gamma_w$

$P_1 = P_2$

$P_3 = P_2 + \gamma_w$

$P_3 = P_{atm} + \gamma_{fl}$

$\gamma_{fl} = 3\gamma_w$

Air

1 ft
(0.305 m)

②

④

Air

1 ft
(0.305 m)

1 ft
(0.305 m)

③

Gage fluid
(sp. gr. =
3.0)

+A

Water

Water

100 cm   3.28 ft

50 cm   1.64 ft

Mercury

PROBLEM 3-27   PROBLEM 3-28

3-28 Determine the gage pressure at the center of pipe A in pounds per square inch gage and kilopascals.

3-29 Mercury is poured into the tube in the figure so that the mercury occupies

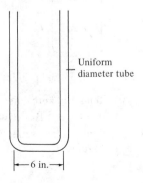

Uniform
diameter tube

6 in.

1 ft (0.30 m) of length of the tube. An equal volume of water is then poured into the left leg. Locate the water and mercury surfaces. Also determine the maximum pressure in the tube.

3-30 What is the specific gravity of the liquid in the left leg of the manometer tube?

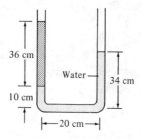

3-31 A U-tube manometer is to be designed to measure the difference in pressure between two points 100 m apart in a horizontal 6-cm pipe. The pipe carries water and the maximum pressure difference is expected to be 60 kPa. Design the manometer and predict the probable degree of accuracy of measurement of Δp for your design.

3-32 Find $p_A - p_B$ if $x = 4.0$ ft, $y = 2.0$ ft, $z = 3.0$ ft, and fluids 1, 2, and 3 are kerosene, water, and kerosene, respectively. Assume sp. gr. $= 0.8$ for kerosene and $T = 60°F$.

3-33 Find $p_A - p_B$ if $x = 3.0$ m, $y = 1.0$ m, $z = 2.0$ m, and the fluids 1, 2, and 3 are kerosene, water, and kerosene, respectively. Assume sp. gr. of kerosene $= 0.8$ and $T = 20°C$.

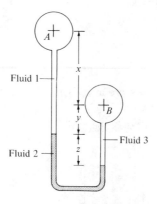

PROBLEMS 3-32, 3-33, 3-34, 3-35, 3-36

3-34 Find z if $p_B - p_A = 2.0$ psi, fluid 1 is kerosene, $\gamma = 180$ lbf/ft³ for fluid 2, and fluid 3 is water. $x = 1$ ft and $y = 3$ ft. Assume $T = 60°F$.

3-35 Find z if $p_B - p_A = 3.0$ psi, fluid 1 is kerosene, fluid 2 is mercury, and fluid 3 is water. $x = 0$, $y = 3$ ft. Assume $T = 60°F$.

3-36   Find z if $p_B - p_A = 10$ kPa, fluid 1 is kerosene, fluid 2 is mercury, and fluid 3 is water. $x = 0$ and $y = 1$ m.

3-37   The pressure intensity in the air space above an oil (sp. gr. $= 0.75$) surface in a tank is 3 psi. What is the pressure 5 ft below the surface of the oil?

3-38   Find the pressure at the center of pipe A. Assume $T = 20°C$.

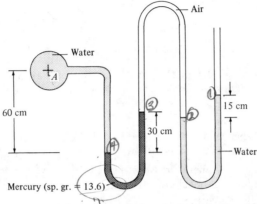

3-39   Find the pressure at the center of pipe A. Assume $T = 10°C$.

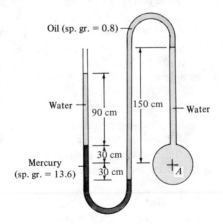

PROBLEMS 3-39, 3-40

3-40   Find the pressure at the center of pipe A if all dimensions given are in inches instead of centimeters.

3-41   Determine (a) the difference in pressure and (b) the difference in piezometric head between points A and B. The elevations z_A and z_B are 10 and 11 m, respectively, $\ell_1 = 1$ m and the manometer deflection ℓ_2 is 50 cm.

3-42   Determine (a) the difference in pressure and (b) the difference in piezometric head between points A and B. The elevations z_A and z_B are 10 and 13 ft, respectively, $\ell_1 = 40$ in. and the manometer deflection ℓ_2 is 25 in.

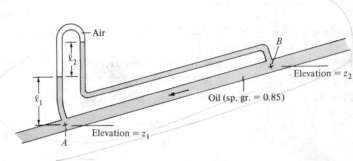

PROBLEMS 3-41, 3-42

3-43   From a depth of 10 m in a lake to an elevation of 4,000 m in the atmosphere, plot the variation of absolute pressure. Assume the lake water surface elevation is at mean sea level and assume standard atmospheric conditions.

3-44   The boiling point of water decreases with elevation because of the pressure change. What is the boiling point of water at 1,500-m elevation and at 3,000-m elevation for standard atmospheric conditions?

3-45   What is the atmospheric pressure at an elevation of 6 km if the pressure and temperature at sea level are 101 kPa and 25°C. Assume the standard lapse rate prevails.

3-46   What is the atmospheric pressure at an elevation of 20,000 ft (6,096 m) if the pressure and temperature at sea level are 14.7 psia (101 kPa) and 60°F (15°C), respectively? Assume standard atmospheric conditions.

3-47   Assume that a man must breathe a constant mass rate of air to maintain his metabolic processes. If he inhales and exhales 16 times per minute at sea level where the temperature is 59°F (15°C) and the pressure is 14.7 psia (101 kPa), what would you expect his rate of breathing at 15,000 ft (4,570 m) to be? Use standard atmospheric conditions.

3-48   A pressure gage in an airplane indicates a pressure of 95 kPa at takeoff where the airport elevation is 1 km and the temperature is 10°C. If the standard lapse rate of 6.5°C/km is assumed, at what elevation is the plane when a pressure of 75 kPa is read? What is the temperature for that condition?

3-49   A pressure gage in an airplane indicates a pressure of 13.6 psia at takeoff where the airport elevation is 2,000 ft and the temperature is 70°F. If the standard lapse rate of 0.003566°F/ft is assumed, at what elevation is the plane when a pressure of 10 psia is read?

3-50   Denver, Colorado, is called the "mile-high" city. What is the pressure, temperature, and density of the air, assuming that standard atmospheric conditions prevail? Give your answer in traditional and SI units.

3-51   A vertical gate 4 ft wide by 3 ft high holds back water to the full height of 3 ft. What is the necessary resisting moment at the bottom to keep the gate in position?

3-52   A vertical gate 4 m wide by 3 m high holds back water to the full height of

3 m. What is the necessary resisting moment at the bottom to keep the gate in position?

3-53   Find the force of the gate on the block below in both SI and traditional units.

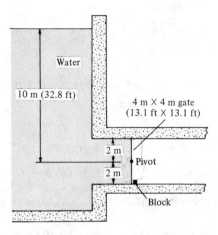

3-54   Assuming that concrete behaves as a liquid ($\gamma = 150$ lbf/ft^3) just after it is poured, determine the force per foot of length exerted on a form by the concrete if the concrete is poured into forms for a wall that is to be 8 ft high. If the forms are held in place as shown with ties between vertical braces spaced every 2 ft apart, what force will be exerted on the bottom tie?

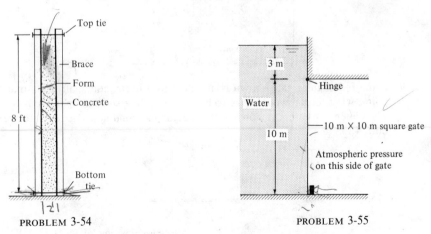

PROBLEM 3-54

PROBLEM 3-55

3-55   Determine the force due to hydrostatic pressure acting on the hinge of the gate shown. Give answers in both traditional and SI units.

3-56   The rectangular gate is 10 m × 4 m in dimension ($\ell = 10$ m) and is pin-connected at point B. If the surface on which the gate rests at A is frictionless and if the water surface is 6 m above point B, what is the reaction at A?

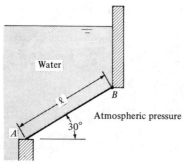

PROBLEMS 3-56, 3-57

3-57 The rectangular gate is 10 ft × 6 ft in dimension ($\ell = 10$ ft) and is pin-connected at point B. If the surface on which the gate rests at A is frictionless and if the water surface is 7 ft above point B, what is the reaction at A?

3-58 In the figure, the gate holding back the oil is 80 cm high by 120 cm long. If it is held in place only along the bottom edge, what is the necessary resisting moment at that edge?

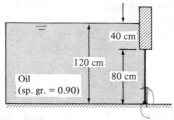

3-59 If the rectangular gate shown is attached to a horizontal shaft at its midpoint, what torque would have to be applied to the shaft to open the gate? The dimensions are $\ell = 5$ m and the rectangular conduit and gate width are both 4 m.

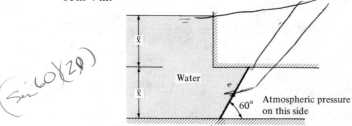

PROBLEMS 3-59, 3-60

3-60 If the rectangular gate shown is attached to a horizontal shaft at its midpoint, what torque would have to be applied to the shaft to open the gate? Here $\ell = 12$ ft and the rectangular conduit and gate are both 5 feet wide.

3-61   If gate AB is rectangular and is 2 m wide, what force F is needed to open the gate if region C is air at atmospheric pressure? $\ell = 5$ m and the water surface lies 2 m above the hinge at B.

3-62   If gate AB is rectangular and is 5 ft wide, what force F is needed to open the gate if region C is air at atmospheric pressure? $\ell = 10$ ft and the water surface lies 10 ft above the hinge at B.

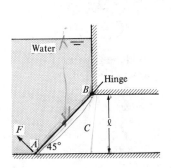

PROBLEMS 3-61, 3-62

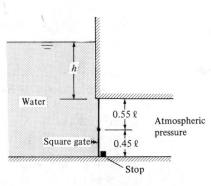

PROBLEM 3-63

3-63   The square gate shown is eccentrically pivoted so that it automatically opens with a certain value of h. What is that value in terms of ℓ?

3-64   Determine the minimum volume of concrete ($\gamma = 23.6$ kN/m³) needed to keep the 1-m wide gate in a closed position. Note the hinge at the bottom of the gate. $\ell = 2$ m.

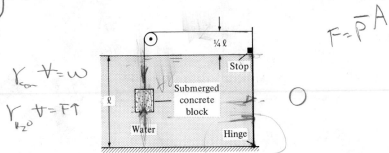

PROBLEMS 3-64, 3-65

3-65   Determine the minimum volume of concrete ($\gamma = 150$ lb-ft³) needed to keep the 2-ft-wide gate in a closed position. $\ell = 5$ ft.

3-66   For this gate $\alpha = 45°$, $y_1 = 1$ m, and $y_2 = 3$ m. Will the gate fall or stay in position under the action of the hydrostatic and gravity forces if the gate itself weighs 90 kN and is 1.0 wide? Assume $T = 10°C$.

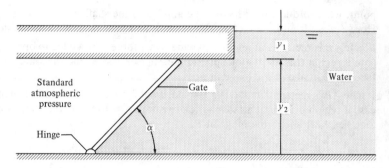

PROBLEMS 3-66, 3-67

3-67   For this gate $\alpha = 45°$, $y_1 = 4$ ft, and $y_2 = 7.07$ ft. Will the gate fall or stay in position under the action of the hydrostatic and gravity forces if the gate itself weighs 18,000 lb and is 3 ft wide? Assume $T = 50°F$.

3-68   The triangular gate ABC is pivoted at the bottom edge AC and closes a triangular opening ABC in the wall of the tank. The opening is 4 m wide ($W = 4$ m) and 9 m high ($H = 9$ m). The depth d of water in the tank is 10 m. Determine the hydrostatic force on the gate and determine the horizontal force P at B required to hold the gate closed.

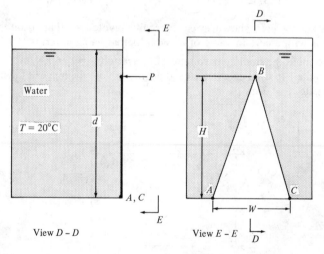

View D – D   View E – E

3-69   A plate that has dimensions W wide, H high, and t thick is completely submerged with vertical orientation in a liquid. The liquid has a variable specific weight given by $\gamma = \gamma_0 \times (1 + kd/d_0)$, where k is a positive constant and d is depth below the liquid surface, and γ_0 is the specific weight at reference depth d_0. Derive a formula for the magnitude of hydrostatic force on one side of the plate. Will the location of the center of pressure be below or above that for a plate located similarly in a constant density liquid?

3-70   The air above the liquid is under a pressure of 2 psi gage and the specific gravity of the liquid in the tank is 0.80. If the rectangular gate is 5 ft wide and if $y_1 = 2$ ft and $y_2 = 10$ ft, what force P is required to hold the gate in place?

3-71   The air above the liquid is under a pressure of 30 kPa gage and the specific gravity of the liquid in the tank is 0.80. If the rectangular gate is 1.0 m wide and if $y_1 = 1.0$ m and $y_2 = 3$ m, what force P is required to hold the gate in place?

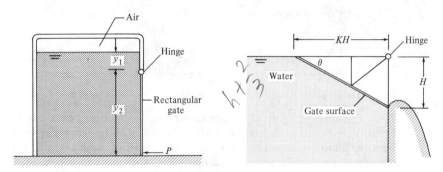

PROBLEMS 3-70, 3-71   PROBLEM 3-72

3-72   If KH is large enough, the gate will be on the verge of opening when the water level is even with the hinge. What is K for this condition? Neglect the weight of the gate.

3-73   Determine the hydrostatic force F on this triangular gate, which is hinged at the bottom edge and held by the reaction R_T at the upper corner. Express F in terms of γ, h, and W. Also determine the ratio R_T/F.

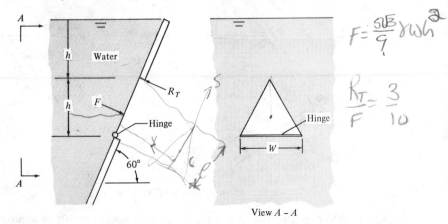

View $A - A$

3-74   For the plane rectangular gate ($\ell \times w$ in size), figure (*a*), what is the magnitude of the reaction at A in terms of γ_w and the dimension ℓ? For the cylindrical gate, figure (*b*), will the magnitude of the reaction at *A* be greater than, less than, or the same as the plane gate?

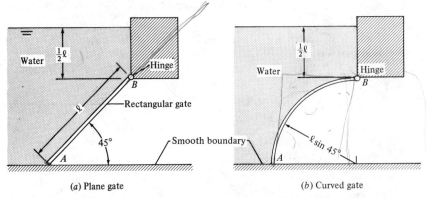

(*a*) Plane gate   (*b*) Curved gate

3-75 When constructing dams, the concrete is poured in lifts of approximately 1.5 m ($y_1 = 1.5$ m) as shown. The forms for the face of the dam are re-used from one lift to the next. The figure shows one such form, which is bolted to the already cured concrete. For the new pour, what moment will occur at the base of the form per meter of length (normal to page)? Assume concrete acts as a liquid when it is first poured and has a specific weight of 24 kN/m³.

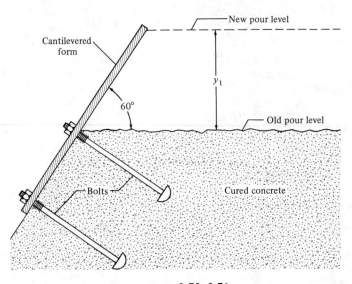

PROBLEMS 3-75, 3-76

3-76 When constructing dams, the concrete is poured in lifts of approximately 4 ft ($y_1 = 4$ ft) as shown. The forms for the face of the dam are reused from one lift to the next. The figure shows one such form, which is bolted to the already cured concrete. For the new pour, what moment will occur at the base of the form per foot of length (normal to page)? Assume concrete acts as a liquid when it is first poured and has a specific weight of 150 lb/ft³.

3-77   Gate *ABC* is hinged at *B* and is 4 ft long (direction perpendicular to the paper). Neglecting the weight of the gate, determine the horizontal force *F* required to open the gate.

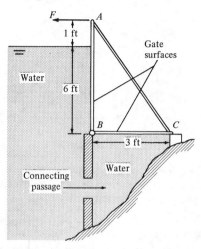

3-78   The wood pole is attached to the wall by a hinge as shown. The pole is in equilibrium under the action of the weight and buoyant forces. Determine the specific weight of the wood of the partially submerged pole.

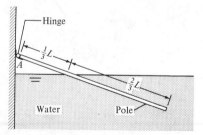

3-79   The wooden pole is hinged at *A* and the free end is partially submerged as shown. If the specific gravity of the wood is 0.56, what will be the magnitude of the angle *α* between the pole and water surface?

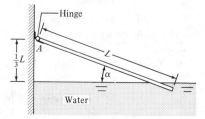

3-80 This dome (half-sphere) is located below the water surface as shown. Determine the magnitude and sign of the force components needed to hold the dome in place and the line of action of the horizontal component of force. Here $y_1 = 3$ ft and $y_2 = 4$ ft. Assume $T = 50°F$.

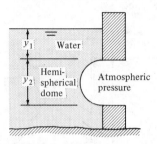

PROBLEMS 3-80, 3-81

3-81 This dome (half-sphere) is located below the water surface as shown. Determine the magnitude and sign of the force components needed to hold the dome in place and the line of action of the horizontal component of force. Here $y_1 = 1$ m and $y_2 = 2$ m. Assume $T = 10°C$.

3-82 A vertical wall at A-A has the same length (normal to paper) as the triangular shape at B-B. Neglecting the weight of the walls, which would require the greatest resisting moment at the base to hold it in place, or would the required moment be the same in both cases?

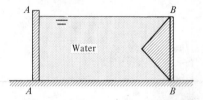

3-83 A block of material of unknown volume is submerged in water and found to weigh 500 N (in water). The same block weighs 650 N in air. Determine the specific weight and volume of the material.

3-84 A block of material of unknown volume is submerged in water and found to weigh 97.6 lbf in the water. The same block weighs 128.8 lbf in air. Determine the specific weight of the material.

3-85 An object weighs 55 N when submerged in oil (sp. gr. = 0.80) and requires a force of 45 N to hold it submerged in mercury. Determine its weight, volume, specific weight, and specific gravity.

3-86 A rock weighs 912 N (205 lbf) in air and 609 N (137 lbf) in water. Find its volume.

3-87 The steel pipe and steel chamber together weigh 600 lbf. What force will all the bolts have to exert on the chamber to hold it in place? The dimension ℓ

is equal to 2 ft. *Note*: There is no bottom on the chamber—only a flange bolted to the floor.

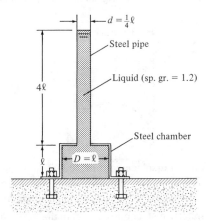

PROBLEMS 3-87, 3-88

3-88 The steel pipe and steel chamber together weigh 5,000 N. What force will all the bolts have to exert on the chamber to hold it in place? The dimension ℓ is equal to 1 m. *Note*: There is no bottom on the chamber—only a flange bolted to the floor.

3-89 Find the vertical component of force in the metal at the base of the spherical dome shown when gage *A* reads 10 psi. Indicate whether the metal is in compression or tension. The specific gravity of the enclosed fluid is 1.5. The dimension *L* is 3 ft. Assume the dome weighs 1,000 lbf.

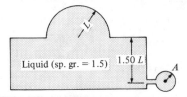

PROBLEMS 3-89, 3-90

3-90 Find the vertical component of force in the metal at the base of the spherical dome shown when gage *A* reads 70 kPa gage. Indicate whether the metal is in compression or tension. The specific gravity of the enclosed fluid is 1.5. The dimension *L* is equal to 1 m. Assume the dome weights 5,000 N.

3-91 A curved gate is to be constructed to conform to the curve $y = kx^3$ between $0 < x < a$ and $0 < y < b$. The liquid surface is along the line $y = b$. Find the total hydrostatic force on the gate in terms of γ on 1 m width ($z = 1$ m) of gate. Also find the lines of action of the vertical and horizontal components.

3-92 The bottom of a cylindrical water tank is a hemispherical steel shell 1 cm thick and is welded to the inside of the vertical tank wall. Steel weighs 77 kN/m³. Determine the resisting shear force per lineal centimeter required of the weld.

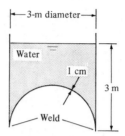

3-93   An 800-ft ship has a displacement of 40,000 tons, and the area defined by the waterline is 40,000 ft². Will the ship take more or less draft when steaming from salt water to fresh water? How much will it settle or rise?

3-94   A submerged spherical steel buoy 1.2 m in diameter and weighing 1600 N is to be anchored in salt water 20 m below the surface. Find the weight of scrap iron to be sealed inside the buoy in order that the force on its anchor chain will not exceed 4.5 kN.

3-95   A balloon is to be used to carry meteorological instruments to an elevation of 15,000 ft where the air pressure is 8.3 psia. The balloon is to be filled with helium and the material from which it is to be fabricated weighs 0.01 lbf/ft². If the instruments weigh 10 lbf, what diameter should the spherical balloon have?

3-96   Determine the magnitude, direction, and line of action of the vertical component of hydrostatic force acting on the surface AB of the figure. Here $\ell = 3$ ft.

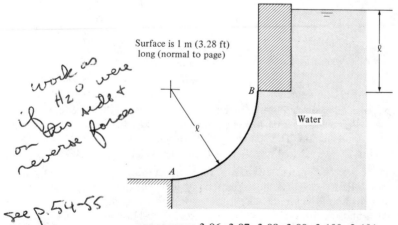

PROBLEMS 3-96, 3-97, 3-98, 3-99, 3-100, 3-101

3-97   Determine the magnitude, direction, and line of action of the vertical component of hydrostatic force acting on the surface *AB* of the figure. Here $\ell = 1$ m.

3-98 Determine the magnitude, direction, and line of action of the horizontal component of hydrostatic force acting on the surface *AB* of the figure given. Here $\ell = 3$ ft.

3-99 Determine the magnitude, direction, and line of action of the horizontal component of hydrostatic force acting on the surface *AB* of the figure given. Here $\ell = 1$ m.

3-100 Determine the resultant hydrostatic force acting on surface *AB* of the figure. Here $\ell = 3$ ft.

3-101 Determine the resultant hydrostatic force acting on surface *AB* of the figure above. Here $\ell = 1$ m.

3-102 What force must be exerted through the bolts to hold the dome in place? *Note*: The metal dome weighs 1,300 lbf and has no bottom. $\ell = 2.0$ ft.

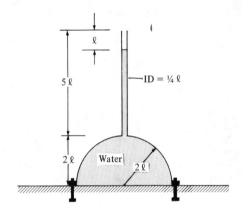

PROBLEMS 3-102, 3-103

3-103 What force must be exerted through the bolts to hold the dome in place? *Note*: The metal dome weighs 6 kN and has no bottom. Here $\ell = 80$ cm.

3-104 Determine the magnitude and direction of the horizontal and vertical components of the hydrostatic force acting on the two-dimensional curved

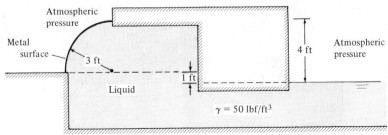

surface per foot of width. Determine the location of the horizontal component of hydrostatic force.

3-105 A hydrometer of the dimensions shown is used to measure the specific gravity of a liquid. The liquid line is 2 in. above the bulb. The hydrometer weighs 0.194 lbf. Calculate the specific gravity of the liquid.

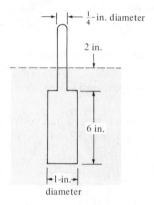

$\frac{1}{4}$-in. diameter

2 in.

6 in.

1-in. diameter

PROBLEMS 3-105, 3-106

3-106 A hydrometer with the configuration shown has a bulb diameter of 2 cm, a bulb length of 8 cm, a stem diameter of 1 cm, a length of 8 cm, and a mass of 35 g. What is the range of specific gravities that can be measured with this hydrometer? (*Hint:* Liquid levels range between bottom and top of stem.)

3-107 A submarine has a displaced volume of 30,000 ft^3 and weighs 965 tons in dry dock. Determine the net weight of the submarine when cruising submerged at sea.

3-108 A barge, 20 ft wide and 50 ft long, is loaded with rock as shown. Assume the center of gravity of the rock and barge is located along the centerline at the top surface of the barge. If the rock and the barge weigh 400,000 lbf, will the barge float upright or will it tip over?

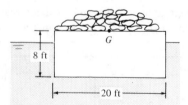

G

8 ft

20 ft

3-109 A cylindrical block of wood 1 m in diameter and 1 m long has a specific weight of 8,000 N/m^3. Will it float in water with its axis vertical?

3-110 A cylindrical block of wood 1 m in diameter and 1 m long has a specific weight of 5,000 N/m^3. Will it float in water with the ends horizontal?

3-111 Will this block be stable in the position shown? Show your calculations. *Note*: The block's specific gravity is equal to 0.5.

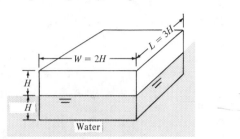

3-112 A cylindrical wooden block of density 500 kg/m³ floats in water at 20°C. The length of the block is 20 cm. If the block is submerged from its equilibrium position and released, it will oscillate about the equilibrium position. Neglecting the effect of hydrodynamic drag on the moving block, calculate the frequency of oscillation.

3-113 Large concrete cylindrical shells sealed at each end are to be used in construction of an off-shore oil drilling rig. These cylinders are floated out to the site and then uprighted to form part of the structure. The cylinders are 20 m in diameter and when floating in the horizontal position sink to a depth of 15 m. When erected, what will be their height above the water?

REFERENCE

1. "Standard Atmosphere, Tables and Data for Altitudes to 65,800 Feet," *NACA Rept. 1235 (1955).*

These plumes of warm liquid in an otherwise quiescent pool originate from the heated plate at the bottom. Flow visualization is accomplished by an electrochemical process in which different degrees of shading occur with local change in pH.

4 FLUIDS IN MOTION

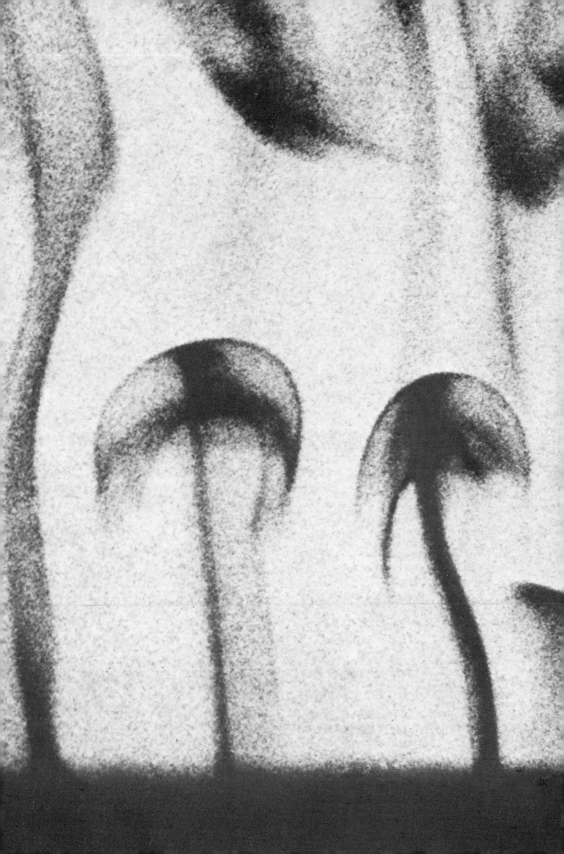

THIS CHAPTER DISCUSSES THE BASIC CONCEPTS paramount in the study of the kinematics of fluids in motion. If one develops a good understanding of velocity, acceleration, and flow visualization, one can more easily grasp the nature of pressure variation in flowing fluids when it is discussed in Chapter 5. Moreover, the control-volume approach introduced here will be used extensively in derivation and application of both the momentum and energy equations that appear in Chapters 6 and 7, respectively. In addition, the control-volume approach is employed to derive the continuity equation, which is basic to all flow problems. This chapter is undoubtedly the most important chapter in the text; hence, students who master the concepts presented will indeed have invested their time wisely.

4-1 VELOCITY AND FLOW VISUALIZATION

Introduction

The velocity of flow for most engineering problems in fluid mechanics is of primary concern. If the application involves flow past structural or machine parts, a knowledge of the velocity allows pressures and forces acting on the structure to be calculated. The information is then used in the design of the structure. In other cases, such as canal design or bridge-pier design, the engineer may be more interested in velocity of flow from the point of view of its scouring action on the channel bottom itself than in the overall pressure or force. In any case, the engineer involved with flow problems must be able to determine the velocity by either experimental or analytical means. The following sections describe methods for making such determinations.

Velocity; Lagrangian and Eulerian viewpoints

There are two ways to express the equations for fluids in motion. One way, the *Lagrangian* viewpoint, has to do with considering an individual fluid particle for all time, which is the familiar approach in dynamics. In this case, the particle velocity is obtained by differentiating the particle's position vector with respect to time. Using the cartesian coordinate system, the position vector is expressed as

$$\mathbf{r}(t) = x\mathbf{i} + y\mathbf{j} + z\mathbf{k} \qquad (4\text{-}1)$$

where **i**, **j**, and **k** are the unit vectors in the x, y, and z directions. Differentiating Eq. (4-1) with respect to time, we obtain the velocity of the fluid particle:

$$\mathbf{V}(t) = \frac{dx}{dt}\,\mathbf{i} + \frac{dy}{dt}\,\mathbf{j} + \frac{dz}{dt}\,\mathbf{k} \tag{4-2}$$

or
$$\mathbf{V}(t) = u\mathbf{i} + v\mathbf{j} + w\mathbf{k} \tag{4-3}$$

where u, v, and w are the component velocities in their respective coordinate directions. Of course, the motion of one fluid particle is inadequate to describe an entire flow field, so the motion of all fluid particles must be considered simultaneously. The motion of the flow field is obtained by solving the equation of motion ($\mathbf{F} = m\mathbf{a}$) for each and every fluid particle in the field of flow.

The other way is to focus on a certain point in space and consider the motion of fluid particles that pass that point as time goes on. This is known as the *Eulerian* approach. In this case, the fluid particle velocity depends on the point in space and time:

$$\begin{aligned}
u &= f_1(x, y, z, t) \\
v &= f_2(x, y, z, t) \\
w &= f_3(x, y, z, t)
\end{aligned} \tag{4-4}$$

In this method we observe the motion of particles passing a specific point in space as opposed to the Lagrangian method in which we track the position of a specific particle as time passes. In order to describe the entire flow field, the fluid motion at all points in the field must be known.

It is an enormous task to keep track of the positions of all particles in a flow field, because unlike the movement of solid bodies, their relative positions continuously change with time. For this reason, the Eulerian approach is generally favored over the Lagrangian approach. However, in some areas—such as rarefied gas dynamics where we are concerned with the motion of individual molecules—the Lagrangian approach is used; but the overwhelming majority of fluid dynamic analyses are based on the Eulerian approach.

Equations (4-4) give the component velocities as a function of space and time in the cartesian coordinate system, but there is also another useful way of expressing velocity by using the Eulerian viewpoint to describe the total velocity as a function of position along a streamline and time. This is given as

$$\mathbf{V} = \mathbf{V}(s, t) \tag{4-5}$$

The streamline will be defined in the next section.

Streamlines and flow patterns

Often it is desirable to construct lines in the flow field to indicate the speed and direction of flow. Such a construction is called a flow pattern, and the lines, called *streamlines,* are defined as lines drawn through the flow field in such a manner that the velocity vector of the fluid at each and every point on the streamline is tangent to the streamline at that instant. Consequently, a tangent to the curve at any point along the streamline gives us the direction of the velocity vector at that point in the flow field. For example, consider flow of water from a slot in the side of a tank, as shown in Fig. 4-1. The velocity vectors have been sketched at three different positions, a, b, and c. One can see that the flow pattern is a very effective way of illustrating the geometry of fluid flow. In Fig. 4-1 it may be noted that the two outer streamlines that bound the free jet also follow the walls inside the tank. This tangency of streamlines with fixed boundaries follows directly from the definition of a streamline. That is, since there is no flow through an impervious boundary, all velocity vectors of the flow adjacent to the boundaries must be parallel to the boundary; therefore, all streamlines directly adjacent to the wall are also parallel and actually follow the contour of the wall. Later, in the section on the continuity equation, we will find that the speed of flow is also indicated by the flow pattern; it will be seen that the speed of flow, V, is inversely proportional to the spacing of the streamlines in two-dimensional flow.

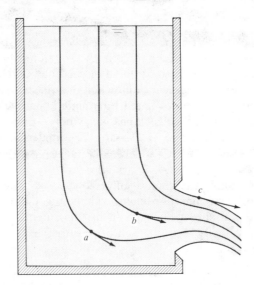

FIGURE 4-1 Flow from a slot.

Now that the general concept of the flow pattern has been introduced, it is convenient to make distinctions between different types of flow. First, we shall consider whether a flow is *uniform* or *nonuniform*. In uniform flow the velocity does not change from point to point along any of the streamlines in the flow field; therefore, it follows that the streamlines depicting such flow must be straight and parallel. If they are not straight, there will be a directional change of velocity. If they are not parallel, there will be a change of speed along the streamlines. Mathematically these conditions of uniform or nonuniform flow are expressed as follows:

$$\frac{\partial \mathbf{V}}{\partial s} = 0 \qquad \text{(uniform flow)}$$

$$\frac{\partial \mathbf{V}}{\partial s} \neq 0 \qquad \text{(nonuniform flow)}$$

Here $\mathbf{V}$ is the total velocity at a given point on a streamline and s is the distance along the streamline measured from an arbitrary point on the streamline. Flow patterns for uniform flow in an open channel and between parallel plates are shown in Fig. 4-2. In nonuniform flow the

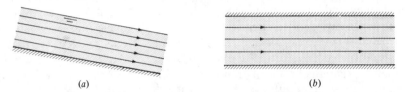

(a)   (b)

FIGURE 4-2   Uniform flow patterns. (a) Open-channel flow. (b) Flow between parallel plates.

velocity changes from point to point along the streamline; therefore, the flow pattern consists of streamlines which are either curving in space or converging or diverging. Flow patterns for cases of nonuniform flow are shown in Figs. 4-1 and 4-3.

Another flow classification is based upon the variation of velocity with respect to time at a given point in a flow field. If at any given point the velocity does not vary in magnitude or direction with time, then the flow is *steady*. Steady or *unsteady* flow conditions are mathematically defined as follows:

$$\frac{\partial \mathbf{V}}{\partial t} = 0 \qquad \text{(steady flow)}$$

$$\frac{\partial \mathbf{V}}{\partial t} \neq 0 \qquad \text{(unsteady flow)}$$

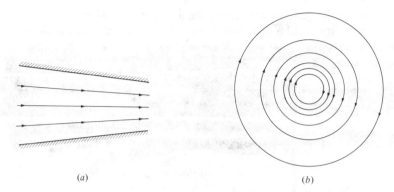

FIGURE 4-3   Flow patterns for nonuniform flow. (*a*) Converging flow. (*b*) Vortex flow.

Here velocity **V** is at a given point in the flow field. An example of steady flow would be a constant rate of flow through a pipe. However, if the discharge were being increased, by the opening of a valve, for example, then the flow would be unsteady. And, if the pipe is straight, one classifies the flow as uniform and unsteady. A simple case of nonuniform-unsteady flow is flow from a nozzle at a changing rate of discharge. Here the flow is nonuniform because of the converging flow passage, consequently converging streamlines, and it is unsteady because of the time change of velocity associated with the change of flow rate.

If the student understands the fundamentals of flow patterns, he or she will appreciate the fact that by simply studying the flow pattern one can ascertain whether or not the flow is uniform. However, the flow pattern itself says nothing with regard to its steady or unsteady state. In general, a flow pattern for unsteady flow is only an instantaneous representation of flow geometry.

Laminar and turbulent flow

Turbulent flow is characterized by a mixing action throughout the flow field, and this mixing is caused by *eddies* of varying size within the flow. Simple observations will reveal this type of flow in rivers and in the atmosphere. Gusts of wind are the result of large-scale eddies that at times reinforce and at other times subtract from the mean wind velocity. Turbulent-flow phenomena can also be observed when smoke from a large stack discharges into the surrounding air.

Laminar flow, on the other hand, is devoid of the intense mixing phenomena and eddies common to turbulent flow. Thus the flow has a very smooth appearance. A typical example is the flow of honey or thick syrup from a pitcher.

To understand the role of turbulence in the flow process, first consider laminar flow in a given situation and then visualize what would happen to the flow if in some hypothetical way one could superimpose a set of eddies onto this flow. The velocity distributions for these two situations are presented in Fig. 4-4 for flow in a straight pipe. For the laminar case, the velocity distribution is parabolic across the section; and at any given distance from the pipe wall, the velocity will be constant with respect to time. In the turbulent-flow case, two effects are readily apparent. First, because the eddies cause the flow to be mixed rather thoroughly, the velocity distribution over most of the section is more uniform than is the case in laminar flow. This results because the turbulent mixing process transports the low-velocity fluid near the wall toward the center, and the higher-velocity fluid in the central region is transported toward the wall.

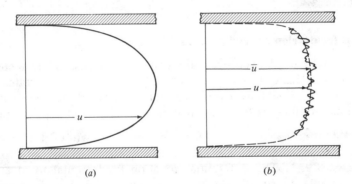

FIGURE 4-4   Laminar and turbulent flow in a pipe. (*a*) Laminar flow, (*b*) Turbulent flow.

The second effect of turbulence is to add continuously fluctuating components of velocity to the flow. At a given instant the distribution of the velocity component in the direction of flow is consequently irregular, as shown by the solid line in Fig. 4-4*b*. However, if we average the velocity over a sufficiently long period of time at various points across the section, the broken line shown in the same figure is obtained. It is thus seen that at a given point the velocity varies with time; therefore, according to our previous definition, the flow is called unsteady. However, if we consider the temporal mean (average with respect to time) velocity at a given point taken over a relatively long period of time, the velocity is virtually constant and the flow is therefore termed steady. For most discussions, this latter point of view (based upon the temporal mean velocity) is commonly used to characterize the flow as steady or unsteady. An index relating to turbulence is the Reynolds number, defined as $Re = VD\rho/\mu$. If the Reynolds number is large ($Re > 2,000$), the flow in a pipe will generally be turbulent; but if the Reynolds number is smaller than 2,000, the flow will be laminar. The significance of the Reynolds number will be considered in more detail in Chapters 8 through 15.

The student may wonder why flow is classified in so many different ways. Of the many reasons for the various classifications, two should serve as valid examples. First, certain types of problems, for example steady-flow problems and unsteady-flow problems, require different methods of solution. The classifications help the engineer to organize his or her thoughts with regard to problem analysis. Sections on acceleration later in this chapter and discussions of pressure distribution in Chap. 5 include examples which will help to illustrate this point. Second, by means of the various classifications, it is much easier for fellow engineers and scientists to communicate—which is why a language exists and what keeps it evolving.

Methods for developing flow patterns

Up to now we have defined streamlines and have discussed how a number of streamlines make up the flow patterns, but we have not stated how these streamlines and flow patterns are obtained. The three basic methods used to predict flow patterns are analytical, analog, and experimental.

ANALYTICAL METHOD   The oldest and best known of the analytic methods derive from ideal-flow theory; that is, the flow of incompressible nonviscous fluids. The differential equation fundamental to ideal-flow theory is Laplace's equation, which is encountered frequently in science and engineering. Solutions of Laplace's equation can be applied to many flow problems such as low-speed flows past airfoils and free-surface flows such as wave motion.

Ideal-flow theory is not applicable to flows where viscous and/or compressibility effects are important. In these cases few closed-form analytic solutions exist and are restricted to simple flow configurations.

The computational capability of the computer has led to new and incredible advances in the analysis of flow fields in recent years. Turbulent two-dimensional flows, unsteady-nonuniform flows, and rarefied gas dynamics are only a few examples for which computer solutions have become available. Numerical fluid mechanics is a field in which new developments are rapidly taking place in step with improved computer capability and capacity.

ANALOG METHOD   Where boundary conditions are complicated, it is often possible to develop flow patterns by analog methods, the most common of which is electrical analogy. In this we take advantage of the fact that the ideal-flow theory serves to describe electrical fields as well as certain fluid-flow fields. To apply the method, an electrical conductor is made in the same shape as the flow passage it represents, and a voltage potential is applied between the ends of the conductor, thus establishing a flow of electrons. By utilizing the data obtained when measuring the voltage at various points of the conductor, it is possible to construct the flow pattern and to predict the pressure distribution in the conduit. The method is valid for irrotational flow cases, a concept that will be covered in more detail in Sec. 4-6.

EXPERIMENTAL METHOD For certain types of turbulent flow, mathematical or analog methods have not been developed to the point where they can reliably predict the flow pattern; therefore, it is necessary to resort to physical models. In other cases, such as basic research studies, experimental methods are also employed to define the pattern of flow. In these experimental setups, dye streams and floating or immersed particles are often used to define the flow pattern. When a photograph is taken of floating or neutrally buoyant particles in a flowing fluid (as in Fig. 4-5, which is the approach channel to a spillway of a model dam), the particles produce light marks, which thus indicate the paths of the particles for the

FIGURE 4-5 Pathlines of floating particles.

period of time of exposure. Each one of these light marks is a segment of a pathline for a given particle. By definition then, a *pathline* is a line drawn through the flow field in such a way that it defines the path that a given particle has taken.

Another technique used to visualize flow patterns is to inject dye or smoke at a given point in the flow field and to observe the dye or smoke trace as it travels downstream. Such a trace is accordingly called a *streakline* (see Fig. 4-6; note the separation zone downstream of the airfoil where turbulence causes the smoke to be diffused).

We have discussed the *streamline, pathline,* and *streakline,* all of which in one way or another are associated with the pattern of flow. For steady flow, all three of these lines are coincident if they originate at the same point. Because streaklines and pathlines also define streamlines for steady flow, it is often thought that they always define streamlines. This is a hazardous assumption, as the following paragraph explains.

Consider a body of fluid that has a velocity of $u = 1.0$ ft/sec, $v = 0$, and $w = 0$ for a period of 4 sec. At the end of 4 sec, hypothetically assume that it will take on a new velocity of $u = 0.707$ ft/sec, $v = 0.707$ ft/sec, and $w = 0$, and assume furthermore that the latter velocity will continue indefinitely. Figure 4-7 shows streamlines, a streakline, and a pathline for the flow at time $t = 4$ sec and at time $t = 10$ sec.

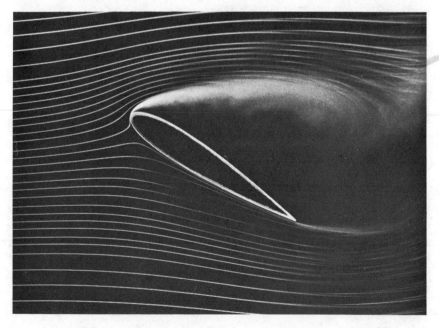

FIGURE 4-6   Smoke traces about an airfoil with a large angle of attack.

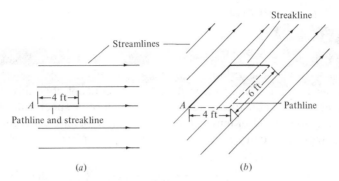

FIGURE 4-7   Streamline, streakline, and pathline. (*a*) *t* = 4 sec. (*b*) *t* = 10 sec.

Note that injection of a single stream of dye was started at point *A* at *t* = 0 and that the particle of fluid for which the path is drawn passed by point *A* at time *t* = 0. For the first period, Fig. 4-7*a*, the pathline and streakline coincide; however, for the entire period there are widely differing traces (see Fig. 4-7*b*). One can conclude from this illustration that streaklines and pathlines *do not* in general define streamlines for *unsteady* flow.

One-, two-, and three-dimensional flow

In general, three coordinate directions are needed to describe the velocity and property changes in a flow field. Such flows are three-dimensional.

For some flow situations there are no changes in one coordinate direction; then two dimensions suffice to describe the flow. For example, the flow between two parallel walls in which there is no velocity component in the direction normal to the wall is a case of two-dimensional flow.

The simplest flow field is the one-dimensional case, in which only one coordinate is needed to relate velocity and property changes. An example is flow in a duct, in which the velocity is constant across each section but varies with distance along the duct. Actually, the velocity is never completely constant across a conduit section. However, for problems in which we are primarily interested in the average velocity parallel to the conduit, this type of flow is also called one-dimensional flow. In this case we are interested in the changes of average velocity and pressure occurring along the length of the conduit rather than across the pipe. Such a concept is used extensively in application of the momentum and energy equations to be introduced in Chapters 6 and 7. Throughout the remainder of the text the term *one-dimensional flow* will generally pertain to cases in which we are primarily interested in the mean, or average, velocity in a conduit.

4-2 RATE OF FLOW

Volume rate of flow

Often simply called *flow rate* or *discharge,* the volume rate of flow refers to the rate at which the volume of fluid passes a given section in the flow stream. Because it is usually easier to grasp the concept of volume rate for a solid, that will be considered first; then we will refer specifically to fluid flow. In Fig. 4-8, a rod of length L and cross-sectional area A is moved to the right with a velocity u. Every second, u feet of rod passes section

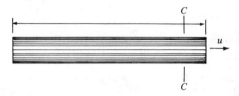

FIGURE 4-8   Motion of a rod past a section.

C-C or, in other words, the volume of rod that passes section C-C every second is uA. Thus the volume rate of passage of the rod past section C-C is uA. In a similar manner, if we have a fluid of constant velocity V flowing in a pipe having a cross-sectional area A (Fig. 4-9), then the volume rate of flow through this pipe will be VA.

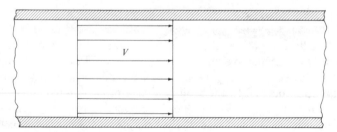

FIGURE 4-9   Flow in a pipe.

Flow rate with variable velocity

The velocity, in general, will be variable across the section through which it flows, such as in Fig. 4-10. Then the rate of flow past a differential area of the section will be $V\,dA$, and the total rate of flow Q will be obtained by integration over the entire flow section:

$$Q = \int_A V\,dA$$

In a similar manner, it can be easily seen that the mass rate of flow past a section would be given by

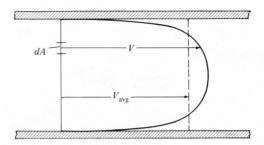

FIGURE 4-10   Flow between parallel boundaries.

$$\dot{m} = \int_A \rho V \, dA$$

In the foregoing developments, the cross-sectional area was always oriented normal to the velocity vector. If other orientations are considered, such as in Fig. 4-11, where flow occurs past section A-A, it can be

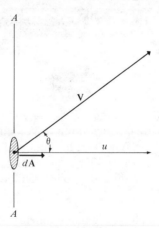

FIGURE 4-11   Velocity not normal to the section.

seen that only the normal component of velocity, the x component in this case, contributes to flow through the section; therefore, to evaluate flow rate one must always consider the area of a section normal to the total velocity or consider the velocity component normal to the given area. Thus the discharge for the case of Fig. 4-11 is given by

$$Q = \int_A u \, dA$$

or $$Q = \int_A V \cos \theta \, dA$$

If we define an area vector as one that has the magnitude of the area in question and that is oriented normal to the area, then by definition, $V \cos \theta \, dA = \mathbf{V} \cdot d\mathbf{A}$, and the discharge can be written as

$$Q = \int_A \mathbf{V} \cdot d\mathbf{A}$$

If the velocity is constant over the area, then the discharge is given as

$$Q = \mathbf{V} \cdot \mathbf{A}$$

In the equations above, because of the scalar, or dot, product, it can be visualized that only the normal component of velocity is multiplied by the area to obtain the discharge.

Average or mean velocity

In many computations, for example one-dimensional flow problems in pipes, we may be given the rate of flow and need to find the average velocity without knowing the actual velocity distribution across the pipe. The average velocity, by definition, is the discharge divided by the total cross-sectional area:

$$\bar{V} = \frac{Q}{A}$$

For turbulent flow in pipes the average velocity may be a fairly close approximation to the actual distribution of velocity over most of the section, as shown in Fig. 4-10; however, for laminar flow the average velocity differs considerably from the velocity over most of the flow section. It is customary to leave the bar off the velocity symbol and to simply indicate the mean velocity with a V.

EXAMPLE 4-1   Air that has a mass density of 1.24 kg/m³ (0.00241 slugs/ft³) flows in a 30-cm-diameter (0.984 ft) pipe at a mass rate of flow of 3 kg/s (0.206 slugs/sec). What is the average velocity of flow in this pipe and what is the volume rate of flow?

Solution

$$\dot{m} = \rho Q \quad \text{or} \quad Q = \frac{\dot{m}}{\rho} = 2.42 \text{ m}^3/\text{s (85.5 cfs)} \quad \blacktriangleleft$$

$$V = \frac{Q}{A}$$

$$= \frac{2.42}{(\frac{1}{4}\pi) \times (0.30)^2} = 34.2 \text{ m/s (112 ft/sec)} \quad \blacktriangleleft$$

EXAMPLE 4-2   Water flows in a channel that has a slope of 30°. If the velocity is assumed to be constant, 12 m/s (39.4 ft/sec), and if the flow is uniform with a depth of 60 cm (1.97 ft) measured along a vertical line, what is the discharge per meter (foot) of width of the channel?

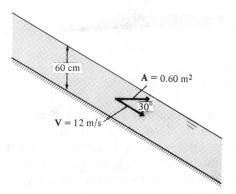

Solution   The discharge per meter (foot) of width is often designated by lowercase q, and in this case it can be solved for as follows:

$$q = \mathbf{V} \cdot \mathbf{A}$$
$$= 12 \text{ m/s} \times \cos 30° \times 0.60 \text{ m} \times 1.0 \text{ m}$$
$$= 6.23 \text{ m}^3/\text{s per meter} \qquad \blacktriangleleft$$
$$= 67.2 \text{ ft}^3/\text{s per foot} \qquad \blacktriangleleft$$

EXAMPLE 4-3   The water velocity in the channel as shown in the figure has a distribution across the vertical section equal to $u/u_{max} = (y/d)^{1/2}$. What is the discharge in the channel if the channel is 2-m deep ($d = 2$m) and 5-m wide and the maximum velocity is 3 m/s?

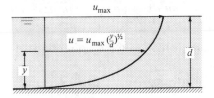

Solution

$$Q = \int_A u \, dA$$

$$= \int_0^2 u_{max}(y/d)^{1/2} \times 5 \, dy$$

$$= \frac{5u_{max}}{d^{1/2}} \int_0^2 y^{1/2} \, dy$$

$$= \frac{5 \times 3}{2^{1/2}} \times \left. \frac{y^{3/2}}{3/2} \right|_0^2$$

$$= 20.0 \text{ m}^3/\text{s} \qquad \blacktriangleleft$$

4-3 ACCELERATION

Basic operations

The acceleration of a fluid particle is the rate of change of the particle's velocity with time. In the Lagrangian formulation, each component of velocity is a function of time only, so the differentiation of each component involves the differentiation of a single-variable function. By the Eulerian approach, the velocity components are functions of both space and time as given by Eq. (4-4). Thus

$$\mathbf{V} = u\mathbf{i} + v\mathbf{j} + w\mathbf{k}$$

where

$$u = f_1(x, y, z, t)$$

$$v = f_2(x, y, z, t)$$

$$w = f_3(x, y, z, t)$$

The acceleration in the x direction is given by

$$a_x = \frac{du}{dt}$$

and, by using the chain rule for differentiation of a multivariable function, can be expressed as

$$a_x = \frac{\partial u}{\partial x} \frac{dx}{dt} + \frac{\partial u}{\partial y} \frac{dy}{dt} + \frac{\partial u}{\partial z} \frac{dz}{dt} + \frac{\partial u}{\partial t} \qquad (4\text{-}6)$$

In time dt the fluid particle moves in the x direction a distance $dx = u \, dt$ so

$$u = \frac{dx}{dt}$$

and similarly

$$v = \frac{dy}{dt} \quad \text{and} \quad w = \frac{dz}{dt}$$

Thus the acceleration component a_x is given by

$$a_x = u \frac{\partial u}{\partial x} + v \frac{\partial u}{\partial y} + w \frac{\partial u}{\partial z} + \frac{\partial u}{\partial t} \qquad (4\text{-}7)$$

Similarly for the y and z components we obtain

$$a_y = u \frac{\partial v}{\partial x} + v \frac{\partial v}{\partial y} + w \frac{\partial v}{\partial z} + \frac{\partial v}{\partial t} \tag{4-8}$$

$$a_z = u \frac{\partial w}{\partial x} + v \frac{\partial w}{\partial y} + w \frac{\partial w}{\partial z} + \frac{\partial w}{\partial t} \tag{4-9}$$

Tangential and normal acceleration

If we solve for the acceleration using the velocity as given by Eq. (4-5) we will obtain

$$a_t = \frac{\partial V_s}{\partial s} \frac{ds}{dt} + \frac{\partial V_s}{\partial t}$$

$$= V_s \frac{\partial V_s}{\partial s} + \frac{\partial V_s}{\partial t} \tag{4-10}$$

This is only the acceleration component tangential to the streamline. However, in addition there will be a component normal to the streamline which is given by

$$a_n = \frac{V_s^2}{r} \tag{4-11}$$

The term on the right-hand side of Eq. (4-11) follows from normal acceleration for curvilinear motions. Here r is the radius of curvature of the streamline and the acceleration is toward the center of curvature of the streamline. Equations (4-10) and (4-11) are valid for those cases where the streamlines do not rotate.

Convective and local acceleration

Inspection of Eqs. (4-7) through (4-11) will reveal that the terms on the right-hand sides of the equations are of two different types: those that include changes of velocity with respect to position, $u \, \partial u/\partial x$, $v \, \partial v/\partial y$, and so on, and those that are changes of velocity with respect to time, $\partial u/\partial t$, $\partial v/\partial t$, $\partial w/\partial t$, and $\partial V_s/\partial t$. Terms of the first type are called *convective* accelerations because they are associated with velocity changes that occur because of changes in position in the flow field. However, for the second type, acceleration results because the velocity changes with respect to time at a given point. These are called *local* accelerations. Obviously, *local* acceleration results when the flow is unsteady; it can also be seen that convective acceleration occurs when the flow is nonuniform, that is, if the velocity changes along a streamline.

EXAMPLE 4-4   A nozzle is designed so that its cross-sectional area changes linearly from the base of the nozzle to the tip of the nozzle. If the inside diameters of the nozzle at the base and tip are 9 cm and 3 cm, respectively, and if the nozzle is 36 cm long, what is the convective acceleration at a section midway between base and tip? Assume one-dimensional flow of water at a constant discharge of 0.02 m³/s.

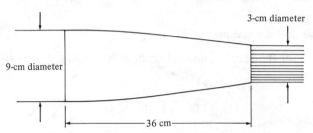

Solution   Using Eq. (4-10) we have the convective acceleration given by

$$a = V_s \frac{\partial V_s}{\partial s}$$

where

$$V_s = \frac{Q}{A}$$

However, because the area is linearly distributed, we can write

$$A = A_0 - \frac{\Delta A}{\Delta s} s$$

where A_0 = area of nozzle at the base = 6.36×10^{-3} m²

   ΔA = difference in area between base and tip of
                       nozzle = 5.65×10^{-3} m²

   Δs = distance between base and tip = 0.36 m

Thus

$$A = 0.00636 \text{ m}^2 - \frac{0.00565 \text{ m}^2}{0.36 \text{ m}} s$$

$$= 0.00636 - 0.0157 \, s$$

The area at midsection is given by

$$A_{\text{mid}} = 0.00636 - 0.0157 \times 0.18$$

$$= 0.00353 \text{ m}^2$$

and

$$V_{\text{mid}} = \frac{Q}{A_{\text{mid}}} = \frac{0.02 \text{ m}^3/\text{s}}{0.00353 \text{ m}^2} = 5.66 \text{ m/s}$$

The velocity gradient is given by

$$\frac{\partial V_s}{\partial s} = \frac{\partial}{\partial s}\frac{Q}{A}$$

$$= \frac{\partial}{\partial s}\frac{Q}{0.00636 - 0.0157\ s}$$

$$= \frac{0.0157\ Q}{(0.00636 - 0.0157\ s)^2}$$

Then, at midsection where $s = 0.18$ m,

$$\frac{\partial V_s}{\partial s} = 25.1\ s^{-1}$$

Then

$$a_{\text{convec}} = V_s\frac{\partial V_s}{\partial s} = 142\ \text{m/s}^2 \qquad \blacktriangleleft$$

4-4 BASIC CONTROL-VOLUME APPROACH

Introduction

As defined in Sec. 2-2, a *system* is a given quantity of matter; therefore, because a given body in solid mechanics is easily identified, it is the system approach that is generally used to solve dynamic problems for solids. For flow of fluids, however, individual particles are not easily identified one from another; therefore, a method is needed that will allow problems to be solved by focusing attention on a given space through which the fluid flows rather than directing attention solely to a given mass of fluid. Such a method is available; it is called the Reynolds transport theorem or, more descriptively, the control-volume approach, and it has wide application to a variety of problems in fluid mechanics. The next few paragraphs include preliminary developments that will be used in the derivation of the control-volume equation. Then the control-volume equation will be derived and applications of it will be given.

Review of extensive and intensive properties

In the derivation of the control-volume approach, we will be concerned with extensive and intensive properties. The extensive properties are mass M, momentum $M\mathbf{V}$, and energy E. The corresponding intensive properties are mass per unit mass, which is unity; momentum per unit mass, which is $\mathbf{V}$; and energy per unit of mass, e. Because we will be considering different intensive and extensive properties at different times when we actually apply the control-volume approach, it is desirable for this derivation that we use a special symbol such as B to represent a general extensive property. We will also let β (beta) be the symbol for the corresponding intensive property. Because the intensive property is given as the extensive property per unit of mass, it can be seen that the relationship between the intensive and extensive properties for a given system is denoted by

$$B = \int \beta \, dm = \int \beta \rho \, d\forall \qquad (4\text{-}12)$$

Here dm and $d\forall$ are differential mass and differential volume, respectively, and the integral is over the volume occupied by the system at a given instant.

Control volume and control surface

The control volume is a region in space that one establishes to aid in the solution of flow problems, and the control surface is the surface surrounding the control volume. In most problems, part of the control surface will coincide with some physical boundary, such as the wall of a pipe. The remaining part of the control surface in such problems is a hypothetical surface through which fluid can flow, such as is shown in Fig. 4-12 for one-dimensional flow in a conduit.

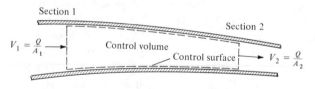

FIGURE 4-12

Flow into and out of a control volume

It has already been shown that the volume rate of flow past a given area A can be written as

$$Q = \mathbf{V} \cdot \mathbf{A}$$

Here $\mathbf{V}$ is the velocity vector of the flow and $\mathbf{A}$ is the area vector that has the magnitude of the area and is directed normal to the area in question. It is obvious that all surface areas have two sides, so we must establish a rule to govern whether the area vector is pointing in one direction or the other. The rule that we use with the control-volume approach is that the area vector always points outward from the control volume. Thus in Fig. 4-13 we show one-dimensional flow in a passage with proper orientation of the area vectors. The cosine of the angle between $\mathbf{V}_A$ and $\mathbf{A}_A$ is -1 and

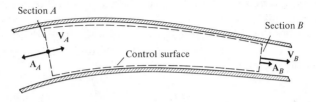

FIGURE 4-13

that between $\mathbf{V}_B$ and $\mathbf{A}_B$ is $+1$, so the flow rate out of the control volume minus the flow rate into the control volume can simply be given as

$$\text{Flow rate out} - \text{flow rate in} = V_B A_B - V_A A_A$$
$$= \mathbf{V}_B \cdot \mathbf{A}_B + \mathbf{V}_A \cdot \mathbf{A}_A$$
$$= \sum_{cs} \mathbf{V} \cdot \mathbf{A} \qquad (4\text{-}13)$$

Equation (4-13) states that if we sum up the $(\mathbf{V} \cdot \mathbf{A})$'s for all flows into and out of a control volume, we will get the net rate of outflow. So if the resulting summation is a positive number, it will mean that the net rate of flow is out of the control volume. On the other hand, if the summation is negative, it will mean a net inflow to the control volume. Of course, if the outflows and inflows are equal, then $\Sigma_{cs}\mathbf{V} \cdot \mathbf{A}$ will be zero.

If we want to evaluate the mass rate of flow out of the control volume, we simply multiply the volume rate by ρ, or

$$\dot{m} = \sum_{cs} \rho \mathbf{V} \cdot \mathbf{A} \qquad (4\text{-}14)$$

In a similar manner, if we want the rate of flow of an extensive property B out of the control volume, then we multiply the mass rate by the intensive property β:

$$\dot{B} = \sum_{cs} \beta \rho \overbrace{\mathbf{V} \cdot \mathbf{A}}^{\dot{m}} \qquad (4\text{-}15)$$

To reinforce the validity of Eq. (4-15), one may consider the dimensions involved. Equation (4-15) states that the flow rate of B is given by

$$\overset{\beta}{\underset{}{\left(\frac{B}{\text{mass}}\right)}} \overset{\dot{m}}{\left(\frac{\text{mass}}{\text{time}}\right)} = \left(\frac{B}{\text{time}}\right) = \dot{B}$$

Equations (4-13), (4-14), and (4-15) are applicable for all one-dimensional flows. If the velocity varies across a flow section, then it would be necessary to integrate the velocity across the section to obtain the rate of flow. A more general expression for the rate of flow of the extensive property from the control volume is thus given as

$$\dot{B} = \int_{cs} \beta\rho\mathbf{V} \cdot d\mathbf{A} \tag{4-15a}$$

Equation (4-15a) will be used in the most general form of the control-volume equations.

Derivation of the control-volume equation

To derive the basic control-volume equation, we first focus attention on a *system* that moves through space; however, in the process of derivation, the control volume becomes significant.

The basic equation for the control-volume approach is derived by considering the rate of change of an extensive property of the *system* of fluid that is flowing through the control volume. In Fig. 4-14 the solid line identifies the control surface that encloses the control volume, and this same

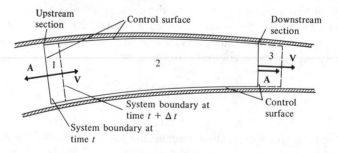

FIGURE 4-14

surface serves to define the *system,* a given mass of fluid, at time t. At time $t + \Delta t$, this *system*, or mass of fluid, is identified by the dashed line in Fig. 4-14, and it has moved with respect to the control surface. The rate of change with respect to time of an arbitrary extensive property B of the system will be given by dB_{sys}/dt, which can be stated according to the fundamental definition of a derivative as

$$\frac{dB_{sys}}{dt} = \lim_{\Delta t \to 0} \left[\frac{B_{t+\Delta t} - B_t}{\Delta t} \right] \tag{4-16}$$

In Eq. (4-16) the derivative dB_{sys}/dt is the rate of change of B of the system as it moves along. So far, Eq. (4-16) is strictly mathematical in nature. To follow the control-volume approach, we must again refer to Fig. 4-14. In this figure, the fluid has been divided into three regions. The fluid of the *system* at time t is given by region 1 plus 2, whereas regions 2 plus 3 identify the same *system* at the time $t + \Delta t$. Utilizing this method of identification, we can write

$$\frac{dB_{sys}}{dt} = \lim_{\Delta t \to 0} \left[\frac{(B_2 + B_3)_{t+\Delta t} - (B_1 + B_2)_t}{\Delta t} \right] \tag{4-17}$$

Rearranging terms, this is expressed as

$$\frac{dB_{sys}}{dt} = \lim_{\Delta t \to 0} \left[\frac{B_{2,t+\Delta t} - B_{2,t}}{\Delta t} \right] + \lim_{\Delta t \to 0} \left[\frac{B_{3,t+\Delta t} - B_{1,t}}{\Delta t} \right] \tag{4-18}$$

Each of the terms on the right-hand side of Eq. (4-18) will be considered separately below.

The first term on the right represents the rate of change of the property B in region 2. However, as Δt approaches zero in the limit, region 2 approaches that of the control volume; therefore, it can be said that this first term on the right of Eq. (4-18) is the rate of change with respect to time of the extensive property B of the fluid inside the control volume at time t. More specifically, we can rewrite the first term as the total derivative of B with respect to time:[1]

$$\lim_{\Delta t \to 0} \left[\frac{B_{2,t+\Delta t} - B_{2,t}}{\Delta t} \right] = \frac{dB_{cv}}{dt} \tag{4-19}$$

or from our preliminary work, Eq. (4-12), we then have

$$\lim_{\Delta t \to 0} \left[\frac{B_{2,t+\Delta t} - B_{2,t}}{\Delta t} \right] = \frac{d}{dt} \int_{cv} \beta \rho \, d\forall \tag{4-20}$$

The second term on the right-hand side of Eq. (4-18) can be analyzed in the following manner. The quantity $B_{3,t+\Delta t}$ represents the amount of the property B that has passed out of the control volume in time Δt, and $B_{1,t}$ represents the amount of B that has passed into the control volume in time Δt. Thus the second term on the right-hand side of Eq. (4-18) is the flow

---

[1] It has been customary to use a partial derivative of B with this term. However, the total derivative is valid because the integral over the control volume is a function only of time; that is, since the integration is over the space of the control volume, it can no longer be a function of the spatial coordinates. Thus it depends only on time.

rate of B out of the control volume minus the flow rate of B into the control volume at time t. But Eq. (4-15), $\dot{B} = \Sigma_{cs}\beta\rho\mathbf{V} \cdot \mathbf{A}$, is precisely equal to this quantity, so we can now make Eq. (4-18) more compact by substituting Eq. (4-15) for the second term on the right-hand side and substituting $d/dt \int_{cv}\beta\rho \, d\Psi$ for the first term on the right-hand side to obtain

$$\frac{dB_{\text{sys}}}{dt} = \frac{d}{dt} \int_{cv} \beta\rho \, d\Psi + \sum_{cs} \beta\rho\mathbf{V} \cdot \mathbf{A} \qquad (4\text{-}21a)$$

The subscript on the summation sign of the second term on the right side of Eq. (4-21a) indicates that we are summing flows across the entire control surface. In the derivation of Eq. (4-21a), we first considered the rate of change of the extensive property B of the *system*, dB_{sys}/dt, then we showed that this could be expressed as the sum of the rate of change of B within the control volume plus the net rate of flow of B out of the control volume. Thus the left-hand side of Eq. (4-21a) is still the rate of change of the extensive property of the *system* (rate of change of the extensive property of the given quantity of matter), but the right-hand side refers to conditions within the control volume and flow across the control surface for the first and second terms, respectively.

More general form of the control-volume equation

In the derivation of Eq. (4-21a) one-dimensional flow was assumed; thus the rate of flow of B at each section was given as $\beta\rho\mathbf{V} \cdot \mathbf{A}$. However, if the velocity is variable across a section, the more general form for the rate of flow of the extensive property must be used, Eq. (4-15a); thus the control-volume equation is given as

$$\frac{dB_{\text{sys}}}{dt} = \frac{d}{dt} \int_{cv} \beta\rho \, d\Psi + \int_{cs} \beta\rho\mathbf{V} \cdot d\mathbf{A} \qquad (4\text{-}21b)$$

In much of our work we are concerned only with steady-flow problems; then, the first term on the right-hand side of Eq. (4-21) drops out. This occurs because the differentiation is taken with respect to conditions within the control volume; if the conditions do not change with time, then the $d(\ )/dt$ term is zero. For steady flow our basic control volume equation thus reduces to

$$\frac{dB_{\text{sys}}}{dt} = \sum_{cs} \beta\rho\mathbf{V} \cdot \mathbf{A} \qquad (4\text{-}22)$$

Depending upon the equation's application, β and B may be either scalar quantities or vector quantities. For example, if we want to derive the general form of the continuity equation, we let B be the mass of the system (which is constant because a system is a given quantity of matter) and then β will be equal to unity. Both these quantities are scalars. However, if we let $B =$ momentum and $\beta = \mathbf{V}$, then we have vector quantities the use of which leads to the momentum equation. In Sec. 4-5, the continuity equation will be derived by utilizing the control-volume equation (Eq. 4-21). Then in Chapters 6 and 7 the control-volume equation will be used to derive the momentum and energy equations, respectively.

Motion of the control volume and steady or unsteady flow

The concepts of steady and unsteady flow were discussed earlier. It is interesting to see how the terms of Eq. (4-21) can be interpreted in light of these concepts. Let us consider the first term on the right-hand side of Eq. (4-21), which is the rate of change of B_{cv} with respect to time; if the flow is steady, there will be no change with respect to time and the term will be zero. Conversely, if the flow process is unsteady, the term must be considered in any control-volume analysis. In this regard, it is most interesting to see how the motion of the control volume itself can, in some cases, dictate whether the flow condition is considered steady or unsteady. Consider the case in which a ship is traveling at a constant speed in shallow water, as shown in Fig. 4-15. If the control surface is fixed to the earth, the velocities within the control volume will continuously change with

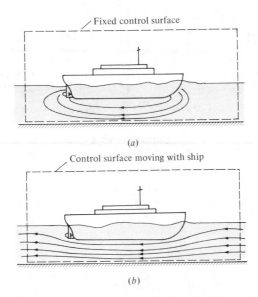

FIGURE 4-15   Change from unsteady to steady flow by change of velocity of control volume. (*a*) Unsteady flow. (*b*) Steady flow.

time, and by definition the flow will be unsteady. However, if the control surface moves at the same speed as the ship, then none of the flow characteristics will change with time, and this would be classified as steady flow. If the control surface is assumed to move in space, the reference system is likewise assumed to have the same motion, which means simply that the velocities considered in the basic equations are measured relative to the control surface—a requirement which the student must be careful to observe in all problems involving the control-volume analysis. Although moving the control surface with the body appears to be a simple way of converting an unsteady-flow process to a steady-flow process, a word of caution should be sounded: If the body itself (and also the control surface) is accelerating and if the control-volume equation is applied to the momentum principle, then it is necessary to reference the velocity to the inertial frame rather than to the control surface itself. Because Eq. (4-21) is involved with rate of change of a flow property with time, it should not be surprising that this basic equation can serve as the starting point for many expressions involving the dynamics of flow as well as for the continuity equation, which is simply kinematic in nature. In the next section we shall see how the control-volume approach can be applied to the continuity principle.

4-5 CONTINUITY EQUATION

The basic concept

The continuity equation is based upon the conservation of mass as it applies to the flow of fluids. In words, the continuity equation states that the mass rate of flow out of a region of space, such as a control volume, minus the mass rate of flow into the region is equal to the rate at which the fluid mass is being evacuated from the region. When the intensive property β of the basic control-volume equation is set equal to unity, the equation reduces to a general form of the continuity equation. This general equation can then be applied to different types of control volumes and different coordinate systems to yield special forms of the continuity equation. The general case and several special forms of the continuity equation will be considered in the paragraphs that follow.

General form of the continuity equation

First we shall write the general control-volume equation with B = mass of the system, which means that β is equal to unity ($\beta = 1$). Then Eq. (4-21b) becomes

$$\frac{d(\text{mass})}{dt} = \frac{d}{dt} \int_{cv} \rho \, d\forall + \int_{cs} \rho \mathbf{V} \cdot d\mathbf{A} \qquad (4\text{-}23)$$

The term on the left is the rate of change of the mass of the system. However, by definition of a system its mass is constant; therefore the left-hand side of the equation is zero and the equation can be written as

$$\int_{cs} \rho \mathbf{V} \cdot d\mathbf{A} = -\frac{d}{dt} \int_{cv} \rho \, d\forall \qquad (4\text{-}24)$$

This is the general form of the continuity equation; it states that the net rate of outflow of mass from the control volume is equal to the rate of decrease of mass within the control volume.

The continuity equation involving flow streams having a constant velocity across the flow section is given as

$$\sum \rho \mathbf{V} \cdot \mathbf{A} = -\frac{d}{dt} \int_{cv} \rho \, d\forall \qquad (4\text{-}25)$$

The following example illustrates the use of the continuity equation.

EXAMPLE 4-5   A jet of water discharges into an open tank and water leaves the tank through an orifice near the bottom at a rate of 0.003 m³/s. If the cross-sectional area of the jet is 0.0025 m² where the velocity of water is 7 m/s, at what rate is water accumulating in (or evacuating from) the tank at this instant?

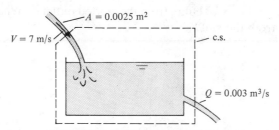

Solution   By drawing the control surface to enclose the entire tank, we can see from the control volume equation that the mass rate at which the tank is emptying $(-d/dt \int \rho \, d\forall)$ is equal to $\sum_{cs} \rho \mathbf{V} \cdot \mathbf{A}$; thus

$$\text{Emptying rate} = \sum_{cs} \rho \mathbf{V} \cdot \mathbf{A}$$

$$\text{Net rate of mass outflow} = \sum_{cs} \rho \mathbf{V} \cdot \mathbf{A}$$

$$= \rho Q_{\text{outflow}} + \rho \mathbf{V}_1 \cdot \mathbf{A}_1$$

Positive because of outflow

$$= 3.00 + 1000(\underset{\uparrow}{-}7.0 \times 0.0025)$$

Minus because area vector
opposes velocity vector

$$= 3.00 - 17.5 \text{ kg/s}$$

Net rate of outflow $= -14.5$ kg/s

Since the net rate of outflow is -14.5 kg/s, we can also say it is filling at 14.5 kg/s. Or, if we want the volume rate at which the water is filling the tank, we divide by ρ to obtain a filling rate of

$$Q = 0.0145 \text{ m}^3/\text{s} \qquad \blacktriangleleft$$

The illustration above is an application of the control-volume approach for a case of unsteady flow. In many problems, however, the flow is steady, and we are left with the following equation:

$$\int_{cs} \rho \mathbf{V} \cdot d\mathbf{A} = 0$$

The example following shows how this can be applied.

EXAMPLE 4-6   Water flows steadily into a tank through the two pipes as shown and discharges at a steady rate out of the tank from a pipe and from a slot where the efflux velocity varies linearly (increases) along the slot. If the velocity of inflow and outflow is 50 ft/sec at points 1, 2, and 3, what is the maximum velocity of efflux from the slotted outlet 4? What is the mass rate of flow from the slot?

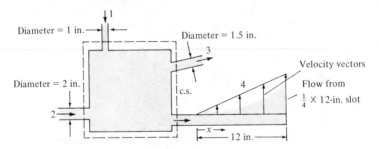

Solution   Since this is steady flow, we use

$$\int_{cs} \rho \mathbf{V} \cdot d\mathbf{A} = 0$$

$$-\rho_1 V_1 A_1 - \rho_2 V_2 A_2 + \rho_3 V_3 A_3 + \int_{A_4} \rho_4 V_4 dA_4 = 0$$

First let us solve for the mass rate of flow from 4.

$$\dot{m}_4 = \int_{A_4} \rho_4 V_4 dA_4 = \rho_1 V_1 A_1 + \rho_2 V_2 A_2 - \rho_3 V_3 A_3$$

For water $\rho = 1.94$ slugs/ft^3; therefore,

$$\dot{m}_4 = \frac{1.94 \times 50}{144} \frac{\pi}{4} (1 + 4 - 2.25) = 1.45 \text{ slugs/sec} \qquad \blacktriangleleft$$

Now we can solve for $V_{4\,\text{max}}$. Since the velocity distribution along the slot is linear, we can write $V_4 = kx$ and at $x = 1$, $V_4 = V_{\text{max}}$; therefore, $V_4 = x V_{\text{max}}$.

$$\int_0^1 1.94 x V_{\text{max}} \overbrace{\left(\frac{1}{48} dx\right)}^{dA} = 1.45$$

or

$$\frac{1.94}{48} V_{\text{max}} \frac{x^2}{2} \Big|_0^1 = 1.45$$

then substituting in the limits yields

$$\frac{1.94}{48} \frac{V_{\text{max}}}{2} = 1.45$$

$$V_{\text{max}} = 71.8 \text{ ft/sec} \qquad \blacktriangleleft$$

or

$$V_{\text{max}} = 21.9 \text{ m/s} \qquad \blacktriangleleft$$

Continuity equation for steady one-dimensional flow in a conduit

Consider the steady-flow case where we have flow in a conduit, Fig. 4-16, in which we are interested in establishing a relationship between the mean velocities at two sections of the conduit. A control volume is drawn such

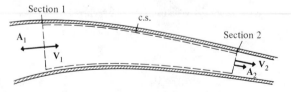

FIGURE 4-16

that two ends of the control volume are the two sections 1 and 2. For flow through this control volume we apply the continuity equation

$$\sum \rho \mathbf{V} \cdot \mathbf{A} = 0$$

When this is expanded for the example under consideration, we obtain

$$-\rho_1 V_1 A_1 + \rho_2 V_2 A_2 = 0$$

or

$$\rho_1 V_1 A_1 = \rho_2 V_2 A_2 \tag{4-26}$$

Furthermore, if we have constant density flow, the ρ's will cancel, leaving

$$V_1 A_1 = V_2 A_2 \tag{4-27}$$

This equation states that the volume rate of flow at section 1 is equal to the volume rate of flow at section 2; therefore, it can also be written as

$$Q_1 = Q_2$$

Equation (4-27) is a very common form of the continuity equation and used in numerous applications where the flow can be considered to be one-dimensional and incompressible.

EXAMPLE 4-7   A 120-cm pipe is in series with a 60-cm pipe. If the rate of flow of water in the system of pipes is 2 m³/s, what is the velocity of flow in each?

Solution

$$Q = V_{120} A_{120}$$

Here $Q = 2 \text{ m}^3/\text{s}$ and $A_{120} = \dfrac{\pi}{4} \times (1.20)^2 = 1.13 \text{ m}^2$

Therefore

$$V_{120} = \frac{Q}{A_{120}}$$

$$= \frac{2 \text{ m}^3/\text{s}}{1.13 \text{ m}^2}$$

$$= 1.77 \text{ m/s} \qquad \blacktriangleleft$$

Also

$$V_{120} A_{120} = V_{60} A_{60}$$

So

$$V_{60} = V_{120} \frac{A_{120}}{A_{60}}$$

$$= V_{120} \left(\frac{120}{60}\right)^2$$

$$= 7.08 \text{ m/s} \qquad \blacktriangleleft$$

EXAMPLE 4-8   The river discharges into the reservoir at a rate of 400,000 ft³/sec (cfs), and the outflow rate from the reservoir through the flow passages in the dam is 250,000 cfs. If the reservoir surface area is 40 miles², what is the rate of rise of water in the reservoir?

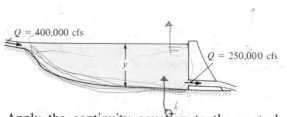

Solution   Apply the continuity equation to the control volume, as shown in the figure. Here flow past the control surface is occurring at three sections 1, 2 and 3. The continuity equation then becomes

$$\sum_{cs} \rho \mathbf{V} \cdot \mathbf{A} = 0$$

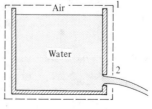

Applied to our example, the equation above becomes

$$-400,000 \text{ ft}^3/\text{sec} + 250,000 \text{ ft}^3/\text{sec} + Q_{\text{rise}} = 0$$

or
$$V_{\text{rise}} A_R = 150,000 \text{ ft}^3/\text{sec}$$

Here A_R is the area of the reservoir. Then we have

$$V_{\text{rise}} = \frac{150,000 \text{ ft}^3/\text{sec}}{40 \text{ mile}^2 \times (5,280)^2 \text{ ft}^2/\text{mile}^2}$$

Rate of rise $= 1.34 \times 10^{-4} \text{ ft/sec}$

or
$$V_{\text{rise}} = 0.484 \text{ ft/hr} \qquad \blacktriangleleft$$

EXAMPLE 4-9   A 10 cm jet of water issues from a 1-m-diameter tank as shown. Assume that the velocity in the jet is $\sqrt{2gh}$ m/s. How long will it take for the water surface in the tank to drop from $h_1 = 2$ m to $h = 0.50$ m?

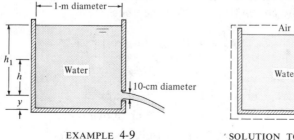

EXAMPLE 4-9   SOLUTION TO EXAMPLE 4-9

Solution   We draw the control surface surrounding the tank as shown in the figure. Then we apply the continuity equation to this flow and let sections 1 and 2 be across the top of the tank and across the jet, respectively:

$$\sum_{cs} \rho \mathbf{V} \cdot \mathbf{A} = -\frac{d}{dt} \int_{cv} \rho \, d\forall$$

or

$$-\rho_1 V_1 A_1 + \rho_2 V_2 A_2 = -\frac{d}{dt} \int_{cv} \rho \, d\forall$$

At section 1, air is flowing into the tank while water is flowing out of the tank at section 2. Thus the continuity equation becomes

$$-\rho_{\text{air}} V_1 A_1 + \rho_{\text{water}} V_2 A_2 = -\frac{d}{dt} \int_{cv} \rho_{\text{air}} \, d\forall_{\text{air}} - \frac{d}{dt} \int_{cv} \rho_{\text{water}} \, d\forall_{\text{water}}$$

or

$$\rho_{\text{air}} \left(-V_{\text{air}} A_1 + \frac{d\forall_{\text{air}}}{dt} \right) + \rho_{\text{water}} V_2 A_2 = -\frac{d}{dt} \int_{cv} \rho_{\text{water}} \, d\forall_{\text{water}}$$

The volume rate of increase of air inside the tank (increase inside control volume) will be the same as the volume rate of flow of air across the control surface at 1; therefore, the first term in the above equation becomes zero and the continuity equation is

$$\rho_{\text{water}} V_2 A_2 = \frac{d}{dt} \int_{cv} \rho_{\text{water}} \, d\forall_{\text{water}}$$

An equally valid approach to setting up the control volume equation for continuity is to assume the control surface moves with the water surface, the volume of the control volume decreasing in size with time. Then the only mass crossing the control surface is at the jet, so the continuity equation becomes directly the above equation. The term

$$\int_{cv} \rho_{\text{water}} \, d\forall_{\text{water}}$$

represents the total mass of water in the control volume, so we write it as $\rho A_T(h + y)$. The continuity equation then becomes

$$\rho_2 V_2 A_2 = -\frac{d}{dt} [\rho A_T(h + y)]$$

where
$$\rho_2 = \rho = \rho_{\text{water}}$$
$$A_T = \text{cross-sectional area of tank}$$
$$h = \text{depth of water in tank above outlet}$$
$$y = \text{depth of water in tank below outlet}$$

Since ρ in the equation above is constant, we cancel it out of the equation; and when the differentiation is carried out (remembering that A_T and y are constant) on the right-hand side, the equation reduces to

$$V_2 A_2 = -A_T \frac{dh}{dt}$$

It was already noted that $V_2 = \sqrt{2gh}$, so substitution for V_2 is made to obtain

$$\sqrt{2gh}\, A_2 = -A_T \frac{dh}{dt}$$

Separating variables, we obtain

$$dt = \frac{-A_T}{\sqrt{2g}\, A_2} \frac{dh}{\sqrt{h}}$$

or

$$dt = \frac{-A_T}{\sqrt{2g}\, A_2} h^{-1/2}\, dh$$

Noting now that $A_T/\sqrt{2g}\, A_2$ is constant, we integrate the differential equation to yield the following:

$$t = \frac{-2A_T}{\sqrt{2g}\, A_2} h^{1/2} + C$$

The constant of integration is evaluated by arbitrarily letting $t = 0$ when $h = h_1$, then

$$C = +\frac{2A_T}{\sqrt{2g}\, A_2} h_1^{1/2}$$

So we have

$$t = \frac{2A_T}{\sqrt{2g}\, A_2} (h_1^{1/2} - h^{1/2})$$

Thus for this particular example the elapsed time for the water level to drop from $h_1 = 2$ m to $h = 0.50$ m will be

$$t = \frac{2A_T}{\sqrt{2g}\, A_2} (2^{1/2} - 0.5^{1/2})$$

But

$$A_T = \frac{\pi}{4} D^2 = \frac{\pi}{4} \times 1^2 = \frac{\pi}{4}\ \text{m}^2$$

$$A_2 = \frac{\pi}{4} (0.10)^2 = 0.01 \left(\frac{\pi}{4}\right)\ \text{m}^2$$

Hence
$$t = \frac{2\pi/4}{\sqrt{2g}(\pi/4 \times 0.01)}(1.414 - 0.707)$$

$$= 31.9 \text{ s} \qquad \blacktriangleleft$$

EXAMPLE 4-10 Methane escapes through a small, 10^{-7} m², hole in a 10-m³ tank. The methane escapes so slowly that the temperature in the tank remains constant at 23°C. The mass flow rate of methane through the hole is given by $\dot{m} = 0.66\ pA/\sqrt{RT}$, where p is the pressure in the tank, A is the area of the hole, R is the gas constant, and T is the temperature in the tank. Calculate the time required for the absolute pressure in the tank to decrease from 500 to 400 kPa.

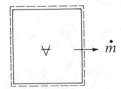

Solution Consider a control surface that just encloses the tank. Writing the continuity equation for the control volume, we have

$$\frac{d}{dt}\int_{cv} \rho\ d\forall + \sum \rho\mathbf{V}\cdot\mathbf{A} = 0$$

Since ρ is uniform throughout the tank and mass crosses the control surface only at the location of the leak, we have

$$\frac{d}{dt}(\rho\forall) + \dot{m} = 0$$

The tank volume is constant, so

$$d\rho/dt = -\dot{m}/\forall$$

The ideal-gas law provides the following relationship between pressure and density: $\rho = p/RT$. Since T is constant, $d\rho = dp/RT$. Substituting this and the mass-flow expression into the equation above gives

$$\frac{dp}{p} = -0.66\frac{A\sqrt{RT}\ dt}{\forall}$$

Integrating, we obtain

$$t = \frac{1.52\forall}{A\sqrt{RT}}\ln\frac{p_0}{p}$$

where p_0 is the initial pressure. Substituting in the appropriate values, we calculate

$$t = \frac{1.52 \ (10 \ \text{m}^3)}{(10^{-7} \ \text{m}^2) \left(518 \ \dfrac{\text{J}}{\text{kg} \cdot \text{K}} \ 300 \ \text{K}\right)^{1/2}} \ln \frac{500}{400} = 8.6 \times 10^4 \ \text{s}$$

or approximately 1 day.   ◄

Continuity at a point

In the analysis of fluid flow, one of the fundamental independent equations that is needed for the solution of problems is a differential equation that is a statement of continuity at a point. This is derived by applying the basic continuity equation, Eq. (4-25), to a control volume of infinitesimal size for which flow is assumed to pass through all surfaces of the element of volume. Consider Fig. 4-17, which shows a cubical element oriented so

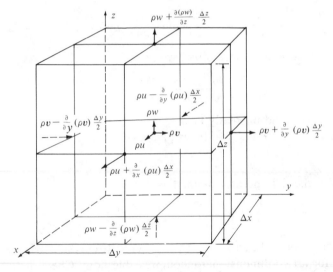

FIGURE 4-17   Continuity for elemental control volume.

that its sides are parallel to the x, y, and z coordinate directions. Because the boundaries of the cubical element are assumed stationary, the control volume continuity equation can be rewritten as

$$\sum \rho \mathbf{V} \cdot \mathbf{A} = - \int_{cv} \frac{\partial \rho}{\partial t} \, d\forall$$

Using the Taylor series expansion (retaining only the first-order terms) for the mass flux on the six faces and applying them to the continuity equation for the cubical element, we obtain the following:

$$\left[\rho u + \frac{\partial(\rho u)}{\partial x}\frac{\Delta x}{2}\right]\Delta y\,\Delta z - \left[\rho u - \frac{\partial(\rho u)}{\partial x}\frac{\Delta x}{2}\right]\Delta y\,\Delta z$$

$$+ \left[\rho v + \frac{\partial(\rho v)}{\partial y}\frac{\Delta y}{2}\right]\Delta x\,\Delta z - \left[\rho v - \frac{\partial(\rho v)}{\partial y}\frac{\Delta y}{2}\right]\Delta x\,\Delta z$$

$$+ \left[\rho w + \frac{\partial(\rho w)}{\partial z}\frac{\Delta z}{2}\right]\Delta x\,\Delta y - \left[\rho w - \frac{\partial(\rho w)}{\partial z}\frac{\Delta z}{2}\right]\Delta x\,\Delta y$$

$$+ \frac{\partial \rho}{\partial t}\Delta x\,\Delta y\,\Delta z = 0 \qquad (4\text{-}28)$$

This reduces to

$$\frac{\partial}{\partial x}(\rho u) + \frac{\partial}{\partial y}(\rho v) + \frac{\partial}{\partial z}(\rho w) = -\frac{\partial \rho}{\partial t} \qquad (4\text{-}29)$$

If the flow is steady, we obtain

$$\frac{\partial}{\partial x}(\rho u) + \frac{\partial}{\partial y}(\rho v) + \frac{\partial}{\partial z}(\rho w) = 0 \qquad (4\text{-}30)$$

And if the fluid is incompressible, we have

$$\frac{\partial u}{\partial x} + \frac{\partial v}{\partial y} + \frac{\partial w}{\partial z} = 0 \qquad (4\text{-}31a)$$

for both steady and unsteady flow.

In vector notation Eq. (4-31a) is given as

$$\nabla \cdot \mathbf{V} = 0 \qquad (4\text{-}31b)$$

where ∇ is the del operator defined as

$$\nabla = \frac{\partial}{\partial x}\mathbf{i} + \frac{\partial}{\partial y}\mathbf{j} + \frac{\partial}{\partial z}\mathbf{k}$$

EXAMPLE 4-11   The expression $\mathbf{V} = 10x\mathbf{i} - 10y\mathbf{j}$ is said to represent the velocity for a two-dimensional incompressible flow. Check it to see if it satisfies continuity.

Solution

$$u = 10x \qquad \text{so} \qquad \frac{\partial u}{\partial x} = 10$$

$$v = -10y \qquad \text{so} \qquad \frac{\partial v}{\partial y} = -10$$

$$\frac{\partial u}{\partial x} + \frac{\partial v}{\partial y} = 10 - 10 = 0$$

Continuity is satisfied.

4-6 ROTATION AND VORTICITY

Concept of rotation

Consider a tank of liquid that is being rotated about a vertical axis. A plan view of such a tank is given in Fig. 4-18. If we focus attention on a given element, it can be seen that this element will rotate but not deform as time

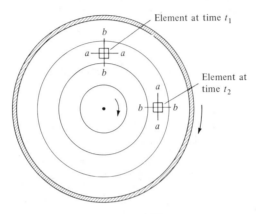

FIGURE 4-18

passes. In this process, all lines drawn through the element, such as *a-a* and *b-b* in Figure 4-18, will rotate at the same rate. This is unquestionably a case of fluid rotation. Now consider fluid flow between two horizontal plates, Fig. 4-19, where the bottom plate is stationary and the top is moving to the right with a velocity V. The velocity distribution is linear; therefore, an element of fluid will deform as shown. Here we see that the element faces that were initially vertical rotate clockwise whereas the horizontal faces do not. It is not so clear whether this is a case of rotational motion or not. The test comes in determining the average rate of rotation of two initially mutually perpendicular lines that follow the motion of the faces of the fluid element. If the average rate of rotation of these lines is zero, then the flow is irrotational; if it is not zero, then by definition rotational flow exists. Thus the flow between the parallel boundaries is also a case of rotational flow. Now we shall derive an expression that

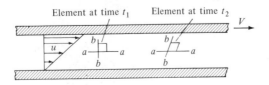

FIGURE 4-19

indicates mathematically the degree of rotation in terms of velocity gradients of flow.

Consider a cubical element of fluid, one face of which is shown in Fig. 4-20. For this element we shall derive an expression that indicates the average rate of rotation of lines AB and AC, which are mutually perpen-

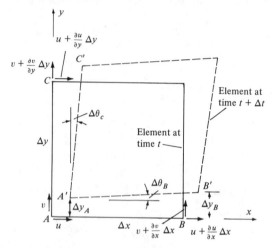

FIGURE 4-20 Deformation of element of fluid.

dicular at time t. Here rotation will be about the z axis; therefore, this rotation will be given as

$$\omega_z = \frac{\omega_{AB} + \omega_{AC}}{2} \tag{4-32}$$

The rate of rotation of AB may be given as the rate of change of θ_B, which can be expressed as a function of the velocity distribution in the following manner:

$$\omega_{AB} = \lim_{\Delta t \to 0} \frac{\Delta \theta_B}{\Delta t}$$

Referring again to Fig. 4-20, we see that $\Delta \theta_B$ may be expressed as $(\Delta y_B - \Delta y_A)/\Delta x$, where $\Delta y_B - \Delta y_A$ is given as

$$\Delta y_B - \Delta y_A = (v + \frac{\partial v}{\partial x} \Delta x - v)\Delta t$$

$$\Delta \theta_B = \frac{\partial v}{\partial x} \frac{\Delta x \, \Delta t}{\Delta x}$$

Then
$$\omega_{AB} = \lim_{\Delta t \to 0} \frac{(\partial v / \partial x) \Delta x \, \Delta t}{\Delta x \, \Delta t}$$

$$= \frac{\partial v}{\partial x} \qquad (4\text{-}33)$$

In a similar manner it can be shown that the negative rate of rotation, clockwise, of line AC is given as

$$-\omega_{AC} = \frac{\partial u}{\partial y}$$

or
$$\omega_{AC} = -\frac{\partial u}{\partial y} \qquad (4\text{-}34)$$

When Eqs. (4-33) and (4-34) are substituted into Eq. (4-32), we obtain

$$\omega_z = \frac{1}{2}\left(\frac{\partial v}{\partial x} - \frac{\partial u}{\partial y}\right)$$

Likewise, the rate of rotations about the x and y axes are found to be

$$\omega_x = \frac{1}{2}\left(\frac{\partial w}{\partial y} - \frac{\partial v}{\partial z}\right)$$

$$\omega_y = \frac{1}{2}\left(\frac{\partial u}{\partial z} - \frac{\partial w}{\partial x}\right)$$

Vorticity

The vorticity is defined as twice the average rate of rotation; consequently, the vorticity vector may be written in the following form:

$$\mathbf{\Omega} = \left(\frac{\partial w}{\partial y} - \frac{\partial v}{\partial z}\right)\mathbf{i} + \left(\frac{\partial u}{\partial z} - \frac{\partial w}{\partial x}\right)\mathbf{j} + \left(\frac{\partial v}{\partial x} - \frac{\partial u}{\partial y}\right)\mathbf{k} \qquad (4\text{-}35)$$

Recall the statement that irrotational flow exists only when the average rates of rotation are zero; thus for irrotational flow each term inside the parentheses of Eq. (4-35) must have a zero value. It then follows that for irrotational flow

$$\frac{\partial w}{\partial y} = \frac{\partial v}{\partial z} \qquad (4\text{-}36)$$

$$\frac{\partial u}{\partial z} = \frac{\partial w}{\partial x} \qquad (4\text{-}37)$$

$$\frac{\partial v}{\partial x} = \frac{\partial u}{\partial y} \qquad (4\text{-}38)$$

The greatest application of these equations is in the area of ideal-flow theory.

EXAMPLE 4-12   Is the flow in Example 4-11 irrotational?

Solution

$$u = 10x \quad \text{so} \quad \frac{\partial u}{\partial y} = 0$$

$$v = -10y \quad \text{so} \quad \frac{\partial v}{\partial x} = 0$$

Then

$$\frac{\partial v}{\partial x} = \frac{\partial u}{\partial y}$$

The flow is irrotational.

4-7   SEPARATION

Flow patterns that are developed by ideal-flow theory, as already indicated, are for irrotational flow. In some applications this is closely approximated; for example, flow of air or water in a region where the streamlines are converging usually approximates irrotational flow quite closely. However, in regions where boundaries turn away from the flow so as to cause the streamlines to diverge, the flow usually "separates" from the boundary and a recirculation pattern is generated in the region.

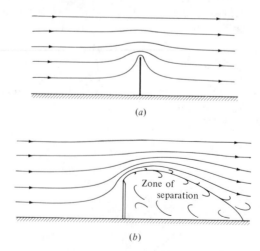

FIGURE 4-21   Flow past a plate. (*a*) Ideal flow past a plate. (*b*) Real flow past a plate.

This phenomenon is called *separation*, a typical case of which is shown schematically in Fig. 4-21*b*. Also shown in Fig. 4-21*a* is the pattern for ideal flow past a similar plate. It can be seen that the two flow patterns are markedly different, especially downstream of the plate; therefore, the engineer must be cautious about applying ideal-flow theory to engineering problems. For some problems it can be very useful, but for others it will give completely fallacious results.

PROBLEMS

4-1   If a flow pattern is indicated by converging streamlines, how would you classify the flow?

4-2   In the system in the figure, the valve at C is gradually opened in such a way that a constant rate of increase in discharge is produced. How would you classify flow at B while the valve is being opened? How would you classify the flow at A?

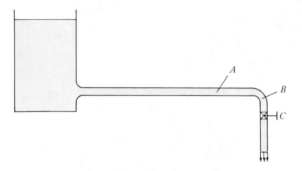

4-3   Select the correct practical example of steady-uniform flow:
(a) Motion of river around bridge piers
(b) Steadily increasing flow through a pipe
(c) Steadily decreasing flow through a reducing bend
(d) Constant discharge through a straight, long pipe

4-4   (a) Water flows in this passage. If the flow rate is decreasing with time, the flow would be classified as: steady, unsteady, uniform, nonuniform (make the correct choice(s)).
(b) In the flow there is: local acceleration, convective acceleration, no acceleration (make the correct choice(s)).

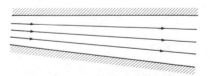

4-5   For a given hypothetical flow the velocity from time $t = 0$ to $t = 10$ s was $u = 2$ m/s, $v = 0$. Then, from time $t = 10$ s to $t = 16$ s, the velocity was $u = -3$ m/s, $v = -4$ m/s. If a dye streak was started at a point in the flow field at time $t = 0$ and the path of a particle in the fluid was also traced from that same point starting at the same time, draw to scale the streakline, pathline of the particle, and streamlines at time $t = 16$ s.

4-6   For a given hypothetical flow the velocity from time $t = 0$ to $t = 10$ sec is $u = 2$ ft/sec, $v = 2$ ft/sec. Then, from time $t = 10$ to $t = 20$ sec, the velocity is $u = 0$ ft/sec, $v = -2$ ft/sec. If a dye streak is started at a point in the flow field at time $t = 0$ and the path of a particle in the fluid is also traced from that same point starting at that same time, draw to scale the streakline, pathline of the particle, and streamlines at time $t = 15$ sec.

4-7   For a given fluid the velocity for the entire field of flow is $u = 4$ m/s, $v = 3t$ m/s. Here t is the time in seconds. If a particle is released at time $t = 0$ and a dye stream is started from the same point at the same time, draw the streakline, pathline of the particle, and the streamlines at time t = 5 s.

4-8   Classify the following according to one-dimensional, two-dimensional, or three-dimensional flows:
(a) Water flow over the crest of a long spillway of a dam
(b) Flow in a straight horizontal pipe
(c) Flow in a constant-diameter pipeline that follows the contour of the ground in hilly country
(d) Air flow from a slit in a plate at the end of a large rectangular duct
(e) Air flow past an automobile
(f) Air flow past a house
(g) Water flow past a pipe that is laid normal to the flow across the bottom of a wide rectangular channel.

4-9   The discharge of water in a 50-cm pipe is 0.40 m³/s. What is the mean velocity?

4-10 A 12-in.-diameter pipe carries water with a velocity of 15 ft/sec. What is the discharge in cubic feet per second and in gallons per minute (1 cfs is equivalent to 449 gpm).

4-11 A 1-m-diameter pipe carries water with a velocity of 3 m/s. What is the discharge in cubic meters per second and in cubic feet per second?

4-12 A 3-cm-diameter pipe transports air with a temperature of 20°C and pressure of 200 kPa absolute at 20 m/s. Determine the mass flow rate.

4-13 Natural gas (methane) flows at 10 m/s through a 1-m-diameter pipe. The temperature of the methane is 15°C and the pressure is 150 kPa gage. Determine the mass flow rate.

4-14 An engineer is designing a system that requires transporting 0.01 m³/s of helium at 15°C and 120 kPa absolute. The velocity in the pipe is limited to 40 m/s. What size of pipe is needed?

4-15 A heating and air-conditioning engineer is designing a system to move 1000 m³ of air per hour at 100 kPa and 20°C. The duct is rectangular with cross-sectional dimensions of 1 m by 20 cm. What will be the air velocity in the duct?

4-16 Flow occurs in a 2-ft-diameter pipe that has the following hypothetical velocity distribution: The velocity is maximum at the centerline and decreases linearly with r to a minimum at the pipe wall. If $V_{max} = 12$ ft/sec and if $V_{min} = 10$ ft/sec, what is the discharge in cubic feet per second and gallons per minute?

4-17 In Prob. 4-16, if $V_{max} = 10$ m/s, if $V_{min} = 8$ m/s, and $D = 2$ m what is the discharge in m³/s and what is the mean velocity?

9.84   9.82

4-18 The velocity at section A-A is 10 m/s and the depth y at the same section is 2 m. If the width of the channel is 10 m what is the discharge in m³/s?

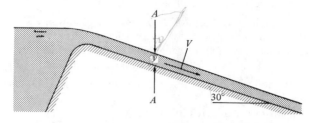

PROBLEMS 4-18, 4-19

4-19 The velocity at section A-A is 15 ft/s and the depth y at the same section is 4 ft. If the width of the channel is 25 ft, what is the discharge in cubic feet per second?

4-20 Water from a pipe is diverted into a weigh tank for exactly 12 min. The increased weight in the tank is 25 kN. What is the discharge in m³/s? Assume $T = 20°C$.

4-21 An empirical equation for the velocity distribution in a horizontal open channel is given by $u = u_{max}(y/d)^{1/7}$, where u is the velocity at a distance y feet above the floor of the channel. If the depth d of flow is 3 ft and $u_{max} = 10$ ft/sec, what is the discharge in cubic feet per second per foot of width of channel? What is the mean velocity?

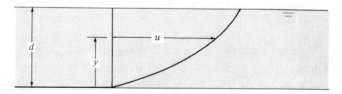

PROBLEMS 4-21, 4-22

4-22 An empirical equation for the velocity distribution in a horizontal open channel is given by $u = u_{max}(y/d)^n$, where u is the velocity at a distance y feet above the floor of the channel. If the depth d of flow is 1.5 m, $u_{max} = 4$ m/s, and $n = 1/7$, what is the discharge in m³/s per meter of width of channel? What is the mean velocity?

4-23 The velocity of flow in a circular pipe varies as

$$\frac{V}{V_C} = \left(\frac{r_0^2 - r^2}{r_0^2}\right)^n$$

where V_C is the centerline velocity, r_0 is the pipe radius, and r is radius measured from the centerline. The exponent n is general and chosen to fit a given profile ($n = 1$ for laminar flow). Determine the mean velocity as a function of V_C and n.

4-24 The equation for the hypothetical velocity distribution in a horizontal open channel is given by $u = 4y^{1/2}$, where u is the velocity at a distance y feet above the floor of the channel. If the depth of flow is 3 ft, what is the discharge for a 10-ft-wide rectangular channel? What is the mean velocity?

4-25 Plot the velocity distribution across the pipe and determine the discharge of a fluid flowing through a 1-ft-diameter pipe that has a velocity distribution given by $V = 10(1 - r^2/r_0^2)$. Here r_0 is the radius of the pipe and r is the radial distance from the centerline. What is the mean velocity?

4-26 One hundred pounds of water per minute flow through a one-inch pipeline. Calculate the mean velocity. Assume $T = 60°F$.

4-27 One thousand kilograms of water per minute flow through a 15-cm pipe. Calculate the mean velocity in meters per second if $T = 20°C$.

4-28 Water from a pipeline is diverted into a weight tank for exactly 10 min. The increased weight in the tank is 4,765 lbf. What is the average flow rate in gallons per minute and in cubic feet per second? Assume $T = 60°F$.

4-29 The mean velocity of water in a 4-in. pipe is 8 ft/sec. Determine the flow in slugs per second, gallons per minute, and cubic feet per second if $T = 60°F$.

4-30 The mean velocity of water in a 10-cm pipe is 3 m/s. Determine the volume rate of flow in cubic meters per second.

4-31 Tests on a sphere are conducted in a wind tunnel at an airspeed of U_0. The velocity of flow toward the sphere along the longitudinal axis is found to be $u = U_0(1 - r_0^3/x^3)$, in which r_0 is the radius of the sphere and x the distance from its center. Determine the acceleration of the air as it passes a point ahead of the center of the sphere in terms of x, r_0 and U_0.

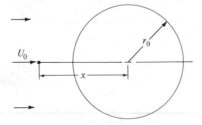

4-32 Two parallel disks of diameter D are brought together each with a normal speed of V. When their spacing is h, what is the convective acceleration at the section just inside the edge of the disk (section A) in terms of V, h, and D? Assume uniform velocity distribution across the section.

Section A

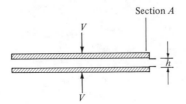

4-33 For the condition in Prob. 4-32, find the local acceleration at A in terms of R, V, and h.

4-34 The velocity along the centerline of the nozzle shown is given by the following expression: $V = 2t/(1 - 0.5x/L)^2$, where V = velocity in feet per second, t = time in seconds, x = distance along the nozzle, and L = length of nozzle = 4 ft. When $x = 0.5L$ and $t = 2$ sec, what is the local acceleration along the centerline? What is the convective acceleration? Assume one-dimensional flow prevails.

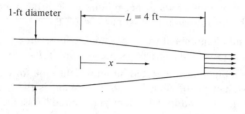

4-35 Liquid flows through this two-dimensional slot at a rate of $q = 2tq_0/t_0$, where q is the discharge per unit of width of slot and q_0 is the discharge when $t = t_0$. Assume that along the line $y = 0$ the velocity is given by $u = q/b$. Then what will be the local acceleration at $x = 2B$ and $y = 0$ in terms of B, t, t_0 and q_0?

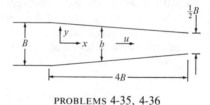

PROBLEMS 4-35, 4-36

4-36 What will be the convective acceleration for the conditions of Prob. 4-35?

4-37 In this flow passage the discharge is varying with time according to the following expression:

$$Q \text{ in cfs} = Q_0 - Q_1 \frac{t}{t_0}$$

At time $t = 0.50$ s it is known that at section A-A the velocity gradient in the S direction is $+2$ m/s per meter. If Q_0, Q_1 and t_0 are constants with values

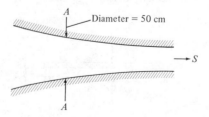

of 0.985 m³/s, 0.4 m³/s, and 1 s, respectively, and assuming that one-dimensional flow applies, answer the following for time $t = 0.5$ s. What is the velocity at A-A? What is the local acceleration at A-A? What is the convective acceleration at A-A?

4-38 Air discharges downward in the pipe and then outward between the parallel disks. Assuming negligible density change in the air, derive a formula for the acceleration of air at point A, which is r distance from the center of the disks. Express the acceleration in terms of the constant air discharge Q, the radial distance r, and disk spacing t. If $D = 10$ cm, $t = 1$ cm and $Q = 0.380$ m³/s, what is the velocity in the pipe and the acceleration at point A where $r = 30$ cm?

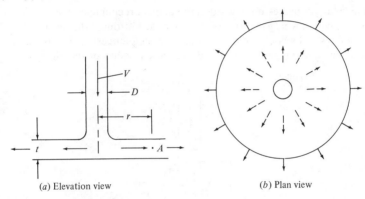

(a) Elevation view   (b) Plan view

4-39 All the conditions of Prob. 4-38 are the same except that the discharge is given as $Q = Q_0(t/t_0)$, where $Q_0 = 0.1$ m³/s and $t_0 = 1$ s. For the additional condition, what will be the acceleration at point A when $t = 2$ s and $t = 3$ s?

4-40 For the conditions in each of the flow cases (a and b) shown, respond to the following questions and statements concerning the application of the control volume equation to the continuity principle. What is the value of β? Determine the value of dB_{syst}/dt. Determine the value of $\Sigma\beta\rho\mathbf{V}\cdot\mathbf{A}$. Determine the value of $d/dt\int_{cv}\beta\rho\,d\forall$.

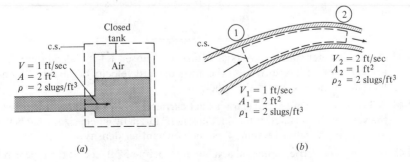

(a)   (b)

4-41 For the conditions shown, respond to the following questions and statements concerning the application of the control volume equation to the continuity principle. What is the value of β? Determine the value of dB_{syst}/dt. Determine the value of $\Sigma\beta\rho\mathbf{V}\cdot\mathbf{A}$. Determine the value of $d/dt \int_{cv}\beta\rho \, d\forall$.

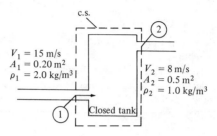

4-42 This plunger moves downward in the conical receptacle, which is filled with oil. At what level (y in terms of d) above the bottom of the receptacle will the mean upward velocity of the oil (between the plunger and receptacle wall) be just the same magnitude as the downward velocity of the plunger?

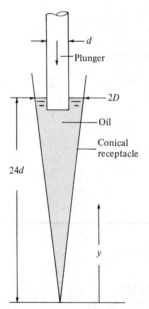

4-43 A 1-ft-diameter sphere falls axially through a closed 1.1-ft-diameter cylinder at a speed of 0.5 ft/sec. What is the mean upward velocity of the surrounding fluid at the midsection of the sphere?

4-44 A 30-cm by 40-cm rectangular air duct carries a flow of 1.5 m³/s. Determine the velocity in the duct. If the duct tapers to 15-cm by 40-cm size, what is the velocity in the latter section? Assume constant air density.

4-45 A 36-in. pipe carries crude oil at a mean velocity of 7 ft/sec. If the pipe is reduced in size to 30 in. in diameter, what will be the velocity in the 30-in. section if the flow rate remains the same?

4-46 A 30-cm pipe divides into a 20-cm branch and a 15-cm branch. If the total flow is 0.5 m³/s and if the same mean velocity occurs in each branch, what is the discharge rate in each branch?

4-47 Water flows in an 8-in. pipe that is connected in series with a 6-in. pipe. If the rate of flow is 449 gpm (gallons per minute), what is the mean velocity in each pipe?

4-48 What is the velocity in leg B of the tee shown in the figure for steady incompressible flow of water?

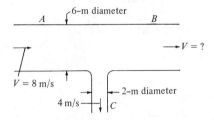

4-49 For the steady flow of gas in the below conduit, what is the mean velocity at section 2?

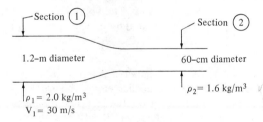

4-50 Is the tank filling or emptying? At what rate is the water level rising or falling in the tank?

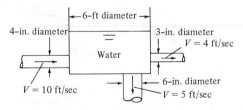

4-51 A stationary nozzle discharges water against a plate moving toward the nozzle at half the jet velocity. When the rate of discharge from the nozzle is 3 cfs, at what rate will the plate deflect water?

4-52 Assuming that complete mixing occurs between the two inflows before the mixture discharges from the pipe at C, what is the mass rate of flow, the velocity and the specific gravity of the mixture in the pipe at C?

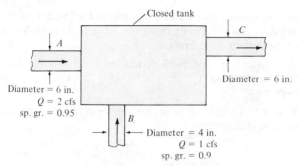

4-53 Oxygen and methane are mixed at 200 kPa absolute pressure and 100°C. The velocity of the gases into the mixer is 1 m/s. The density of the gas leaving the mixer is 2.2 kg/m³. Determine the exit velocity of the gases.

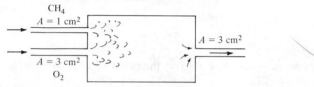

4-54 A slow leak develops in an inner tube, in which it takes 4 hr for the pressure to decrease from 30 psig to 25 psig. The volume of the inner tube is 0.5 ft³, and temperature remains constant at 60°F. The mass-flow rate of air is given by $\dot{m} = 0.68pA/\sqrt{RT}$. Calculate the area of the hole in the tube. Atmospheric pressure is 14 psia.

4-55 Oxygen leaks slowly through a small orifice from an oxygen bottle. The volume of the bottle is 0.1 m³ and the diameter of the orifice is 0.1 mm. The temperature in the tank remains constant at 18°C and the mass-flow rate is given by $\dot{m} = 0.68pA/\sqrt{RT}$. How long will it take the absolute pressure to decrease from 10 to 5 MPa?

4-56 How long will it take the water surface in the tank shown to drop from an $h = 3$ m to $h = 30$ cm?

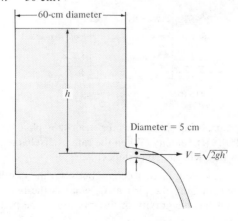

4-57 Water enters the tank through a pipe at point A at a rate of 0.10 ft^3/sec. What length of time will it take for the water surface to drop from $h = 9$ ft to $h = 4$ ft?

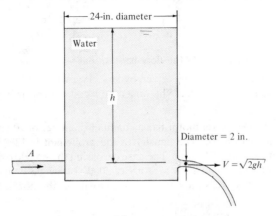

4-58 For this type of tank the tank diameter is given as $D = d + C_1 h$, where d is the bottom diameter and C_1 is a constant. Derive a formula for the time of fall of liquid surface from $h = h_0$ to $h = h$ in terms of d_j, d, h_0, h, and C_1. Solve for t if $h_0 = 1$ m, $h = 20$ cm, $d = 20$ cm, $C_1 = 0.4$, and $d_j = 6$ cm.

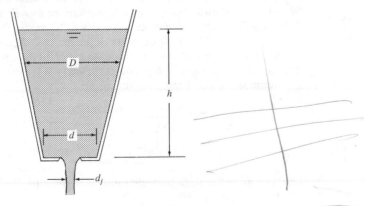

4-59 The velocity components for a two-dimensional flow are

$$u = \frac{C(y^2 - x^2)}{(y^2 + x^2)^2} \qquad v = \frac{-2Cxy}{(x^2 + y^2)^2}$$

where C is a constant. Does this description of the flow field satisfy continuity? Is the flow irrotational?

4-60 It is predicted that a flow field will have the following velocity components:

$$u = U(x^3 + xy^2) \qquad v = U(y^3 + yx^2) \qquad w = 0$$

Is such a flow field possible (does it satisfy continuity)?

4-61 The velocity variation for flow near a wall is given by $u = Uy$, $v = Cy$, and $w = 0$, where U and C are constants. Is the flow irrotational?

4-62 The velocity components of a flow field are given by

$$u = \frac{y}{(x^2 + y^2)^{3/2}} \qquad v = \frac{-x}{(x^2 + y^2)^{3/2}}$$

Is continuity satisfied? Is the flow irrotational?

4-63 A u component of velocity is given by $u = Axy$, where A is a constant. What is a possible v component? What must the v component be if the flow is irrotational?

4-64 An end-burning rocket motor has a chamber diameter of 10 cm and a nozzle exit diameter of 8 cm. The density of the propellant is 1800 kg/m³ and regresses at the rate of 1 cm/s. The gases crossing the nozzle exit plane have a pressure of 10 kPa and a temperature of 2000°C. The gas constant of exhaust gases is 415 J/kg K. Calculate the gas velocity at the nozzle exit plane.

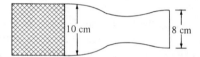

4-65 A cylindrical-port rocket motor has a grain design consisting of a cylindrical shape as shown. The curved internal surface as well as both ends burn. The propellant surface regresses uniformly at 1.2 cm/s. The propellant density is 2000 kg/m³. The inside diameter of the motor is 20 cm. The propellant grain is 40 cm long and has an inside diameter of 12 cm. The diameter of the nozzle exit plane is 20 cm. The gas velocity at the exit plane is 2000 m/s. Determine the gas density at the exit plane.

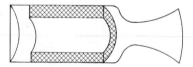

4-66 The mass flow rate through a nozzle is given by

$$\dot{m} = 0.65 \frac{p_c A_t}{\sqrt{RT_c}}$$

where p_c and T_c are pressure and temperature in the rocket chamber and R is the gas constant of the gases in the chamber. The propellant burning rate (surface regression rate) can be expressed as $\dot{r} = ap_c^n$ where a and n are two empirical constants. Show, by application of the continuity equation, that the chamber pressure can be expressed as

$$p_c = \left(\frac{a\rho_p}{0.65}\right)^{1/(1-n)} \left(\frac{A_g}{A_t}\right)^{1/(1-n)} (RT_c)^{1/[2(1-n)]}$$

where ρ_p is the propellant density and A_g is the grain surface burning area. If the operating chamber pressure of a rocket motor is 3.5 MPa and $n = 0.3$, how much will the chamber pressure increase if a crack develops in the grain, increasing the burning area by 20%?

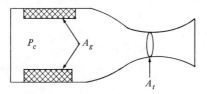

4-67 A piston is moving up during the exhaust stroke of a four-cycle engine. Mass escapes through the exhaust port at a rate given by

$$\dot{m} = 0.65 \, \frac{p_c A_v}{\sqrt{RT_c}}$$

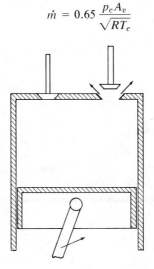

where p_c and T_c are the cylinder pressure and temperature, A_v is the valve opening, and R is gas constant of the exhaust gases. The bore of the cylinder is 10 cm and the piston is moving upward at 30 m/s. The distance between the piston and head is 10 cm. The valve opening is 1 cm², the chamber pressure is 300 kPa, the chamber temperature is 500°C and the gas constant is 350 J/kg K. Applying the continuity equation, determine the rate at which the gas density is changing in the cylinder. Assume the density and pressure are uniform in the cylinder and the gas is ideal.

REFERENCES

1. Bradshaw, P. *Experimental Fluid Mechanics*. Pergamon Press, New York, 1964.
2. Macagno, E. O. "Flow Visualization in Liquids." *Iowa Inst. Hydraulics Resh. Rept.* 114 (1969).
3. Maltby, R. L. (comp.). "Flow Visualization in Wind Tunnels Using Indicators." *AGARDograph* 70 (April 1962).
4. National Committee for Fluid Mechanics Films. *Illustrated Experiments in Fluid Mechanics*. The M.I.T. Press, Cambridge, Mass., 1972.
5. Prandtl, L., and Tietjens, O. G. *Fundamentals of Hydro and Aero Mechanics*. McGraw-Hill Book Company, New York, 1934.
6. Rouse, H. *Elementary Mechanics of Fluids*. John Wiley & Sons, Inc., New York, 1946.

FILMS

7. Kline, S. J. *Flow Visualization*. National Committee for Fluid Mechanics Films, distributed by Encyclopaedia Britannica Educational Corporation.
8. Lumley, John L. *Eulerian and Lagrangian Descriptions in Fluid Mechanics*. National Committee for Fluid Mechanics Films, distributed by Encyclopaedia Britannica Educational Corporation.
9. Stewart, R. W. *Turbulence*. National Committee for Fluid Mechanics Films, distributed by Encyclopaedia Britannica Educational Corporation.

Leonhard Euler (1707–1783) (left), Professor of Physics and Mathematics. Euler had a greater interest in mathematics than did Bernoulli and, in fact, Euler formalized the equation we now call Bernoulli's. Euler developed the basic equations of fluid motion and published hundreds of papers on this and other subjects. In fact, 50 pages were required in his eulogy to list only the titles of his works!

Daniel Bernoulli (1700–1782) (right). Professor of Mathematics. Bernoulli published works dealing with the statics and dynamics of fluids and made the first observations and notes relating to the equation that bears his name. Bernoulli and Euler both studied under Bernoulli's father, Johann, at Basel, Switzerland, and they both held professorships at St. Petersburg in Russia for a number of years.

5 PRESSURE VARIATION IN FLOWING FLUIDS

P RESSURE VARIATION IS IMPORTANT to the engineer for several reasons. In certain cases, as in the design of tall structures, the pressure variation resulting from the wind must be considered in the design of individual parts, such as windows, as well as in the design of the basic structure to resist the overall wind load. In the design of pump impellers, hydrofoils, or even pipelines, the engineer must avoid the possibility of *cavitation*, the phenomenon of boiling in a flowing liquid at normal temperatures, which results from low pressure. In this case, pressure variation is the most significant aspect of the problem. In aircraft design, the pressure variation around the wing produces lift, but it also contributes to the drag of the aircraft. Thus pressure plays a major role in many areas of engineering design and analysis.

Even in our daily lives, many phenomena that affect us are related to pressures caused by flowing fluids. For example, an indicator of our health, blood pressure, relates directly to the flow of the blood in our veins and arteries. On a global scale the atmospheric pressure readings, reported in weather forecasts, relate to the vortex strength (tangential velocity of the wind) in cyclonic storms and the relative position of the observer to the storm center. The drag that we experience with gusts of wind is a pressure-related phenomenon. Even the vortex motion as we stir a cup of coffee or the rise velocity of bubbles in our favorite beverage is a result of pressure variations in flowing fluids. The basic principles introduced in this chapter will help to explain natural phenomena such as are mentioned above and will also set the stage for a more thorough treatment of pressure-related topics later in the text.

5-1   BASIC CAUSES OF PRESSURE VARIATION IN A FLOWING FLUID

Introduction

In Chapter 3 we saw that gravity causes pressure to vary with elevation; now we shall consider other causes of pressure variation. In fluid flow there are basically two causes of pressure variation in addition to the weight effect—these are acceleration and viscous resistance. To accelerate a mass of fluid in a given direction there must be a net force in the direction of acceleration; therefore, the pressure must decrease in the direction of acceleration. The fact that a net force on the fluid acts in the direction of decreasing pressure is illustrated in Fig. 5-1 for the case of flow in a pipe. When we isolate a mass of fluid such as this, we see that the

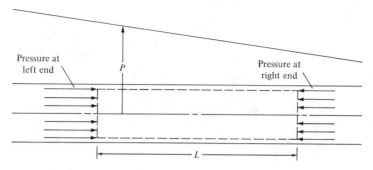

FIGURE 5-1   Pressure variation in a pipe.

greater pressure on the left end acts to the right in the direction of decreasing pressure, and the smaller pressure at the right end acts in the opposite direction. Since the areas are the same, the net force on the fluid acts to the right in the direction of decreasing pressure. In addition to acceleration, pressure variation is needed to overcome the viscous resistance which, like friction in solids, acts in opposition to the motion of the fluid. In the foregoing discussion, it has been assumed that the net force that produces acceleration and overcomes viscous resistance is a result of the pressure distribution. It should be pointed out, however, that gravity may also enter the problem. Therefore, it is seen that fluid mechanics is an extension of basic mechanics with added variations in the way that forces are applied.

Pressure variation due to weight and acceleration

Consider the cylindrical element of fluid shown in Fig. 5-2. Here the element is being accelerated in the ℓ direction and is acted upon by pressure and weight forces only. Note also that the coordinate axis z is vertically upward and that it is assumed that the pressure varies along the length of the element. Applying Newton's second law in the ℓ direction and using the system approach, we have

$$\sum F_\ell = Ma_\ell$$

or
$$p\,\Delta A - \left(p + \frac{\partial p}{\partial \ell}\,\Delta\ell\right)\Delta A - W\sin\alpha = \rho\,\Delta\ell\,\Delta A\,a_\ell \qquad (5\text{-}1)[1]$$

However, $W = \gamma\,\Delta\ell\,\Delta A$; therefore, Eq. (5-1) reduces to

$$-\frac{\partial p}{\partial \ell} - \gamma\sin\alpha = \rho a_\ell \qquad (5\text{-}2)$$

---

[1] The partial derivative of p with respect to ℓ, $\partial p/\partial \ell$, is used here because pressure may also vary with time.

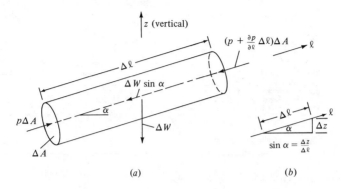

(a)   (b)

FIGURE 5-2   Pressure and weight forces acting on an accelerating fluid element. (a) Fluid element. (b) Trigonometric relation.

In Fig. 5-2b it is seen that sin α is equal to $\partial z/\partial \ell$ in the limiting condition as Δz approaches zero. Thus, when this substitution is made in Eq. (5-2) we obtain

$$-\frac{\partial p}{\partial \ell} - \gamma \frac{\partial z}{\partial \ell} = \rho a_\ell$$

or, taking γ as a constant,

$$-\frac{\partial}{\partial \ell}(p + \gamma z) = \rho a_\ell \qquad (5\text{-}3)$$

Equation (5-3) is Euler's equation of motion for a fluid. It is of interest to note that when the acceleration is zero, Eq. (5-3) reduces to $d/d\ell(p + \gamma z) = 0$, which gives the familiar expression for hydrostatics, $p + \gamma z = C$, when integrated. In other words, along a path of zero acceleration the pressure distribution must be hydrostatic. Again, this assumes that gravity and pressure forces are the only forces acting.

5-2   EXAMPLES OF PRESSURE VARIATION RESULTING FROM ACCELERATION

Uniform acceleration of a tank of liquid

Assume that the open tank of liquid shown in Fig. 5-3 is accelerated to the right, positive x direction, at a rate of a_x. For this to occur, a net force must act on the liquid in the x direction; this is accomplished when the liquid redistributes itself in the tank as shown by $A'B'CD$. Under this condition the hydrostatic force at the left end is greater than the hydrostatic force at the right, which is consistent with the requirement of $F = Ma$.

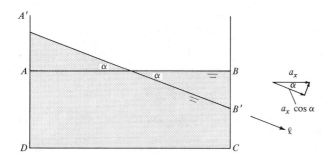

FIGURE 5-3   Uniform acceleration of a tank of liquid.

Further quantitative analysis of the acceleration of the tank of liquid is made with Eq. (5-3). First consider application of the equation along the liquid surface $A'B'$. Here the pressure is constant, $p = p_{atm}$; consequently, $\partial p/\partial \ell = 0$. The acceleration along $A'B'$ is given by $a_\ell = a_x \cos \alpha$; hence, Eq. (5-3) reduces to

$$\frac{d}{d\ell}(\gamma z) = -\rho a_x \cos \alpha \qquad (5\text{-}4)$$

where the total derivative is used because the variables do not change with time. The specific weight in Eq. (5-4) is constant; therefore, Eq. (5-4) becomes

$$\frac{dz}{d\ell} = -\frac{a_x \cos \alpha}{g}$$

But $dz/d\ell = -\sin \alpha$; thus we obtain

$$\sin \alpha = \frac{a_x \cos \alpha}{g}$$

or
$$\tan \alpha = \frac{a_x}{g} \qquad (5\text{-}5)$$

Still further analysis can be made if Eq. (5-3) is applied along a horizontal plane in the liquid, such as at the bottom of the tank. Now z is constant and Eq. (5-3) reduces to $\partial p/\partial x = -\rho a_x$, which shows that the pressure must decrease in the direction of acceleration. The change in pressure is consistent with the change in depth of liquid because hydrostatic pressure variation prevails in the vertical, since there is no component of acceleration in that direction. Thus as the depth decreases in the direction of acceleration, the pressure along the bottom of the tank must also decrease. Another case of uniform acceleration is given in the following example.

EXAMPLE 5-1   The tank on a tank truck is filled completely with gasoline, which has a specific weight of 42 lbf/ft³ (6.60 kN/m³).

a. If the tank on the trailer is 20 ft (6.1 m) long and if the pressure at the top rear end of the tank is atmospheric, what is the pressure at the top front when the truck decelerates at a rate of 10 ft/sec² (3.05 m/s²)?
b. If the tank is 6 ft (1.83 m) high, what is the maximum pressure in the tank?

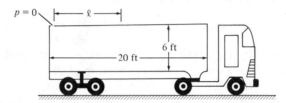

Solution   Apply Eq. (5-3) along the top of the tank. Here z = constant and the pressure does not vary with time during this phase of deceleration; therefore, one may write

$$\frac{dp}{d\ell} = -\rho a_\ell$$

Integrating, one obtains

$$p = -\rho \ell a_\ell + C$$

When $\ell = 0$, $p = 0$; hence, $C = 0$ and $p = -\rho \ell a_\ell$.

Now substituting -10 ft/sec² (-3.05 m/s²) for a_ℓ, 20 ft (6.1 m) for ℓ, and 1.30 slugs/ft³ for ρ, which is equal to γ/g, one obtains

$$p = -1.30 \text{ slugs/ft}^3 \times (-10 \text{ ft/sec}^2) \times 20 \text{ ft}$$
$$= 260 \text{ psfg} \qquad \blacktriangleleft$$

SI units $p = -672 \text{ kg/m}^3 \times (-3.05 \text{ m/s}^2) \times 6.1 \text{ m}$
$$= 12{,}500 \text{ N/m}^2 = 12{,}500 \text{ Pa gage} \qquad \blacktriangleleft$$

The maximum pressure in the tank will occur at the front end of the tank bottom. Since the pressure variation is hydrostatic in the vertical direction, one obtains $p + \gamma z$ = constant or

$$p_{\text{bottom}} + \gamma z_{\text{bottom}} = p_{\text{top}} + \gamma z_{\text{top}}$$

Solving, $p_{\text{bottom}} = 260 + (42)(6)$

$$p_{\max} = p_{\text{bottom}} = 512 \text{ psfg} \qquad \blacktriangleleft$$

SI units $p_{\max} = p_{\text{bottom}} = 12{,}500 \text{ N/m}^2 + 6.6 \text{ kN/m}^3 \times 1.83 \text{ m}$
$$= 24.6 \text{ kPa gage} \qquad \blacktriangleleft$$

Rotation of tank of liquid

In Sec. 5-3 it will be shown that the variation in piezometric head $(p/\gamma + z)$ can be easily determined by use of the Bernoulli equation if the flow is irrotational. However, for the case of pure rotation, such as a tank of liquid in rotation, one must use Eq. (5-3) at the outset.

Consider a cylindrical tank of liquid rotating at a constant rate ω, as shown in Fig. 5-4. Here AA depicts the liquid surface before rotation, and surface $A'A'$ shows how it appears after a period of time when a steady

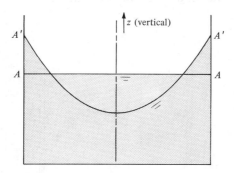

FIGURE 5-4   Rotating vessel of liquid.

state has been established. When Eq. (5-3) is applied in a radial direction for this tank, one obtains

$$\frac{d}{dr}(p + \gamma z) = -\left(-\rho\,\frac{V^2}{r}\right) \tag{5-6}$$

Here, Eq. (5-3) is written for the r direction, so r is used as the space variable instead of ℓ. We know from mechanics that the acceleration in the r direction is negative (acceleration is toward the center of rotation); hence, a negative sign is given to the acceleration term. Substituting $r\omega$ for V and clearing the negative signs then yields

$$\frac{d}{dr}(p + \gamma z) = \rho r \omega^2 \tag{5-7}$$

Integrating Eq. (5-7) with respect to r then gives us

$$p + \gamma z = \frac{\rho r^2 \omega^2}{2} + \text{constant}$$

but $V = r\omega$ and $\rho = \gamma/g$, so we have

$$\frac{p}{\gamma} + z - \frac{V^2}{2g} = \text{constant} \tag{5-8}$$

Replacing V by $r\omega$ in Eq. (5-8) reveals that the liquid surface is in the form of a paraboloid of revolution. *Note*: Although Eq. (5-8) at first glance appears similar to Bernoulli's equation (see next section)—the terms p/γ, z, and $V^2/2g$ are the same as in Bernoulli's equation—the similarity is not complete because the sign on the $V^2/2g$ term is different from that on the other two terms. In the Bernoulli equation the signs are all the same.

EXAMPLE 5-2   When the U-tube is not rotated, the water stands in the tube as shown. If the tube is rotated about the eccentric axis at a rate of 8 rad/s, what are the new levels of water in the tube?

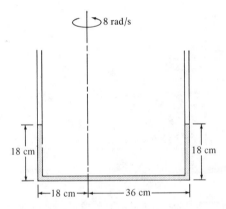

Solution   Solution of this problem is based upon Eq. (5-8) and also upon the fact that the water occupies a given volume of the tube, which may be expressed in terms of a given length of tube. Let the elevation reference be at the level of the horizontal part of the tube; then, by considering a point at the water surface in the left tube where $p = 0$ and also a point at the surface in the right tube, Eq. (5-8) can be written as follows:

$$z_l - \frac{r_l^2 \omega^2}{2g} = z_r - \frac{r_r^2 \omega^2}{2g}$$

Another independent equation involving the volume of tube which the liquid occupies may be written as

$$z_l + z_r = 0.36 \text{ m}$$

When $r_l = 0.18$ m, $r_r = 0.36$ m, and $\omega = 8$ rad/s are substituted into the equation above and the equations are solved for z_l and z_r, one obtains

$$z_l = 2.1 \text{ cm} \qquad \blacktriangleleft$$
$$z_r = 33.9 \text{ cm} \qquad \blacktriangleleft$$

5-3 THE BERNOULLI EQUATION

The Bernoulli equation along a streamline

When Euler's equation, Eq. (5-3), is written for flow along a streamline (in the s direction) using the general expression for acceleration along a streamline [Eq. (4-10)] the equation becomes

$$-\frac{\partial}{\partial s}(p + \gamma z) = \rho \left(V_s \frac{\partial V_s}{\partial s} + \frac{\partial V_s}{\partial t} \right) \tag{5-9}$$

By integrating Euler's equation for steady flow between points in the field of flow, we derive the Bernoulli equation. In steady flow the changes are with respect to position only, so the differentials of Eq. (5-9) become total differentials. Euler's equation for steady flow along a streamline, Eq. (5-9), then becomes

$$-\frac{d}{ds}(p + \gamma z) = \rho V_s \frac{dV_s}{ds}$$

or

$$-\frac{d}{ds}(p + \gamma z) = \rho \frac{d}{ds}\left(\frac{V_s^2}{2}\right) \tag{5-10}$$

When Eq. (5-10) is integrated with respect to s for incompressible flow, we get

$$p + \gamma z + \rho \frac{V_s^2}{2} = C \tag{5-11}$$

Equation (5-11) is Bernoulli's equation, which can also be written as

$$\frac{p}{\gamma} + z + \frac{V_s^2}{2g} = C_1 \tag{5-12}$$

Here p/γ, z, and $V_s^2/2g$ are called pressure head, elevation, and velocity head, respectively.

The Bernoulli equation for irrotational flow

To derive the Bernoulli equation for irrotational-incompressible-steady flow we utilize the cartesian coordinate system in two dimensions.[1] First, we write Euler's equation for the x coordinate direction.

$$-\frac{\partial}{\partial x}(p + \gamma z) = \rho a_x$$

---

[1] The same type of derivation can be easily used to derive Bernoulli's equation for three-dimensional flow. The two-dimensional case is presented here for brevity.

If we divide through by γ we have

$$-\frac{\partial}{\partial x}\left(\frac{p}{\gamma} + z\right) = \frac{a_x}{g}$$

By definition $(p/\gamma) + z = h$. Therefore, we obtain

$$-\frac{\partial h}{\partial x} = \frac{a_x}{g} \tag{5-13}$$

Then from Sec. 4-3 the acceleration in the x direction for steady flow is given as

$$a_x = u\frac{\partial u}{\partial x} + v\frac{\partial u}{\partial y} \tag{5-14}$$

When Eq. (5-14) is substituted into Eq. (5-13) we get

$$-\frac{\partial h}{\partial x} = \frac{1}{g}\left(u\frac{\partial u}{\partial x} + v\frac{\partial u}{\partial y}\right) \tag{5-15}$$

Similarly, for the y direction we have

$$-\frac{\partial h}{\partial y} = \frac{1}{g}\left(u\frac{\partial v}{\partial x} + v\frac{\partial v}{\partial y}\right) \tag{5-16}$$

The condition for irrotationality of flow in the x-y direction was given by Eq. (4-39), which is

$$\frac{\partial v}{\partial x} = \frac{\partial u}{\partial y} \tag{4-39}$$

Therefore, when $\partial u/\partial y$ in Eq. (5-15) is replaced by $\partial v/\partial x$ one obtains

$$-\frac{\partial h}{\partial x} = \frac{1}{g}\left(u\frac{\partial u}{\partial x} + v\frac{\partial v}{\partial x}\right) \tag{5-17}$$

However, note that $u\ \partial u/\partial x = \partial/\partial x\ (u^2/2)$ and $v\ \partial v/\partial x = \partial/\partial y\ (v^2/2)$. Therefore, one can write Eq. (5-17) as

$$\frac{\partial}{\partial x}\left(h + \frac{u^2 + v^2}{2g}\right) = 0$$

The sum $u^2 + v^2$ is the total velocity squared, V_s^2, so we have

$$\frac{\partial}{\partial x}\left(h + \frac{V_s^2}{2g}\right) = 0 \tag{5-18}$$

Thus when the flow is steady, incompressible, and irrotational there is no change in $h + V_s^2/2g$ in the x direction according to Eq. (5-18).

In a similar manner, if we take Eq. (5-16) and replace $\partial v/\partial x$ by $\partial u/\partial y$ because of the irrotationality condition it can be shown that

$$\frac{\partial}{\partial y}\left(h + \frac{V_s^2}{2g}\right) = 0 \tag{5-19}$$

This shows that there is no change in $h + V_s^2/2g$ in the y direction for the given conditions. Therefore, we have shown that $h + V_s^2/2g$ is constant throughout the flow field, or

$$h + \frac{V_s^2}{2g} = C \tag{5-20}$$

Also, $h = p/\gamma + z$ so we have

$$\frac{p}{\gamma} + z + \frac{V_s^2}{2g} = C \tag{5-21}$$

Equation (5-21) is exactly the same as Eq. (5-12). However, the limitations in the derivations are different. In the first derivation the equation could be applied along a streamline, but irrotational flow was not stipulated. In the latter derivation it was shown that $h + V_s^2/2g$ is constant in both the x and y directions; therefore, $h + V_s^2/2g$ is constant throughout the field of flow. However, to get this solution the irrotational condition was required. Therefore, if the flow is irrotational, incompressible, and steady, the Bernoulli equation can be applied between any points in the flow field.

In most applications the subscript s is omitted from the velocity symbol and the equation is written between two points in the field of flow as follows:

$$\frac{p_1}{\gamma} + \frac{V_1^2}{2g} + z_1 = \frac{p_2}{\gamma} + \frac{V_2^2}{2g} + z_2 \tag{5-22}$$

The Bernoulli equation relates the pressure, velocity, and elevation between any two points in the flow field for flow which is *steady, irrotational, nonviscous*, and *incompressible*. The nonviscous limitation follows from the inherent limitation of Euler's equation from which Bernoulli's equation is derived. Note that V is the speed of the fluid; thus component velocities are not valid in Bernoulli's equation. Bernoulli's equation can be used to predict the pressure distribution within the fluid or the pressure distribution on a body if the flow pattern about the body is known.

5-4   APPLICATION OF THE BERNOULLI EQUATION

Stagnation tube

Consider a curved tube such as that shown in Fig. 5-5. When the Bernoulli equation is written between points 1 and 2, we obtain

$$\frac{p_1}{\gamma} + \frac{V_1^2}{2g} + z_1 = \frac{p_2}{\gamma} + \frac{V_2^2}{2g} + z_2 \tag{5-23}$$

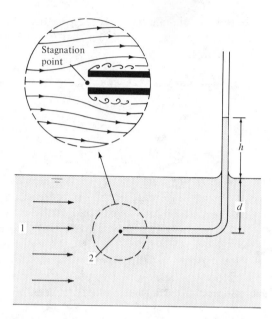

FIGURE 5-5 Stagnation tube.

However, $z_2 = z_1$ and the velocity at point 2 is zero (a stagnation point); hence, Eq. (5-23) reduces to

$$\frac{V_1^2}{2g} = \frac{p_2}{\gamma} - \frac{p_1}{\gamma} \qquad (5\text{-}24)$$

By the equations of hydrostatics (there is no acceleration normal to the streamlines where the streamlines are straight and parallel), $p_1 = \gamma d$ and $p_2 = \gamma(h + d)$; therefore, Eq. (5-24) can now be written as

$$\frac{V_1^2}{2g} = \frac{\gamma(h + d) - \gamma d}{\gamma}$$

which then reduces to $\boxed{V_1 = \sqrt{2gh}}$ (5-25)

Thus it is seen that a very simple device such as this curved tube can be used to measure the velocity of flow.

Pitot tube

The Pitot tube, named after the eighteenth century French hydraulic engineer who invented it, is based upon the same principle as the stagnation tube; however, it is much more versatile than the stagnation tube. The Pitot tube has a pressure tap at the upstream end of the tube for sensing the *stagnation* pressure. There are also ports several tube diameters

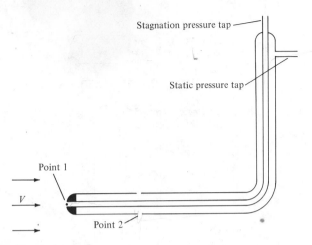

FIGURE 5-6   Pitot tube.

downstream of the front end of the tube where the flow is essentially undisturbed for sensing the *static* pressure in the fluid. When the Bernoulli equation is applied between points 1 and 2, Fig. 5-6, we have

$$\frac{p_1}{\gamma} + \frac{V_1^2}{2g} + z_1 = \frac{p_2}{\gamma} + \frac{V_2^2}{2g} + z_2$$

But $V_1 = 0$, so solving for V_2 from the equation above gives

$$V_2 = \left\{ 2g \left[\left(\frac{p_1}{\gamma} + z_1 \right) - \left(\frac{p_2}{\gamma} + z_2 \right) \right] \right\}^{1/2}$$

Here $V_2 = V$ and $p/\gamma + z = h$; hence we obtain

$$V = \sqrt{2g(h_1 - h_2)} \qquad (5\text{-}26)$$

where V = velocity of the stream and h_1 and h_2 are the piezometric heads at points 1 and 2, respectively.

By connecting a pressure gage or manometer between taps that lead to points 1 and 2, the flow velocity can be easily measured with the Pitot tube. A major advantage of the Pitot tube is that it can be used to measure velocity in a pressure pipe; a simple stagnation tube is not convenient to use in such a situation. In gas-flow measurement, where a single differential pressure gage is connected across the taps, Eq. (5-25) simplifies to $V = \sqrt{2\,\Delta p/\rho}$, where Δp is the pressure difference across the taps.

EXAMPLE 5-3   A mercury-kerosene manometer is connected to the Pitot tube as shown. If the deflection on the manometer is 7 in. (17.8 cm), what is the kerosene velocity in the pipe? Assume that the specific gravity of the kerosene is 0.81.

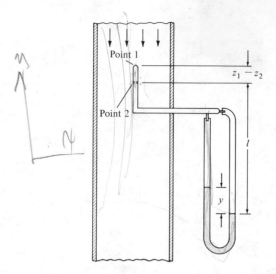

Solution   We need to know $h_1 - h_2$, the difference in piezometric head, between points 1 and 2, so we evaluate this first by applying the principles of hydrostatics through the manometer:

$$p_1 + (z_1 - z_2)\gamma_k + l\gamma_k - y\gamma_{Hg} - (l - y)\gamma_k = p_2$$

This then reduces to

$$p_1 - p_2 = y(\gamma_{Hg} - \gamma_k) - (z_1 - z_2)\gamma_k$$

or

$$\frac{p_1 - p_2}{\gamma_k} + z_1 - z_2 = \frac{y(\gamma_{Hg} - \gamma_k)}{\gamma_k}$$

Then for

$$h_1 - h_2 = \left(\frac{p_1}{\gamma} + z_1\right) - \left(\frac{p_2}{\gamma} + z_2\right)$$

we have

$$h_1 - h_2 = y\left(\frac{\gamma_{Hg}}{\gamma_k} - 1\right) = \frac{7}{12}(16.7 - 1)$$

The velocity is then

$$V = \sqrt{2g\left[\frac{7}{12}(16.7 - 1)\right]} = 24.3 \text{ ft/sec} \quad \blacktriangleleft$$

SI units $V = \sqrt{(19.63 \text{ m/s}^2)(0.178 \text{ m})(15.7)} = 7.41 \text{ m/s} \quad \blacktriangleleft$

Note   The -1 of the quantity $(16.7 - 1)$ reflects the effect of the column of kerosene in the right leg of the manometer, which tends to counterbalance the mercury in the left leg. Thus if we have a gas-liquid manometer, the counterbalancing effect will be nil.

EXAMPLE 5-4   A differential pressure gage is connected across the taps of a Pitot tube. When this Pitot tube is used in a wind tunnel test the gage indicates a Δp of 730 Pa. What is the air velocity in the tunnel? The pressure and temperature in the tunnel are 98 kPa absolute and 20°C, respectively.

Solution

$$V = \sqrt{2 \, \Delta p / \rho}$$

where

$$\rho = \frac{p}{RT} = \frac{98 \times 10^3 \text{ N/m}^2}{(287 \text{ J/kg K}) \times (20 + 273 \text{ K})}.$$

$$\rho = 1.16 \text{ kg/m}^3$$

$$\Delta p = 730 \text{ Pa}$$

therefore

$$V = \sqrt{(2 \times 730 \text{ N/m}^2)/(1.16 \text{ kg/m}^3)}$$

$$= 35.5 \text{ m/s} \qquad \blacktriangleleft$$

Pressure variation near curved boundaries

If flow passages are converging, such as is shown in Fig. 5-7, then irrotational flow will be approximated for low-viscosity fluids such as water or air. Hence the Bernoulli equation can be used to obtain the pressure variation between points in the flow field, including points adjacent to the boundaries. A common procedure for such an application is to use one

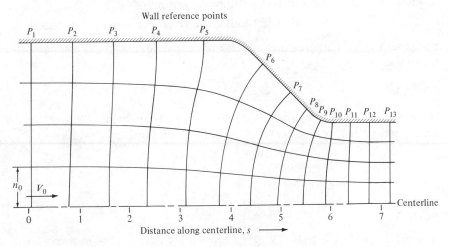

FIGURE 5-7   Flow net for a transition (half-section).

point as a reference (for example, point 0 far upstream in Fig. 5-7). When we write Bernoulli's equation between the reference point and any other point, we have

$$\frac{p}{\gamma} + \frac{V^2}{2g} + z = \frac{p_0}{\gamma} + \frac{V_0^2}{2g} + z_0 \qquad (5\text{-}27)$$

where p_0, V_0, and z_0 = pressure, velocity, and elevation, respectively, at the reference point

p, V, and z = pressure, velocity, and elevation at any other given point.

By simple rearrangement, Eq. (5-27) is written as

$$p - p_0 = \gamma(z_0 - z) + \frac{\rho}{2}(V_0^2 - V^2) \qquad (5\text{-}28)$$

Equation (5-28) expresses the pressure change in terms of the change in hydrostatic pressure (the first term on the right) and the change in kinetic pressure (the second term on the right). Thus the dynamic effect is given by

$$\left(\frac{p}{\gamma} + z\right) - \left(\frac{p_0}{\gamma} + z_0\right) = \frac{\rho}{2\gamma}(V_0^2 - V^2) \qquad (5\text{-}29)$$

Now Eq. (5-29) expresses the change in piezometric head as a function of the difference in the velocities squared, and it reduces to

$$h - h_0 = \frac{V_0^2 - V^2}{2g} \qquad (5\text{-}30)$$

where h = piezometric head at a given point

h_0 = piezometric head at the reference point

For gases in which hydrostatic effects are negligible, Eq. (5-28) can be expressed as

$$p - p_0 = \frac{\rho}{2}(V_0^2 - V^2) \qquad (5\text{-}31)$$

When we nondimensionalize Eqs. (5-30) and (5-31) by dividing through by $V_0^2/2g$ and $\rho V_0^2/2$, respectively,[1] we obtain the following equations:

$$\frac{h - h_0}{V_0^2/2g} = 1 - \left(\frac{V}{V_0}\right)^2 \qquad (5\text{-}32)$$

$$\frac{p - p_0}{\rho V_0^2/2} = 1 - \left(\frac{V}{V_0}\right)^2 \qquad (5\text{-}33)$$

[1] Because the terms $V_0^2/2g$ and $\rho V_0^2/2$ occur so often in hydraulics and fluid mechanics calculations, they have been given special names: velocity head and kinetic pressure, respectively. Another term, dynamic pressure, which is closely associated with kinetic pressure, is equal to the difference between the total pressure (pressure at a point of stagnation) and the static pressure. Under conditions where Bernoulli's equation applies, kinetic and dynamic pressure are equal; however, in high-speed gas flow, where compressibility effects are important, the two may have significantly different values.

These are dimensionless forms of the Bernoulli equation, the latter being for the case where the hydrostatic variation of pressure is negligible.

Because the velocity is inversely proportional to the cross-sectional area through which flow occurs in a flow passage, $V/V_0 = A_0/A$, we can express the relative pressure distribution or piezometric-head distribution in terms of the dimensions of the flow passage. For two-dimensional flow, the streamline spacing is directly proportional to the flow area; thus we have $V/V_0 = n_0/n$ for the relationship between the relative-velocity distribution and the relative-streamline spacing. Here n is the distance between two adjacent streamlines measured along the line (possibly curved) perpendicular to both streamlines. These perpendicular or normal lines are shown in Fig. 5-7. We can now express the relative-piezometric-head distribution or relative-pressure distribution for the case of gases, where the γz term is negligible, in terms of the streamline spacing for a two-dimensional-irrotational flow pattern:

$$\frac{h - h_0}{V_0^2/2g} = 1 - \left(\frac{n_0}{n}\right)^2 \tag{5-34a}$$

$$\frac{p - p_0}{\rho V_0^2/2} = 1 - \left(\frac{n_0}{n}\right)^2 \tag{5-34b}$$

Furthermore, since both sides of Eqs. (5-34) are dimensionless, application of the equations is not a function of the density of the fluid or the absolute size of the passage that controls the flow. Consequently, tests can be made on a small-scale structure, a model, and the results may be applied to a larger-scale structure. This is the principle of model testing, which will be covered in more detail in Chapter 8. The left side of Eqs. (5-34) is often called the *pressure coefficient* C_p. It is the change in piezometric head (or pressure) between two points in the flow field relative to the velocity head (kinetic pressure) of the reference velocity.

For the conduit of Fig. 5-7, the relative pressure (pressure coefficient) along the centerline and along the boundary at various sections is plotted in Fig. 5-8. Because there are greater variations of velocity near the boundary, the pressure variations are also greater along the boundary than along the centerline.

EXAMPLE 5-5   If air ($\rho = 1.2$ kg/m^3) flows through the two-dimensional passage of Fig. 5-7, what is the difference in pressure between points P_5 and P_{10} when the flow rate is 60 m^3/s per meter of width? If water flows through the passage, what is the difference in pressure between the same two points? Assume the half-size of the passage at the reference section is 2 m and the downstream size is 1 m. The view shown in Fig. 5-7 is an elevation view.

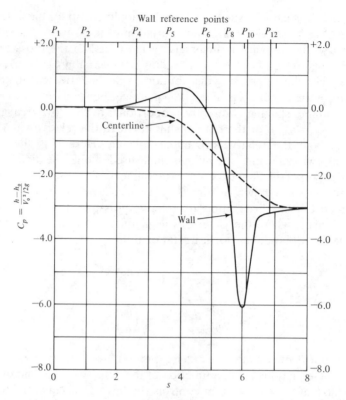

FIGURE 5-8   Relative piezometric head along the wall and centerline of transition in Figure 5-7.

Solution To solve for the pressure difference, we use Fig. 5-8, which shows that at point P_5

$$\frac{h_5 - h_0}{V_0^2/2g} = 0.5 \quad \text{or} \quad h_5 - h_0 = 0.5 \frac{V_0^2}{2g}$$

while at point P_{10}

$$h_{10} - h_0 = -5.6 \frac{V_0^2}{2g}$$

Then

$$h_5 - h_{10} = \frac{V_0^2}{2g} [0.5 - (-5.6)]$$

$$= 6.1 \frac{V_0^2}{2g}$$

But

$$h = \frac{p}{\gamma} + z$$

so $$\frac{p_5}{\gamma} + z_5 - \frac{p_{10}}{\gamma} - z_{10} = 6.1 \frac{V_0^2}{2g}$$

$$p_5 - p_{10} = \gamma \left[(z_{10} - z_5) + 6.1 \frac{V_0^2}{2g} \right]$$

For air flow,

$$p_5 - p_{10} = 1.2g \left[(1 - 2) + 6.1 \left(\frac{V_0^2}{2 \times 9.81} \right) \right]$$

But $$V_0 = \frac{60}{4} = 15 \text{ m/s}$$

hence $$p_5 - p_{10} = 11.77 \, (-1.0 + 70.0) = 812 \text{ Pa} \quad \blacktriangleleft$$

For water flow,

$$p_5 - p_{10} = 9{,}810 \, (1 - 2 + 70.0)$$

$$p_5 - p_{10} = 677 \text{ kPa} \quad \blacktriangleleft$$

Pressure distribution around a circular cylinder—ideal fluid

If a fluid is nonviscous (an ideal fluid) and if the flow of such a fluid is initially incompressible and irrotational, then the flow will be irrotational throughout the entire flow field.[1] Then, in addition, if the flow is steady Bernoulli's equation will apply because all the restrictions for Bernoulli's equation will have been satisfied. The flow pattern about a circular cylinder with such restrictions is shown in Fig. 5-9.

Because the flow pattern is symmetrical with either the vertical or hori-

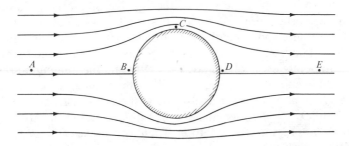

FIGURE 5-9 Irrotational flow past a cylinder.

[1] This can be seen in a qualitative sense if one visualizes a small spherical mass of fluid within a nonviscous flow field. Since pressure forces will act normal to the surface of the spherical mass, the mass will deform if the pressure is not of equal intensity over the entire surface. However, the mass cannot rotate (irrotational situation) because there is no shear stress (viscosity is zero) on the surface of the sphere to possibly cause rotation.

zontal axis through the center of the cylinder, the pressure distribution on the surface of the cylinder, obtained by application of the Bernoulli equation, will also be symmetrical. In Fig. 5-10 the relative pressure C_p is

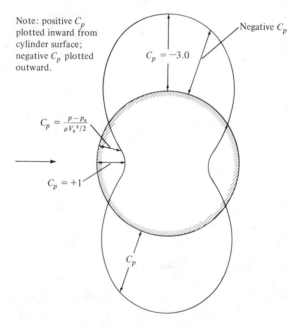

Note: positive C_p plotted inward from cylinder surface; negative C_p plotted outward.

Negative C_p

$C_p = -3.0$

$$C_p = \frac{p - p_0}{\rho V_0^2 / 2}$$

$C_p = +1$

C_p

FIGURE 5-10   Pressure distribution on a cylinder—irrotational flow.

plotted outward (negative) or inward (positive) from the surface of the cylinder depending upon the sign of the relative pressure and on a line normal to the surface of the cylinder. It should also be noted that p_0 and V_0 are the pressure and velocity of the free stream far upstream or downstream of the body. Thus we see that the points at the front and rear of the cylinder are points of stagnation ($C_p = +1.0$) and the minimum pressure ($C_p = -3.0$) occurs at the midsection where the velocity is highest. If we visualize a fluid particle as it travels around the cylinder from A to B to C and finally to D, Fig. 5-9, we first see that it decelerates, consistent with the increase in pressure from A to B. Then as it passes from B to C it is accelerated to its highest speed by the action of the pressure gradient; that is, the pressure decreases over the entire path from B to C. Next, as the particle travels from C to D, its momentum at C is sufficient to allow it to travel to D against the adverse pressure gradient (pressure increases in the direction of flow here). Finally, the particle accelerates to the freestream velocity in its passage from D to E. This qualitative description of how the fluid travels from one point to another will be helpful in under-standing the phenomenon of separation, which is explained in the next section.

5-5 SEPARATION AND ITS EFFECT ON PRESSURE VARIATION

Separation

In Chapter 4 we introduced the concept of separation, and emphasized that separation usually occurs where the physical boundaries turn away from the main stream of flow. We will now consider the basic cause of separation and its consequences.

Consider the flow of a real (viscous) fluid past a circular cylinder, as shown in Fig. 5-11. The flow pattern upstream of the midsection of the

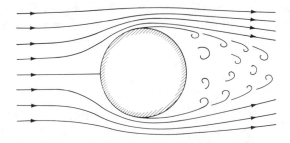

FIGURE 5-11 Flow of a real fluid past a circular cylinder.

cylinder is quite similar to the flow pattern of irrotational flow about the cylinder, except very close to the boundary surface. Here, due to the viscous resistance, a thin layer of fluid has its velocity reduced from that predicted by irrotational theory. In fact, the fluid particles directly adjacent to the surface have zero velocity (this "no-slip" condition at a boundary is characteristic of all fluids). The normal tendency is for the layer of reduced velocity (called the *boundary layer*) to grow in thick-ness in the direction of flow; however, because the main stream of fluid out-side the boundary layer is accelerating in the same direction, the bound-ary layer remains quite thin up to approximately the midsection.

Downstream of the midsection (from C to D, Fig. 5-9), the normal irrotational-flow pattern shows a significant deceleration of fluid next to the boundary with a corresponding increase in pressure. For real flow, however, deceleration of the fluid next to the boundary is limited because its velocity is already small (much smaller than for irrotational flow) because of the viscous resistance. Therefore, the fluid near the boundary can proceed only a very short distance against the adverse pressure gradient before stopping completely. Once the fluid next to the boundary is halted, it causes the main stream of flow to be diverted away or to be "separated" from the boundary; thus the process of separation is produced. Downstream of this point of separation the fluid outside the surface of separation has a high velocity and the fluid inside the surface of separation has a relatively low velocity. Because of the steep velocity gradient along the surface of separation, eddies are generated which through viscous action are finally dissipated into heat.

Since the location of the point of separation on a rounded body, such as the cylinder, depends upon the character of the flow in the boundary layer, it is not surprising that roughness of the surface or turbulence in the approach flow has an effect on the location of the separation point. These effects will be considered in more detail in Chapter 11. For angular-type bodies, however, the point of separation occurs at the sharp break in boundary configuration. Thus in Fig. 5-12 we observe flow separation at the boundary discontinuity for flow past a square rod and a disk and through a sharp-edged orifice.

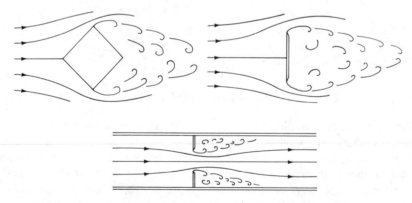

FIGURE 5-12   Flow past a square rod and disk and through an orifice.

We have already indicated that the point of separation may be related to the shape and roughness of a body. Because separation is closely associated with the viscous resistance of the fluid, it is not surprising that the Reynolds number, the value of which is inversely proportional to the relative viscous resistance, is an indicator of the onset of separation. For example, in flow past a circular cylinder, separation will occur for a Reynolds number ($VD\rho/\mu$) greater than 50. For Reynolds numbers less than 50, the entire flow field is dominated by relatively large viscous stresses that inhibit the onset of eddy motion in the fluid.

The effect of separation on pressure distribution

When separation occurs, the flow pattern is changed from that of irrotational flow; therefore, one would expect corresponding changes in the pressure distribution. Indeed, this occurs. For flow past a blunt body, the slight change (due to viscosity) of the flow pattern next to the forward part of the body changes the pressure distribution only slightly. However, in the zone of separation, marked changes in pressure result. It is a rule of thumb that the pressure that prevails at the point of separation also prevails over the body within the zone of separation. This is borne out for flow past the cylinder and disk, as shown in Fig. 5-13. Note that for both the disk and cylinder, the pressure on the rear half of the body is much less than the pressure on the front half; consequently, a net force in the downstream direction is imposed on the body. This force is the drag of the body, which will be considered in more detail in Chapter 11.

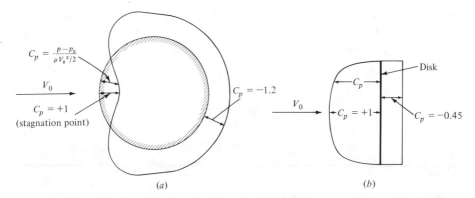

FIGURE 5-13   Pressure distribution on a circular cylinder and a disk. (a) Circular cylinder, Re = 10^5; after Fage and Warsap (2). (b) Circular disk, Re = 10^5; after Rouse (4).

5-6   CAVITATION

Cavitation occurs in liquid systems when the pressure at any point in the system is reduced to the vapor pressure of the liquid. Under such conditions, vapor bubbles form (boiling occurs) and then collapse (condense), thereby producing dynamic effects that can often lead to decreased efficiency and/or equipment failure.

Consider water flow through the pipe restriction shown in Fig. 5-14. In Fig. 5-14a, the physical configuration along with plots of piezometric head along the wall of the conduit for different flows are shown; and in Fig. 5-14b, the dimensionless plot of piezometric head along the wall is

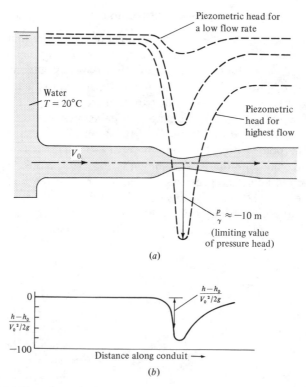

(b)

FIGURE 5-14   Flow through pipe restriction. (*a*) Variation of piezometric head. (*b*) Relative piezometric head.

shown. Here, the reference point is taken at the center of the pipe. For low and medium rates of flow, there will be a relatively small drop in pressure at the constriction; therefore, the pressure in the water will be well above the vapor pressure, and cavitation will not occur. This is indicated in Fig. 5-14*a* for the low rate where the piezometric head is everywhere above the conduit itself. However, if the rate of flow is high, the piezometric-head line will actually drop below the pipe, thereby indicating a less-than-atmospheric pressure for the liquid in the constriction; but the pressure can drop no lower than the vapor pressure of the liquid, because at this pressure the liquid will boil. Such a condition is depicted in Fig. 5-15*a*, where vapor bubbles are shown forming at the restriction, growing in size, then collapsing as they move into a region of higher pressure as they are swept downstream with the flow. Experiments reveal that very high intermittent pressures result in the vicinity of the bubbles when they collapse. These pressures may be as high as 689 MPa (100,000 psi) (5).

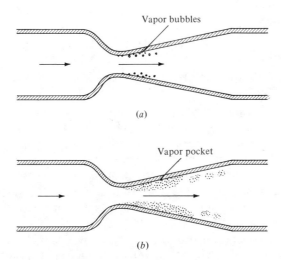

FIGURE 5-15 Formation of vapor bubbles in the process of cavitation. (*a*) Cavitation. (*b*) Cavitation—higher flow rate.

Therefore, if the bubbles collapse close to a physical boundary such as pipe walls, pump impellers, or valve casings, they can cause damage.[1] Usually this damage occurs in the form of a fatigue failure by the action of millions of bubbles impacting (in effect, imploding) against the metal over a long period of time, thus producing pitting of the metal in the vicinity of the zone of cavitation.

If the flow is increased even more than indicated above, the minimum pressure will still be restricted to the vapor pressure of the water; however, the zone of vaporization will increase, as shown in Fig. 5-15*b*. For such a condition the entire vapor pocket may intermittently grow and collapse, thereby producing serious vibration problems. Needless to say, cavitation is a phenomenon that should be avoided by proper design of equipment and by proper operation thereof.

---

[1] Cavitation in an enclosed pipe or machine can often be detected by the characteristic sound generated. In large structures, it sounds like large rocks are being carried through the system and are hitting the sides of the conduit.

PROBLEMS

5-1   What pressure gradient is required to accelerate water in a horizontal pipe at a rate of 3 m/s²?

5-2   What pressure gradient is required to accelerate water in a horizontal pipe at 6 ft/sec²?

5-3   What pressure gradient is required to accelerate kerosene (sp. gr. = 0.8) upward in a 6-in. vertical pipe at a rate of 5 ft/sec²?

5-4   Water is accelerated from rest in a 100-m-long 30-cm-diameter horizontal pipe. If the acceleration rate is 4 m/s², what is the pressure at the upstream end if the pressure at the downstream end is 70 kPa gage?

5-5   Water stands at a depth of 10 ft in a vertical pipe that is closed at the bottom end by a piston. What is the upward acceleration of the piston necessary to create a pressure of 10 psi immediately above it?

5-6   Water stands at a depth of 3 m in a vertical pipe that is closed at the bottom end by a piston. What is the upward acceleration of the piston necessary to produce a pressure of 60 kPa immediately above it?

5-7   Water stands at a depth of 10 ft in a vertical pipe that is closed at the bottom end by a piston. Assuming $p_v = 0$ psia, what is the maximum downward acceleration that can be given the piston without causing the water immediately above it to vaporize?

5-8   In this two-dimensional conduit, which discharges water into the atmosphere, the discharge is given by $q = q_0 t/t_0$, where q_0 and t_0 are 1 cfs/ft and 1 sec, respectively. What is the rate of change of pressure with respect to x at point A when $t = 2$ sec? $B = 12$ in., $b = 6$ in., and $L = 3$ ft; $T = 60°$F.

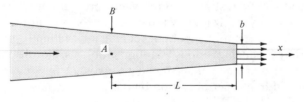

PROBLEMS 5-8, 5-9

5-9   In this two-dimensional conduit, which discharges water into the atmosphere, the discharge is given by $q = q_0 t/t_0$, where q_0 and t_0 are 0.1 m³/s per meter and 1 s, respectively. $B = 30$ cm, $b = 20$ cm, and $L = 1$ m; and $T = 10°$C. What is the rate of change of pressure with respect to x at point A when $t = 2$ s?

5-10   The flow rate in cubic feet per second for the nozzle shown is given by the following equation: $Q = 0.25t$. Here the time t is given in seconds and the fluid is a nonviscous liquid with a specific gravity of 1.6. $D = 2$ in., $d = 1$ in., $L_1 = 30$ in., and $L_2 = 15$ in.

(a) If the simplifying assumption is made that the velocity is constant across

any given section, what will be the convective acceleration of a fluid particle at point A when $t = 2$ sec?
- (b) With the same conditions given above, what will be the local acceleration at point A?
- (c) With the same conditions as above, what is the pressure gradient in the z direction at point A?
- (d) What is the pressure at point A if the pressure surrounding the free jet is atmospheric?

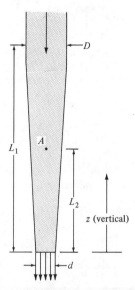

PROBLEMS 5-10, 5-11

5-11 The flow rate in m³/s for the nozzle shown is given by the following equation: $Q = 0.02t$. Here the time t is given in seconds and the fluid is a nonviscous liquid with a specific gravity of 1.6. $D = 10$ cm, $d = 4$ cm, $L_1 = 60$ cm, and $L_2 = 30$ cm.
- (a) If the simplifying assumption is made that the velocity is constant across any given section, what will be the convective acceleration of a fluid particle at point A when $t = 2$ s?
- (b) With the same conditions given above, what will be the local acceleration at point A?
- (c) With the same conditions as above, what is the pressure gradient in the z direction at point A?
- (d) What is the pressure at point A if the pressure surrounding the free jet is atmospheric?

5-12 In Prob. 4-38 assume the air density is 1.5 kg/m³. What is the pressure gradient at point A? If the maximum disk radius is 50 cm, what is the pressure at point A? Neglect frictional effects (viscous effects) in this problem and assume the atmospheric pressure (pressure at outlet) is 100 kPa.

5-13 It is desired to design a vertically downward-discharging nozzle so that the pressure between the base of the nozzle and the nozzle tip is zero gage for a given discharge of water. Derive a formula for d as a function of V_0, z_0, z, and d_0 to achieve the desired objective.

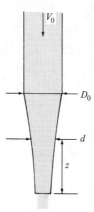

5-14 This tank moves in the x direction in a way such that the liquid surface does not change slope. What is the acceleration of the tank?

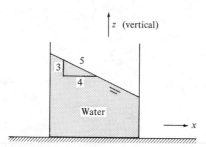

5-15 A truck carries a tank that is open at the top. The tank is 20 ft long, 6 ft wide, and 7 ft high. Assuming that the driver will not accelerate or decelerate the truck at a rate greater than 8.02 ft/sec², to what maximum depth may the tank be filled so that water will not be spilled?

5-16 A truck carries a tank that is open at the top, 6 m long, 2 m wide, and 3 m deep. Assuming that the driver will not accelerate or decelerate the truck at a rate greater than 3 m/s², what is the maximum depth to which the tank may be filled to prevent spilling?

5-17 The closed tank shown, which is full of liquid, is accelated downward at 48.3 ft/sec² and to the right at 32.2 ft/sec². $L = 3$ ft, $H = 4$ ft, and the specific gravity of the liquid is 1.1. Determine $p_C - p_A$ and $p_B - p_A$.

5-18 The closed tank shown, which is full of liquid, is accelerated downward at 6 m/s² and to the right at 9.81 m/s². $L = 2$ m and $H = 3$ m, and the liquid has a specific gravity of 1.5. Determine $p_C - p_A$ and $p_B - p_A$.

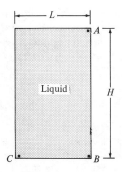

PROBLEMS 5-17, 5-18

5-19 The tank shown is 4 m long, 3 m high, and 3 m wide and closed except for a small opening at the right end. It contains oil (sp. gr. = 0.8) to a depth of 2 m in a static situation. If the tank is uniformly accelerated to the right at a rate of 9.81 m/s², what will be the maximum pressure intensity in the tank during acceleration?

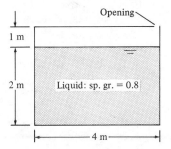

5-20 A tank of liquid (sp. gr. = 0.80) that is 1 ft in diameter and 1 ft high ($h = 1$ ft) is rigidly fixed (as shown) to a rotating arm having a 2-ft radius. The arm rotates so that the speed at point A is 20 ft/sec. If the pressure intensity at A is 30 psf, what is the pressure intensity at B?

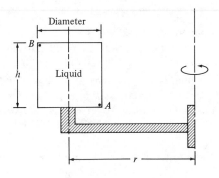

PROBLEMS 5-20, 5-21

5-21 A tank of liquid (sp. gr. = 0.80) that is 40 cm in diameter and 30 cm high (h = 30 cm) is rigidly fixed (as shown) to a rotating arm having a 1-m radius. The arm rotates so that the speed at point A is 10 m/s. If the pressure intensity at A is 200 kPa, what is the pressure intensity at B?

5-22 The open tank truck carries a load of sludge that has a specific gravity of 1.1. If the truck travels at a speed of 60 km/h, what is the minimum radius of un-banked turn it can make without spilling the sludge over the side of the tank?

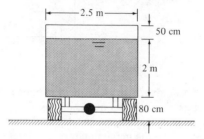

5-23 If the U-tube is rotated about the axis 0-0 at a rate of 16.06 rad/sec, determine the new position of the water surface in the outside leg during rotation. The values of L_1, L_2, and L_3 are 50 in., 6 in., and 10 in., respectively.

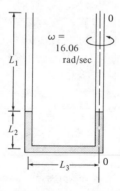

PROBLEMS 5-23, 5-24

5-24 If the U-tube is rotated about the axis 0-0 at a rate of 16.06 rad/sec, determine the new position of the water surface in the outside leg during rotation. The values of L_1, L_2, and L_3 are 1 m, 20 cm, and 30 cm, respectively.

5-25 This closed tank, which is 4 ft in diameter, is filled with water and is spun around its vertical centroidal axis at a rate of 10 rad/sec. An open piezometer is connected to the tank as shown so it is also rotating with the tank. For these conditions, what is the pressure at the center of the bottom of the tank?

5-26 Water stands in these tubes as shown when no rotation occurs. If the entire system is rotated about axis A-A, at what angular speed will water just begin to spill out of the small tube?

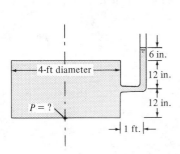

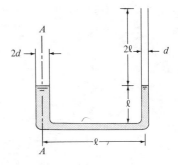

PROBLEM 5-25   PROBLEM 5-26

5-27 Water fills a slender tube $\frac{1}{2}$ in. in diameter, 18 in. long, and closed at one end. If the tube is rotated in a horizontal plane about its open end at a constant speed of 62.8 rad/sec, what force is exerted on the closed end?

5-28 Water fills a slender tube 1 cm in diameter, 50 cm long and closed at one end. If the tube is rotated in the horizontal plane about its open end at a constant speed of 62.8 rad/s, what force is exerted on the closed end?

5-29 Water fills a slender tube 1 cm in diameter, 50 cm long and closed at one end. If the tube is rotated in the horizontal plane about its open end at a constant speed of 30 rpm, what force is exerted on the closed end?

5-30 Mercury maintains the position in a rotating tube as shown. What is the rate of rotation in terms of g and ℓ?

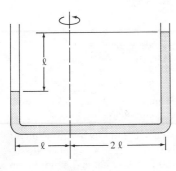

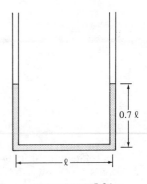

PROBLEM 5-30   PROBLEM 5-31

5-31 The U-tube is shown with water standing in it when there is no rotation.
   (a) If the tube is rotated about the left leg at a rate of 5 rad/s, where will the water stand in the tube for $\ell = 30$ cm?
   (b) Where will the water surfaces be if the rate of rotation is increased to 12 rad/s for $\ell = 30$ cm?

5-32 This tube and liquid is rotated about a vertical axis as indicated. If the rate of rotation is 10 rad/sec, what will be the pressure in the liquid at point A (at the axis of rotation) and at point B? Now, if the tube is rotated at a rate of 20 rad/sec, what will be the pressure at points A and B?

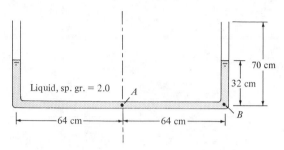

5-33 A tank of liquid is rotated about a vertical axis at a rate of 10 rad/sec. On the horizontal bottom inside the tank, what will be the pressure difference between points A and B if point B is 1.5 ft from the axis of rotation and point A is 1.0 ft from the axis? $\rho = 2.0$ slugs/ft³.

5-34 A closed cylindrical tank of water is rotated about its horizontal axis as shown. The water inside the tank rotates with the tank ($v = r\omega$). Derive an equation for dp/dz along a vertical-radial line through the center of rotation. What is dp/dz along this line for $z = -1$ m, $z = 0$, and $z = +1$ m when $\omega = 5$ rad/s? Here $z = 0$ at the axis.

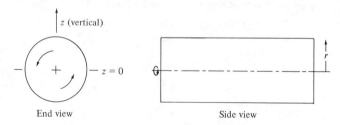

End view   Side view

PROBLEMS 5-34, 5-35

5-35 For the conditions of Prob. 5-34 derive an equation for the maximum pressure difference in the tank as a function of the significant variables.

5-36 The 4-ft-diameter by 10-ft-long tank shown is closed and filled with water. It is rotated about its horizontal centroidal axis to produce the velocity distribution as shown. If the maximum velocity is 20 ft/sec, what is the maximum difference in pressure in the tank? Where is the point of minimum pressure?

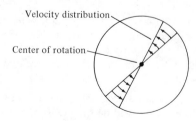

Velocity distribution

Center of rotation

5-37 The velocity in the outlet pipe from this reservoir is 20 ft/sec and h = 20 ft. Due to the rounded transition, the flow is assumed to be irrotational. With these conditions, what is the pressure at A?

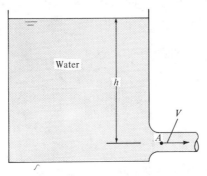

PROBLEMS 5-37, 5-38

5-38 The velocity in the outlet pipe from this reservoir is 10 m/s and h = 15 m. Due to the rounded entrance to the pipe, the flow is assumed to be irrotational. With these conditions, what is the pressure at the section where the velocity is 10 m/s?

5-39 Water flows from this large orifice at the bottom of the tank. Point B is at zero elevation and point A is at 1 ft elevation. If V_A = 10 ft/sec at an angle of 45° with the horizontal and if V_B = 20 ft/sec vertically downward, what is the value of $p_A - p_B$?

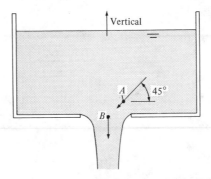

5-40 Water is forced out of this cylinder by the piston. If the piston is driven at a speed of 4 ft/sec, what will be the speed of efflux of the water from the

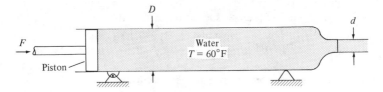

nozzle if $d = 2$ in. and $D = 4$ in.? Neglecting friction, what will be the force F required to drive the piston?

5-41 For the contraction $p_1 = 160$ kPa, $p_2 = 100$ kPa and water is flowing. If $d = 40$ cm and $D = 60$ cm, what is the velocity V_2?

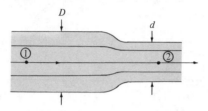

PROBLEMS 5-41, 5-42

5-42 For the contraction $p_1 = 24$ psi, $p_2 = 18$ psi and water is flowing. If $d = 4$ in. and $D = 6$ in. what is the velocity V_2?

5-43 The flow pattern through the pipe contraction is as indicated and the discharge of water is 60 cfs. For $d = 2$ ft, $D = 6$ ft, what will be the pressure at point B if the pressure at point A is 2,500 psf?

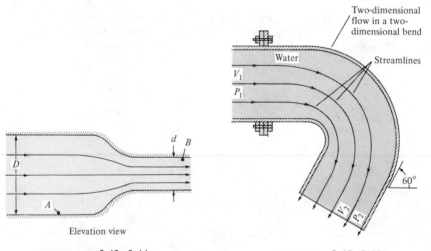

Elevation view

PROBLEMS 5-43, 5-44   PROBLEMS 5-45, 5-46

5-44 The flow pattern through the pipe contraction is as indicated and the discharge of water is 20 m³/s. For $d = 160$ cm and $D = 200$ cm what will be the pressure at point B if the pressure at point A is 10 kPa?

5-45 Water flows in this two-dimensional bend, which turns in a horizontal plane. Here $V_1 = V_2 = 40$ ft/sec, $p_1 = 200$ psfg, and $p_2 = 0$ psfg. Would you expect cavitation to occur any place within the bend if the atmospheric pressure is 14.7 psia? *Hint*: The streamlines are drawn to scale.

5-46 Water flows in this two-dimensional bend, which turns in a horizontal plane. Here $V_1 = V_2 = 13$ m/s, $p_1 = 10$ kPa gage, and $p_2 = 0$ gage. Would you expect cavitation to occur any place within the bend if the atmospheric pressure is 100 kPa absolute? *Hint*: The streamlines are drawn to scale.

5-47 When gage A indicates a pressure of 100 kPa gage, then cavitation just starts to occur in the venturi meter. If $D = 39$ cm and $d = 10$ cm what is the water discharge in the system for this condition of incipient cavitation? The atmospheric pressure is 100 kPa and the water temperature is 10°C.

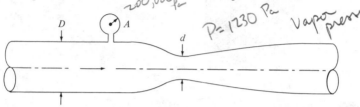

5-48 An atomizer consists of a constriction in an air duct that reduces the pressure and draws liquid from the reservoir into the airstream. The pressure on the liquid and at the exit of the atomizer is atmospheric. The velocity of the air at the exit is 5 m/s and the air density is 1.2 kg/m³. Assuming the air flow is steady, incompressible and inviscid, calculate the maximum permissible area ratio (A_c/A_e) for which the atomizer will operate.

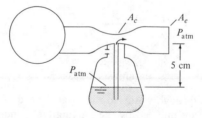

5-49 A spherical probe is used for finding gas velocity by measuring the pressure difference between the upstream and downstream points A and B. The pressure coefficient at points A and B are 1.0 and -0.5. The pressure difference $p_A - p_B$ is 5 kPa and the gas density is 1.50 kg/m³. Calculate the gas velocity.

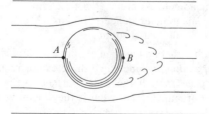

5-50 Refer to Fig. 5-7. If in this flow field air is flowing ($\rho = 0.0024$ slugs/ft³) and the maximum pressure in the field is 100 psf at a point where the velocity is 30 ft/sec, what is the minimum pressure in the field?

5-51 Refer to Fig. 5-7. If in this flow field air is flowing ($\rho = 1.2$ kg/m³) and the maximum pressure in the field is 1.40 kPa at a point where the velocity is 20 m/s, what is the minimum pressure in the field?

5-52 Water in a flume is shown for two conditions. If the depth d is the same for each case, will gage A read greater or less than gage B? Explain.

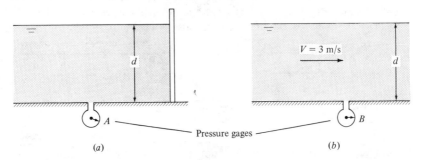

(a)   (b)

5-53 Body A travels through water at a constant speed of 30 ft/sec. Velocities at points B and C are induced by the moving body and are observed to have magnitudes of 10 ft/sec and 5 ft/sec, respectively. What is $p_B - p_C$?

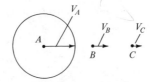

PROBLEMS 5-53, 5-54

5-54 Body A travels through water at a constant speed of 10 m/s. Velocities at points B and C are induced by the moving body and are observed to have magnitudes of 4 m/s and 2 m/s, respectively. What is $p_B - p_C$?

5-55 An air-water manometer is connected to a Pitot tube used to measure air velocity. If the manometer deflects 2.0 in., what is the velocity? Assume $T = 60°F$ and $p = 15$ psia.

5-56 An air-water manometer is connected to a Pitot tube used to measure air velocity. If the manometer deflects 4.5 cm what is the velocity? Assume $T = 20°C$ and $p = 100$ kPa absolute.

5-57 A glass tube with a 90° bend is open at both ends. It is inserted into a flowing stream of water so that one opening is directed upstream and the other is vertical (Fig. 5-5 page 152). If the water surface in the vertical leg is 15 cm higher than the stream surface, what is the stream velocity?

5-58 A glass tube with a 90° bend is open at both ends. It is inserted into a flowing stream of water so that one opening is directed upstream and the other is vertical (see Fig. 5-6). If the water surface in the vertical tube is 10 in. higher than the stream surface, what is the velocity?

5-59 A Pitot tube is used to measure the velocity at the center of a 12-in. pipe. If kerosene at 33°F is flowing and the deflection on a mercury-kerosene manometer connected to the Pitot tube is 8 in., what is the velocity?

5-60 A Pitot tube is used to measure the velocity at the center of a 40-cm pipe. If kerosene at 20°C is flowing and the deflection on a mercury-kerosene manometer connected to the Pitot tube is 20 cm, what is the velocity?

5-61 A Pitot tube used to measure air velocity is connected to a differential pressure gage. If the air temperature is 20°C at standard atmospheric pressure at sea level and if the differential gage reads a pressure difference of 3 kPa, what is the air velocity?

5-62 A Pitot tube used to measure air velocity is connected to a differential pressure gage. If the air temperature is 60°F at standard atmospheric pressure at sea level and if the differential gage reads a pressure difference of 10 psf, what is the air velocity?

5-63 Ideal-flow theory will yield a flow pattern past an airfoil similar to that shown. If the approach air velocity V_0 is 100 ft/sec, what is the pressure difference between the bottom and top of this airfoil at the point where the velocities are $V_1 = 200$ ft/sec and $V_2 = 90$ ft/sec? Assume ρ_{air} is constant at 0.0022 slugs/ft^3.

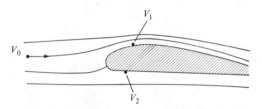

PROBLEMS 5-63, 5-64

5-64 Ideal-flow theory will yield a flow pattern past an airfoil similar to that shown. If the approach air velocity V_0 is 50 m/s, what is the pressure difference between the bottom and top of this airfoil at the point where the velocities are $V_1 = 100$ m/s and $V_2 = 80$ m/s? Assume ρ_{air} is constant at 1.1 kg/m^3.

5-65 When testing the hydrofoil shown, the minimum pressure on the surface of the foil was found to be 80 kPa absolute when the foil was submerged 1.80 m

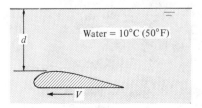

PROBLEMS 5-65, 5-66, 5-67, 5-68

and towed at a speed of 7 m/s. At the same depth, at what speed will cavitation first occur? Assume irrotational flow for both cases and $T = 10°C$.

5-66 For the hydrofoil of Prob. 5-65 at what speed will cavitation begin if the depth is increased to 3 m?

5-67 When testing the hydrofoil shown, the minimum pressure on the surface of the foil was found to be 2 psi vacuum when the foil was submerged 4 ft and towed at a speed of 20 ft/sec. At the same depth, at what speed will cavitation first occur? Assume irrotational flow for both cases and $T = 60°F$.

5-68 For the conditions of Prob. 5-67, at what speed will cavitation begin if the depth is increased to 10 ft?

5-69 The minimum pressure on a cylinder moving horizontally in water $(T = 10°C)$ at 5 m/s at a depth of 2 m is 70 kPa absolute. At what velocity will cavitation begin? Atmospheric pressure is 100 kPa absolute.

5-70 A rugged instrument used frequently for monitoring gas velocity in smoke stacks consists of two open tubes oriented to the flow direction, as shown below, and connected to a manometer. The pressure coefficient is 1.0 at A and -0.4 at B. Assume water, at 20°C, is used in the manometer and a 0.5-cm deflection is noted. The pressure and temperature of the stack gases is 101 kPa and 250°C. The gas constant of the stack gases is 200 J/kg K. Determine the velocity of the stack gases.

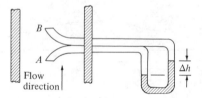

5-71 An airspeed indicator used frequently on World War II air-craft consisted of a converging-diverging duct as shown below. The diameter of the duct at ② is three-quarters of the entrance diameter at ①. The differential pressure gage records a pressure of 1.5 kPa and the density of the air is 1 kg/m³. The air flow is steady, incompressible, inviscid and irrotational. Determine the airspeed U.

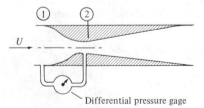

Differential pressure gage

5-72 A Pitot tube is used to measure the airspeed of an airplane. The Pitot tube is connected to a pressure-sensing device calibrated to indicate the correct airspeed when the temperature is 17°C and the pressure is 101 kPa. The airplane flies at 3,000 m altitude where the pressure and temperature are 70 kPa and −5°C. The indicated airspeed is 60 m/s. What is the true airspeed?

REFERENCES

1. Bernoulli, D., *Hydrodynamics,* and Bernoulli, J., *Hydraulics.* Trans. Thomas Carmody and Helmut Kobus. Dover Publications, Inc., New York, 1968.

2. Fage, A., and Warsap, J. H. "The Effects of Turbulence and Surface Roughness on the Drag of a Circular Cylinder." *Aero. Res. Comm., London, Rept. Mem.* 1283 (1929).

3. Knapp, R. T., and Hollander, A. "Laboratory Investigations on the Mechanism of Cavitation." *Trans. ASME,* 70 (1948).

4. Rouse, H. *Elementary Mechanics of Fluids.* John Wiley & Sons, Inc., New York, 1946.

5. Streeter, V. L. (ed.). *Handbook of Fluid Dynamics.* McGraw-Hill Book Company, New York, 1961.

6. Thomas, H. A., and Schuleen, E. P. "Cavitation in Outlet Conduits of High Dams." *Trans. Am. Soc. Civil Eng.,* 107 (1942).

FILMS

7. Eisenberg, Philip. *Cavitation.* National Committee for Fluid Mechanics Films, distributed by Encyclopaedia Britannica Educational Corporation.

8. Shapiro, A. H. *Pressure Fields and Fluid Acceleration.* National Committee for Fluid Mechanics Films, distributed by Encyclopaedia Britannica Educational Corporation.

Penstock "anchors," Shasta Dam, California. The massive concrete structures at the upper ends of these 15-ft- (4.57-m-) diameter penstock sections are anchor blocks designed to withstand the dynamic forces (pressure and momentum) acting on the bend from a water flow in each penstock of 2,700 cfs (76.4 m³/s) under a pressure of 140 psi (965 kPa). The penstocks' plate thickness is 2 in. and each anchor weighs over 2 million lb (9 MN). (Courtesy U.S. Bureau of Reclamation.)

6 MOMENTUM PRINCIPLE

IN SOLID MECHANICS you will recall that the impulse, $\int \mathbf{F}\, dt$, applied to a body was equal to the change of momentum, $M\mathbf{V}_2 - M\mathbf{V}_1$, for a given time interval. The directly analogous situation for fluid flow is the case where liquid in a pipe is accelerated by means of a pressure differential along the pipe. Both these cases may be termed simple unsteady-state cases, for which the basic equation of linear momentum is applicable. When we have nonuniform flow, which may be either steady or unsteady, the process becomes more complicated, and we turn to the basic control-volume approach to develop the momentum equations.

6-1 THE MOMENTUM EQUATION

Derivation of basic equation

First, let us consider Eq. (4-21a), which is rewritten here for convenience:

$$\frac{dB_{\text{sys}}}{dt} = \frac{d}{dt} \int_{cv} \beta \rho \, d\forall + \sum_{cs} \beta \rho \mathbf{V} \cdot \mathbf{A} \qquad (4\text{-}21a)$$

We let B equal the momentum of the *system,* or, in other words, the momentum of a given quantity of matter. Then dB/dt will be the rate of change of momentum of the system with respect to time $d(\text{momentum})/dt$. Momentum, by definition, is the product of mass and velocity. Therefore, β, the corresponding intensive property—or momentum per unit mass—is simply the velocity $\mathbf{V}$. We now substitute $d(\text{momentum})/dt$ for dB/dt and $\mathbf{V}$ for β in Eq. (4-21a) to obtain

$$\frac{d(\text{momentum})}{dt} = \sum_{cs} \mathbf{V}\rho\mathbf{V} \cdot \mathbf{A} + \frac{d}{dt} \int_{cv} \mathbf{V}\rho \, d\forall \qquad (6\text{-}1)$$

According to Newton's second law, the summation of all external forces on a *system* is equal to the rate of change of momentum of that *system,* $\Sigma \mathbf{F} = d(\text{momentum})/dt$; thus, when the appropriate substitution is made in Eq. (6-1) we have

$$\sum \mathbf{F} = \sum_{cs} \mathbf{V}\rho\mathbf{V} \cdot \mathbf{A} + \frac{d}{dt} \int_{cv} \mathbf{V}\rho \, d\forall \qquad (6\text{-}2)$$

The force term on the left may include various types of forces. For example, the two types of forces usually considered are surface forces and body forces. When these are so designated, Eq. (6-2) can be written as

$$\sum \mathbf{F}_S + \sum \mathbf{F}_B = \sum_{cs} \mathbf{V}\rho\mathbf{V} \cdot \mathbf{A} + \frac{d}{dt}\int_{cv} \mathbf{V}\rho \; d\forall \qquad (6\text{-}3a)$$

This is the basic form of the momentum equation when there is a uniform velocity in the streams crossing the control surface, and it is a very powerful tool for solving many problems in fluid mechanics.

The more general form of the momentum equation, when the velocity in general is variable across the control surface, is given by

$$\sum \mathbf{F}_S + \sum \mathbf{F}_B = \int_{cs} \mathbf{V}\rho\mathbf{V} \cdot d\mathbf{A} + \frac{d}{dt}\int_{cv} \mathbf{V}\rho \; d\forall \qquad (6\text{-}3b)$$

There are no limitations for use of the momentum equation taken as a whole; however, there are certain limitations and assumptions for each term of the equation that the student should be aware of. These are considered in detail below.

EXTERNAL FORCES Let us first consider the terms on the left of Eqs. (6-3). These represent the external forces acting on the system or, in other words, acting on the mass inside the control volume at the instant the equation is applied. The surface forces $\sum\mathbf{F}_S$ may be in the form of pressure forces transmitted through the fluid, such as those shown acting at section 1 in Fig. 6-1; or they may be forces transmitted through a solid, such as the force transmitted by the metal in the joint which joins the pipe to the nozzle of Fig. 6-1. The surface force may also include the force of the physical boundary on the fluid if the control surface is so drawn. In other words, whenever forces are transmitted across the control surface, they must be considered in $\sum\mathbf{F}_S$ of Eqs. (6-3). The body forces $\sum\mathbf{F}_B$ for most engineering applications consist of the forces of gravity which act on the mass inside the control volume; however, more advanced studies, such as magnetohydrodynamics, will also include electromagnetic forces in this category.

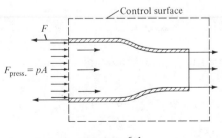

FIGURE 6-1

It is important to recognize that the term $\sum\mathbf{F}$ in Eq. (6-2) can be considered in exactly the same manner that one considers the forces on a free-body in basic engineering mechanics. In our case the body under consideration is the mass that at the instant is inside the control volume; therefore, the control surface delineates the body. One should keep this free-body concept in mind when solving flow problems using the momentum equation.

VELOCITY REFERENCE   Recall that in the basic derivation of the control-volume equation, the velocity in the $\mathbf{V} \cdot \mathbf{A}$ term is always referenced to the control volume itself because it represents the discharge across the control surface; therefore, in its application in the momentum equation, this same requirement holds. If the control volume itself is not accelerating (it can be stationary or moving with a constant velocity), then the velocity that is used for β will also be referenced to the same reference frame to which the control volume is fixed. Most of your problems will be of this type. However, if the problem is one for which it would be desirable to let the control volume accelerate—for example, to let the control volume accelerate with a body that is also accelerating—then the velocity used for β must be with respect to the inertial frame of reference. This is required to satisfy Newton's second law, which is used in the derivation of the momentum equation and is valid only in an inertial frame of reference.

UNSTEADINESS   When conditions at a point change with respect to time, we have unsteady flow. In Eqs. (6-3) this is taken into account by the last term on the right. A simple case of unsteady flow in a straight pipe illustrates the applicability of this term, as is shown in Example 6-1.

EXAMPLE 6-1   A straight, horizontal 20-cm-diameter pipe discharges water into the air. If the rate of flow is 0.01 m³/s and is increasing at the rate of 0.15 m³/s², what will be the pressure 100 m from the outlet end? Neglect viscous effects.

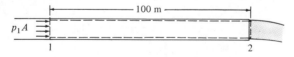

Solution   If we write Eq. (6-3a) for the x direction we have

$$p_1A_1 - p_2A_2 = \sum_{cs} V\rho\mathbf{V} \cdot \mathbf{A} + \frac{d}{dt}\int_{cv} V\rho \, d\forall$$

Here the term $\Sigma_{cs} V\rho\mathbf{V} \cdot \mathbf{A} = \rho V(-V_1A_1 + V_2A_2) = 0$. The last term on the right-hand side of the equation represents the rate of change of momentum inside the control volume, which can be expressed as

$$\frac{d}{dt} \int_{cv} V\rho \, d\forall = \frac{d}{dt} \left(\frac{Q}{A} \rho \forall \right)$$

$$= \frac{d}{dt} \left(\frac{Q}{A} \rho L A \right)$$

$$= \rho L \frac{dQ}{dt}$$

In the statement of the example it was given that the discharge rate was increasing at a rate of 0.15 m³/s/s ($dQ/dt = +0.15$); hence, this substitution is made along with values for ρ and L in the expression above:

$$\frac{d}{dt} \int_{cv} V\rho \, d\forall = 1{,}000 \times 100 \times 0.15 = 15 \text{ kN}$$

Then $\qquad (p_1 - p_2)A = 15 \text{ kN}$

Since $p_2 = 0$ gage

$$p_1 = p_{100} = \frac{15 \text{ kN}}{0.0314} = 478 \text{ kPa} \qquad \blacktriangleleft$$

In many problems the flow inside the control volume is steady and the unsteadiness term drops out of the equation; however, the student should always be aware of this term and include it whenever it is significant.

NONUNIFORMITY OF FLOW   The first term on the right-hand side of Eqs. (6-3) is the change in the flow of momentum across a given control volume; therefore, if the flow through the control volume is uniform, as in Example 6-1, there will be no change in the flow of momentum and this term will be zero. However, if the flow is nonuniform, a change in the flow of momentum will undoubtedly exist between the inflow and outflow sections and the magnitude of the term must be evaluated. Common applications involving nonuniform flow are flow in bends and flow through nozzles. These will be covered in detail later in this chapter.

Advantages of the momentum equation

The momentum equation can be used in the solution of numerous problems; however, a common application is the determination of the force exerted on a piece of equipment such as a nozzle or bend in a pipe given a certain discharge and pressure. One way to solve for such a force is to evaluate the pressure and shear stress, τ, over the surface of the device in question and carry out the integration of $p \, dA$ and $\tau \, dA$ for a particular direction in question. In theory this can be done, but in practice it is usually too difficult to solve for the exact pressure and shear stress distribution over the area; therefore, we turn to the momentum equation for a simple means of evaluating the force. If the flow is steady, we need only know the pressure and velocity where the flow crosses the control surface to apply the momentum equation. This is of great significance, because for many problems we can draw the control surface such that it crosses one-dimensional flow zones where the velocity and piezometric head across each section are essentially constant, thus avoiding complicated flow patterns. Indeed, it is by making a judicious choice of control surface that we are often justified in using the mean velocity in our calculations instead of the more exact variable velocity, which would require integration. It is often advantageous to use both the momentum and Bernoulli equations to obtain a solution to the problem.

Momentum equation in the cartesian coordinate system

It is often convenient to use separate momentum equations in the x, y, and z coordinate directions instead of a single vector equation; thus, Eq. (6-2) is written as follows in these respective directions:

x direction:

$$\sum F_x = \sum_{cs} u(\rho \mathbf{V} \cdot \mathbf{A}) + \frac{d}{dt} \int_{cv} u\rho \, d\forall \qquad (6\text{-}4)$$

y direction:

$$\sum F_y = \sum_{cs} v(\rho \mathbf{V} \cdot \mathbf{A}) + \frac{d}{dt} \int_{cv} v\rho \, d\forall \qquad (6\text{-}5)$$

z direction:

$$\sum F_z = \sum_{cs} w(\rho \mathbf{V} \cdot \mathbf{A}) + \frac{d}{dt} \int_{cv} w\rho \, d\forall \qquad (6\text{-}6)$$

Several examples are given in the next section to illustrate the methods of solution and to show the wide range of application of the momentum equation.

EXAMPLE 6-2   A rocket motor is tested on a test stand (see page 187). During a test, fuel is fed in at the rate of 1 kg/s and LOX (liquid oxygen) at 10 kg/s. The flexible couplings and supporting rollers offer negligible horizontal force. The gases exit at atmospheric pressure and with a velocity of 2,000 m/s. Under these conditions, what force will be exerted on the thrust stand?

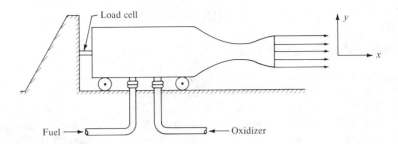

Solution First, for computational purposes, we isolate the rocket with a control surface as shown below and apply the steady-flow momentum equation in the x direction:

$$\sum F_x = \sum u\rho \mathbf{V} \cdot \mathbf{A}$$

Since the pressure is uniform around the control surface except for the fuel and LOX lines, there is no net force due to pressure in the x direction. The flexible couplings and rollers offer no force in the x direction. The single surface force is F, the force of the test stand on the rocket motor. The body force is due to gravity, so there is no component in the x direction. Thus the momentum equation becomes

$$F = \sum u\rho \mathbf{V} \cdot \mathbf{A}$$

Mass enters the control surface through the fuel and oxidizer connections. However, the flow in these lines is vertical, so u is zero. Thus the momentum equation reduces to

$$F = u_{\text{exit}}\rho \mathbf{V} \cdot \mathbf{A}$$
$$= u_{\text{exit}}\dot{m}$$
$$= (2{,}000 \text{ m/s})(1 + 10) \text{ kg/s}$$
$$= 22{,}000 \text{ N}$$

This force is the force of the thrust stand on the rocket motor; thus the reaction to this force is the force on the thrust stand and will be in the negative x direction:

$$F_{\text{on stand}} = -22{,}000 \text{ N} \qquad \blacktriangleleft$$

6-2 APPLICATIONS OF THE MOMENTUM EQUATION

Jet deflected by a plate or vane

Consider a jet of water turned through a horizontal angle, such as is shown in Fig. 6-2. Here it is assumed that the pressure in the stream leaving the vane is the same as that entering the vane, and it is also assumed that the speed of the jet is not appreciably reduced by the surface resistance of the vane. Then, if we draw the control surface so that it includes only the jet as shown in Fig. 6-3, we can easily apply the mo-

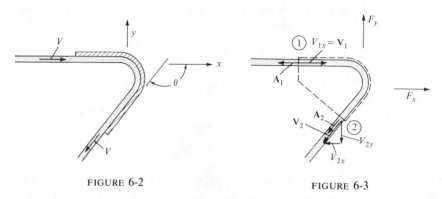

FIGURE 6-2   FIGURE 6-3

mentum equation to solve for the force components F_x and F_y, which are the forces exerted by the vane on the jet. We initially assume that these forces have a positive sense as shown; however, if they are in fact negative, a negative sign will result when we solve for them. The momentum equation for the x direction is as follows:

$$F_x = \sum_{cs} u\rho \mathbf{V} \cdot \mathbf{A} + \frac{d}{dt} \int_{cv} u\rho \, d\forall \qquad (6\text{-}7)$$

If the flow is steady, the second term on the right of Eq. (6-7) will be zero, which leaves the following:

$$F_x = \sum_{cs} u\rho \mathbf{V} \cdot \mathbf{A} \qquad (6\text{-}8)$$

At section 1 the velocity is constant over the section and the area vector $\mathbf{A}_1$ is in the reverse direction of the velocity vector $\mathbf{V}$. Therefore, for this part of the control surface we have

$$\sum_{cs} u\rho \mathbf{V} \cdot \mathbf{A} = V_{1x}\rho(-V_1 A_1)$$

By a similar analysis for section 2 (in this case, the velocity $\mathbf{V}_2$ and area $\mathbf{A}_2$

have the same sense), we get

$$\sum_{cs} u\rho \mathbf{V} \cdot \mathbf{A} = V_{2x}\rho V_2 A_2$$

However, $V_1 A_1 = V_2 A_2 = Q$, so when these substitutions are made, Eq. (6-8) becomes

$$F_x = \rho Q(V_{2x} - V_{1x}) \tag{6-9}$$

In a similar manner the force in the y direction, F_y, will be

$$F_y = \rho Q(V_{2y} - V_{1y}) \tag{6-10}$$

and for this particular case, Fig. 6-3, $F_y = \rho Q(V_{2y})$. Since the forces given by Eqs. (6-9) and (6-10) are the forces exerted by the vane on the system (on the jet, in this case), we obtain the forces of the jet on the vane by simply reversing the sign on F_x and F_y.

Equations (6-9) and (6-10) are very useful for cases such as this deflected jet, where a single-steady stream passes through the control volume; however, it is recommended that Eq. (6-3) be utilized in most applications because of its generality. For example, if a single stream enters a control volume but then divides and leaves as two streams with different velocities, then Eqs. (6-9) and (6-10) do not apply.

EXAMPLE 6-3   Consider a jet that is deflected by a stationary vane, such as is given in Fig. 6-2. If the jet speed and diameter are 100 ft/sec and 1 in., respectively, and the jet is deflected 60°, what force is exerted on the vane by the jet?

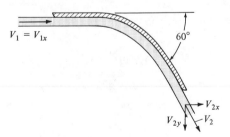

$$V_1 = V_{1x}$$ 60°

$$V_{2x}$$
$$V_{2y} \quad V_2$$

Solution   First solve for F_x, the x component of force of the vane on the jet, by using Eq. (6-9).

$$F_x = \rho Q(V_{2x} - V_{1x})$$

Here, the final velocity in the x direction is given as

$$V_{2x} = 100 \cos 60°$$

Hence, $$V_{2x} = 100 \times 0.500 = 50 \text{ ft/sec}$$

also, $V_{1x} = 100$ ft/sec

and $Q = V_1 A_1 = 100 \dfrac{0.785}{144} = 0.545$ cfs

Therefore, $F_x = 1.94$ lbf-sec^2/ft^4 $\times$ 0.545 ft^3/sec $\times$ (50 $-$ 100) ft/sec

$= -52.9$ lbf

Similarly determined, the y component of force on the jet is

$F_y = 1.94$ lbf-sec^2/ft^4 $\times$ 0.545 ft^3/sec $\times$ (-86.6 ft/sec $-$ 0)

$= -91.6$ lbf

Then the force on the vane will be the reactions to the forces of the vane on the jet, or

$$F_x = +52.9 \text{ lbf}$$
$$F_y = +91.6 \text{ lbf}$$

If the vane were moving, such as on a turbine runner, we could carry out the same type of analysis as given above, except that it would be convenient to let the control volume move with the vane. By this choice, the flow field is steady so the momentum equation becomes

$$\sum F_x = \int u(\rho \mathbf{V} \cdot \mathbf{A})$$

In addition, because the vane is moving at a constant velocity, the vane represents an inertial reference frame for the velocities. The control volume for this application is shown in Fig. 6-4, where the velocities are referenced to the moving control volume. The force applied to the vane in the x direction is then

$$F_v = -\rho(V - V_v)A[(V - V_v)_{2x} - (V - V_v)_{1x}] \qquad (6\text{-}11)$$

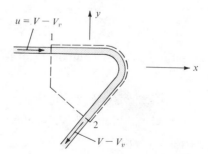

FIGURE 6-4   Control volume for jet deflected by moving vane.

The power delivered by the vane is equal to the product of the force on the vane and the speed of the vane. The resulting power is then $F_v V_v$. Obviously, no power results unless the vane speed is greater than zero and less than the velocity of the approaching jet. Other interesting and useful relations can be developed, such as the optimum speed and θ for most efficient generation of power. Such topics are discussed in more detail in Chapter 14.

Forces on bends

Consider the bend in the pipe of Fig. 6-5. Here it is assumed that the flow is steady and that the velocity is uniform across sections 1 and 2. Assume also that the bend is oriented so that y is vertically upward. In the design of such a bend, one of the primary design considerations is the force required to hold the bend in place. For small- or medium-sized pipes, such a force is supplied by the flange bolts or by welds if it happens to be a welded connection; however, if the pipe is large, specially designed anchor blocks may be required to hold the bend. In the latter case, we are primarily interested in the reactions R_x and R_y needed to hold the bend in place.

The analysis is carried out by isolating the bend by means of a control

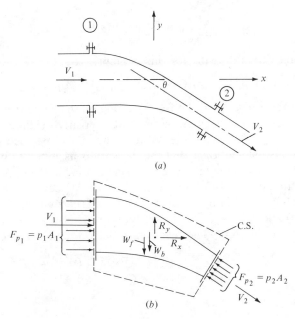

FIGURE 6-5   Flow in a bend. (*a*) Bend. (*b*) Control volume.

surface as shown in Fig. 6-5b, and the momentum equation is applied for the x, y, and z directions with due care to be certain that all relevant forces such as pressure and gravity are included in the solution.

EXAMPLE 6-4   Water flows through a 180° vertical reducing bend, as shown. The discharge is 0.25 m³/s and the pressure at the center of the inlet of the bend is 150 kPa. If the bend volume is 0.10 m³ and it is assumed the Bernoulli equation is valid, what reaction is required to hold the bend in place? Assume that the metal in the bend weighs 500 N.

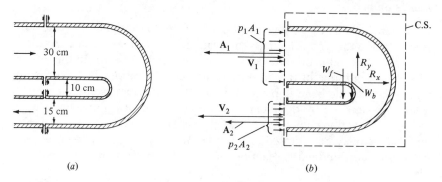

(a)   (b)

Solution   The momentum equation is applied with the control volume shown. For the x direction we have

$$\sum F_x = \sum_{cs} V_x \rho \mathbf{V} \cdot \mathbf{A}$$

which becomes the following for this example:

$$R_x + p_1 A_1 + p_2 A_2 = u_1 \rho(-V_1 A_1) + u_2 \rho(V_2 A_2)$$

Note that in the x direction, the forces acting on the system are the pressure forces $p_1 A_1$ and $p_2 A_2$, both acting in the positive x direction for this example, and R_x. Here R_x is the net force acting across the metal at the flange. Note also at section 1 that the velocity is assumed to be constant across the section and the area vector $\mathbf{A}_1$ is pointing in the opposite direction to $\mathbf{V}_1$; therefore, $u_1 \rho \mathbf{V} \cdot \mathbf{A} = V_1 \rho(-V_1 A_1)$. At section 2 the area and velocity vectors point in the same direction, so the sign on $\mathbf{V} \cdot \mathbf{A}$ is positive. However $u_2 = -V_2$; consequently, there is also a negative sign on the last term on the right. The reaction in the x direction is then given as

$$R_x = -p_1 A_1 - p_2 A_2 - \rho V_1^2 A_1 - \rho V_2^2 A_2$$

However, $V_1 A_1 = V_2 A_2 = Q$

so this reduces to

$$R_x = -p_1 A_1 - p_2 A_2 - \rho Q(V_1 + V_2)$$

The velocities in the foregoing equation are determined by the continuity equation:

$$V_1 = \frac{Q}{A_1} = \frac{0.25}{(\pi/4) \times 0.30^2} = 3.54 \text{ m/s}$$

$$V_2 = \frac{Q}{A_2} = \frac{0.25}{(\pi/4) \times 0.15^2} = 14.15 \text{ m/s}$$

The initial pressure p_1 is given; however, p_2 will have to be obtained by the Bernoulli equation written between the center of section 1 and the center of section 2.

$$\frac{p_1}{\gamma} + \frac{V_1^2}{2g} + z_1 = \frac{p_2}{\gamma} + \frac{V_2^2}{2g} + z_2$$

Here $z_2 - z_1 = -0.325$ m, $V_1 = 3.54$ m/s, $V_2 = 14.15$ m/s, so we obtain

$$\frac{p_2}{\gamma} = \frac{150 \times 10^3}{9,810} + \frac{3.54^2}{2 \times 9.81} + 0.325 - \frac{14.15^2}{2 \times 9.81}$$

$$p_2 = 59.3 \text{ kPa}$$

Now we can solve for R_x as follows:

$$\begin{aligned}
R_x &= -150 \times 10^3 \times (\pi/4) \times 0.3^2 - 59.3 \times (\pi/4) \\
&\quad \times 0.15^2 - 1,000 \times 0.25(14.15 + 3.54) \\
&= -10,602 - 1,048 - 4,422 \\
&= -16.07 \text{ kN} \qquad \blacktriangleleft
\end{aligned}$$

Because there are no velocities in the y direction where the flow passes across the control surface, R_y is obtained by static equilibrium: $\Sigma F_y = 0$ or

$$R_y - 500 - 0.1 \times 9,810 = 0$$

$$R_y = 500 + 981$$

$$= 1.48 \text{ kN} \qquad \blacktriangleleft$$

The determination of the sign on the unknown reaction R is determined from one basic rule. In writing out the momentum equation, the direction of R is designated and the appropriate sign is given to it. Solving for R and finding the sign of the answer is positive means that the direction chosen is the correct choice. However, if the assumption is incorrect, the sign that will be obtained will be negative. To yield the proper sign automatically, most engineers prefer always to write the initial momentum equation with a positive sense on the unknown; then the resulting sign will be the correct one. This latter procedure is the one that was followed in Example 6-4.

EXAMPLE 6-5   A 1-m-diameter pipe has a 30° horizontal bend in it, as shown above, and carries crude oil (sp. gr. = 0.94) at a rate of 2 m³/s. If the pressure in the bend is assumed to be constant at 75 kPa gage, if the volume of the bend is 1.2 m³, and if the metal in the bend weighs 4 kN, what forces must be applied to the bend to hold it in place?

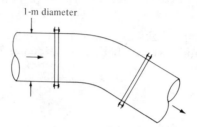

1-m diameter

Solution   Isolate the bend by a control surface as shown below and solve for the unknown forces R_x, R_y, and R_z using the momentum equation. First solve for the forces in the x direction. The flow is steady; therefore, the relevant equation is

$$\sum F_x = \sum \rho u \mathbf{V} \cdot \mathbf{A}$$

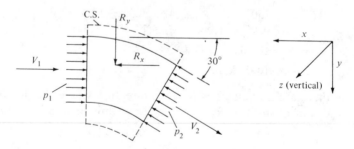

which for this example becomes

$$R_x - p_1 A_1 + 0.866 p_2 A_2 = V_{1x}\rho(-V_1 A_1) + V_{2x}\rho(V_2 A_2)$$

$$V_{1x} = \frac{-Q}{A_1} = -2.54 \text{ m/s} \qquad V_{2x} = \frac{-Q}{A_2}\cos 30° = -2.20 \text{ m/s}$$

$$R_x = +p_1 A_1 - 0.866 p_2 A_2 + \rho Q(-V_{1x} + V_{2x})$$

$$= 59 \times 10^3 - 51 \times 10^3 + 0.94 \times 10^3 \times 2(2.54 - 2.20)$$

$$= 8 \times 10^3 + 0.64 \times 10^3 \text{ N}$$

$$= 8.64 \text{ kN}$$

$\sum F_x = Q\rho(V_2 - V_1)$

For the y direction we have

$$\sum F_y = \sum v\rho \mathbf{V} \cdot \mathbf{A}$$

$$R_y - p_2 A_2 \sin 30° = V_{1y}\rho(-V_1 A_1) + V_{2y}\rho(V_2 A_2)$$

But $V_{1y} = 0$ and $V_{2y} = +\dfrac{Q}{A_2} \sin 30° = +1.27 \text{ m/s}$

$$R_y = +p_2 A_2 \sin 30° + V_{2y}\rho Q$$
$$= 29.5 \text{ kN} + 1.27 \times 0.94 \times 10^3 \times 2 \text{ N}$$
$$= 31.9 \text{ kN}$$

There are no components of velocity in the z direction (normal to the xy plane); hence, the momentum equation reduces to

$$\sum F_z = 0$$
$$R_z - W_B - W_f = 0$$
$$R_z = W_B + W_f$$

where W_B = weight of bend and W_f = weight of the fluid in bend.

$W_B = 4 \text{ kN}$ and $W_f = \rho g \forall$

where $\forall$ = volume of fluid in bend

so $W_f = 0.94 \times 1,000 \times 9.81 \times 1.2 = 11.1 \text{ kN}$

Thus $R_z = 4 \text{ kN} + 11.1 \text{ kN}$

$R_z = 15.1 \text{ kN}$

$\mathbf{R} = (8.64\mathbf{i} + 31.9\mathbf{j} + 15.1\mathbf{k}) \text{ kN}$ ◀

Motion of a rocket

A further interesting application of the momentum equation is the derivation of the equation of motion for a rocket accelerating along a vertical rectilinear path. Consider the rocket illustrated in Fig. 6-6. The exhaust gases exit through a nozzle with area A_e with a velocity V_e and a pressure in the exhaust stream of p_e. The pressure of the atmosphere is p_0. Let us take our control surface around the rocket and allow it to accelerate with the rocket. In regard to the accelerating control volume, it should be noted that our applications to this point have all been for nonaccelerating control volumes; i.e., the inertial frame to which velocities were referenced, to give momentum per unit mass, was the same reference frame for the control volume. However, if the control volume is accelerating, it simply means that we must go back to the basic control-volume equation,

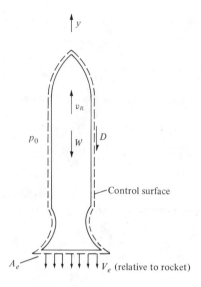

FIGURE 6-6

Eq. 4-21*a*, and make certain that the velocity (momentum per unit mass) used for β is referenced to the inertial frame of reference. Thus, when we apply the momentum equation in the vertical direction to the accelerating control volume, we have

$$\sum F_y = \sum_{cs} \mathbf{v}\rho\mathbf{V} \cdot \mathbf{A} + \frac{d}{dt}\int_{cv} \mathbf{v}\rho \, d\forall$$

where $\mathbf{v}$ refers to the velocity with respect to the vertical reference frame and $\mathbf{V}$ represents velocities measured with respect to the control volume, i.e., rocket. It is convenient to express the velocity $\mathbf{v}$ as a sum of the velocity of the rocket with respect to inertial space $\mathbf{v}_R$ and the velocity relative to the rocket $\mathbf{V}$:

$$\mathbf{v} = \mathbf{v}_R + \mathbf{V}$$

Substituting into the momentum equation above, we have for the y direction

$$\sum F_y = (v_R - V_e)\rho V_e A_e + \frac{d}{dt}\int_{cv} v_R\rho \, d\forall + \frac{d}{dt}\int_{cv} V\rho \, d\forall \quad (6\text{-}12)$$

where V_e is the exit velocity of the exhaust gases and the negative sign is taken because it is in the negative sense. Since v_R is constant over the space of the control volume, we have

$$\frac{d}{dt}\int_{cv} v_R\rho \, d\forall = \frac{d}{dt}\left[v_R \int_{cv} \rho \, d\forall \right] = \frac{d}{dt}(v_R M)$$

where M is the instantaneous mass of the rocket. The last term in Eq. (6-12) is zero because the relative velocity of the solid elements in the rocket is zero and the density and relative velocity of the gas flow in the rocket do not vary with time. The momentum equation then reduces to

$$\sum F_y = -\rho V_e^2 A_e + v_R \left(\rho_e V_e A_e + \frac{dM}{dt} \right) + M \frac{dv_R}{dt}$$

Applying the continuity equation to the control volume we have

$$\sum \rho \mathbf{V} \cdot \mathbf{A} + \frac{d}{dt} \int \rho \, d\forall = 0$$

$$\rho_e V_e A_e + \frac{dM}{dt} = 0$$

which further simplifies the momentum equation to

$$\sum F_y + \rho V_e^2 A_e = M \frac{dv_R}{dt}$$

The external forces on the rocket accrue from the pressure difference between the atmosphere and nozzle at the exit plane, the drag on the forward part of the rocket (upstream of the exit plane), and the force due to gravity.

$$-D - W + (p_e - p_0)A_e + \rho_e V_e^2 A_e = M \frac{dv_R}{dt}$$

The combination $(p_e - p_0)A_e + \rho_e V_e^2 A_e$ is the thrust T of the rocket engine, so the equation of motion becomes

$$-D - W + T = M \frac{dv_R}{dt} \qquad (6\text{-}13)$$

EXAMPLE 6-6 Neglecting the drag force, calculate the velocity a rocket will achieve after a 100-s burn if the thrust is 200 kN and the initial and final masses are 10,000 and 1,000 kg, respectively. Assume that the rocket's mass decreases linearly with time.

Solution Using Eq. (6-13), we have

$$T - Mg = M \frac{dv_R}{dt} \qquad (6\text{-}14)$$

$$\frac{T}{M} - g = \frac{dv_R}{dt}$$

The mass decreases linearly with time, so

$$M = M_0 - \lambda t$$

where λ has a value of 90 kg/s.

Substituting into Eq. (6-14) and integrating, we have

$$\int_0^t \frac{T\,dt}{M_0 - \lambda t} - \int_0^{t_b} g\,dt = v_R$$

or

$$v_R = \frac{T}{\lambda}\ln\frac{M_0}{M_f} - gt_b$$

The final result is

$$v_R = \frac{2 \times 10^5 \text{ N}}{90 \text{ kg/s}}\ln\frac{10{,}000 \text{ kg}}{1{,}000 \text{ kg}} - (9.81 \text{ m/s}^2)(100 \text{ s})$$

$$= 4{,}135 \text{ m/s} \qquad \blacktriangleleft$$

Force on rectangular sluice gate

A gate across a channel under which the water flows is called a sluice gate. In Fig. 6-7 the pressure distribution near the upper part of the sluice gate is approximately hydrostatic because the velocity of flow in this region is very low. However, near the bottom of the sluice gate the velocity is quite large; consequently, there is a large deviation from the hydrostatic pressure distribution. Because of this variable pressure on the sluice gate, it is not easy to determine the force on the gate by integration of the pressure over the gate surface. The simplest, although indirect, method to obtain the force is to apply the momentum equation to the liquid in the control volume shown in Fig. 6-7. Here the control surface is identified by the dashed line and the flow across the control surface on the upstream side and downstream side is assumed to be uniform and steady.

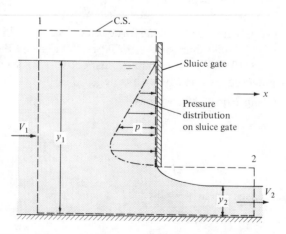

FIGURE 6-7   Flow under a sluice gate.

Hence the pressure distribution across both sections 1 and 2 will be hydrostatic. The upstream hydrostatic pressure will act against the water in the control volume in the positive x direction (to the right) and the downstream hydrostatic pressure will act to the left or with a negative sign. Then, for a sluice gate in a horizontal channel, the momentum equation for the x direction will yield the following:

$$\sum F_x = \sum_{cs} V_x \rho \mathbf{V} \cdot \mathbf{A}$$

$$\frac{\gamma y_1^2 b}{2} - \frac{\gamma y_2^2 b}{2} + F_{GW} - F_{visc} = V_1 \rho(-V_1 y_1 b) + \rho V_2(V_2 y_2 b) \quad (6\text{-}15)$$

where F_{GW} = force of gate acting on the water in control volume

F_{visc} = viscous force from bottom of channel

b = width of channel and gate

For most applications of this type the viscous force is negligible; therefore, it is easy to solve for the gate force given the discharge plus upstream and downstream depths. The procedure is to solve for F_{GW} and then change the sign on F_{GW} to obtain the force on the gate F_G:

$$F_G = \rho Q(V_1 - V_2) + \tfrac{1}{2} b \gamma(y_1^2 - y_2^2) \quad\quad\quad (6\text{-}16)$$

Water hammer—physical description

Whenever a valve is closed in a pipe, a positive pressure wave is created upstream of the valve that travels up the pipe at the speed of sound. In piping systems this pressure wave may be great enough to cause pipe failure; therefore, a basic understanding of this process, called water hammer, is necessary for the proper design and operation of such systems. The simplest case of water hammer will be considered here. For a more comprehensive treatment of the subject, the reader is referred to Streeter and Wylie (3).

Consider flow in the pipe of Fig. 6-8. Initially the valve at the end of the pipe is only partially open (Fig. 6-8a); consequently, an initial velocity V and initial pressure p_0 exist in the pipe. At time $t = 0$ it is assumed that the valve is instantaneously closed, thus creating the pressure wave that travels toward the reservoir with the speed of sound c. All the water between the pressure wave and the upper end of the pipe will have the initial velocity V, but all the water on the other side of the pressure wave (between the wave and the valve) will be at rest. This condition is shown in Fig. 6-8b. Once the pressure wave reaches the upper end of the pipe (after time $t = L/c$), it can be visualized that all of the water in the pipe will be under a pressure $p_0 + \Delta p$; however, the pressure in the reservoir

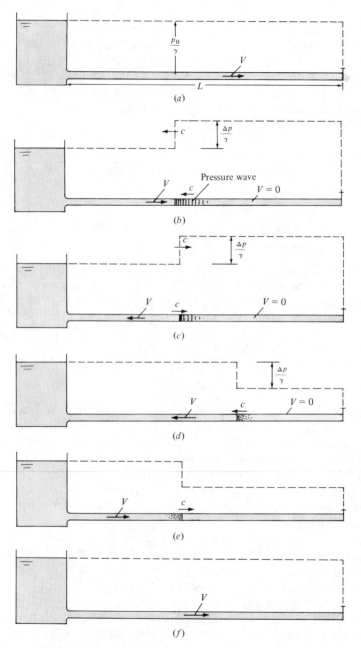

FIGURE 6-8 Water-hammer conditions. (*a*) Water-hammer process—initial condition. (*b*) Condition during time $0 < t < L/c$. (*c*) Condition during time $L/c < t < 2L/c$. (*d*) Condition during time $2L/c < t < 3L/c$. (*e*) Condition during time $3L/c < t < 4L/c$. (*f*) Condition at time $t = 4L/c$.

at the end of the pipe is only p_0. This imbalance of pressure at the reservoir end causes the water to flow from the pipe back into the reservoir with a velocity V. Thus a new pressure wave is formed that travels toward the valve end of the pipe (Fig. 6-8c). When this wave finally reaches the valve, all the water in the pipe is flowing toward the reservoir with a velocity V. However, such a condition is only momentary, because the closed valve prevents any sustained flow.

Next, $2L/c < t < 3L/c$, a rarefied wave of pressure travels up to the reservoir, as shown in Fig. 6-8d. When the wave reaches the reservoir, all the water in the pipe has a pressure less than that in the reservoir; therefore, this imbalance of pressure causes flow to be established again in the entire pipe, as shown in Fig. 6-8f, and the condition is exactly the same as in the initial condition (Fig. 6-8a); hence, the process will repeat itself in a periodic manner.

From the description given above it may be seen that the pressure in the pipe immediately upstream of the valve will be alternately high and low as shown in Fig. 6-9a. A similar observation for the pressure at the midpoint of the pipe will reveal a more complex variation of pressure with time as shown in Fig. 6-9b. Obviously, a valve cannot be closed instantaneously, and viscous effects, which were neglected here, will have a damping effect on the process; therefore, a more realistic pressure-time trace for the point just upstream of the valve is given in Fig. 6-9c. The finite time of closure erases the sharp discontinuities in the pressure trace that were

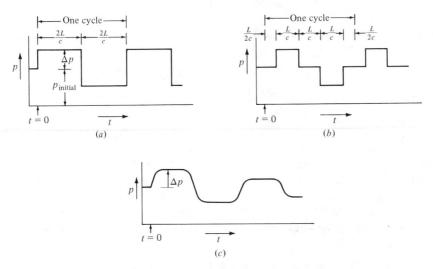

FIGURE 6-9 Variation of water-hammer pressure with time at two points in a pipe. (a) Location: adjacent to value. (b) Location: at midpoint of pipe. (c) Actual variation of pressure near valve.

present in Fig. 6-9a; however, it should be noted that the maximum pressure developed at the valve will be virtually the same as for instantaneous closure if the time of closure is less than $2L/c$. That is, the change in pressure will be the same for a given change in velocity unless the negative wave from the reservoir mitigates the positive pressure, and it takes a time $2L/c$ before this negative wave can reach the valve.

Magnitude of water-hammer pressure and speed of pressure wave

The quantitative relations for water hammer can be analyzed with the momentum equation by letting the control volume either move with the pressure wave, and thus creating steady motion, or be fixed, thus retaining the inherently unsteady character of the process. To illustrate the use of the momentum equation with unsteady motion the latter approach will be used. Consider a rigid pipe that includes the pressure wave, Fig. 6-10. The density, pressure, and velocity of the fluid on the reservoir side of the

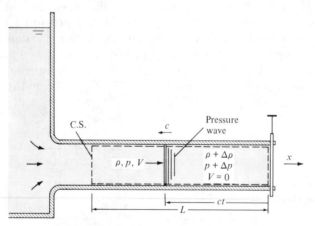

FIGURE 6-10   Pressure wave in a pipe.

pressure wave are ρ, p, and V, respectively, and the similar quantities on the valve side of the wave are $\rho + \Delta\rho$, $p + \Delta p$, and 0. Because the wave in this case is traveling from the valve to the reservoir, its distance from the valve at any time t will be given as ct. We can now apply the momentum equation to the flow in the control volume. Let the x direction be along the pipe. Then the momentum equation in the x direction will be given as

$$\sum F_x = \sum_{cs} V_x \rho \mathbf{V} \cdot \mathbf{A} + \frac{d}{dt} \int_{cv} V_x \rho \, d\forall$$

where
$$\sum F_x = pA - (p + \Delta p)A \qquad (6\text{-}17)$$

$$\sum_{cs} V_x \rho \mathbf{V} \cdot \mathbf{A} = V\rho(-VA) \qquad (6\text{-}18)$$

$$\frac{d}{dt} \int_{cv} V\rho \, d\forall = \frac{d}{dt}[V\rho(L - ct)A]$$
$$= -V\rho cA \qquad (6\text{-}19)$$

When Eqs. (6-17), (6-18), and (6-19) are substituted into the basic momentum equation, the following is obtained:

$$pA - (p + \Delta p)A = -\rho V^2 A - \rho VcA$$

which reduces to

$$\Delta p = \rho V^2 + \rho Vc$$

In the above equation the first term on the right side is usually negligible with respect to the second term on the right because for liquids c is much greater than V; consequently, the equation reduces to

$$\Delta p = \rho Vc \qquad (6\text{-}20)$$

We will now determine the speed of the pressure wave by applying the continuity equation to the control volume in Fig. 6-10. The basic continuity equation is given as

$$0 = \sum_{cs} \rho \mathbf{V} \cdot \mathbf{A} + \frac{d}{dt} \int_{cv} \rho \, d\forall$$

and when this is applied to Fig. 6-10 we obtain

$$0 = \rho(-VA) + \frac{d}{dt}[\rho(L - ct)A + (\rho + \Delta\rho)ctA]$$

which reduces to

$$\frac{\Delta\rho}{\rho} = \frac{V}{c}$$

or
$$c = \frac{V}{\Delta\rho/\rho} \qquad (6\text{-}21)$$

However, by definition $E_v = \Delta p/(\Delta\rho/\rho)$; therefore,

$$\frac{\Delta\rho}{\rho} = \frac{\Delta p}{E_v} \qquad (6\text{-}22)$$

Now when $\Delta\rho/\rho$ is eliminated between Eqs. (6-21) and (6-22) we have

$$c = \frac{VE_v}{\Delta p} \tag{6-23}$$

From Eq. (6-20), $\Delta p = \rho Vc$; therefore, Eq. (6-23) becomes

$$c = \sqrt{\frac{E_v}{\rho}} \tag{6-24}$$

Thus, by application of the momentum and continuity equations we have derived the equations for both Δp and c.

EXAMPLE 6-7   If the initial water velocity in a rigid pipe is 4 ft/sec (1.22 m/s) and if the initial pressure is 40 psi ($p = 276$ kPa gage), what maximum pressure will result with rapid closure of a valve in the pipe? Assume that $T = 60°F$ (16°C).

   Solution

$$\Delta p = \rho Vc$$

Here $\rho = 1.94$ slugs/ft³ (1,000 kg/m³)

                  $V = 4$ ft/sec (1.22 m/s)

and $c = \sqrt{\dfrac{E_v}{\rho}} = \sqrt{\dfrac{320,000 \times 144}{1.94}} = 4,800$ ft/sec

   SI units $E_v = 2.20$ GN/m²

so $c = \sqrt{2.20 \times (10^9 \text{ N/m}^2)/(10^3 \text{ kg/m}^3)} = 1.48 \times 10^3$ m/s

Therefore, $\Delta p = 1.94 \times 4 \times 4,800 = 37,200$ psf

                  $= 259$ psi

   SI units $\Delta p = 1000$ kg/m³ $\times 1.22$ m/s $\times 1.48 \times 10^3$ m/s

                  $= 1.81$ MN/m² $= 1.81$ MPa

Then the maximum pressure will be the initial pressure plus the pressure change, which is

                  $p_{max} = 40 + 259 = 299$ psi   ◄

   SI units $p_{max} = 0.276$ MPa $+ 1.81$ MPA $= 2.08$ MPa   ◄

EXAMPLE 6-8   If the pipe of Example 6-7 leads from the reservoir and is 4,000 ft long, what is the maximum time of closure for the generation of the maximum pressure of 299 psi, as given in Example 6-7?

Solution

$$t_c = \frac{2L}{c}$$

where t_c = maximum time of closure (critical time)
to produce maximum pressure

L = length of pipe = 4,000 ft

c = speed of pressure wave = 4,800 ft/sec

Then $t_c = 2 \times \dfrac{4,000}{4,800} = 1.67$ sec   ◄

As indicated by Example 6-7, water-hammer pressures can be quite large; therefore, engineers must design piping systems to keep the pressure within acceptable design limits. This is done by installing an accumulator near the valve and/or operating the valve so that rapid closure is prevented. Accumulators may be in the form of air chambers for relatively small systems, or surge tanks (large open tank connected by a branch pipe to the main pipe) for cases such as large hydropower systems. Another way to eliminate excessive water-hammer pressures is to install pressure-relief valves at critical points in the pipe system. These valves are pressure actuated so that water is automatically diverted out of the system when the water-hammer pressure reaches excessive levels.

6-3 MOMENT-OF-MOMENTUM EQUATION

The moment-of-momentum equation relates the moment applied to a system with the rate of change of angular momentum of the system and the flow of angular momentum across the control surface. Because the phenomenon involves the rate of change of a physical property, we can employ the basic control-volume approach. The basic control-volume equation is

$$\frac{dB_{sys}}{dt} = \sum_{cs} \beta\rho\mathbf{V} \cdot \mathbf{A} + \frac{d}{dt}\int_{cv} \beta\rho\, d\forall \qquad (4\text{-}21a)$$

Here, if we let $\beta = \mathbf{r} \times \mathbf{V}$ where $\times$ denotes the vector product of $\mathbf{r}$ and $\mathbf{V}$, then the extensive property B_{sys} will be

$$B_{sys} = \int (\mathbf{r} \times \mathbf{V})\rho\, d\forall = \text{angular momentum of the system} \quad (6\text{-}25)$$

When Eq. (6-25) and $\beta = \mathbf{r} \times \mathbf{V}$ are substituted back into Eq. (4-21a), we obtain

$$\frac{d[\int (\mathbf{r} \times \mathbf{V})\rho \, d\forall]}{dt} = \sum_{cs} (\mathbf{r} \times \mathbf{V})\rho \mathbf{V} \cdot \mathbf{A} + \frac{d}{dt} \int_{cv} (\mathbf{r} \times \mathbf{V})\rho \, d\forall \quad (6\text{-}26)$$

Equation (6-26) then states that the rate of change with respect to time of the angular momentum of the system is equal to the net rate of angular momentum from the control volume plus the rate of change of angular momentum within the control volume. Here the origin for the angular momentum is the origin of the position vector $\mathbf{r}$. From basic mechanics it is known that the net moment $\Sigma \mathbf{M}$ applied to a system is equal to the rate of change of angular momentum of the system; therefore, the left side of Eq. (6-26) will be replaced by $\Sigma \mathbf{M}$. We then have

$$\sum \mathbf{M} = \sum_{cs} (\mathbf{r} \times \mathbf{V})\rho \mathbf{V} \cdot \mathbf{A} + \frac{d}{dt} \int_{cv} (\mathbf{r} \times \mathbf{V})\rho \, d\forall \quad (6\text{-}27a)$$

Equation (6-27a) is the basic moment-of-momentum equation for cases where the velocity distribution is uniform across the flow section. When the velocity is variable across the flow section, we then use the more general form:

$$\sum \mathbf{M} = \int_{cs} (\mathbf{r} \times \mathbf{V})\rho \mathbf{V} \cdot \mathbf{dA} + \frac{d}{dt} \int_{cv} (\mathbf{r} \times \mathbf{V})\rho \, d\forall \quad (6\text{-}27b)$$

EXAMPLE 6-9   In Example 6-4, if the bend is to be supported from above by an external system of supports, what moment about an axis that is level with the top of the pipe and 30 cm to the right of the flange must the support system be designed for? Assume that the center of mass of the bend and the water taken together is 30 cm below the top of the pipe and 50 cm to the right of the flanges of the bend.

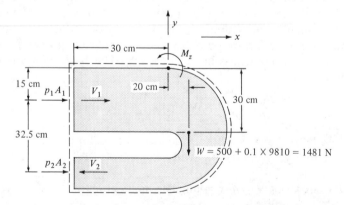

Solution   We draw the control surface as shown and apply Eq. (6-27a). We have steady flow so the last term of Eq. (6-27a) drops out. Then we have

$$\sum M_{z\,\text{axis}} = \sum_{cs} rV\rho\mathbf{V} \cdot \mathbf{A}$$

Here the moments will be from the pressure forces, weight of the bend, weight of the water, and external moment M_z applied at the axis in question. When these moments, the velocities, and their moment arms, are inserted in the equation above we get

$$
\begin{aligned}
M_z + 0.15p_1A_1 &+ 0.475p_2A_2 - 0.2 \times 1{,}481 = (0.15 \times 3.54)(1{,}000) \\
&\times [-3.54 \times 0.3^2 \times (\pi/4)] + (-0.475 \times 14.15)(1{,}000) \\
&\qquad\qquad\times [14.15 \times 0.15^2 \times (\pi/4)] \ \text{N} \cdot \text{m}
\end{aligned}
$$

Here $p_1 = 150$ kPa and $p_2 = 59.3$ kPa, so we solve for M_z:

$$
\begin{aligned}
M_z &= -0.15 \times 150 \times 10^3 \times 0.3^2 \times (\pi/4) - 0.475 \times 59.3 \times 10^3 \\
&\quad \times 0.15^2 \times (\pi/4) + 0.2 \times 1{,}481 - 133 - 1{,}681 \\
&= -1{,}590 - 498 + 296 - 133 - 1{,}681 \\
&= -3.61 \ \text{kN} \cdot \text{m} \qquad\qquad\blacktriangleleft
\end{aligned}
$$

Thus a moment of 3,610 N · m applied in the clockwise sense about the axis in question is needed. Stated differently, the support system must be designed to withstand a counterclockwise moment of 3,610 N · m.

EXAMPLE 6-10 Determine the power produced by a simple sprinkler-like turbine that rotates in a horizontal plane at 500 rpm. The radius of the turbine is 0.5 m. Water enters the turbine from a vertical pipe that is coaxial with the axis of rotation and the fluid exits through 10-cm² nozzles with a velocity of 50 m/s with respect to the nozzle. The water density is 1,000 kg/m³ and the pressure at the exit is atmospheric.

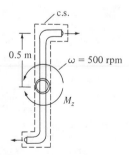

Solution We draw a control surface around the turbine as shown taking the z axis as the axis of rotation. Mass enters the control volume at the hub and exits at the nozzle. Applying the moment of momentum equation to the control volume gives

$$M_z = \sum (r \times V)\rho\mathbf{V} \cdot \mathbf{A}$$

since the flow is steady. The moment of momentum associated with the flow into the hub is zero because $r = 0$. Thus the moment of momentum equation reduces to

$$M_z = (r \times V)\dot{m}$$

where $\dot{m}$ is the mass efflux from the nozzles. The velocity V must be referenced to an inertial frame so

$$r \times V = -r(V_e - \omega r)$$

where r is the radius and V_e is the exit velocity with respect to the nozzle. Thus the moment applied to the system (that is, to the arms) by the generator is $-r(V_e - \omega r)\dot{m}$ but the moment applied to the generator is the negative of the moment of the generator on the arms, or

$$M_z = (0.5)[50 - 500 \times 2\pi \times (0.50/60)](50)(10^{-4})(1,000) \times 2$$
$$= 119 \text{ N} \cdot \text{m}$$

The power produced by the turbine-generator is the product of the moment and angular velocity:

$$P = M_z\omega = \frac{(500)(2\pi)(119)}{60}$$

$$= 6.24 \text{ kW} \qquad \blacktriangleleft$$

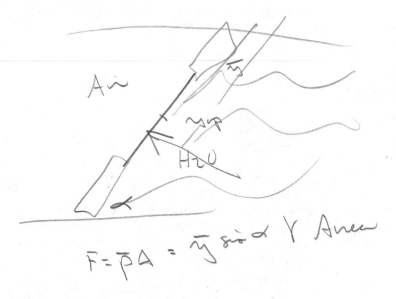

PROBLEMS

6-1 A horizontal water jet impinges upon a vertical-perpendicular plate. The discharge is 0.05 m³/s. If the external force required to hold the plate in place is 400 N, what is the water speed?

6-2 A horizontal water jet impinges upon a vertical-perpendicular plate. The discharge is 2 cfs. If the external force required to hold the plate in place is 120 lbf, what is the velocity of the water?

6-3 A horizontal water jet issues from a circular orifice in a large tank. The jet strikes a vertical plate that is normal to the plane of the jet. A force of 500 lbf is needed to hold the plate in place against the action of the jet. If the pressure in the tank is 9 psi, what is the diameter of the jet just downstream of the orifice?

6-4 A horizontal water jet issues from a circular orifice in a large tank. The jet strikes a vertical plate that is normal to the plane of the jet. A force of 2.0 kN is needed to hold the plate in place against the action of the jet. If the pressure in the tank is 70 kPa, what is the diameter of the jet just downstream of the orifice?

6-5 A horizontal water jet issues from a circular orifice in a large tank. The discharge in the jet is 0.3 m³/s and it strikes a vertical plate that is normal to the plane of the jet. What horizontal force is required to hold the plate in position if the pressure in the tank is 50 kPa?

6-6 A horizontal water jet issues from a circular orifice in a large tank. The discharge in the jet is 10 cfs and it strikes a vertical plate that is normal to the plane of the jet. What horizontal force is required to hold the plate in position if the pressure in the tank is 6 psi?

6-7 A jet of water is discharging at a constant rate of 1.20 cfs from the upper tank. If the jet diameter at section 1 is 4 in., what forces will be measured by scales A and B? Assume the empty tank weighs 200 lbf and the cross-sectional area of the tank is 4 ft². $h = 9$ ft and $H = 1$ ft.

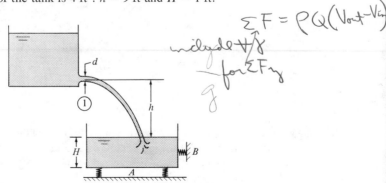

PROBLEMS 6-7, 6-8

6-8 A jet of water is discharging at a constant rate of 0.04 m³/s from the upper

tank. If the jet diameter at section 1 is 4 in., what forces will be measured by
by scales *A* and *B*? Assume the empty tank weighs 800 N and the cross-
sectional area of the tank is 1 m². $h = 3$ m and $H = 30$ cm.

6-9 Determine the external reactions in the *x* and *y* directions needed to hold this
fixed vane, which turns the oil jet in a horizontal plane. Here V_1 is 90 ft/sec,
$V_2 = 85$ ft/sec, and $Q = 3$ cfs.

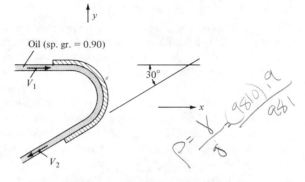

PROBLEMS 6-9, 6-10

6-10 Determine the external reaction in the *x* and *y* directions needed to hold the
fixed vane that turns the oil jet in a horizontal plane. Here $V_1 = 25$ m/s,
$V_2 = 23$ m/s, and $Q = 0.1$ m³/s.

6-11 A 3-in.-diameter jet of water having a velocity of 50 ft/sec is deflected by the
vane as shown. If the vane is moving at a rate of 10 ft/sec in the *x* direction,
what components of force are exerted on the vane by the water in the *x* and *y*
directions?

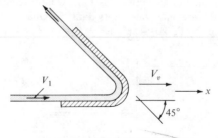

PROBLEMS 6-11, 6-12

6-12 A 6-cm-diameter jet of water having a velocity of 20 m/s is deflected by the
vane as shown. If the vane is moving at a rate of 4 m/s in the *x* direction,
what components of force are exerted on the vane by the water in the *x* and *y*
directions?

6-13 The water in this 3-in.-diameter jet has a speed of 100 ft/sec to the right and
is deflected by a cone that is moving to the left with a speed of 40 ft/sec. Determine the external horizontal force needed to move the cone.

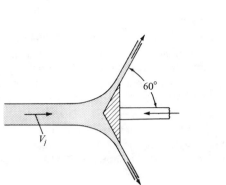

PROBLEMS 6-13, 6-14

6-14 The water in this 10-cm-diameter jet has a speed of 30 m/s to the right and is deflected by a cone that is moving to the left with a speed of 15 m/s. Determine the external horizontal force needed to move the cone.

6-15 This two-dimensional liquid jet impinges upon the horizontal floor. Derive formulas for d_2 and d_3 as a function of b_1 and θ. Assume that the jet speed is large enough to cause the gravitational effects to be nil. Also assume that the speed of the liquid is the same at sections 1, 2, and 3.

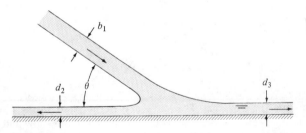

6-16 Assume that a scoop as shown, which is 8 in. wide, is used as a braking device for studying deceleration effects, such as those on space vehicles. If the scoop is attached to a 2,000 lbm sled that is initially traveling horizontally at the rate of 350 ft/sec, what will be the initial deceleration of the sled? The scoop dips into the water 3 in. ($d = 3$ in.).

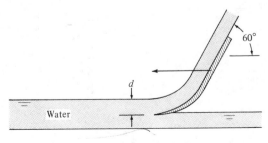

PROBLEMS 6-16, 6-17

6-17 Assume that the scoop as shown, which is 20 cm wide, is used as a braking device for studying deceleration effects, such as those on space vehicles. If the scoop is attached to a 1000-kg sled that is initially traveling horizontally at the rate of 100 m/s, what will be the initial deceleration of the sled? The scoop dips into the water 8 cm (d = 8 cm).

6-18 This snowplow "cleans" a swath of snow that is 3 in. deep (d = 3 in.) and 2 ft wide (B = 2 ft). The snow leaves the blade in the direction indicated in the sketches. Neglecting friction between the snow and blade, estimate the power required for just the snow removal if the speed of the snowplow is 40 ft/sec.

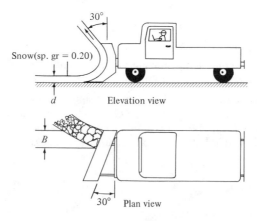

Snow(sp. gr = 0.20)

d Elevation view

B

30° Plan view

PROBLEMS 6-18, 6-19

6-19 This snowplow "cleans" a swath of snow that is 8 cm deep (d = 8 cm) and 60 cm wide (B = 60 cm). The snow leaves the blade in the direction indicated in the sketches. Neglecting friction between the snow and blade, estimate the power required for just the snow removal if the speed of the snowplow is 12 m/s.

6-20 A large tank of liquid is resting on a frictionless plane as shown. If the cap is removed from the short pipe, explain in a qualitative way what will happen thereafter.

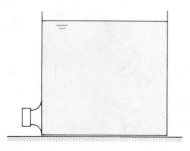

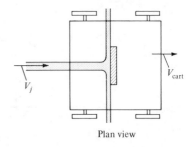

V_j

V_{cart}

Plan view

PROBLEM 6-20

PROBLEMS 6-21, 6-22

6-21 A 2-in.-diameter water jet with a velocity $V_j = 50$ ft/sec causes the cart to move at a constant rate of speed of 10 ft/sec. What is the rolling resistance of the cart?

6-22 A 6-cm-diameter water jet with a velocity $V_j = 15$ m/s causes the cart to move at a constant speed of 3 m/s. What is the rolling resistance of the cart?

6-23 A discharge Q of a very low viscosity liquid drops vertically into a short horizontal-rectangular channel of width B as shown. The depth at section 2 is y_2. Derive an equation that gives y_1 in terms of y_2, Q, B, and γ.

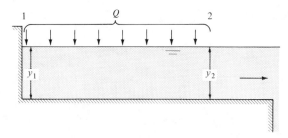

6-24 A disk that is held stable by a wire is free to move in the vertical and has a jet of water striking it from below. The disk weighs 6 lbf. The initial speed of the jet is 40 ft/sec and the initial jet diameter is 1 in. Find the height to which the disk will rise and remain stationary. *Note:* The wire is only for stability and should not enter into your calculations.

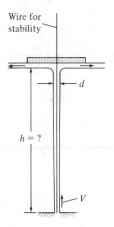

PROBLEMS 6-24, 6-25

6-25 A disk that is held stable by a wire is free to move in the vertical and has a jet of water striking it from below. The disk weighs 30 N. The initial speed of the jet is 10 m/s and the initial jet diameter is 3 cm. Find the height to which the disk will rise and remain stationary. *Note:* The wire is only for stability and should not enter into your calculations.

6-26 A 6-in.-horizontal pipe has a 180° bend in it. If the rate of flow of water in the bend is 5 cfs and the pressure therein is 20 psi, what external force in the original direction of flow is required to hold the bend in place?

6-27 The diameter D is 1 ft and the pressure at the center of the upper pipe is 10 psi. If the flow in the bend is 12 cfs, what external force will be required to hold the bend in place against the action of the water? The bend weighs 100 lbf and the volume of the bend is 3 ft³. Neglect head losses.

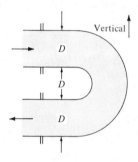

PROBLEMS 6-27, 6-28

6-28 The diameter D is 1 m, the spacing between the upper and lower pipe is also 1 m, and the pressure at the center of the upper pipe is 70 kPa. If the flow in the bend is 10 m³/s, what external force in newtons and pounds will be required to hold the bend in place against the action of the water? Determine both the horizontal and vertical components. The bend weighs 5,000 N and the volume in the bend is 1 m³. Neglect losses.

6-29 The gage pressure in the horizontal 90° pipe bend is 300 kPa. If the pipe diameter is 1 m and the water flow rate is 10 m³/s, what x component of force must be applied to the bend to hold it in place against the water action?

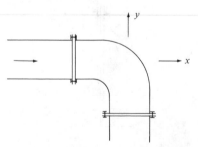

6-30 This 30° vertical bend in a 2-ft-diameter pipe carries water at a rate of 31.4 cfs. If the pressure p_1 at the lower end of the bend is 10 psi where the elevation is 100 ft, and p_2 at the upper end is 8 psi where the elevation is 103 ft, what will be the vertical component of force that must be exerted by the "anchor" on the bend to hold it in position? The bend itself weighs 300 lb and the length L is 4 ft.

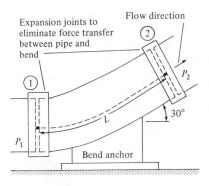

Expansion joints to eliminate force transfer between pipe and bend

Flow direction

Bend anchor

PROBLEMS 6-30, 6-31

6-31 This 30° vertical bend in a 60-cm-diameter pipe carries water at a rate of 1.0 m³/s. If the pressure p_1 at the lower end of the bend is 70 kPa gage where the elevation is 100 m, and p_2 at the upper end is 56 kPa gage where the elevation is 101 m, what will be the vertical component of force that must be exerted by the "anchor" on the bend to hold it in position? The bend itself weighs 1.2 kN and the length L is 130 cm.

6-32 A 90° horizontal bend reduces from 2-ft-diameter upstream to 1-ft-diameter downstream. If the bend is discharging water into the atmosphere and the pressure upstream is 25 psi, what is the magnitude of the component of external force exerted on the bend in the x direction (direction parallel to the initial flow direction) required to hold the bend in place?

6-33 A 90° horizontal bend reduces from 60-cm-diameter upstream to 30-cm-diameter downstream. If the bend is discharging water into the atmosphere and the pressure upstream is 140 kPa gage, what is the magnitude of the component of external force exerted on the bend in the x direction (direction parallel to the initial flow direction) required to hold the bend in place?

6-34 This is a plan view of a horizontal bend. If water is flowing at a rate of 15 cfs and if $D_1 = 12$ in. and $D_2 = 10$ in., estimate the x component of force (through the bolts in the flange) needed to hold it in place. *Note*: The jet from the bend discharges into the atmosphere.

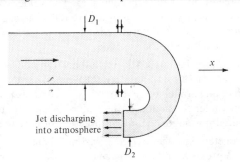

Jet discharging into atmosphere

PROBLEMS 6-34, 6-35

6-35 This is a plan view of a horizontal bend. If water is flowing at a rate of 4 m³/s and if $D_1 = 50$ cm, and $D_2 = 40$ cm, estimate the x component of force (through the bolts in the flange) needed to hold it in place. *Note*: The jet from the bend discharges into the atmosphere.

6-36 This bend discharges water into the atmosphere. Determine the magnitude and direction of the external force components at the flange required to hold the bend in place. The bend shown lies in a horizontal plane and water is flowing. Assume that the interior volume of the bend is 7 ft³. $D_1 = 2$ ft, $D_2 = 1$ ft, and $V_2 = 15$ ft/sec.

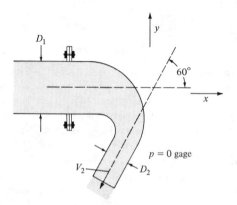

PROBLEMS 6-36, 6-37

6-37 Determine the magnitude and direction of the external force components at the flange required to hold the bend in place. The bend shown lies in a horizontal plane and water is flowing. Assume that the interior volume of the bend is 0.25 m³. $D_1 = 60$ cm, $D_2 = 30$ cm, and $V_2 = 10$ m/s.

6-38 A 1-ft-diameter pipe bends through an angle of 135°. The velocity of flow of gasoline (sp. gr. = 0.8) is 20 ft/sec and the pressure is 10 psi in the bend. What external force is required to hold the bend against the action of the gasoline? Neglect the gravitational force.

6-39 A 30-cm-diameter pipe bends through 135°. The velocity of flow of gasoline (sp. gr. = 0.8) is 8 m/s and the pressure is 100 kPa gage in the bend. Neglecting gravitational force, determine the external force required to hold the bend against the action of the gasoline.

6-40 A 1-ft-diameter pipe has a 135° horizontal bend in it. The pipe carries water under a pressure of 20 psi at a rate of 7.85 cfs. What external force component in a direction parallel to the initial flow direction is necessary to hold the bend in place under the action of the water? What horizontal force component is required normal to the initial direction of flow to hold the bend?

6-41 A 40-cm-diameter pipe has a 135° horizontal bend in it. The pipe carries water under a pressure of 80 kPa gage at a rate of 0.40 m³/s. What external force component in a direction parallel to the initial flow direction is necessary to hold the bend in place under the action of the water? What horizontal force component is required normal to the initial direction of flow to hold the bend?

6-42 A horizontal reducing bend turns a flow of water through 90°. The inlet diameter is 20 cm and the outlet diamter is 10 cm. The flow rate is 0.1 m³/s and the upstream pressure is 100 kPa gage. Neglecting body forces and assuming irrotational flow calculate the external force required to hold the bend in place.

6-43 This 130-cm overflow pipe from a small hydroelectric plant conveys water from the 70-m elevation to the 40-m elevation. The pressure in the water at the bend entrance and exit is 20 kPa and 25 kPa, respectively. The bend interior volume is 3 m³ and the bend itself weighs 10 kN. Determine the force that a thrust block must exert on the bend to secure it if the discharge is 15 m³/s.

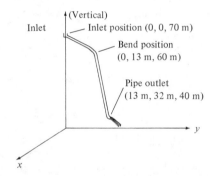

6-44 For this Y fitting, which lies in a horizontal plane, the cross-sectional areas at sections 1, 2, and 3 are 1 ft², 1 ft², and 0.3 ft², respectively. At these same respective sections pressures are 1,000 psfg, 900 psfg, and 0, and the water discharges are 20 cfs to the right, 11 cfs to the right, and 9 cfs. What forces will have to be applied by the flange bolts to hold it in place?

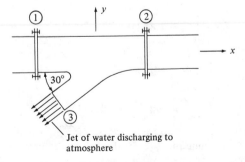

PROBLEMS 6-44, 6-45

6-45 For this Y fitting, which lies in a horizontal plane, the cross-sectional areas at sections 1, 2, and 3 are 0.1 m², 0.1 m², and 0.03 m³, respectively. At these same respective sections the pressures are 50 kPa gage, 45 kPa gage, and zero gage, and the water discharges are 0.6 m³/s to the right, 0.3 m³/s to the right, and 0.3 m³/s. What forces will have to be applied by the flange bolts to hold it in place?

6-46 For this horizontal tee through which water is flowing, the following data are given: $Q_1 = 0.25$ m³/s, $Q_2 = 0.15$ m³/s, $p_1 = 100$ kPa, $p_2 = 70$ kPa, $p_3 = 80$ kPa, $D_1 = 15$ cm, $D_2 = 10$ cm, and $D_3 = 15$ cm. For these conditions what external force in the x-y plane (through the bolts or other supporting devices) is needed to hold the tee in place?

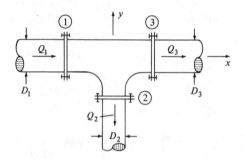

PROBLEMS 6-46, 6-47

6-47 For this horizontal tee through which water is flowing, the following data are given: $Q_1 = 7.85$ cfs, $Q_2 = 4.00$ cfs, $p_1 = 15$ psig, $p_2 = 10$ psig, $p_3 = 12$ psig, $D_1 = 6$ in., $D_2 = 4$ in., and $D_3 = 6$ in. For these conditions what external force in the x-y plane (through bolts or other supporting devices) is needed to hold the tee in place?

6-48 Water flows through this nozzle at a rate of 10 cfs and discharges into the atmosphere. Determine the force at the flange required to hold the nozzle in place. Assume irrotational flow. $D_1 = 12$ in. and $D_2 = 6$ in.

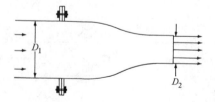

PROBLEMS 6-48, 6-49

6-49 Water flows through this nozzle at a rate of 0.3 m³/s and discharges into the atmosphere. Determine the force at the flange required to hold the nozzle in place. Assume irrotational flow. $D_1 = 30$ cm and $D_2 = 15$ cm.

6-50 This "double" nozzle discharges water into the atmosphere at a rate of 15.7 cfs. If the nozzle is lying in a horizontal plane, what x component of force

acting through the flange bolts is required to hold the nozzle in place? *Note*:
Assume irrotational flow and the water speed in each jet to be the same. Jet
A is 4 in. in diameter, jet B is 4.5 in. in diameter, and the pipe is 1 ft in diameter.

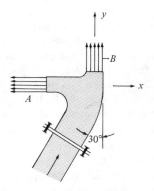

PROBLEMS 6-50, 6-51

6-51 This "double" nozzle discharges water into the atmosphere at a rate of 0.50
m³/s. If the nozzle is lying in a horizontal plane, what x component of force
acting through the flange bolts is required to hold the nozzle in place? *Note*:
Assume irrotational flow and the water speed in each jet to be the same. Jet
A is 10 cm in diameter, jet B is 12 cm in diameter, and the pipe is 30 cm in
diameter.

6-52 Plate A is 1 ft in diameter and has a sharp-edged orifice at its center. A water
jet strikes the plate concentrically with a speed of 100 ft/sec. With the plate
held stationary, what external force is needed to hold the plate in place if the
jet issuing from the orifice also has a speed of 100 ft/s? The jet diameters are
D = 4 in. and d = 2 in.

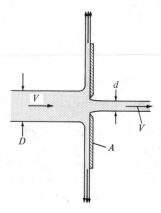

PROBLEMS 6-52, 6-53

6-53 Plate A is 50 cm in diameter and has a sharp-edged orifice at its center. A water jet strikes the plate concentrically with a speed of 30 m/s. With the plate held stationary, what external force is needed to hold the plate in place if the jet issuing from the orifice also has a speed of 30 m/s? The jet diameters are $D = 10$ cm and $d = 5$ cm.

6-54 A 1-ft horizontal pipe carries water to three jets discharging into the atmosphere from the end of the pipe. The velocity of water in the jets is 40 ft/sec. One jet discharges parallel to the pipe and has a 2-in. diameter. The second jet is inclined upward and in the direction of flow in the pipe at an angle of 30° with respect to the pipe and has a 1-in. diameter. The third jet discharges vertically upward and has a diameter of 3 in. If an upstream section in the pipe is considered, what horizontal component of force must be transmitted through the metal at this section to hold the end in place?

6-55 A 30-cm horizontal pipe carries water to three jets discharging into the atmosphere from the end of the pipe. The velocity of water in the jets is 15 m/s. One jet discharges parallel to the pipe and has a 5-cm diameter. The second jet is inclined upward and in the direction of flow in the pipe at an angle of 30° with respect to the pipe and has a 2-cm diameter. The third jet discharges vertically upward and has a diamter of 7 cm. If an upstream section in the pipe is considered, what horizontal component of force must be transmitted through the metal at this section to hold the end in place?

6-56 This spray head discharges water a rate of 10 cfs. Assuming irrotational flow and an efflux speed of 50 ft/sec in free jet, what force is needed to keep the spray head on the 1-ft-diameter pipe? Neglect gravitational forces in this problem.

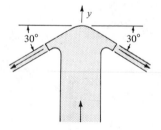

PROBLEMS 6-56, 6-57

6-57 This spray head discharges water at a rate of 0.3 m³/s. Assuming irrotational flow and an efflux speed of 20 m/s in the free jet, what force is needed to keep the spray head on the 30-cm pipe? Neglect gravitational forces in this problem.

6-58 Two circular water jets of 1-in. diameter ($d = 1$ in.) issue from this unusual nozzle. If the efflux speed is 80.2 ft/sec, what force is required at the flange to hold the nozzle in place? The pressure in the 4 in. ($D = 4$ in.) pipe is 43 psig.

6-59 Two circular water jets issue from this unusual nozzle. The jet diameters, d,

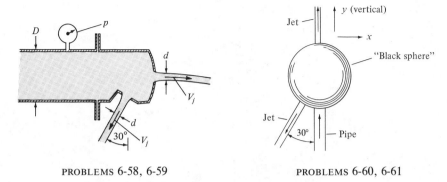

PROBLEMS 6-58, 6-59   PROBLEMS 6-60, 6-61

are 5 cm each, and the pipe diameter, D, is 20 cm. If the efflux speed, v_j, is 30 m/s, what force is required at the flange to hold the nozzle in place? Assume irrotational flow conditions prevail. Neglect the force of gravity.

6-60 Liquid (sp. gr. = 1.5) enters the "black sphere" through a 2-in. pipe with velocity of 40 ft/sec and a pressure of 60 psig. It leaves the sphere through two jets as shown. The velocity in the vertical jet is 100 ft/sec and its diameter is 1 in. The other jet diameter is also 1 in. What force through the 2-in. pipe wall is required in the x and y directions to hold the sphere in place? Assume the sphere plus liquid inside it weighs 100 lb.

6-61 Liquid (sp. gr. = 1.5) enters the "black sphere" through a 5-cm pipe with velocity of 10 m/s and a pressure of 400 kPa. It leaves the sphere through two jets as shown. The velocity in the vertical jet is 30 m/s and its diameter is 25 mm. The other jet diameter is also 25 mm. What force through the 5-cm pipe wall is required in the x and y directions to hold the sphere in place? Assume the sphere plus liquid inside it weighs 500 N.

6-62 What force is required to hold this "black box" in place if gravitational forces are neglected and if the liquid's specific gravity is 2.0?

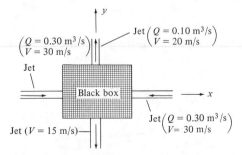

6-63 The "black box" and associated equipment weighs 100 lbf and when operating as shown contains 3 ft³ of water. Assuming that steady flow prevails, what external force must be applied to the black box in the y direction (vertical) to keep it in equilibrium?

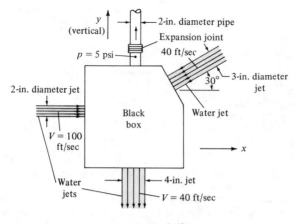

PROBLEM 6-63

6-64 Neglecting viscous resistance, determine the force of the water per unit of width acting on a sluice gate for which the upstream depth is 3 ft and the downstream depth is 0.6 ft.

6-65 Neglecting viscous resistance, determine the force of the water per unit of width acting on a sluice gate for which the upstream depth is 1 m and the downstream depth is 15 cm.

6-66 A 15-cm nozzle is bolted with six bolts to the flange of a 30-cm pipe. If water discharges from the nozzle into the atmosphere, calculate the tension load in each bolt when the pressure in the pipe is 280 kPa. Assume irrotational flow.

6-67 A nozzle at the end of a 3-in. hose produces a jet $1\frac{1}{2}$ in. in diameter. Determine the longitudinal force in the joint at the base of the nozzle required to hold the nozzle when it is discharging 300 gpm of water.

6-68 A nozzle on a 3-cm-diameter garden hose has an outlet diameter of 15 mm. Determine the longitudinal force at the nozzle-hose joint required to hold the nozzle for a discharge of 0.005 m³/s.

6-69 If water is discharged from the two-dimensional slot shown at the rate of 4 cfs per foot of slot, determine the pressure p at the gage and determine the

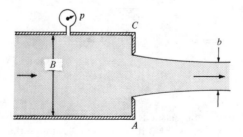

PROBLEMS 6-69, 6-70

water force per foot on the vertical end plates AC. The slot and jet dimensions B and b are 8 in. and 3 in., respectively.

6-70 If water is discharged from the two-dimensional slot shown at the rate of 0.40 m^3/s per meter of slot, determine the pressure p at the gage and determine the water force per meter on the vertical end plates AC. The slot and jet dimensions B and b are 20 cm and 7 cm, respectively.

6-71 The propeller on a swamp boat produces a slipstream 3 ft in diameter with a velocity relative to the boat of 80 ft/sec. If the air temperature is 80°F, what is the propulsive force when the boat is not moving and also when its forward speed is 20 ft/sec? *Hint*: Assume that the pressure, except in the immediate vicinity of the propeller, is atmospheric.

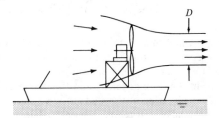

PROBLEMS 6-71, 6-72

6-72 The propeller on a swamp boat produces a slipstream 1 m in diameter with a velocity relative to the boat of 30 m/s. If the air temperature is 30°C, what is the propulsive force when the boat is not moving and also when its forward speed is 10 m/s?

6-73 A torpedolike device is tested in a wind tunnel with air density of 1 kg/m^3. The tunnel is 1 m in diameter, the upstream pressure is 5 kPa gage, and the downstream pressure is 1 kPa gage. If the mean air velocity, V, is 30 m/s, what is the mass rate of flow and what is the maximum velocity at the downstream section at C? If the pressure is assumed to be constant across the sections at A and C, what is the drag of the device and supporting vane? Assume viscous resistance at the wall is negligible.

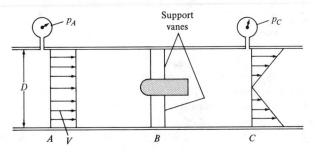

PROBLEMS 6-73, 6-74

6-74 A torpedolike device is tested in a wind tunnel with air density of 0.0030 slugs/ft³. The tunnel is 3 ft in diameter, the upstream pressure is 0.24 psig, and the downstream pressure is 0.10 psig. If the mean air velocity, V, is 100 ft/sec, what is the mass rate of flow and what is the maximum velocity at the downstream section at C? If the pressure is assumed to be constant across the sections at A and C, what is the drag of the device and supporting vane? Assume viscous resistance at the wall is negligible.

6-75 Consider a tank of water in a container that rests on a sled. A high pressure is maintained by a compressor so that the water leaving the tank through the orifice does so at a constant speed of 50 ft/sec relative to the tank. If there is 3 ft³ of water in the tank at time t and the area of the jet is 0.5 in.², what will be the acceleration of the sled at time t if the empty tank and compressor have a weight of 70 lbf and the coefficient of friction between the sled and the ice is 0.05?

6-76 Consider a tank of water in a container that rests on a sled. A high pressure is maintained by a compressor so that the water leaving the tank through the orifice does so at a constant speed of 20 m/s relative to the tank. If there is 0.10 m³ of water in the tank at time t and the diameter of the jet is 15 mm, what will be the acceleration of the sled at time t if the empty tank and compressor have a weight of 350 N and the coefficient of friction between the sled and the ice is 0.05?

6-77 The hemicircular nozzle sprays a sheet of liquid through 180° of arc as indicated. The velocity is V at the efflux section where the sheet thickness is t. Derive a formula for the external force F (in the y direction) required to hold the nozzle system in place. This force should be a function of ρ, V, and t.

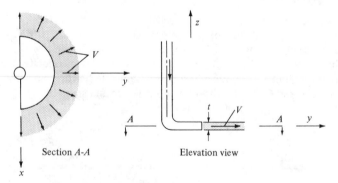

Section A-A Elevation view

6-78 A 3-in. orifice is located in a plate at the end of a 6-in. pipe. Water flows through the pipe and orifice at a rate of 3 cfs. If the water jet downstream of the orifice is 2.5 in. in diameter, what is the external force required to hold the orifice in place?

6-79 A 4-cm orifice is located in a plate at the end of a 10-cm pipe. Water flows through the pipe and orifice at 0.05 m³/s. The diameter of the water jet downstream of the orifice is 3.5 cm. Calculate the external force required to hold the orifice in place.

6-80 Water is discharged from the slot in the pipe as shown. If the resulting two-dimensional jet is 100 cm long and 15 mm thick, and if the pressure at section *A-A* is 30 kPa, what is the reaction at section *A-A*? In this calculation do not consider the weight of the pipe.

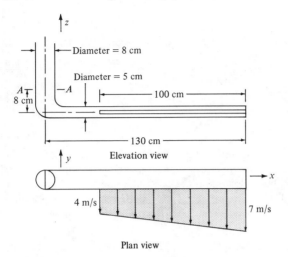

Elevation view

Plan view

6-81 What is the reaction at section 1? Water is flowing in the system. Neglect gravitational forces.

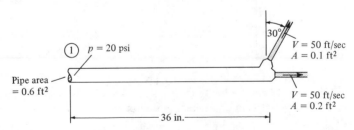

6-82 What is the reaction at section 1? Water is flowing and the axes of the two jets lie in a vertical plane. The pipe and nozzle system weigh 90 N.

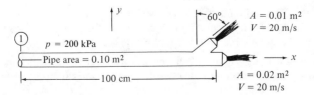

6-83 A reducing pipe bend is held in place by a pedestal as shown. There are expansion joints at sections 1 and 2 so that no force is transmitted through the pipe past these sections. If the pressure at section 1 is 20 psi and if the rate of flow of water is 2 cfs, what is the resultant force system which must be transmitted through the base of the pedestal at section 3? That is, determine the twisting and bending moments on the pedestal as well as the force components at section 3 due to the dynamic effect of the water. Assume that the flow is irrotational.

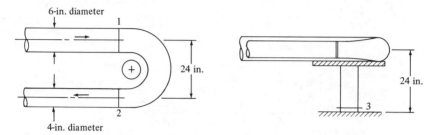

6-84 The figure illustrates the principle of the jet pump. Derive a formula for $p_2 - p_1$ as a function of d_j, V_j, D_0, V_0, and ρ. Assume that the fluid from the jet and initially flowing in the pipe have the same density and assume they are completely mixed at section ② so that the velocity is uniform across that section. Also assume that the pressures are constant across both sections ① and ②.

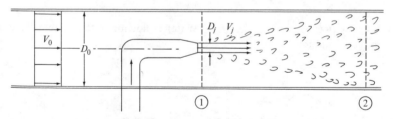

6-85 Estimate the maximum water-hammer pressure that is generated in a rigid pipe if the initial water velocity is 2 m/s and the pipe is 10 km long with a valve at the downstream end that is closed in 10 s.

6-86 The length of a 20-cm rigid pipe carrying 0.1 m³/s of water is estimated by instantaneously closing a valve at the downstream end and noting the time required for the pressure fluctuation to complete a cycle. If the time interval is 3 s, what is the pipe length?

6-87 Estimate the maximum water-hammer pressure that is generated in a rigid pipe if the initial water velocity is 5 ft/sec and the pipe is 5 miles long with a valve at the downstream end, which is closed in 10 sec.

6-88 A 4-km-long 12-cm rigid pipe discharges water at the rate of 0.02 m³/s. If a valve at the end of the pipe is closed in 3 s, what is the maximum force resulting from the pressure rise which will be exerted on the valve? Assume that the water temperature is 10°C.

6-89 By letting the control volume move with the water-hammer wave, steady-flow conditions are established. By use of the momentum and continuity equations and using the steady-flow approach, derive Eq. (6-20).

6-90 The 60-cm pipe carries water with an initial velocity, V_0, of 0.10 m/s. If the valve at C is instantaneously closed at time $t = 0$, then what does the pressure versus time trace look like at point B for the next 5 s? Graph your results and indicate significant quantitative relations or values from $t = 0$ to $t = 5$ s. What does the pressure versus position along the pipe look like at $t = 1.5$ s. Plot your results and indicate the velocity or velocities in the pipe.

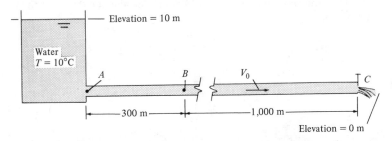

6-91 Steady flow initially occurs in this 1-m steel pipe. There is a rapid-acting valve at the end of the pipe at point B and there are pressure transducers at both points A and B. If the valve is closed at B and the p versus t traces are made as shown, estimate the initial discharge and length from A to B.

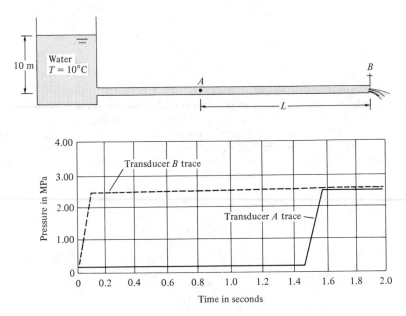

6-92 It is common practice in rocket-trajectory analyses to neglect the body-force term and drag, so the velocity at burnout is given by

$$v_{bo} = \frac{T}{\lambda} \ln \frac{M_0}{M_f}$$

Taking a thrust-to-mass-flow ratio of 3,000 N · s/kg and a final mass of 50 kg, calculate the initial mass needed to establish the rocket in an earth orbit at a velocity of 7,200 m/s.

6-93 A rocket is designed to have four nozzles, each canted at 30° with respect to the rocket's centerline. The gases exit at 2,000 m/s through an exit area of 1 m². The density of the exhaust gases is 0.3 kg/m³, and exhaust pressure is 50 kPa. The atmospheric pressure is 10 kPa. Determine the thrust on the rocket in newtons.

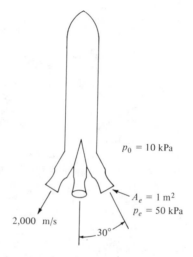

6-94 A rocket-nozzle designer is concerned about the force required to hold the nozzle section on the body of a rocket. The nozzle section is shaped as shown in the figure. The pressure and velocity at the entrance to the nozzle are 1.5 MPa and 100 m/s. The exit pressure and velocity are 80 kPa and 2,000 m/s. The mass flow through the nozzle is 220 kg/s. The atmospheric pressure is 100 kPa. The rocket is not accelerating. Calculate the force on the nozzle-chamber connection.

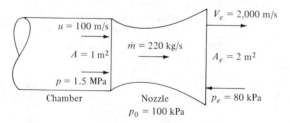

6-95 The expansion section of rocket nozzles are often conical in shape; and because the flow diverges, the thrust derived from the nozzle is less than would be obtained if the exit velocity were everywhere parallel to the nozzle axis. By considering the flow through the spherical section subtended by the cone and assuming that the exit pressure is equal to the atmospheric, show that the thrust is given by

$$T = \dot{m} V_e \frac{(1 + \cos \alpha)}{2}$$

where $\dot{m}$ is the mass flow through the nozzle, V_e is the exit velocity, and α is the nozzle half-angle.

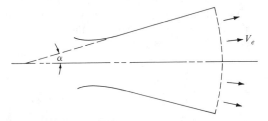

6-96 Show how the momentum equation can be applied to derive Euler's equation for the flow of inviscid fluids (Eq. 5-3). *Hint:* Select an arbitrary control volume of length Δs enclosed by a stream tube in an unsteady, nonuniform flow as shown. The volume of the control volume is $[A + (\partial A/\partial s)(\Delta s/2)]\Delta s$. First derive the continuity equation by applying the continuity principle to the flow through the control volume. Then apply the momentum equation along the stream-tube direction and use the continuity equation to reduce it to Euler's equation.

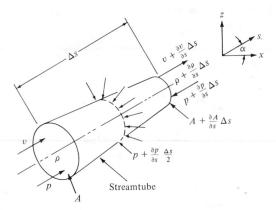

6-97 A ramjet can be represented by the duct shown below. Air enters from the

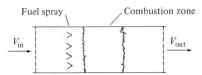

left. A fuel spray is mixed with the air and burned in the combustion zone. Hot gas exits at high speed to the right. Calculate the thrust if the diameter of the duct is 50 cm, the inlet velocity is 200 m/s, the inlet density is 1 kg/m³, and the exit velocity is 500 m/s.

6-98 Two small liquid-propellant rocket motors are mounted at the tips of a helicopter rotor to augment power under emergency conditions. The diameter of the helicopter rotor is 8 m and rotates at 1 rev/s. The air enters at the tip speed of the rotor and exhaust gases exit at 500 m/s with respect to the rocket motor. The intake area of each motor is 20 cm² and the air density is 1.2 kg/m³. Calculate the power provided by the rocket motors. Neglect the mass rate of flow of fuel in this calculation.

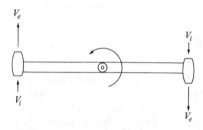

6-99 A windmill is operating in a 10 m/s wind with a density of 1.2 kg/m³. The diameter of the windmill is 4 m. The constant pressure (atmospheric) streamline has a diamter of 3 m upstream of the windmill and 4.5 downstream. Assume the velocity distributions are uniform and the air is incompressible. Determine the thrust on the windmill.

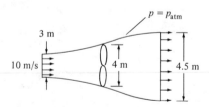

REFERENCES

1. Daily James W., and Harleman, D. R. F. *Fluid Dynamics*. Addison-Wesley Publishing Company, Inc., Reading, Mass., 1966.

2. Shames, Irving. *Mechanics of Fluids*. McGraw-Hill Book Company, New York, 1962.

3. Streeter, V. L., and Wylie, E. B. *Hydraulic Transients*. McGraw-Hill Book Company, New York, 1967.

4. Sutton, G. P. *Rocket Propulsion Elements*. John Wiley & Sons, Inc., New York, 1963.

7 ENERGY PRINCIPLE

To THIS POINT, we have been concerned with the mechanical forces (pressure, gravity, shear stress) on a fluid. The energy equation allows us to incorporate the thermal energies, as well. In the early nineteenth century, J. P. Joule carried out a number of tests that verified the general principle of the conservation of energy that had been previously hypothesized. From this was developed the *first law of thermodynamics*, which can be written for a given system (given quantity of matter) as follows:

$$\Delta E = Q - W$$

Here Q is the heat transferred to the system in a given time t and W is the work done by the system on its surroundings in this same interval of time.[1] The energy E of a system can take a variety of forms, such as *kinetic* and *potential* energy of the system as a whole and energy associated with motion of the molecules. The latter includes energy involved with the structure of the atom, chemical energy, and electrical energy. It is convenient to consider kinetic energy E_k and potential energy E_p separately and lump all other energies into a single term called *internal* energy E_u. Thus the total energy of the system is

$$E = E_u + E_k + E_p$$

Now, if we want the rate of change of E with time, the differential form of the first law of the thermodynamics is given as follows:

$$\frac{dE}{dt} = \dot{Q} - \dot{W}$$

In the next section, the first law of thermodynamics along with the control-volume approach will be used to develop the energy equation for fluid flow.

7-1 DERIVATION OF THE ENERGY EQUATION

Control-volume approach applied to the first law of thermodynamics

The energy E introduced above refers to the total energy of the system; thus E is an extensive property of the system. Then the corresponding intensive property (energy per unit of mass) is given by e, which is made up

---

[1] Heat transferred to the system and work done by the system are defined, by convention, to be positive quantities. Heat transferred from the system and work done on the system are negative quantities.

of e_k, e_p, and u. dE/dt is the rate of change of energy of a system with respect to time; consequently, the basic control-volume equation may be applied. In applying the control-volume equation, Eq. (4-21a), we let $B_{sys} = E$ and $\beta = e$ to obtain

$$\frac{dE}{dt} = \frac{d}{dt} \int_{cv} e\rho \, d\forall + \sum_{cs} e\rho \mathbf{V} \cdot \mathbf{A} \tag{7-1}$$

However, from the first law of thermodynamics, $dE/dt = \dot{Q} - \dot{W}$; consequently, this substitution is made for dE/dt in Eq. (7-1) to yield

$$\dot{Q} - \dot{W} = \frac{d}{dt} \int_{cv} e\rho \, d\forall + \sum_{cs} e\rho \mathbf{V} \cdot \mathbf{A} \tag{7-2}$$

When e is replaced by its equivalent, $e_k + e_p + u$, the following is obtained:

$$\dot{Q} - \dot{W} = \frac{d}{dt} \int_{cv} (e_k + e_p + u)\rho \, d\forall + \sum_{cs} (e_k + e_p + u)\rho \mathbf{V} \cdot \mathbf{A} \tag{7-3}$$

Several terms on the right of Eq. (7-3) are still too general for practical application; therefore, let us examine these carefully and express them in terms of variables associated with the flow of fluids.

The kinetic energy per unit of mass e_k is given by the total kinetic energy of mass having velocity V divided by the mass itself, or

$$e_k = \frac{\Delta M V^2 / 2}{\Delta M} = \frac{V^2}{2} \tag{7-4}$$

The potential energy per unit of mass e_p is given by $E_p / \Delta M$, where E_p is the product of weight and elevation of the centroid of the incremental mass. In this case, the potential energy is referenced to the datum from which elevation is measured. Then

$$e_p = \frac{\gamma \, \Delta \forall z}{\Delta M} = \frac{\gamma \, \Delta \forall z}{\rho \, \Delta \forall} = gz \tag{7-5}$$

When Eqs. (7-4) and (7-5) are substituted into Eq. (7-3) we obtain

$$\dot{Q} - \dot{W} = \frac{d}{dt} \int_{cv} \left(\frac{V^2}{2} + gz + u \right) \rho \, d\forall$$

$$+ \sum_{cs} \left(\frac{V^2}{2} + gz + u \right) \rho \mathbf{V} \cdot \mathbf{A} \tag{7-6}$$

On the left side of this equation we have the two terms $\dot{Q}$ and $\dot{W}$, which are the rate of flow of heat into the system and the rate of work done by the system on its surroundings, respectively. For convenience of analysis, work is divided into shaft work W_s and flow work W_f. These are discussed next.

Flow work

Flow work is the work done by pressure forces as the system moves through space. Let us consider the basic figure (Fig. 7-1) depicting the system and control volume to get a better understanding of flow work.

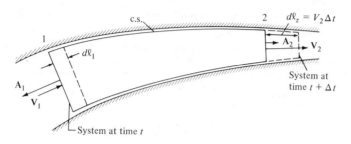

FIGURE 7-1

Consider the area A_2, which is the right end of the fluid system. The force acting upon the surrounding fluid will be $p_2 A_2$, and the distance traveled by the area in the time Δt will be $\Delta \ell = V_2 \, \Delta t$. The work done on the surrounding fluid from this force in time Δt will be the product of the component of force in the direction of motion ($p_2 A_2$) and the distance traveled by the area ($V_2 \, \Delta t$). Hence the flow work done by the system on the surrounding fluid in time Δt by the downstream end of the system will be

$$\Delta W_{f_2} = V_2 p_2 A_2 \, \Delta t$$

and the rate at which flow work is done on the area will be obtained by dividing through by Δt:

$$\dot{W}_{f_2} = V_2 p_2 A_2 \qquad (7\text{-}7)$$

It can also be seen from Fig. 7-1 that $V_2 A_2 = \mathbf{V}_2 \cdot \mathbf{A}_2$; consequently, Eq. (7-7) reduces to

$$\dot{W}_{f_2} = p_2 \mathbf{V}_2 \cdot \mathbf{A}_2$$

In a similar manner it can be shown that the flow work done on the surrounding fluid by the upstream end of the system will be $-p_1 V_1 A_1$. A negative sign occurs here because the pressure force on the surrounding fluid acts in a direction opposite to the motion of the system boundary. The rate at which work is done on the surrounding fluid by the upstream end of the system can also be given in terms of the scalar product:

$$\dot{W}_{f_2} = p_1 \mathbf{V}_1 \cdot \mathbf{A}_1$$

A negative rate of work results from this product because the velocity vector $\mathbf{V}_1$ and the area vector $\mathbf{A}_1$ have opposite sense; thus the scalar product has a negative sign. Hence all system surfaces that are moving (represented by streams of fluid passing across the control surface) do work on the surrounding fluid according to the expression

$$\dot{W}_f = p\mathbf{V} \cdot \mathbf{A} \tag{7-8}$$

Then the rate at which flow work is done on the system's surroundings is obtained by summing Eq. (7-8) for all streams passing through the control surface:

$$\dot{W}_f = \sum_{cs} p\mathbf{V} \cdot \mathbf{A} \tag{7-9}$$

Later in this derivation, simplification will result if the quantity under the summation symbol, Eq. (7-9), is multiplied and divided by ρ; hence we obtain

$$\dot{W}_f = \sum_{cs} \frac{p}{\rho} \rho\mathbf{V} \cdot \mathbf{A} \tag{7-10}$$

Shaft work

Shaft work is defined as any work other than flow work. This is usually in the form of a shaft (from which the term originates) which either takes energy out of the system or puts energy into the system. In the first case, the system does work on a mechanism (such as a turbine blade) attached to the shaft. In the second case, the shaft is attached to a mechanism (such as a pump impeller) that does work on the system. In this latter case, to retain our original terminology for W, we say that the fluid system is doing negative work on its surroundings.

Basic form of the energy equation

If we now substitute for $\dot{W}$ in Eq. (7-6) the sum of the shaft-work rate $\dot{W}_s$ and flow-work rate, Eq. (7-10), the following form of the energy equation for fluid flow is obtained:

$$\dot{Q} - \dot{W}_s - \sum_{cs} \frac{p}{\rho} \rho\mathbf{V} \cdot \mathbf{A}$$

$$= \frac{d}{dt} \int_{cv} \left(\frac{V^2}{2} + gz + u \right) \rho \, d\forall + \sum_{cs} \left(\frac{V^2}{2} + gz + u \right) \rho\mathbf{V} \cdot \mathbf{A} \tag{7-11}$$

It may be seen that the Σ term on the left side is the same form as the second term on the right side. Hence, these two terms may be combined as follows:

$$\dot{Q} - \dot{W}_s = \frac{d}{dt} \int_{cv} \left(\frac{V^2}{2} + gz + u \right) \rho \, d\forall$$

$$+ \sum_{cs} \left(\frac{p}{\rho} + \frac{V^2}{2} + gz + u \right) \rho \mathbf{V} \cdot \mathbf{A} \quad (7\text{-}12a)$$

Equation (7-12a) is the basic form of the energy equation.

More general form of the energy equation

If the velocity distribution across a flow section is not uniform, then it is necessary to use the more general form of the control-volume equation, Eq. (4-21b), in the derivation. The steps in the derivation are essentially the same as given above except that integration of the flow past the control surface results in the following more general form of the energy equation:

$$\dot{Q} - \dot{W}_s = \frac{d}{dt} \int_{cv} \left(\frac{V^2}{2} + gz + u \right) \rho \, d\forall$$

$$+ \int_{cs} \left(\frac{p}{\rho} + \frac{V^2}{2} + gz + u \right) \rho \mathbf{V} \cdot d\mathbf{A} \quad (7\text{-}12b)$$

7-2   DISCUSSION OF THE ENERGY EQUATION

Even though we have derived Eq. (7-12a) for particular applicability to fluids, it is also valid for a solid mass. Consider a block of metal at rest that is being heated. For this case, $\dot{Q}$ will be a positive quantity, and all quantities on the right-hand side will drop out except $d/dt \int_{cv} u\rho \, d\forall$. In other words, the internal energy will increase at the same rate as heat is supplied to the mass; and this will of course be reflected in the temperature rise of the metal.

Now let us take the case in which there is no heat transfer and the rate of flow is increasing in a straight pipe of constant diameter. Let us also assume that an incompressible nonviscous fluid is flowing. Figure 7-2 illustrates the physical situation. Since there is no heat flow and no shaft work involved, then the left side of Eq. (7-12a) will be zero.

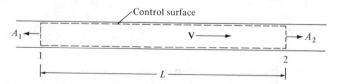

FIGURE 7-2   Acceleration of fluid in a pipe.

The velocity, elevation, and internal energy are assumed to be uniform throughout the control volume. The pressure is the only quantity that changes along the pipe at a given instant; thus Eq. (7-12a) reduces to

$$0 = \sum_{cs} \frac{p}{\rho} \rho \mathbf{V} \cdot \mathbf{A} + \frac{d}{dt} \int_{cv} \frac{V^2}{2} \rho \, d\forall \qquad (7\text{-}13)$$

Note that the first term on the right reduces to $-p_1 V_1 A_1 + p_2 V_2 A_2$ and the last term can be expressed as

$$\frac{d}{dt} \left(\frac{V^2}{2} \rho A L \right) = \rho A L V \frac{dV}{dt} \qquad (7\text{-}14)$$

But $V_1 A_1 = V_2 A_2 = VA$, so when the appropriate substitutions are made in the basic equation and it is simplified, we get

$$F_1 - F_2 = M \frac{dV}{dt} \qquad (7\text{-}15)$$

which is Newton's second law. In other words, in the absence of thermal energies the energy equation reduces to a form of the momentum equation.

7-3 SIMPLIFIED FORMS OF THE ENERGY EQUATION

Steady-flow energy equation for a fixed control volume

When Eq. (7-12a) is written for steady flow, we have

$$\dot{Q} - \dot{W}_s = \sum_{cs} \left(\frac{p}{\rho} + \frac{V^2}{2} + gz + u \right) \rho \mathbf{V} \cdot \mathbf{A} \qquad (7\text{-}16)$$

An alternative form of Eq. (7-16) is the following:

$$\dot{Q} - \dot{W}_s = \sum_{cs} \left(\frac{V^2}{2} + gz + h \right) \rho \mathbf{V} \cdot \mathbf{A} \qquad (7\text{-}17)$$

where $h = p/\rho + u$ is defined as the enthalpy of the fluid.

Equations (7-16) and (7-17) are much simpler than Eqs. (7-12) because in their application we are no longer concerned with the fluid mass inside the control surface; Eqs. (7-16) and (7-17) pertain only to the rate of transfer of heat and energy across the control surface plus the rate at which work is done by the system on its surroundings.

EXAMPLE 7-1   A steam turbine receives superheated steam at 1.4 MPa absolute and 400°C, which corresponds to a specific enthalpy of 3,121 kJ/kg. The steam leaves the turbine at 101 kPa absolute and 100°C, for which the specific enthalpy is 2,676 kJ/kg. The steam enters the turbine at 15 m/s and exits at 60 m/s. The elevation between entry and exit ports is negligible. The heat loss through the turbine wall is 7,600 kJ/h. Calculate the power output if the mass flow through the turbine is 0.5 kg/s.

Solution   First sketch the general layout of the turbine indicating the inlet and outlet velocities, shaft work, and heat transfer.

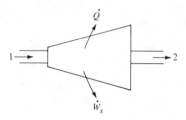

Writing down the energy equation for this turbine and neglecting the elevation terms gives

$$\dot{Q} - \dot{W}_s = \left(\frac{V_2^2}{2} + h_2\right)\rho_2 V_2 A_2 - \left(\frac{V_1^2}{2} + h_1\right)\rho_1 V_1 A_1$$

However, $$\rho_1 V_1 A_1 = \rho_2 V_2 A_2 = \dot{m}$$

from the conservation of mass, so we have

$$\dot{W}_s = \dot{Q} + \dot{m}\left(\frac{V_1^2}{2} - \frac{V_2^2}{2} + h_1 - h_2\right)$$

We must now be careful to check units before substituting numbers into the equation and evaluating $\dot{W}_s$. The units for enthalpy are joules per kilogram and for velocity squared, meters squared per second squared. One finds that meters squared per second squared is equivalent to joules per kilogram:

$$\frac{m^2}{s^2} = \frac{kg \cdot m^2}{kg \cdot s^2} = \frac{kg \cdot m}{s^2}\frac{m}{kg} = \frac{N \cdot m}{kg} = \frac{J}{kg}$$

Substituting the numbers into the energy equation and using the correct units gives

$$W_s = \frac{-7,600}{3,600}\frac{kJ}{h}\frac{h}{s} + 0.5\frac{kg}{s}\left[\frac{15^2 - 60^2}{2 \times 10^3}\frac{kJ}{kg} + (3,121 - 2,676)\frac{kJ}{kg}\right]$$

$$= -2.11 + 0.5(-1.69 + 445)$$

$$= 220\frac{kJ}{s} = 220\ kW \qquad \blacktriangleleft$$

Typically, in this type of problem the kinetic-energy term is negligible compared with the enthalpy difference.

Energy equation for steady one-dimensional flow in a pipe

Consider flow through the pipe system shown in Fig. 7-3. Here the magnitude of the velocity is variable across the flow sections; thus we use the more general form of the energy equation [Eq. (7-12b)]. When Eq. (7-12b) is written between sections 1 and 2 and when the flow quantities relating to section 1 are transferred to the left side of the equation, we obtain

$$\dot{Q} - \dot{W}_s + \int \left(\frac{p_1}{\rho} + gz_1 + u_1\right) \rho_1 V_1 \, dA_1 + \int \frac{\rho_1 V_1^3}{2} \, dA_1$$

$$= \int \left(\frac{p_2}{\rho} + gz_2 + u_2\right) \rho_2 V_2 \, dA_2 + \int \frac{\rho_2 V_2^3}{2} \, dA_2 \quad (7\text{-}18)$$

At sections 1 and 2 where the flow is uniform, hydrostatic conditions prevail across the section; therefore, $(p/\rho + gz)$ is constant across the section.[1] At any section the internal energy is also virtually constant across the section; therefore, $(p/\rho + gz + u)$ can be taken outside the integral sign to yield

$$\dot{Q} - \dot{W}_s + \left(\frac{p_1}{\rho_1} + gz_1 + u_1\right) \int \rho_1 V_1 \, dA_1 + \int \rho_1 \frac{V_1^3}{2} \, dA_1$$

$$= \left(\frac{p_2}{\rho_2} + gz_2 + u_2\right) \int \rho_2 V_2 \, dA_2 + \int \rho_2 \frac{V_2^3}{2} \, dA_2 \quad (7\text{-}19)$$

It can be seen that $\int \rho V \, dA = \rho \bar{V} A = \dot{m}$, the mass rate of flow; consequently, some simplification will result if we divide through by $\dot{m}$. How-

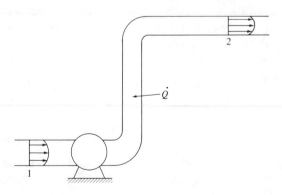

FIGURE 7-3

[1] The pressure is hydrostatically distributed across the section because the flow is assumed to be uniform at each section. For uniform flow, the streamlines are straight and parallel; hence there is no acceleration normal to the streamlines and the pressure is thus hydrostatically distributed. For hydrostatic conditions $(p/\gamma + z) = $ const, see page 36; therefore, if we multiply $(p/\gamma + z)$ by the constant g we get $(p/\rho + gz)$, which is also constant.

ever, $\dot{m}$ does not appear as a factor of $\int(\rho V^3/2)dA$; so it is common to express $\int(\rho V^3/2)dA$ as $\alpha(\rho \bar{V}^3/2)A$. Then when we factor out $\dot{m}$ from each of these terms, Eq. (7-19) becomes

$$\dot{Q} - \dot{W}_s + \left(\frac{p_1}{\rho_1} + gz_1 + u_1 + \alpha_1 \frac{\bar{V}_1^2}{2}\right) \dot{m}$$

$$= \left(\frac{p_2}{\rho_2} + gz_2 + u_2 + \alpha_2 \frac{\bar{V}_2^2}{2}\right) \dot{m} \quad (7\text{-}20)$$

When we divide through by $\dot{m}$, we get

$$\frac{1}{\dot{m}}(\dot{Q} - \dot{W}_s) + \frac{p_1}{\rho_1} + gz_1 + u_1 + \alpha_1 \frac{\bar{V}_1^2}{2} = \frac{p_2}{\rho_2} + gz_2 + u_2 + \alpha_2 \frac{\bar{V}_2^2}{2} \quad (7\text{-}21)$$

The coefficients α_1 and α_2 are kinetic-energy-correction factors and are evaluated by the original expressions in which they were first introduced:

$$\alpha \frac{\rho \bar{V}^3 A}{2} = \int_A \frac{\rho V^3 \, dA}{2} \quad (7\text{-}22)$$

or

$$\alpha = \frac{1}{A} \int_A \left(\frac{V}{\bar{V}}\right)^3 dA \quad (7\text{-}23)$$

Thus $\alpha = 1$ when the velocity is constant across the section and it is greater than 1 for nonuniform velocity distributions. Computations show that $\alpha = 2$ for laminar flow in a pipe where the velocity has a parabolic distribution across the section. For most cases of turbulent flow, $\alpha \approx 1.05$; and because this is quite close to unity, it is common practice in engineering applications to let $\alpha_1 = \alpha_2 = 1$. A similar correction factor is used with the momentum flux terms in the momentum equation for one-dimensional flow; however, it deviates even less from unity than does α for a given velocity distribution.

EXAMPLE 7-2   The velocity distribution for laminar flow in a pipe is given by the equation

$$V = V_{\max} \left[1 - \left(\frac{r}{r_0}\right)^2\right]$$

Here r_0 is the radius of the pipe and r is the radial distance from the center. Determine $\bar{V}$ in terms of $V_{\max}$ and evaluate the kinetic energy correction factor α.

Solution   A sketch for the velocity distribution is shown in the figure. The discharge is given by $Q = \int V \, dA$, or

$$Q = \bar{V}A = \int_0^{r_0} V_{\max} \left(1 - \frac{r^2}{r_0^2}\right) 2\pi r \, dr$$

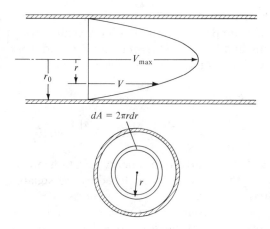

$dA = 2\pi r\,dr$

or

$$\bar{V} = \frac{-\pi}{A}\, r_0^2\, V_{\max} \left.\frac{(1 - r^2/r_0^2)^2}{2}\right|_0^{r_0}$$

$$= \frac{1}{2}\, V_{\max} \qquad \blacktriangleleft$$

The mean velocity is one-half the maximum velocity. To evaluate α we apply Eq. (7-23):

$$\alpha = \frac{1}{\pi r_0^2}\int_0^{r_0} \frac{V_{\max}^3(1 - r^2/r_0^2)^3}{(\frac{1}{2})^3 V_{\max}^3}\, 2\pi r\, dr$$

$$= 2 \qquad \blacktriangleleft$$

The shaft-work term in Eq. (7-21) is usually the result of a turbine or pump in the flow system. When fluid passes through a turbine, the fluid system is doing shaft work on the surroundings; on the other hand, a pump does work on the fluid. It is convenient, then, to represent the shaft-work term as

$$\dot{W}_s = \dot{W}_t - \dot{W}_p$$

where $\dot{W}_t$ and $\dot{W}_p$ are magnitudes of power (work per unit time) delivered by a turbine or supplied to a pump. Substituting the expression above for shaft work into Eq. (7-21) and dividing by g results in

$$\frac{\dot{W}_p}{\dot{m}g} + \frac{p_1}{\gamma} + z_1 + \alpha_1 \frac{\bar{V}_1^2}{2g} = \frac{\dot{W}_t}{\dot{m}g} + \frac{p_2}{\gamma} + z_2$$

$$+ \alpha_2 \frac{\bar{V}_2^2}{2g} + \frac{u_2 - u_1}{g} - \frac{\dot{Q}}{\dot{m}g} \qquad (7\text{-}24)$$

Now all of the terms of Eq. (7-24) have the single dimension, length. Hence, we designate the first term involving shaft work as h_p, head supplied by a pump, and the second term involving shaft work h_t, head given up to a turbine. Equation (7-24) is now written as

$$\frac{p_1}{\gamma} + \alpha_1 \frac{\bar{V}_1^2}{2g} + z_1 + h_p = \frac{p_2}{\gamma} + \alpha_2 \frac{\bar{V}_2^2}{2g} + z_2$$

$$+ h_t + \left[\frac{1}{g}(u_2 - u_1) - \frac{\dot{Q}}{\dot{m}g} \right] \quad (7\text{-}25)$$

At this point we have separated the energy equation into a mechanical and a thermal part, with the last term enclosed by the square brackets representing the thermal energy.

In the flow process, some of the system's mechanical energy is converted to thermal energy through viscous action between fluid particles; consequently, there may be a finite increase in internal energy of the fluid between sections 1 and 2, or some of the heat that is generated through energy dissipation may escape from the pipe $(-\dot{Q})$. In any event, the bracketed terms in Eq. (7-25) represent a loss of mechanical energy due to viscous stresses. This lost energy is usually simply lumped together in a single term called *head loss* and identified by the symbol h_L.[1] Thus, Eq. (7-25) becomes

$$\frac{p_1}{\gamma} + \alpha_1 \frac{\bar{V}_1^2}{2g} + z_1 + h_p = \frac{p_2}{\gamma} + \alpha_2 \frac{\bar{V}_2^2}{2g} + z_2 + h_t + h_L \quad (7\text{-}26)$$

In Chapter 10, we shall consider specific practical ways to estimate h_L; however, for the present we shall use it only in a general way.

In most applications of the energy equation it is understood that the kinetic energy term $\alpha \bar{V}^2/2g$ involves the mean velocity $\bar{V}$; hence, the bar over the V is usually omitted. The coefficient α is often also omitted because it usually has a value very near unity since most flows are turbulent. In addition, it is often necessary to convert the total power of a pump or turbine to h_p or h_t, respectively, or vice versa. This is done through the original definition of h_p and h_t. That is, $h_p = \dot{W}_p/\dot{m}g = \dot{W}_p/Q\rho g = \dot{W}_p/Q\gamma$. Similarly, $h_t = \dot{W}_t/Q\gamma$.

It must be cautioned that Eq. (7-26) is not valid when compressibility effects are significant. The energy equation for compressible flow will be covered in Chapter 12.

---

[1] It might appear from Eq. (7-25) that with sufficient heat transfer to the conduit (positive $\dot{Q}$) h_L could be eliminated or even made negative. However, increasing $\dot{Q}$ will also increase $u_2 - u_1$, maintaining positive head loss. The fact that h_L is always positive can be shown by application of the entropy principle of thermodynamics.

EXAMPLE 7-3   A horizontal pipe carries cooling water for a thermal power plant from a reservoir as shown. The head loss in the pipe is given as

$$\frac{0.02(L/D)V^2}{2g}$$

where L = length of pipe from reservoir to point in question
   V = mean velocity in pipe
   D = diameter of pipe

If the pipe diameter is 20 cm and the rate of flow is 0.06 m³/s, what is the pressure in the pipe at L = 2,000 m?

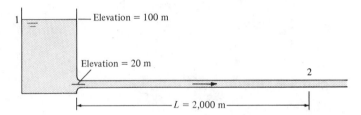

Solution   Write the one-dimensional energy equation, Eq. (7-26), between the water surface in the reservoir and a section in the pipe:

$$\frac{p_1}{\gamma} + \frac{V_1^2}{2g} + z_1 = \frac{p_2}{\gamma} + \frac{V_2^2}{2g} + z_2 + h_L$$

where $p_1 = 0$
$V_1 = 0$
$z_1 = 100$ m
$z_2 = 20$ m
$V_2 = Q/A = 0.06/[(\pi/4) \times 0.2^2] = 1.91$ m/s
$h_L = 0.02(L/0.2) \times 1.91^2/(2 \times 9.81) = 0.0186L$ m

Then

$$\frac{p_2}{\gamma} = 100 - 20 - 0.186 - 0.0186L$$

$$= 79.8 - 0.0186L$$

at L = 2,000 m

$$\frac{p_2}{\gamma} = 79.8 - 37.2 = 42.6 \text{ m}$$

$$p_2 = 417.9 \text{ kPa}$$

EXAMPLE 7-4   The pipe in Fig. 7-3 is 50 cm in diameter and carries water at a rate of 0.5 m³/s. Also, $z_2 = 40$ m, $z_1 = 30$ m, and $p_1 = 70$ kPa gage. What power in kilowatts and horsepower must be supplied to the flow by the pump if the gage pressure at section 2 is to be 350 kPa? Assume $h_L = 3$ m of water and $\alpha_1 = \alpha_2 = 1$.

Solution   Write the energy equation, Eq. (7-26), between sections 1 and 2:

$$\frac{p_1}{\gamma} + \frac{V_1^2}{2g} + z_1 + h_p = \frac{p_2}{\gamma} + \frac{V_2^2}{2g} + z_2 + h_L$$

where $p_1 = 70$ kPa

$\gamma = 9.80$ kN/m³

$V_1 = Q/A_1 = (0.50 \text{ m}^3/\text{s})/[\pi(0.25^2 \text{ m}^2)] = 2.55$ m/s

$V_2 = 2.55$ m/s

$z_1 = 30$ m

$z_2 = 40$ m

$p_2 = 350$ kPa

$h_L = 3$ m

$g = 9.81$ m/s²

then

$$h_p = \frac{p_2 - p_1}{\gamma} + \frac{V_2^2 - V_1^2}{2g} + z_2 - z_1 + h_L$$

$$= \frac{(350 - 70) \text{ kN/m}^2}{9.80 \text{ kN/m}^3} + \frac{(2.55^2 - 2.55^2) \text{ m}^2/\text{s}^2}{9.81 \text{ m/s}^2} + 40 \text{ m} - 30 \text{ m} + 3 \text{ m}$$

$$= 28.6 + 0 + 10 + 3 = 41.6 \text{ m}$$

$$= 41.6 \text{ m} \cdot \text{N/N}$$

The energy supplied by the pump is 41.6 m · N/N of fluid that is flowing; therefore, we obtain the total power supplied by the pump by taking the product of h_p and the weight rate of flow:

$$P = \dot{W}_p = Q\gamma h_p = 0.5 \text{ m}^3/\text{s} \times 9.80 \text{ kN/m}^3 \times 41.6 \text{ m} = 204 \text{ m} \cdot \text{kN/s}$$

$$= 204 \text{ kJ/s} = 204 \text{ kW} \qquad \blacktriangleleft$$

$$= 273 \text{ hp} \qquad \blacktriangleleft$$

EXAMPLE 7-5   This hydroelectric power plant at maximum rate of power generation takes a discharge of 141 m³/s. If the head loss through the intakes, penstock, and outlet works is 1.52 m, what is the rate of power generation?

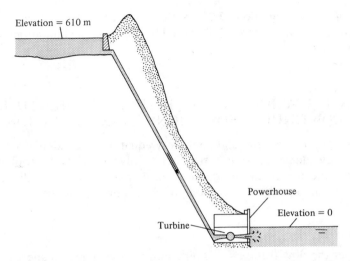

Solution Assume $\alpha_1 = \alpha_2 = 1$. Then Eq. (7-26) for this application reduces to

$$\frac{p_1}{\gamma} + \frac{V_1^2}{2g} + z_1 = \frac{p_2}{\gamma} + \frac{V_2^2}{2g} + h_L + h_t$$

If section 1 is at the upstream reservoir and section 2 is at sea level, then $p_1 = 0$, $V_1 = 0$, $z_1 = 610$ m, $p_2 = 0$, $V_2 = 0$, and $h_L = 1.52$ m.

Then
$$h_t = 610 - 1.52 = 608.5 \text{ m}$$
$$P = \dot{W}_t = Q\gamma h_t = 141 \times 9{,}810 \times 608.5$$
$$= 842 \text{ MW} \qquad \blacktriangleleft$$

Energy equation for incompressible-nonviscous one-dimensional flow

Nonviscous flow means that there will be no head loss in the system; consequently, Eq. (7-26) reduces to

$$\frac{p_1}{\gamma} + \alpha_1 \frac{V_1^2}{2g} + z_1 + h_p = \frac{p_2}{\gamma} + \alpha_2 \frac{V_2^2}{2g} + z_2 + h_t \qquad (7\text{-}27)$$

In addition, if we apply the equation to a stream tube (a filament of fluid of infinitesimal cross section bounded by streamlines) without any shaft work involved, then Eq. (7-27) becomes

$$\frac{p_1}{\gamma} + \frac{V_1^2}{2g} + z_1 = \frac{p_2}{\gamma} + \frac{V_2^2}{2g} + z_2 \qquad (7\text{-}28)$$

Note here that α_1 and α_2 have been dropped because for a stream tube the velocities at the end sections are point velocities; hence, α has its limiting value of unity. Equation (7-28) is the same as the Bernoulli equation derived in Chapter 5. Indeed, it should be, because the limitations are exactly the same.

7-4   APPLICATION OF THE ENERGY, MOMENTUM, AND CONTINUITY EQUATIONS IN COMBINATION

The energy, momentum, and continuity equations are independent equations. They can hence be used together to solve a variety of problems. To illustrate the use of these equations in combination, we shall consider flow through an abrupt expansion and forces on transitions.

Abrupt expansion

Consider the flow from a small pipe into a larger pipe, as shown in Fig. 7-4. We want to solve for the head loss due to the expansion as a function of the flow velocities in the two pipes. Normally such a problem is not amenable to analytic solution; however, one can solve this problem with a reasonable assumption about the pressure distribution at the change in section. Experience tells us that when flow occurs past an abrupt enlargement such as this, the flow will separate from the boundary. Hence, in effect, a jet of fluid from the smaller pipe discharges into the larger pipe.

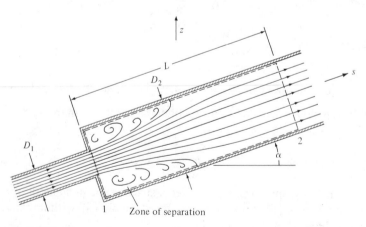

FIGURE 7-4   Flow in an abrupt expansion.

Because the streamlines in the jet are initially straight and parallel, the pressure distribution across the jet will be simply hydrostatic. Because the fluid in the zone of separation at section 1 has such low velocity, it can be likewise assumed that the pressure in this zone is the same as in the jet except for the hydrostatic variation. With this basic assumption for the pressure at section 1, we can now proceed to the application of the momentum and energy equations to solve for the head loss due to the expansion. The one-dimensional energy equation written between sections 1 and 2 is

$$\frac{p_1}{\gamma} + \alpha_1 \frac{V_1^2}{2g} + z_1 = \frac{p_2}{\gamma} + \alpha_2 \frac{V_2^2}{2g} + z_2 + h_L \tag{7-29}$$

We are assuming turbulent flow conditions here; hence we may assume $\alpha_1 = \alpha_2 = 1$. The momentum equation for the fluid in the large pipe between section 1 and section 2 written for the s direction will be as follows:

$$\sum F_s = \sum_{cs} V \rho \mathbf{V} \cdot \mathbf{A}$$

$$p_1 A_2 - p_2 A_2 - \gamma A_2 L \sin \alpha = \rho V_2^2 A_2 - \rho V_1^2 A_1$$

or $$\frac{p_1}{\gamma} - \frac{p_2}{\gamma} - (z_2 - z_1) = \frac{V_2^2}{g} - \frac{V_1^2}{g} \frac{A_1}{A_2} \tag{7-30}$$

The continuity equation, $V_1 A_1 = V_2 A_2$, is also a valid independent equation for this problem; therefore, when Eqs. (7-29) and (7-30) along with the continuity equation are used to solve for h_L, the following is obtained:

$$h_L = \frac{(V_1 - V_2)^2}{2g} \tag{7-31}$$

Hence we see that the head loss for an abrupt expansion is given by the square of the difference in mean velocities in the two pipes divided by $2g$. *Note*: The head loss for an abrupt expansion is *not* $(V_1^2 - V_2^2)/2g$!

Forces on transitions

The method for determining the forces required to hold a transition or any other flow passage in place will be presented in the form of the example given below.

EXAMPLE 7-6   Water flows through the contraction at a rate of 0.707 m^3/s. The head loss due to this particular contraction is given by the empirical equation

$$h_L = 0.2 \frac{V_2^2}{2g}$$

Here V_2 is the velocity in the 20-cm pipe. What horizontal force is required to hold the transition in place if $p_1 = 70$ kPa?

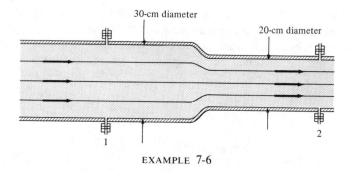

30-cm diameter

20-cm diameter

1

2

EXAMPLE 7-6

Solution Write the momentum equation for the transition between sections 1 and 2. The control surface is drawn so that it encloses the fluid and transition itself; consequently, the control volume is as shown. The pressures are gage pressures, so that the pressure on the exterior of the transi-

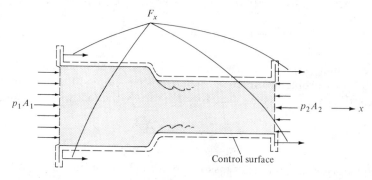

F_x

$p_1 A_1$

$p_2 A_2$ x

Control surface

tion is zero; hence, the force on the exterior is zero. Then, for this control volume the momentum equation is

$$p_1 A_1 - p_2 A_2 + F_x = \sum_{cs} V_x \rho \mathbf{V} \cdot \mathbf{A}$$

$$p_1 A_1 - p_2 A_2 + F_x = \rho V_2^2 A_2 - \rho V_1^2 A_1$$

$$F_x = \rho Q(V_2 - V_1) + p_2 A_2 - p_1 A_1$$

Here Q and the velocities are known and F_x is the unknown. Also, p_2 is still unknown. We obtain p_2 by applying the energy equation between sections 1 and 2:

$$\frac{p_1}{\gamma} + \frac{V_1^2}{2g} + z_1 = \frac{p_2}{\gamma} + \frac{V_2^2}{2g} + z_2 + h_L$$

Here we have assumed $\alpha_1 = \alpha_2 = 1$. For our problem, $z_1 = z_2$; hence, the foregoing equation reduces to

$$\frac{p_1}{\gamma} + \frac{V_1^2}{2g} = \frac{p_2}{\gamma} + \frac{V_2^2}{2g} + h_L$$

$$p_2 = p_1 - \gamma \left(\frac{V_2^2}{2g} - \frac{V_1^2}{2g} + h_L\right)$$

Substituting this expression for p_2 into the equation for the force on the transition yields

$$F_x = \rho Q(V_2 - V_1) + A_2 \left[p_1 - \gamma \left(\frac{V_2^2}{2g} - \frac{V_1^2}{2g} + h_L\right)\right] - p_1 A_1$$

We can obtain the velocities in this equation by use of the continuity equation:

$$V_1 = \frac{Q}{A_1} = \frac{0.707}{(\pi/4) \times 0.3^2} = 10 \text{ m/s}$$

$$V_2 = \frac{Q}{A_2} = \frac{0.707}{(\pi/4) \times 0.2^2} = 22.5 \text{ m/s}$$

$$h_L = \frac{0.2 V_2^2}{2g} = \frac{0.2 \times 22.5^2}{2 \times 9.81} = 5.16 \text{ m}$$

Then $$F_x = 1,000 \times (0.707)(22.5 - 10) + \frac{\pi}{4} \times 0.2^2$$

$$\times \left[70,000 - 9,810 \left(\frac{22.5^2}{2 \times 9.81} - \frac{10^2}{2 \times 9.81} + 5.16\right)\right]$$

$$- 70,000 \times 0.785 \times 0.3^2$$

$$= -1.88 \text{ kN} \qquad \blacktriangleleft$$

Thus it is seen that a force of -1.88 kN must be applied in the negative x direction to hold the transition in place for the given conditions.

7-5 CONCEPT OF THE HYDRAULIC AND ENERGY GRADE LINES

The units of Eq. (7-26) are meters, and we can attach a useful physical relationship to them. Consider the flow in the pipe of Fig. 7-5. The sum of the terms on the left side of Eq. (7-26) represent the total mechanical energy (stated in energy per unit weight of flowing fluid, $N \cdot m/N$) plus the flow-work term in the fluid at the upstream section plus the energy supplied by a pump, whereas the sum on the right-hand side is the total energy per unit weight at the downstream section plus the head loss and energy per unit weight given up to a turbine between the two sections. For

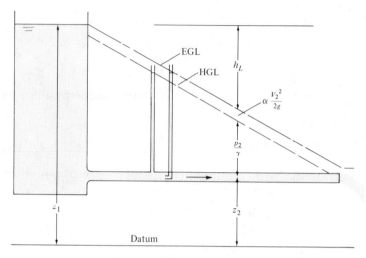

FIGURE 7-5   EGL and HGL in a straight pipe.

the case shown in Fig. 7-5, the velocity of flow in the reservoir is negligible; hence, the total energy at the surface is potential energy, and in terms of energy per unit weight is simply z. At the downstream section the liquid is at a different elevation than in the reservoir and it has significant velocity; therefore, the "energy per unit weight" here is given in terms of p_2/γ, z_2, and $\alpha_2 V_2^2/2g$. It is common practice to lump all these terms together as *total head* in feet or meters. Note that the sum of these terms plus the h_L between section 1 and 2 is equal to the total head at section 1. By analyzing a graph such as Fig. 7-5 it is possible at a glance to ascertain the pressure at any point in the pipeline: One simply picks off values of p/γ and multiplies by γ to obtain p.

If one were to tap a piezometer into the pipe of Fig. 7-5, the liquid would rise to a height p/γ above the pipe; hence, the height of the p/γ line is called the *hydraulic grade line* (HGL). The total head $(p/\gamma + \alpha V^2/2g + z)$ in the system is greater than $p/\gamma + z$ by an amount $\alpha V^2/2g$; consequently, the *energy grade line* (EGL) is above the HGL a distance $\alpha V^2/2g$. The engineer who develops a visual concept of the energy equation as explained above will find it much easier to sense trouble spots in the system (usually points of low pressure) and to ascertain ways of solving the problem.

Other helpful hints for drawing hydraulic grade lines and energy grade lines are enumerated below:

1. By definition, the EGL is positioned above the HGL an amount $\alpha V^2/2g$. Thus if the velocity is zero as in a lake or reservoir, the HGL and EGL will concide (see Fig. 7-5) with the liquid surface.

2. Head loss for flow in a pipe or channel always means that the EGL will slope downward in the direction of flow (see Fig. 7-5). The only exception to this rule occurs when a pump supplies energy (and pressure) to the flow. Then an abrupt rise in the EGL (and HGL) occurs from the upstream side to the downstream side of the pump (see Fig. 7-6).

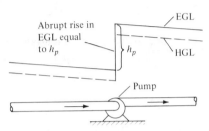

FIGURE 7-6   Rise in EGL and HGL due to pump.

3. In point 2 above, it was noted that a pump can cause an abrupt rise in the EGL (and HGL) because energy is introduced into the flow by the pump. Similarly, if energy is abruptly taken out of the flow by a turbine, for example, then the EGL and HGL will drop abruptly as in Fig. 7-7. In Fig. 7-7 and other figures (Figs. 7-8 through Fig. 7-10) it is assumed that $\alpha = 1.0$. Figure 7-7 also shows that much of the kinetic energy can be converted to pressure if there is a gradual expansion such as at the outlet. Thus the head loss at the outlet is reduced, making the turbine installation more efficient. If the outlet to

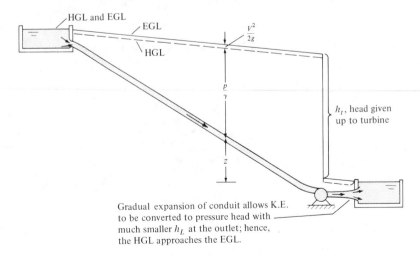

FIGURE 7-7   Drop in EGL and HGL due to turbine.

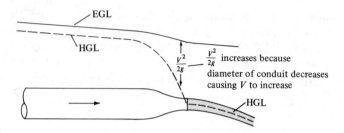

FIGURE 7-8   Change in HGL due to change in conduit diameter.

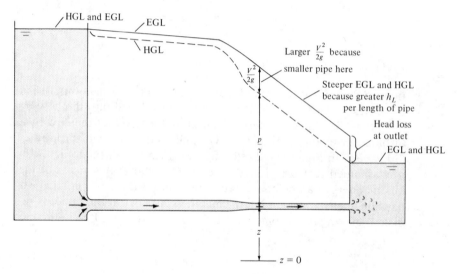

FIGURE 7-9   Change in EGL and HGL due to change in pipe diameter.

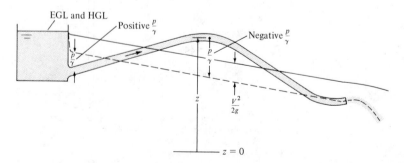

FIGURE 7-10   Subatmospheric pressure when pipe is above the HGL.

a reservoir is an abrupt expansion as in Fig. 7-9, all the kinetic energy is lost; thus the EGL will drop an amount $\alpha V^2/2g$ at the outlet.

4. In a pipe or channel where the pressure is zero, the HGL is coincident with the system because $p/\gamma = 0$ at these points. This fact can be used to locate the HGL at certain points in the physical system, such as at the outlet end of a pipe where the liquid discharges into the atmosphere or at the upstream end where the pressure is zero in the reservoir (see Fig. 7-5).

5. For steady flow in a pipe that has uniform physical characteristics (diameter, roughness, shape, and so on) along its length, the head loss per unit of length will be constant; thus the slope $(\Delta h_L/\Delta L)$ of the EGL and HGL will be constant along the length of pipe (see Fig. 7-5).

6. If a flow passage changes diameter, such as in a nozzle or through a change in pipe size, the velocity therein will also change; hence, the distance between the EGL and HGL will change (see Fig. 7-8). In addition, the slope on the EGL will change because the head loss per length will be larger in the conduit with the larger velocity (see Fig. 7-9). You will learn more about this latter point in Chapter 10 when head loss in pipes is considered in more detail.

7. If the HGL falls below the pipe, then p/γ is negative, thereby indicating subatmospheric pressure (see Fig. 7-10).

EXAMPLE 7-7 A pump draws water from a reservoir, water surface elevation = 520 ft (158.5 m), and forces the water through a 5,000-ft-long (1,524-m), 1-ft-diameter (30.5-cm) pipe. This pipe then discharges the water into a reservoir, with water-surface elevation = 620 ft (189.0 m). If the flow rate is 7.85 cfs (0.222 m³/s) and if the head loss in the pipe is given by $0.01(L/D)(V^2/2g)$, determine the head supplied by the pump h_p, the power supplied to the flow, and draw the hydraulic and energy grade lines for the system. Assume that the pipe is horizontal and is 510 ft (155.4 m) in elevation.

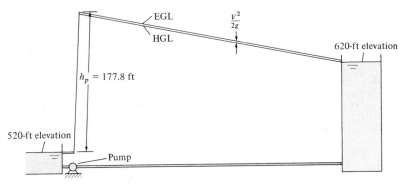

Solution   First solve for h_p by use of the energy equation written (one-dimensional flow assumed) from the water surface in the lower reservoir to the water surface in the upper reservoir:

$$\frac{p_1}{\gamma} + \frac{V_1^2}{2g} + z_1 + h_p = \frac{p_2}{\gamma} + \frac{V_2^2}{2g} + z_2 + h_L$$

In this example, p_1/γ, p_2/γ, V_1, and V_2 are all zero, but $z_1 = 520$ ft, $z_2 = 620$ ft ($z_1 = 158.5$ m, $z_2 = 189.0$ m).

$$h_L = 0.01 \times \frac{5{,}000}{1}\frac{V_p^2}{2g} \quad \text{and} \quad V_p = \frac{Q}{A_p} = 10 \text{ ft/sec}$$

SI units $$h_L = 0.01 \times \frac{1{,}524}{0.305}\frac{V_p^2}{2g} \quad \text{and} \quad V_p = \frac{Q}{A_p} = 3.05 \text{ m/s}$$

Then, $$h_p = 620 - 520 + 0.01 \times \frac{5{,}000}{1}\frac{100}{64.4} \text{ ft-lbf/lbf}$$

$$= 178 \text{ ft} \qquad \blacktriangleleft$$

SI units $$h_p = 189.0 - 158.5 + 0.01 \times \frac{1{,}524}{0.305}\frac{3.05^2}{2 \times 9.81}$$

$$= 54.2 \text{ m} \qquad \blacktriangleleft$$

However, the product of the flow rate Q and specific weight will give us the weight rate of flow. Then the power supplied by the pump will be $h_p \times$ weight rate of flow, or

$$P = h_p Q \gamma \text{ ft-lbf/sec}$$

$$= \frac{h_p Q \gamma}{550} = 158 \text{ hp} \qquad \blacktriangleleft$$

SI units $$P = 54.2 \text{ m} \times 0.222 \text{ m}^3/\text{s} \times 9.80 \text{ kN/m}^3$$

$$= 117.9 \text{ m} \cdot \text{kN/s} = 117.9 \text{ kW} \qquad \blacktriangleleft$$

PROBLEMS

7-1   A turbine receives steam at 300 psia, $T = 700°F$ (enthalpy = 1,268 Btu/lbm). The entrance velocity to the turbine is 50 ft/sec. The steam exits the turbine at 10 psia, $T = 193°F$ as a gas-liquid mixture with an enthalpy of 1,098 Btu/lbm. The exit velocity is 200 ft/sec. The heat loss from the turbine is 2,500 Btu/hr and the mass flow rate of steam through the turbine is 5,800 lbm/hr. Calculate the power output. The elevation difference between the inlet and exit ports is negligible.

7-2   A turbine receives steam at 2.0 MPa, 500°C (enthalpy = 3,062 kJ/kg) at a velocity of 10 m/s. The steam leaves the turbine as gas-liquid mixture at 101 kPa, 373 K with an enthalpy of 2,621 kJ/kg. The exit velocity is 50 m/s. Thermal energy is lost through the turbine walls at a rate of 5 kJ/h. The potential energy due to elevation difference between the inlet and exit port can be neglected. Calculate the power if 4,000 kg/h of steam pass through the turbine.

7-3   A turbine receives steam at 300 psia, $T = 800°F$ (enthalpy = 1,422 Btu/lbm), and the inlet velocity is 20 ft/sec. The steam exits at 150 ft/sec with an enthalpy of 1,100 Btu/lbm. Steam flows through the turbine at 2 lbm/sec and the turbine develops 900 hp. Calculate the rate of heat loss from the turbine. Neglect the potential energy due to elevation difference between inlet and exit ports.

7-4   A turbine receives steam at 1.80 MPa, 500°C (enthalpy = 3,470 kJ/kg) at a velocity of 5 m/s. The steam exits at an enthalpy of 2,630 kJ/kg with a velocity of 70 m/s. The steam flows through at a rate of 1 kg/s and the turbine develops 830 kW. Calculate the rate of heat loss from the turbine. Neglect the potential energy due to elevation difference.

7-5   Air flows in a 1,000-m-long 8-cm-diameter pipe at a rate of 0.5 kg/s. The pressure and temperature at the upstream end of the pipe are at 50 kPa gage and 20°C, respectively. If the pipe is well insulated so that a negligible amount of heat is transferred to or from the air in the pipe, what is the velocity and temperature in the stream of air at the outlet as it discharges into the atmosphere? Assume that atmospheric pressure is 100 kPa. Assume that air is an ideal gas and that the specific enthalpy is given by h (kJ/kg) = $1.004T$ (K).

7-6   A hypothetical velocity distribution in a pipe has a maximum velocity of V_{max} at the center and a minimum velocity of $0.8V_{max}$ at the wall. If the velocity is linearly distributed from the center to the wall, what is α? What is the mean velocity V in terms of V_{max}?

7-7   For this hypothetical velocity distribution in a wide rectangular channel (two-dimensional flow), evaluate the kinetic energy correction factor α.

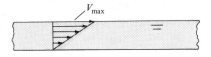

7-8 For these velocity distributions in a pipe, indicate whether the kinetic energy correction factor α is greater than, equal to, or less than unity.

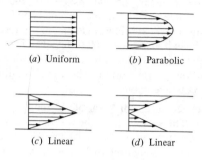

(a) Uniform   (b) Parabolic

(c) Linear   (d) Linear

PROBLEMS 7-8, 7-9, 7-10

7-9 Calculate α for case (<i>c</i>) in problem 7-8.

7-10 Calculate α for case (<i>d</i>) in problem 7-8.

7-11 An approximate equation for the velocity distribution in a pipe with turbulent flow is given by

$$\frac{u}{u_{max}} = \left(\frac{y}{r_0}\right)^m$$

where u_{max} is the centerline velocity, y is the distance from the wall of the pipe, and r_0 is the radius of the pipe. m is an exponent that depends upon the Reynolds number but varies between $\frac{1}{6}$ and $\frac{1}{8}$ for most applications. Derive a formula for α as a function of m. What is α if $m = \frac{1}{7}$?

7-12 An approximate equation for the velocity distribution in a rectangular channel with turbulent flow is given by

$$\frac{u}{u_{max}} = \left(\frac{y}{d}\right)^m$$

where u_{max} is the velocity at the surface, y is the distance from the floor of the channel and d is the depth of flow. m is an exponent that varies from about $\frac{1}{6}$ to $\frac{1}{8}$ depending upon the Reynolds number. Derive a formula for α as a function of m. What is the value of α for $m = \frac{1}{8}$?

7-13 If $D_A = 8$ in., $D_B = 4$ in., and $L = 4$ ft, and if water is flowing at a rate of 3 cfs, determine the difference in pressure between sections A and B. Assume $\alpha_1 = \alpha_2 = 1.0$, and $h_L = 0$.

7-14 If $D_A = 20$ cm, $D_B = 10$ cm, and $L = 1$ m, and if crude oil (sp. gr. = 0.90) is flowing at a rate of 0.06 m³/s, determine the difference in pressure between sections A and B. Neglect head losses.

7-15 If the pipe has constant diameter, is this type of HGL possible? If so, under what additional conditions? If not, why not?

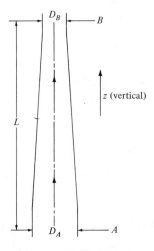

PROBLEMS 7-13, 7-14

PROBLEM 7-15

7-16 Given steady viscous flow of a liquid in a constant-diameter pipe, the pressure along the pipe is observed to increase in the direction of the flow. What sort of orientation can cause the increase in pressure along the pipe?

7-17 Liquid in this pipe has a density of 1,100 kg/m³. The acceleration of the liquid is zero. Is the liquid flowing upward, not moving, or flowing downward in the pipe?

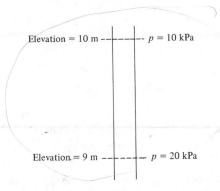

7-18 In the system, $d = 15$ cm, $D = 30$ cm, $\Delta z_1 = 2$ m, and $\Delta z_2 = 4$ m. If the discharge of water in the system is 0.06 m³/s, what are the pressures at points A and B? Is the machine a pump or a turbine? Neglect head losses.

7-19 In the system $d = 6$ in., $D = 12$ in., $\Delta z_1 = 4$ ft, and $\Delta z_2 = 12$ ft. If the discharge of water in the system is 10 cfs, is the machine a pump or a turbine? What are the pressures at points A and B? Neglect head losses.

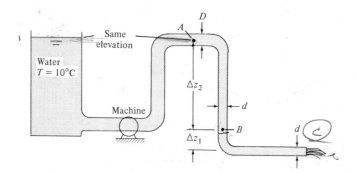

PROBLEMS 7-18, 7-19

7-20 A pipe drains a tank as shown. If $x = 10$ ft, $y = 6$ ft, and head losses are neglected, what is the pressure intensity at point A?

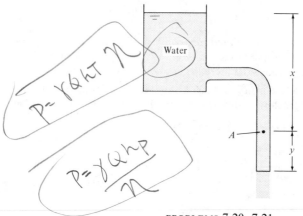

PROBLEMS 7-20, 7-21

7-21 A pipe drains a tank as shown. If $x = 4$ m, $y = 2$ m, and head losses are neglected, what is the pressure intensity at point A?

7-22 Determine the discharge in the pipe and the pressure at point B. Neglect head losses.

7-23 A simple siphon is operating as shown. The head loss between the inlet and B is 2 ft. Between B and C the head loss is 4 ft. Determine the velocity in the pipe and the pressure in the pipe at B if $L_1 = 5$ ft, $L_2 = 10$ ft, and $D = 6$ in.

7-24 A simple siphon is operating as shown. The head loss between the inlet and B is 1 m. Between B and C the head loss is 1.5 m. Determine the velocity in the pipe and the pressure in the pipe at B if $L_1 = 2$ m, $L_2 = 3$ m, and $D = 30$ cm.

7-25 For this siphon the elevations at A, B, C, and D are 30 m, 32 m, 27 m, and 26 m, respectively. The head loss between the inlet and point B is $\frac{3}{4}$ of the

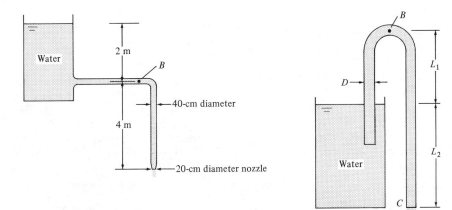

PROBLEM 7-22   PROBLEMS 7-23, 7-24

velocity head, and the head loss in the pipe itself between point B and the
end of the pipe is $\frac{1}{4}$ of the velocity head. For these conditions what is the discharge and what is the pressure at point B?

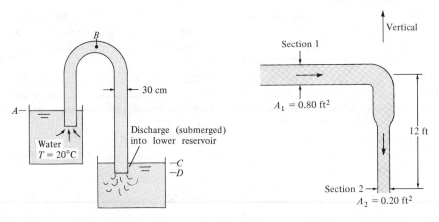

PROBLEM 7-25   PROBLEM 7-26

7-26 Gasoline having a specific gravity of 0.8 is flowing in the pipe shown at a rate
of 5 cfs. What is the pressure at section 2 when the pressure at section 1 is 10
psig and the head loss is 5 ft between the two sections?

7-27 A pump draws water through an 8-in. suction pipe and discharges through a
6-in. pipe in which the velocity is 12 ft/sec. The 6-in. pipe discharges horizontally into air at C. To what height h above A can the water be raised if
30 hp is delivered to the pump? Assume that the pump operates at 70% efficiency and that the head loss in the pipe between A and C is equal to $2V_C^2/2g$.

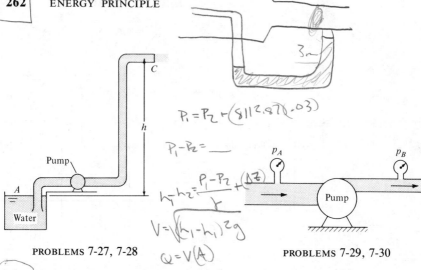

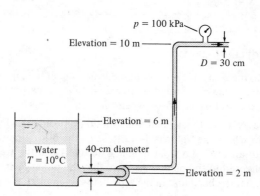

C

h

Pump

A

Water

p_A

Pump

p_B

$P_1 = P_2 + (811^{2.87})(.03)$

$P_1 - P_2 = \underline{\quad}$

$h_1 - h_2 = \dfrac{P_1 - P_2}{r} + (\Delta z)$

$V = \sqrt{(h_1 - h_1)^2 g}$

$Q = V(A)$

PROBLEMS 7-27, 7-28   **PROBLEMS 7-29, 7-30**

7-28 A pump draws water through a 20-cm suction pipe and discharges it through a 15-cm pipe in which the velocity is 4 m/s. The 15-cm pipe discharges horizontally into air at point C. To what height h above the water surface at A can the water be raised if 25 kW is delivered to the pump? Assume that the pump operates at 70% efficiency and that the head loss in the pipe between A and C is equal to $2V_C^2/2g$.

7-29 As shown in the accompanying figure, the pump supplies energy to the flow such that the upstream pressure (12-in. pipe) is 10 psi and the downstream pressure (6-in. pipe) is 30 psi when the flow of water is 3.92 cfs. What horsepower is delivered by the pump to the flow?

7-30 As shown in the figure, the pump supplies energy to the flow such that the upstream pressure (30-cm pipe) is 7.0 kPa and the downstream pressure (15-cm pipe) is 14.0 kPa when the flow of water is 0.2 m³/s. What power in kilowatts is delivered by the pump to the flow?

7-31 If water is flowing at a rate of 0.20 m³/s and it is assumed that $h_L = 1.5V^2/2g$, where V is the velocity in the 30-cm pipe, what power must the pump supply?

$p = 100$ kPa

Elevation = 10 m

D = 30 cm

Elevation = 6 m

Water
T = 10°C

40-cm diameter

Elevation = 2 m

7-32 For the system, the discharge of water is 1.5 cfs, $x = 2$ ft, $y = 3$ ft, $z = 20$ ft, and the pipe diameter is 1 ft. Neglect head losses for all computations. What is the pressure head at point 2 in feet of water if the nozzle diameter is 6 in.?

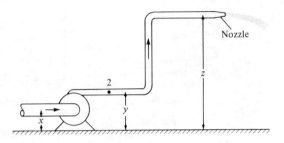

PROBLEMS 7-32, 7-33

7-33 For the system, the discharge of water is 0.20 m³/s, $x = 1.0$ m, $y = 2.0$ m, $z = 7.0$ m, and the pipe diameter is 30 cm. Neglecting head losses, what is the pressure head at point 2 if the jet from the nozzle is 15 cm in diameter?

7-34 In the pump test shown, the rate of flow is 5 cfs of oil (sp. gr. = 0.88). Calculate the horsepower which the pump supplies to the fluid if there is a differential reading of 36 in. of mercury in the U-tube manometer.

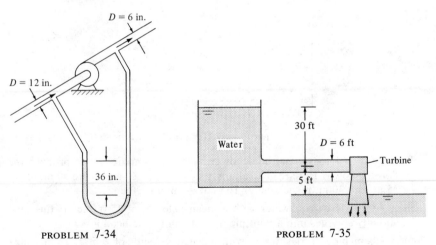

PROBLEM 7-34   PROBLEM 7-35

7-35 If the flow rate is 250 cfs, what power output may be expected from the turbine? Assume that the turbine efficiency is 80% and the overall head loss is $V^2/2g$, where V is the velocity in the 6-ft penstock.

7-36 The discharge of water through this hydroturbine is 30 m³/s. What power is generated if the turbine efficiency is 80%, the head difference H is 30 m, and the overall head loss is 1 m?

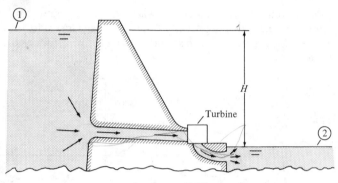

PROBLEM 7-36

7-37 Neglecting losses, what horsepower must the pump deliver to produce the flow as shown? Here the elevations at points A, B, C, and D are 120 ft, 200 ft, 110 ft, and 90 ft, respectively. The nozzle area is 0.10 ft^2.

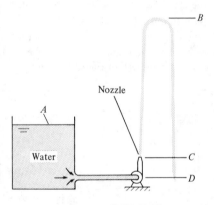

PROBLEMS 7-37, 7-38

7-38 Neglecting losses, what horsepower must the pump deliver to produce the flow as shown? Here the elevations at points A, B, C, and D are 40 m, 65 m, 35 m, and 30 m, respectively. The nozzle area is 30 cm².

7-39 A 4-in. pipe carries water with a mean velocity of 15 ft/sec. If this pipe abruptly enlarges into a 6-in. pipe, what will be the head loss?

7-40 A 10-cm pipe carries water with a mean velocity of 6 m/s. If this pipe abruptly enlarges into a 15-cm pipe, what will be the head loss?

7-41 A 6-in. pipe abruptly expands to a 12-in. size. If the flow rate of water in the pipes is 5 cfs, what is the head loss due to abrupt expansion?

7-42 An 18-in. pipe abruptly expands to 24-in. size. These pipes are horizontal and the rate of flow of water from the smaller size to the larger is 25 cfs. What horizontal force is required to hold the transition in place if the pressure in the 18-in. pipe is 10 psi?

7-43 A 40-cm pipe abruptly expands to 60-cm size. These pipes are horizontal, and the rate of flow of water from the smaller size to the larger is 0.8 m³/s. What horizontal force is required to hold the transition in place if the pressure in the 40-cm pipe is 70 kPa gage?

7-44 The 6-in.-diameter rough aluminum pipe weighs 2 lb per foot of length of pipe and the length L is 50 ft. If the discharge of water is 3 cfs and if the head loss due to friction from section 1 to the end of the pipe is 10 ft, what is the longitudinal force transmitted across section 1 through the pipe wall?

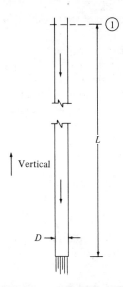

7-45 Water flows in this bend at a rate of 5 m³/s and the pressure at the inlet is 650 kPa. If the head loss in the bend is 10 m, what will be the pressure at the outlet of the bend? Also estimate the force of the anchor block on the bend in the x direction required to hold the bend in place.

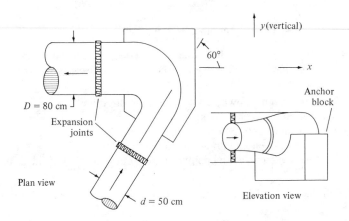

7-46 The pipe diameter D is 30 cm, d is 15 cm, and the atmospheric pressure is 100 kPa. What is the maximum allowable discharge before cavitation occurs at the throat of the venturi meter if $H = 5$ m?

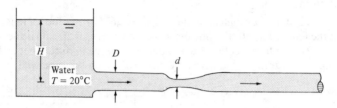

7-47 In this pipe and venturi system $d = 20$ cm, $D = 40$ cm, and the head loss from the venturi meter to the end of the pipe is given by $h_L = 0.5\ V^2/2g$, where V is the velocity in the pipe. Neglecting all other head losses, what head H will first initiate cavitation if the atmospheric pressure is 100 kPa? What will be the discharge at incipient cavitation?

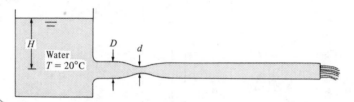

7-48 What is the head loss at the outlet of the pipe that discharges into the reservoir if the diameter of the pipe is 12 in. and the discharge of water is 10 cfs?

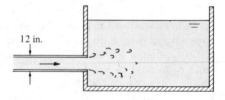

PROBLEMS 7-48, 7-49

7-49 What is the head loss at the outlet of the pipe that discharges into the reservoir if the diameter of the pipe is 50 cm and the discharge of water is 0.60 m³/s?

7-50 (a) For the system shown, what is the flow direction?
(b) What kind of machine is at A?
(c) Do you think both pipes, AB and CA, are the same size?
(d) Sketch in the energy grade line for the system.
(e) Is there a vacuum at any point or region of the pipe? If so, identify the location.

7-51 Sketch the hydraulic grade line and the energy grade line for Example 7-3.

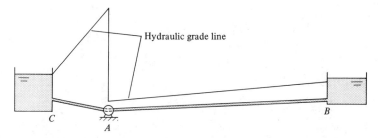

PROBLEM 7-50

7-52 Sketch the hydraulic grade line and the energy grade line for Example 7-5.

7-53 Carefully sketch the hydraulic and energy grade lines for the flow system of Prob. 7-47.

7-54 Sketch the hydraulic and energy grade lines for this conduit, which tapers uniformly from the left end to the right end.

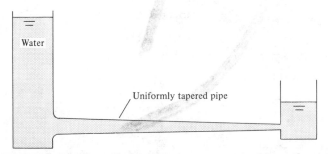

7-55 (*a*) Shown on page 268 are the HGL and EGL for a pipeline. Indicate which is the HGL and which is the EGL.

    (*b*) Are all pipes the same size? If not, which is the smallest?

    (*c*) Is there any region in the pipe where the pressure is below atmospheric pressure? If so, where?

    (*d*) Where is the point of maximum pressure in the system?

    (*e*) Where is the point of minimum pressure in the system?

    (*f*) What do you think is located at the end of the pipe at point *E*?

    (*g*) Is the pressure in the air in the tank above or below atmospheric pressure?

    (*h*) What do you think is located at point *B*?

7-56 Water flows from reservoir *A* to *B* (see page 268). The water temperature in the system is 10°C, the pipe diameter *D* is 1 m and the pipe length *L* is 300 m. If *H* = 16 m, *h* = 2 m, and if the pipe head loss is given by $h_L = 0.01 (L/D)$ $(V^2/2g)$, where *V* is the velocity in the pipe, what will be the discharge in the pipe? In your solution include the head loss at the pipe outlet and also sketch hydraulic and energy grade lines. What will be the pressure at point *P* halfway between the two reservoirs?

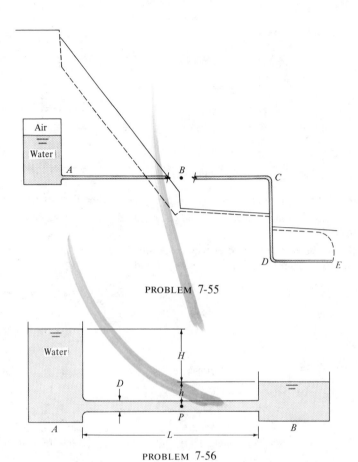

PROBLEM 7-55

PROBLEM 7-56

7-57 What power is required to pump water at a rate of 2 m³/s from the lower to
the upper reservoir? Assume the pipe head loss is given by $h_L = 0.014\,(L/D)$
$(V^2/2g)$, where L is the length of pipe and D is pipe diameter and V is the
velocity in the pipe. The water temperature is 10°C and the water surface elevation in the lower reservoir is 150 m; the surface elevation in the upper

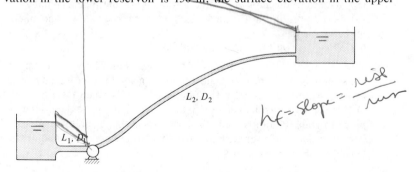

reservoir is 250 m. The pump elevation is 100 m, $L_1 = 100$ m, $L_2 = 1,000$ m, $D_1 = 1$ m, $D_2 = 50$ cm. Assume the pump and motor efficiency is 74%. In your solution include the head loss at pipe outlet and sketch hydraulic and energy grade lines.

7-58 What horsepower must be supplied to the water to pump 2.5 cfs at 68°F from the lower to the upper reservoir? Assume that the head loss in the pipes is given by $h_L = 0.015(L/D)(V^2/2g)$, where L is the length of pipe in feet and D is the pipe diameter in feet. Sketch the hydraulic and energy grade lines.

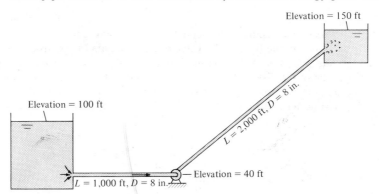

Elevation = 150 ft

Elevation = 100 ft

$L = 2,000$ ft, $D = 8$ in.

— Elevation = 40 ft

$L = 1,000$ ft, $D = 8$ in.

7-59 Assume that the head loss in the pipe is given by $h_L = 0.014(L/D)(V^2/2g)$, where L is the length of pipe in feet and D is the pipe diameter in feet.

(a) Determine the discharge of water through this system.

(b) Draw the hydraulic and energy grade line for the system.

(c) Locate the point of maximum pressure.

(d) Locate the point of minimum pressure.

(e) Calculate the maximum and minimum pressures in the system.

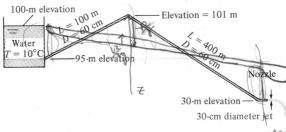

100-m elevation

$L = 100$ m

$D = 60$ cm

Water
$T = 10°C$

—95-m elevation

Elevation = 101 m

$L = 400$ m

$D = 60$ cm

Nozzle

30-m elevation —

30-cm diameter jet

7-60 In Prob. 6-99, what power is developed by the windmill?

7-61 An engineer is designing a subsonic wind tunnel. The test section is to have a cross-sectional area of 4 m² and a speed of 50 m/s. The air density is 1.2 kg/m³. The area of the tunnel exit is 10 m². The head loss through the tunnel is given by

$$h_L = (0.02) \frac{V_T^2}{2g}$$

where V_T is the speed in the test section. Calculate the power needed to operate the wind tunnel. *Hint*: Assume negligible energy loss for the flow approaching the tunnel in region A. Also assume atmospheric pressure at the outlet section of the tunnel.

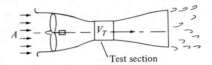

Test section

7-62 Fluid flowing along a pipe of diameter D accelerates around a disc of diameter d as shown in the figure. The velocity far upstream of the disk is U and the fluid density is ρ. Assuming incompressible flow and that the pressure downstream of the disk is the same as at the plane of separation, develop an expression for the force required to hold the disk in place in terms of U, D, d, and ρ. Using the expression developed, determine the force when $U = 10$ m/s, $D = 5$ cm, $d = 4$ cm, and $\rho = 1.2$ kg/m^3.

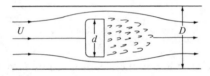

REFERENCES

1. Shames, I. *Mechanics of Fluids*. McGraw-Hill Book Company, New York, 1962.
2. Van Wylen, G. J., and Sonntag, R. E. *Fundamentals of Classical Thermodynamics*. John Wiley & Sons, Inc., New York, 1965.
3. Wark, K. *Thermodynamics*. McGraw-Hill Book Company, New York, 1966.

This model of downtown Montreal is typical of boundary layer wind-tunnel models used for determining design-wind forces on major buildings.

8 DIMENSIONAL ANALYSIS AND SIMILITUDE

THE SOLUTIONS OF MOST ENGINEERING PROBLEMS involving fluid mechanics rely on data acquired by experimental means. In many cases the empirical data are general enough so that engineers have need for them in their normal design practice; consequently, they are made available by publication in journals and textbooks. Examples of such data are the resistance coefficients for pipes and the drag coefficients for blunt bodies.[1] For many problems, however, either the geometry of the structure that guides the flow or the flow conditions themselves are so unique that special tests on a replica at different scale of the structure are required to predict the flow patterns and pressure variation. When such tests are performed, the replica of the structure on which the tests are made is called the *model* and the full-scale structure employed in the actual engineering design is called the *prototype*. The model is usually made much smaller than the prototype for economic reasons.

8-1   THE NEED FOR DIMENSIONAL ANALYSIS

Fluid mechanics is more heavily involved with empirical work than is structural engineering, machine design, or electrical engineering because the analytical tools presently available are not capable of yielding exact solutions to many of the problems in fluid mechanics. It is true that exact solutions are obtainable for all hydrostatic problems and for many laminar-flow problems; however, the most general equations solved on the largest computers yield only fair approximations for turbulent-flow problems—thus the need for experimental evaluation and verification.

For analyzing model studies and for correlating the results of experimental research, it is essential that the researchers employ *dimensionless parameters*. To appreciate the advantages of using dimensionless parameters, let us consider the flow of water through the unusual orifice illustrated in Fig. 8-1. Actually, this is much like a flow nozzle, which will be presented in Chapter 13, except that the flow is in the opposite direction to a nozzle operating in normal fashion. It is true that an orifice operating in such a manner would have a much different performance than a flow nozzle; however, it is not unlikely that a firm or city water department might have such a situation where the flow may occur the "right way" most of the time and the "wrong way" part of the time. Hence the need for such knowledge. The test procedure involves testing several orifices

---

[1] These coefficients will be presented in Chapters 10 and 11.

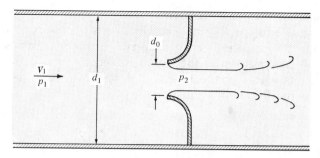

FIGURE 8-1   Flow through inverted flow nozzle.

each with different throat diameters, d_0. Thus we want to measure the pressure difference $p_1 - p_2$ as a function of the velocity V_1, density ρ, and diameter d_0. We think that by carrying out numerous measurements at different values of V_1, d_0, and ρ we can plot our data as shown in Fig. 8-2a. We soon realize that our first plan will involve a terrific amount of work, so we look for a better scheme. We then think that the Bernoulli equation must have relevance here in a manner similar to that shown in Chapter 5. It was noted in Chapter 5 that the conditions for application of the Bernoulli equation are closely approximated if the fluid has fairly low viscosity such as water, if the streamlines converge in the direction of flow, and if the flow is steady. These conditions prevail between sections 1 and 2 in our problem. In Sec. 5-4 we knew by the character of the flow passages how V_1 and V_2 were related; thus we could write the Bernoulli equation between two points in the flow field and solve for Δp directly. However, in our orifice, Fig. 8-1, it should be expected that separation

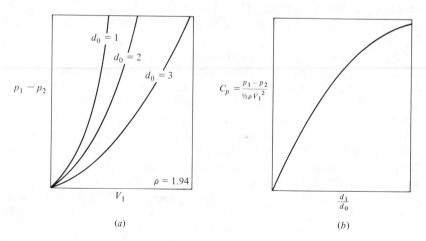

FIGURE 8-2   Pressure-velocity-diameter relations.

will occur downstream of the throat; thus the smallest flow section is smaller than the throat section. Therefore, it is necessary to determine experimentally the relation between Δp and the other variables.

We use a general form of the Bernoulli equation much as we did in Chapter 5, but we insert coefficients for unknown relations. In other words, we write Bernoulli's equation as

$$p_1 + \rho \frac{V_1^2}{2} = p_2 + \rho \frac{V_2^2}{2}$$

or in a dimensionless form

$$\frac{p_1 - p_2}{\rho V_1^2 / 2} = \frac{V_2^2}{V_1^2} - 1 \tag{8-1}$$

In Eq. (8-1) we know that $V_2/V_1 = A_1/A_2$; however, we do not know the value of A_2, and we are most interested in relating the velocity ratio to the ratio of A_1 to the orifice area A_0; therefore, let us simply express A_1/A_0 in functional form: $V_2/V_1 = f(A_1/A_0) = f_1(d_1/d_0)^2$. Then Eq. (8-1) can be rewritten as

$$\frac{p_1 - p_2}{\rho V_1^2 / 2} = [f_1(d_1/d_0)^2]^2 - 1 \tag{8-2}$$

Here the left side of Eq. (8-2) is a form of pressure coefficient C_p, and since the right side is solely a function of d_1/d_0, we can write

$$C_p = f_2 \left(\frac{d_1}{d_0} \right) \tag{8-3}$$

If we plot the pressure coefficient C_p (instead of pressure difference) versus d_1/d_0, we will get a single curve as shown in Fig. 8-2b. Inspection will convince the reader that everything given on Fig. 8-2a is included on Fig. 8-2b.

Thus it is seen that by considering a nondimensional form of Bernoulli's equation we will have made a tremendous reduction in experimental work from that required before considering the nondimensional form. The process of nondimensionalizing the equation reduced the correlating parameters from five (Δp, ρ, V, d_1, d_0) to two $[\Delta p/(\rho V^2/2)$, $d_1/d_0]$. To collect the data for Fig. 8-2a would require, say, 20 data points for each d_0 value, and if five d_0's were tested, then 100 data points would be needed. On the other hand, only about 20 would be required to yield the curve in Fig. 8-2b, thus, effecting a considerable saving of time.

In the foregoing development we had a clue about the governing equation; by considering a dimensionless form of that equation, we were able to obtain a set of dimensionless parameters with which to correlate our data. In many cases, however, the governing equation is not available and the dimensionless parameters must be sought using a formal procedure called *dimensional analysis*. Sections 8-2 and 8-3 present some basic material leading up to dimensional analysis and in Sec. 8-4 the actual procedure of dimensional analysis is explained.

8-2   DIMENSIONS AND EQUATIONS

All variables used in science or engineering are expressed in terms of a limited number of basic dimensions. For most engineering problems the basic dimensions are either force, length, and time; or mass, length, and time—either group being equally as acceptable as the other. Thus we can designate the dimensions of pressure as follows:

$$[p] = \frac{F}{L^2} \tag{8-4}$$

Here the brackets mean "dimension of." Therefore, Eq. (8-4) reads as follows: "the dimensions of p equal force per length squared." In this case, L^2 represents the dimension of area.

It goes without saying that all equations must balance in magnitude; however, all rational equations (those developed from basic laws of physics) must also be dimensionally homogeneous. That is, the left side of the equation must have the same dimensions as the right. Moreover, each term in the equation must have the same dimension.

8-3   THE BUCKINGHAM Π THEOREM

In 1915 Buckingham (3) showed that the number of independent dimensionless groups of variables (dimensionless parameters) needed to correlate the variables in a given process is equal to $n - m$, where n is the number of variables involved and m is the number of basic dimensions included in the variables. Thus, if the drag force F of a fluid flowing past a sphere is known to be a function of the velocity V, mass density ρ, viscosity μ, and diameter D, then we have five variables (F, V, ρ, μ, and D) and three basic dimensions (L, F, and T) involved. Then we will have $5 - 3 = 2$ basic groupings of variables that can be used to correlate experimental results. Note that the same reduction in correlating parameters occurred in our discussion concerning the reverse-flow nozzle in Sec. 8-1. In the next section we will show in a step-by-step process how variables can be combined to form the dimensionless parameters.

8-4   DIMENSIONAL ANALYSIS

The basic method

Several methods may be used to carry out the process of dimensional analysis, but the step-by-step approach, very clearly presented by Ipsen (6), is the easiest and reveals the most about the process. The basic objective in dimensional analysis, as already noted, is to reduce the number of separate variables involved in a problem to a smaller number of independent dimensionless groups of variables (dimensionless parameters). The rationale for this is that all rational equations can be nondimensionalized, as illustrated in Sec. 8-1, and all rational equations have a certain number of independent terms in them. Thus by the procedure of dimensional analysis we will be simply arranging variables into a dimensionless equation. In the process of combining the variables to form dimensionless groups or terms, the number of independent groups thus obtained is less than the number of original variables.

We start the process by identifying only those variables that are significant to the problem. Then we include these variables in a functional equation and note their dimensions. For example, let us consider a simple problem we are already familiar with so that we can get a feel for the process.

EXAMPLE 8-1   Suppose that we want to consider the fall velocity of a body in a vacuum and we know that this velocity V is a function of the acceleration due to gravity g and the distance through which it falls, h. However, let us say that we do not know how to derive the equation for the fall velocity but want to obtain the proper relationship by dimensional analysis and experimentation. Thus in this example we want to determine the dimensionless parameter(s) that apply.

Solution   We include the significant variables in a functional equation as

$$V = f(g, h)$$

where

$$[V] = L/T$$
$$[g] = L/T^2$$
$$[h] = L$$

By referring to our list of variables and their dimensions above, we note that the velocity V has the dimension of time in the denominator. The only variable on the right-hand side of the equation with the dimension of time is g, which has T^2 in its denominator. Thus it can be seen that $\sqrt{g}$ will have T to the first power in the denominator. Now, if we divide the left side by $\sqrt{g}$, that side will be dimensionless in T. The only way we can make the right side dimensionless in T is to combine g with itself; that is, to divide g by g, thereby deleting it from the right side. We then have

$$\frac{V}{\sqrt{g}} = f_1(h)$$

Note that we have reduced the number of variables on the right-hand side by one, namely, g, which is the essence of the technique of dimensional analysis.

Now, in a similar manner we combine h with $V/\sqrt{g}$ to make the entire combination dimensionless in L as well as T, which gives

$$\frac{V}{\sqrt{gh}} = C$$

Of course, our experience in basic physics tells us that $C = \sqrt{2}$. In the illustration above of the step-by-step method of dimensional analysis, it should be noted that the remaining single dimensionless parameter was consistent with the number of dimensionless parameters, $n - m$, that the Buckingham Π theorem predicts. That is, we had three variables ($n = 3$) and two dimensions [L and $T(m = 2)$]; hence, we should have one ($3 - 2$) dimensionless parameter, which we do. In fact, the step-by-step method of dimensional analysis will always yield the correct number of parameters.[1]

Next we will take an example that is somewhat more involved than the first; however, the procedure is the same as before.

EXAMPLE 8-2 If the drag F_D of a sphere in a fluid flowing past the sphere is a function of the viscosity μ, the mass density ρ, the velocity of flow V, and the diameter of the sphere D, what dimensionless parameters would be applicable to the flow process?

Solution
$$F_D = f_1(V, \rho, \mu, D) \tag{8-5}$$

where
$$[F_D] = F$$
$$[V] = L/T$$
$$[\rho] = FT^2/L^4$$
$$[\mu] = FT/L^2$$
$$[D] = L$$

---

[1] Note that in rare instances, the number of parameters may be one more than predicted by the Buckingham Π theorem. This anomaly can occur because it is possible that two-dimensional categories can be eliminated when dividing (or multiplying) by a given variable. See Ipsen (6, page 172) for an example of this.

Eliminate F as a dimension by combining ρ with all variables that have the force dimension in them in such a way that F is canceled:

$$\frac{F_D}{\rho} = f_2(V, \frac{\mu}{\rho}, D)$$

where
$$[F_D/\rho] = FL^4/FT^2 = L^4/T^2$$
$$[V] = L/T$$
$$[\mu/\rho] = FTL^4/L^2FT^2 = L^2/T$$
$$[D] = L$$

Now eliminate the time dimension T by combining V with the remaining groups of variables that include the time dimension:

$$\frac{F_D}{\rho V^2} = f_3\left(\frac{\mu}{\rho V}, D\right)$$

where
$$\left[\frac{F_D}{\rho V^2}\right] = L^2$$

$$\left[\frac{\mu}{\rho V}\right] = L$$

$$[D] = L$$

We now have groupings of variables that have only the length dimension left. We can make these dimensionless by combining the length variable D with them. We finally obtain

$$\frac{F_D}{\rho V^2 D^2} = f_4\left(\frac{\mu}{\rho VD}\right) \tag{8-6}$$

Although Eq. (8-6) still has as many variables as Eq. (8-5), for purposes of correlation the number has been reduced from five to two because the two dimensionless parameters can now be treated as separate variables when one is determining the functional relationship between the variables.

The actual functional relationship between the parameters is determined by running a series of tests to observe values of $F_D/\rho V^2 D^2$ for different values of the ratio $\mu/\rho VD$. The results of these tests are then plotted to yield a usable functional relationship between the two parameters. Since the relationship is experimentally determined, it is just as valid to express Eq. (8-6) with the right-hand side inverted:

$$\frac{F_D}{\rho V^2 D^2} = f_5\left(\frac{VD\rho}{\mu}\right) \tag{8-7}$$

Now the combination on the right is the Reynolds number, which was first introduced in Chapter 4. Hence the parameter $F_D/\rho V^2 D^2$ is a function of the Reynolds number, $VD\rho/\mu$.

The solution of Example 8-2 was found by combining ρ, then V, and finally D with other variables to produce the dimensionless groupings. The question arises as to whether these variables or this order must be used to produce dimensionless groupings. The answer is that this is not the only way it can be done. For example, let us solve Example 8-2 over again with a different order of attack:

We start by noting as before that

$$F_D = f_1(V, \rho, \mu, D)$$

where

$$[F_D] = F$$
$$[V] = L/T$$
$$[\rho] = FT^2/L^4$$
$$[\mu] = FT/L^2$$
$$[D] = L$$

Now, we eliminate the length dimension first by combining D with the other variables. Therefore, we have

$$F_D = f_2 \left(\frac{V}{D}, \rho D^4, \mu D^2 \right)$$

where

$$[F_D] = F$$
$$[V/D] = T^{-1}$$
$$[\rho D^4] = FT^2$$
$$[\mu D^2] = FT$$

Next, we eliminate T by the combination D/V, which itself has the dimension of T. We then get

$$F_D = f \left(\frac{VD}{DV}, \rho D^4 \frac{V^2}{D^2}, \mu D^2 \frac{V}{D} \right)$$

The combination VD/DV is a pure number (unity); it is not a variable and is hence dropped from the groups of variables. We then have

$$F_D = f_3(\rho D^2 V^2, \mu D V)$$

where $[\rho D^2 V^2] = F$ and $[\mu DV] = F$. Finally, use the combination $\rho V^2 D^2$, the dimension of which is F, to eliminate F as a dimension. We then obtain

$$\frac{F_D}{\rho V^2 D^2} = f_4 \left(\frac{\rho D^2 V^2}{\rho D^2 V^2}, \frac{\mu D V}{\rho V^2 D^2} \right)$$

or

$$\frac{F_D}{\rho V^2 D^2} = f_5 \left(\frac{\mu}{\rho V D} \right) = f_6 \left(\frac{VD\rho}{\mu} \right)$$

Therefore, it is seen that we end up with the same result that we had before. We could have gotten different combinations if we had, for example, used μ, V, and D as combining variables instead of ρ, V, and D or combinations thereof.[1] However, it is the usual practice in fluid mechanics to use ρ, V, and a length variable for combining variables. Note that when the variables are combined in this order (first ρ, then V, and finally a length variable) to eliminate F, T, and L, respectively, the process of dimensional analysis becomes quite simple, inasmuch as after the force dimension is eliminated it is not reintroduced when V is combined with the group(s) of variables to eliminate T because the F dimension is not included in V. This is also the case when the length variable is combined with the group(s), because neither F nor T are included in a length variable. If different choices of variables (other than ρ, V, and L) are used for combining variables, it is still desirable to combine them in order from the more complex (the one which has the greatest number of dimensions) to the simplest (the one with merely a single dimension).

Recapitulation of the process of dimensional analysis

In the foregoing paragraphs the concept and procedure of dimensional analysis have been presented. For review purposes and handy reference, the procedure of dimensional analysis can be summarized as follows:

1. Identify all significant variables associated with the problem and include these in a functional equation, such as

$$Z = f(U, V, W, X, Y)$$

2. Combine first one variable that includes a given dimension with other variables having that same dimension in such a way as to make the resulting combination(s) dimensionless in that category. Then choose a second variable to combine with other variables and groups of variables to nondimensionalize the equation in a second dimension. Repeat the process until the entire equation consists of dimensionless groups or parameters.

3. Note that the final form will appear something like the following:

$$\pi_1 = f(\pi_2, \pi_3, \ldots, \pi_{n-m})$$

where $\pi_1, \pi_2, \ldots, \pi_{n-m}$ represent the dimensionless parameters obtained, n is the number of variables, and m is the number of dimensions included in the list of variables.

---

[1] For example, if μ were first combined with F_D and ρ to cancel the F dimension we would have obtained $F_D/\mu = f_1(\rho/\mu, V, D)$. Then, upon combining V and D to cancel the T and L dimensions, respectively, we would have obtained $F_D/\mu DV = f_4(VD\rho/\mu)$.

Limitations of dimensional analysis

All the foregoing procedures deal with straightforward situations; there are some problems that do occur, however. In order to apply dimensional analysis we must first decide which variables are significant. If the problem is not understood well enough to make a good initial choice of variables, dimensional analysis will seldom provide clarification.

One error might be inclusion of variables whose influence is already accounted for. For example, one might tend to include two or three length variables in a scale-model test where only one length variable may be all that is needed.

Another serious error might be the omission of a significant variable. If this is done, one of the significant dimensionless parameters will likewise be missing. In this regard, we may sometimes identify a list of variables that we think are significant to a problem and find that one dimensional category such as F or L or T may be included in only one variable. If such is the case, then at the outset we know that an error in choice of variables has been made because it will not be possible to combine two variables to eliminate the lone dimension. Either the variable with the lone dimension should not have been included in the first place (it is not significant) or another one should have been included (a significant variable was omitted).

How do we know if a variable is significant or not for a given problem? Probably the truest answer is by experience. After one works in the field of fluid mechanics for several years he or she develops a feel for the significance of variables to certain kinds of applications. However, even the inexperienced engineer will appreciate the fact that free-surface effects have no significance in closed-conduit flow; consequently, surface tension, σ, would not be included as a variable. In closed-conduit flow, if the velocity is less than approximately one-third the speed of sound, compressibility effects are usually negligible. Thus these guidelines, which have been observed by previous experimenters, will help the beginning engineer develop confidence in application of dimensional analysis and similitude.

8-5 COMMON DIMENSIONLESS NUMBERS

Derivation of the dimensionless numbers

In Example 8-2, one of the dimensionless parameters obtained was a form of the Reynolds number. This parameter and others recur repeatedly in fluid-flow studies. By considering all the variables that might be significant in a general flow situation, we can derive these parameters or *dimensionless numbers* by means of dimensional analysis. Visualize a hypothetical flow condition in which the pressure difference between two points in the flow field is expected to be a function of V, L, ρ, μ, E_v, σ, and $\Delta\gamma$. Respectively, these variables are velocity, characteristic length involved in the fluid flow, mass density of the fluid, viscosity of the fluid, the bulk modulus of elasticity of the fluid, surface tension of the fluid, and the difference in specific weight between the flowing fluid and the fluid above the free surface. The latter variable, $\Delta\gamma$, is significant when free-surface phenomena such as waves exist. It would never be applied to flow in completely enclosed conduits with the absence of free surfaces. The functional relationship for these variables is given as

$$\Delta p = f(V, L, \rho, \mu, E_v, \sigma, \Delta\gamma)$$

where
$$[\Delta p] = F/L^2$$
$$[V] = L/T$$
$$[L] = L$$
$$[\rho] = FT^2/L^4$$
$$[\mu] = FT/L^2$$
$$[E_v] = F/L^2$$
$$[\sigma] = F/L$$
$$[\Delta\gamma] = F/L^3$$

When a dimensional analysis is made on these variables using ρ, V, and L as combining variables, one can obtain the following functional relationship between the dimensionless parameters:

$$\frac{\Delta p}{\rho V^2} = f\left(\frac{VL\rho}{\mu}, \frac{V}{\sqrt{E_v/\rho}}, \frac{\rho L V^2}{\sigma}, \frac{V^2}{L\,\Delta\gamma/\rho}\right)$$

Without affecting the results with respect to dimensional considerations, we can rewrite the left side of this equation as $\Delta p/\frac{1}{2}\rho V^2$. We do this to put the term in the same form as the pressure coefficient that was introduced in Chapter 5. The speed of sound c is given by $\sqrt{E_v/\rho}$. We then have

$$\frac{\Delta p}{\frac{1}{2}\rho V^2} = f_1\left(\frac{VL\rho}{\mu}, \frac{V}{c}, \frac{\rho L V^2}{\sigma}, \frac{V^2}{L\,\Delta\gamma/\rho}\right) \tag{8-8}$$

The parameter on the left side of Eq. (8-8) is the pressure coefficient, and it is seen to be dependent on the dimensionless parameters on the right side of the equation. On the right side, the first parameter is the Reynolds number already referred to, the second parameter is the Mach number, the third is the Weber number, and the last is the Froude (rhymes with food) number squared. The Froude number is defined as the square root of this last parameter:

$$F = \frac{V}{\sqrt{(\Delta\gamma/\rho)L}}$$

In this particular form it is called the densimetric Froude number, and it is applied in studying the motion of fluids in which there is density stratification such as between salt water and fresh water in an estuary. It also has application in the study of thermal plumes from stacks or from heated water effluents associated with thermal power plants. However, in most applications of the Froude number, the flowing fluid is usually a liquid such as water and the adjacent fluid at the interface is a gas such as air, so that for all practical purposes, $\Delta\gamma = \gamma_{\text{liquid}}$. Then $\Delta\gamma/\rho$ reduces to $\gamma_{\text{liquid}}/\rho_{\text{liquid}} = g$. Thus the Froude number is most often given as

$$F = \frac{V}{\sqrt{gL}} \tag{8-9}$$

We can then write Eq. (8-8) as

$$\frac{\Delta p}{\frac{1}{2}\rho V^2} = f\left(\frac{VL\rho}{\mu}, \frac{V}{c}, \frac{\rho L V^2}{\sigma}, \frac{V}{\sqrt{gL}}\right) \tag{8-10}$$

or
$$C_p = f(\text{Re}, M, W, F) \tag{8-11}$$

In Chapter 5 it was noted that C_p was essentially a dimensionless form of Bernoulli's equation. Thus C_p represents a relative change of pressure; the pressure change is relative to the inertial reaction (mass times acceleration) of the fluid. Similarly, the other dimensionless numbers can be viewed as relative quantities—that is, inertial-force reaction relative to forces relating to fluid characteristics. The next paragraph will demonstrate these relations.

The Reynolds number can be viewed as a ratio of inertial to viscous forces. The viscous force acting on a surface is $F_v = \tau A = (\mu \, du/dy)A$. However, we are primarily interested in dimensional considerations; hence F_v can be written as

$$F_v \propto \mu \frac{V}{L} A \propto \mu \frac{V}{L} L^2$$

$$F_v \propto \mu V L$$

The inertial, or Ma, force is expressed as

$$F_i \propto Ma$$

$$F_i \propto \rho L^3 V \frac{dV}{ds} \propto \rho L^2 V^2$$

where M represents the mass of the fluid and $V\,dV/ds$ is acceleration in the form of convective acceleration. Thus the ratio of inertial to viscous force in terms of the basic variables is

$$\frac{F_i}{F_v} \propto \frac{\rho L^2 V^2}{\mu VL} = \frac{VL\rho}{\mu} = \text{Re}$$

Since we have shown that the Reynolds number is inversely proportional to the shear force, it follows that very low Reynolds numbers imply relatively large viscous shear forces and very large Reynolds numbers imply relatively low viscous shear forces. This concept of the Reynolds number and its relation to viscous stresses is often very helpful in understanding certain flow phenomena.

Expressed in terms of basic variables involved, the other forces (elastic, surface tension, and gravity) which influence the flow are as follows:

$$F_c \propto \rho VcL^2 \qquad \text{(elastic force)}$$

$$F_\sigma \propto \sigma L \qquad \text{(surface-tension force)}$$

$$F_g \propto \Delta\gamma L^3 \qquad \text{(gravity force)}$$

Thus it is seen that the Mach number is a ratio of inertial to elastic force,

$$M = \frac{\rho L^2 V^2}{\rho VcL^2} = \frac{V}{c}$$

the Weber number is a ratio of inertial to surface-tension force,

$$W = \frac{\rho L^2 V^2}{\sigma L} = \frac{\rho V^2 L}{\sigma}$$

and the Froude number squared is the ratio of inertial to gravity force,

$$F^2 = \frac{\rho L^2 V^2}{\Delta\gamma L^3} = \frac{V^2}{L\,\Delta\gamma/\rho}$$

With respect to the Froude number, it should be noted that when gravity simply causes a hydrostatic pressure distribution such as in a closed conduit, then its effect is of no significance in controlling the pattern of flow. However, if the gravitational force influences the pattern of flow, such as flow over a spillway or the motion of a ship in rough seas, then the Froude number is a most significant parameter.

The Mach number is an indicator of how important compressibility effects are in a fluid flow. If the Mach number is low, then the inertial force associated with the fluid motion does not cause a significant density change, and the flow can be treated as incompressible (constant density). On the other hand, a large Mach number often results in appreciable density changes that must be considered in model studies. The significance of the Mach number in compressible flow studies will be discussed in more detail in Chapter 12.

The Weber number is an important parameter in liquid atomization. The surface tension of the liquid at the surface of a droplet is responsible for maintaining the droplet's shape. If a droplet is subjected to an air jet and there is a relative velocity between the droplet and the gas, inertial forces due to this relative velocity cause the droplet to deform. If the Weber number is too high, the inertial force overcomes the surface-tension force to the point that the droplet shatters into even smaller droplets. Thus a Weber-number criterion can be useful in predicting the droplet size to be expected in liquid atomization. Here the size of the droplets resulting from liquid atomization is a very significant parameter in gas-turbine and rocket combustion.

8-6   SIMILITUDE

Scope of similitude

Whenever it is necessary to perform tests on a model to obtain information that cannot be obtained by analytical means alone, the rules of similitude must be applied. Hence similitude is the theory and art of predicting prototype conditions from model observations. We shall see that the theory of similitude involves the application of dimensionless numbers, such as the Reynolds number or the Froude number, to predict prototype performance from model tests. The art of similitude enters the problem when the engineer must make decisions about the model design, model construction, performance of tests, or analysis of results that are not included in the basic theory.

Present engineering practice makes use of model tests much more than most people realize. For example, whenever a new airplane is being designed, tests are made not only on the general scale model of the prototype airplane but also on various components of the plane. Numerous tests are made on individual wing sections as well as on the engine pods and tail sections.

Models of automobiles and high-speed trains are also tested in wind tunnels to predict the drag and flow patterns for the prototype. Information derived from these model studies often indicates potential problems that can be corrected before the prototype is built, thereby saving considerable time and expense in system development.

In civil engineering, model tests are always used to predict flow conditions for the spillways of large dams. In addition, river models assist the engineer in the design of flood-control structures as well as in the analysis of sediment movement in the river. Marine engineers make extensive tests on model ship hulls to predict the drag of the ships. Much of this type of testing is done at the Naval Ship Research and Development Center near Washington, D.C. Tests are also regularly performed on models of tall buildings to help predict the wind loads on the buildings, the stability characteristics of the buildings, and the air-flow patterns in their vicinity. The latter information is used by the architects to design walkways and passageways that are safer and more comfortable for pedestrian use.

Geometric similitude

The basic and perhaps the most obvious requirement of similitude is that the model be an exact geometric replica of the prototype.[1] Consequently, if a $1:10$ scale model is specified, this means that all linear dimensions of the model shall be $\frac{1}{10}$ that of the prototype. In Fig. 8-3 if the model and prototype are geometrically similar, the following equalities hold:

$$\frac{\ell_m}{\ell_p} = \frac{w_m}{w_p} = \frac{c_m}{c_p} = L_r \qquad (8\text{-}12)$$

Here ℓ, w, and c are specific linear dimensions associated with the model and prototype, and L_r is the scale ratio between model and prototype. It then follows that the ratio of corresponding areas between model and prototype will be the square of the length ratio: $A_r = L_r^2$. The ratio of corresponding volumes will be given by $\forall_m/\forall_p = L_r^3$.

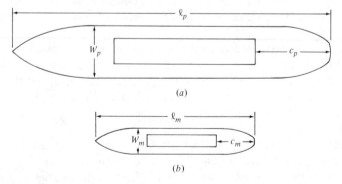

(a)

(b)

FIGURE 8-3   (a) Prototype. (b) Model.

---

[1] For most model studies this is a basic requirement; however, for certain types of problems such as river models, distortion of the vertical scale is often necessary to obtain meaningful results.

and finally
$$\frac{M_m a_m}{M_p a_p} = \frac{F_{p_m}}{F_{p_p}}$$
(8-16)

where
$$F_p \propto \Delta p L^2$$

which yields
$$C_{p_m} = C_{p_p}$$

The foregoing development then means that dynamic similitude (similarity of pressure coefficients) will be completely achieved for flow over a spillway if the Froude numbers and the Reynolds numbers are the same in the model as in the prototype. As will be shown in Sec. 8-7, the latter requirement can be relaxed if the model is reasonably large. In the force polygon of Fig. 8-4, it can be seen that the polygon can be completed with only three of the forces; hence, one of them is dependent on the others. Thus, if we think of the pressure force as a dependent one, it also follows that the pressure coefficient is dependent on the other parameters. In other words, if we have equality of F and Re, then we will automatically have equality of C_p between the model and prototype. This is exactly the conclusion we also reached in Sec. 8-5, but in a different manner. If other forces, such as surface tension or elastic forces, were significant in establishing the flow pattern, then equality of additional dimensionless parameters, such as the Weber number and the Mach number, respectively, would be required.

We then conclude from the foregoing developments *that the requirement for similarity of flow between model and prototype is that the significant dimensionless parameters must be equal for model and prototype.*

For example, if we were model-testing a valve (enclosed flow), the only forces that might affect the flow pattern would be the shear forces produced by viscous effects; consequently, dynamic similarity would prevail with equality of Reynolds number. That is, the model Reynolds number would have to be the same as the prototype Reynolds number. For flow over a spillway, the gravitational force is predominant in establishing the flow pattern. Therefore, it is required that the Froude number in the model be the same as the Froude number in the prototype for dynamic similarity in this type of application. The significance of these requirements and their actual use in predicting prototype performance will be considered in the next two sections.

8-7 MODEL STUDIES FOR FLOWS WITHOUT FREE-SURFACE EFFECTS

Free-surface effects are absent in the flow of liquids or gases in closed conduits, including control devices such as valves, or for the flow about bodies (for example, aircraft) that travel through air or are deeply submerged in a liquid such as water (submarines). Free-surface effects also are absent where a structure such as a building is stationary and wind flows past the structure. In all these cases assuming relatively low Mach numbers, it is the Reynolds-number criterion that must be used for dynamic similarity. That is, the Reynolds number in the model must equal the Reynolds number in the prototype. The following examples will illustrate the application.

EXAMPLE 8-3   The drag characteristics of a 5-m-diameter by 60-m-long blimp is to be studied in a wind tunnel. If the speed of the blimp through still air is 10 m/s, what airspeed in the wind tunnel is needed for dynamically similar conditions if a 1:10 scale model is to be tested in the tunnel? Assume the same air pressure and temperature for both model and prototype.

Solution   For dynamic similarity, the Reynolds number of the model must equal the Reynolds number of the prototype, or

$$\text{Re}_m = \text{Re}_p$$

Hence

$$\frac{V_m L_m \rho_m}{\mu_m} = \frac{V_p L_p \rho_p}{\mu_p}$$

From this we can solve for V_m, which is

$$V_m = V_p \frac{L_p}{L_m} \frac{\rho_p}{\rho_m} \frac{\mu_m}{\mu_p}$$

then

$$V_m = 10 \times 10 \times 1 \times 1$$

$$= 100 \text{ m/s} \quad \blacktriangleleft$$

Here, we get the result that the air speed in the wind tunnel will have to be 100 m/s for true Reynolds number similitude. This speed is quite large and in fact Mach number effects may start to become important at such a speed. However, it will be shown in Sec. 8-9, pg. 296, that it is not always necessary to operate models at true Reynolds number criteria to obtain useful results.

EXAMPLE 8-4   The valve shown is the type used in the control of water in large conduits. Model tests are to be done, using water as the fluid, to determine how the valve will operate under wide-open conditions. If the prototype size is 6 ft in diameter at the inlet, what flow rate is required for the model if the prototype flow is 700 cfs? Assume that temperature for model and prototype is 60°F and that the model inlet diameter is 1 ft.

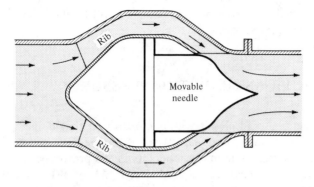

Solution   Again we use the Reynolds-model criterion; therefore

$$\text{Re}_m = \text{Re}_p$$

$$\frac{V_m L_m}{\nu_m} = \frac{V_p L_p}{\nu_p}$$

$$\frac{V_m}{V_p} = \frac{L_p}{L_m} \frac{\nu_m}{\nu_p}$$

However, we are interested in the total rate of flow; therefore, if we multiply both sides of the equation above by the area ratio A_m/A_p, the resulting product on the left-hand side will yield the desired discharge ratio:

$$\frac{A_m}{A_p} \frac{V_m}{V_p} = \frac{A_m}{A_p} \frac{L_p}{L_m} \frac{\nu_m}{\nu_p}$$

or

$$\frac{Q_m}{Q_p} = \frac{A_m}{A_p} \frac{L_p}{L_m} \frac{\nu_m}{\nu_p}$$

Since $A_m/A_p = L_r^2 = (L_m/L_p)^2$, we obtain

$$\frac{Q_m}{Q_p} = \frac{L_m^2}{L_p^2} \frac{L_p}{L_m} \frac{\nu_m}{\nu_p}$$

But $\nu_m = \nu_p$, or $\nu_m/\nu_p = 1$ because water is used for both model and prototype, so

$$\frac{Q_m}{Q_p} = \frac{L_m}{L_p}$$

Finally

$$Q_m = \tfrac{1}{6} Q_p = \tfrac{1}{6} \times 700$$

$$= 116.7 \text{ cfs} \qquad \blacktriangleleft$$

Note: This discharge is very large indeed, and in fact it points up the fact that very few model studies are made that completely satisfy the

Reynolds similitude criterion. This subject will be discussed more completely in the next sections.

8-8   SIGNIFICANCE OF THE PRESSURE COEFFICIENT

In the foregoing examples it was demonstrated that dynamic similarity between model and prototype exists if the significant dimensionless parameters are the same in the model as in the prototype. Since none of the parameters we have considered (Re, F, M, or W) explicitly include Δp, one may wonder how Δp in the model is related to the change in pressure in the prototype. This is accomplished by means of the pressure coefficient. If we refer to Eq. (8-10),

$$\frac{\Delta p}{\frac{1}{2}\rho V^2} = f\left(\frac{VL\rho}{\mu}, \frac{V}{c}, \frac{\rho LV^2}{\sigma}, \frac{V}{\sqrt{gL}}\right) \tag{8-10}$$

we see that the pressure coefficient, $\Delta p/\frac{1}{2}\rho V^2$, is a function of the basic parameters of similitude. Consequently, if dynamic similarity exists, that is, if the significant dimensionless numbers are the same in the model and prototype, then it follows that the pressure coefficient will also be the same in the model and prototype. Thus, when we have dynamic similarity, the following holds:

$$C_{p\,\text{model}} = C_{p\,\text{prototype}} \tag{8-17}$$

or

$$\frac{\Delta p_m}{\frac{1}{2}\rho_m V_m^2} = \frac{\Delta p_p}{\frac{1}{2}\rho_p V_p^2} \tag{8-18}$$

The pressure coefficient can be used like any of the other basic parameters for model analysis. It is useful not only in relating pressure changes in the model to those in the prototype but also in relating total forces in model and prototype. The latter is accomplished by multiplying the pressure ratio by the area ratio.

EXAMPLE 8-5   For the given conditions of Example 8-3, if the pressure difference between two points on the surface of the blimp is measured to be 17.8 kPa in the model, what will be the pressure difference in the prototype for dynamically similar conditions?

Solution   By using Eq. (8-18), the prototype pressure difference will be given as

$$\Delta p_p = \Delta p_m \frac{\rho_p}{\rho_m} \frac{V_p^2}{V_m^2}$$

However, $\qquad \rho_p = \rho_m \quad$ and $\quad \dfrac{V_p}{V_m} = \dfrac{L_m}{L_p}$

Therefore, $\qquad \Delta p_p = \Delta p_m (\tfrac{1}{10})^2$

$$= \frac{1}{100}\,\Delta p_m$$

$$= \frac{17{,}800}{100} = 178 \text{ Pa} \qquad \blacktriangleleft$$

EXAMPLE 8-6 For the given conditions of Example 8-3, if the drag force on the model blimp is measured to be 1,530 N, what corresponding force would be expected in the prototype?

Solution Again using Eq. (8-18), we can write

$$\frac{\Delta p_p}{\Delta p_m} = \frac{\rho_p}{\rho_m}\frac{V_p^2}{V_m^2}$$

Also $\qquad\qquad\qquad \text{Re}_p = \text{Re}_m$

$$\frac{V_p L_p}{\nu_p} = \frac{V_m L_m}{\nu_m}$$

or $\qquad\qquad\qquad \dfrac{V_p}{V_m} = \dfrac{\nu_p}{\nu_m}\dfrac{L_m}{L_p}$

However, for this example, $\rho_p = \rho_m$ and $\nu_p = \nu_m$; thus, when we let $\rho_p/\rho_m = 1$, we obtain from the first equation in this example the following:

$$\frac{\Delta p_p}{\Delta p_m} = \frac{V_p^2}{V_m^2}$$

From the Reynolds-number relationship, when we set $\nu_p/\nu_m = 1$, we get the following:

$$\frac{V_p}{V_m} = \frac{L_m}{L_p} \qquad \frac{V_p^2}{V_m^2} = \frac{L_m^2}{L_p^2}$$

Thus, when we substitute L_m^2/L_p^2 for V_p^2/V_m^2, we get

$$\frac{\Delta p_p}{\Delta p_m} = \frac{L_m^2}{L_p^2}$$

Now multiplying both sides of the foregoing equation by the area ratio A_p/A_m, which is the same as L_p^2/L_m^2, we obtain

$$\frac{\Delta p_p}{\Delta p_m}\frac{L_p^2}{L_m^2} = \frac{L_m^2}{L_p^2}\frac{L_p^2}{L_m^2}$$

The left-hand side of this equation is the ratio of the prototype force to the model force, and the right-hand side is unity. Hence we arrive at the interesting result that the model force is the same as the prototype force. When we use the Reynolds-similarity criterion and use the same fluid for the model and prototype, the forces on the model will always be the same as on the prototype. Consequently, for this example the force on the prototype blimp will be 1,530 N.   ◄

8-9   APPROXIMATE SIMILITUDE AT HIGH REYNOLDS NUMBERS

The primary justification for model tests is that it is more economical to get answers needed for engineering design by such tests than by any other means. However, as revealed by Examples 8-3, 8-4, and 8-6, true similarity according to the Reynolds-similarity criterion yields quantities for the model that would require very costly model setups. Consider the size and power required for wind-tunnel tests of the blimp of Example 8-3. The wind tunnel would probably require at least a 2-m by 2-m section to accommodate the model blimp; and with a 100 m/s airspeed in the tunnel, it can be shown that the power required for producing continuously a stream of air of this size and velocity is in the order of 4 MW. Such a test is not prohibitive, but it is very expensive. It is also conceivable that the 100 m/s airspeed would introduce Mach-number effects not encountered with the prototype, thus generating concern over the validity of the model data. Furthermore, a force of 1,530 N is indeed larger than that usually associated with model tests. Therefore, especially when studying problems involving non-free-surface flows, it is desirable to perform model tests so that the large magnitudes of forces or pressures are not encountered.

For many cases, it is possible to obtain all the needed information from abbreviated tests because often the Reynolds-number effect (relative viscous effect) either becomes insignificant at high Reynolds numbers or it becomes independent of the Reynolds number. The point where testing can be stopped often can be detected by inspection of a graph of the pressure coefficient C_p versus Reynolds number Re. Such a graph for a venturi meter in a pipe is shown in Fig. 8-5. In this meter, Δp is the pressure difference between the points shown, and V is the velocity in the restricted section of the venturi meter. Here it is seen that viscous forces affect the value of C_p below a Reynolds number of approximately 50,000; however, for higher Reynolds numbers, C_p is virtually constant. Physically this means that at low Reynolds numbers (relatively high viscous forces) a significant part of the change in pressure comes from viscous resistance, and the remainder comes from the inertial reaction (change in

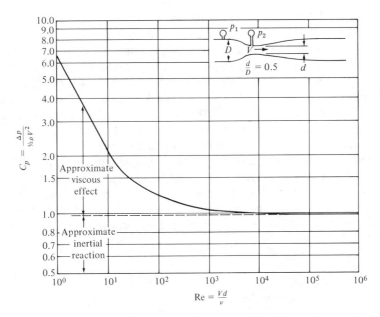

FIGURE 8-5 C_p for a venturi meter as a function of the Reynolds number.

kinetic energy) of the fluid as it passes through the venturi meter. However, with high Reynolds numbers (equivalent to either small viscosity or large product of V, D, and ρ), the viscous resistance is negligible compared with that required to overcome the inertial reaction of the fluid. Since the ratio of Δp to the inertial reaction does not change (constant C_p) with the high Reynolds numbers, there is no need to carry out tests at higher Reynolds numbers. This in general is true so long as the flow pattern does not change with the Reynolds number.

In a practical sense, the one who is in charge of the model test will try to predict from previous works approximately what maximum Reynolds number will be needed to reach the point of insignificant Reynolds number effect, and then the person will design the model accordingly. After a series of tests have been made on the model, C_p versus Re will be plotted to see if, in fact, the range of constant C_p has been reached. If so, then no more data are needed to predict the prototype performance. However, if C_p has not reached a constant value, the test program will have to be expanded or results will have to be extrapolated. Thus it is seen that the results of some model tests can be used to predict prototype performance even though the Reynolds numbers are not the same for the model and for the prototype. This is especially valid for angular-shaped bodies such as model buildings tested in wind tunnels.

EXAMPLE 8-7   Tests were made by Roberson and Crowe (10) on model-building shapes such as shown here to determine the effects of free-stream turbulence and angle of incidence on the surface pressure distribution. Figure (*b*) is an example of the temporal-mean pressure distribution obtained at a relative elevation on the building of $z/H = 0.7$ for an angle of incidence α of 8° and a wind speed V of 20 m/s. The air temperature in the wind tunnel was 20°C. If it is assumed that C_p does not change with higher Re's, what would be the maximum and the minimum temporal mean pressures on a similar 100-m-high building in a 50-m/s wind (10°C) with the same angle of incidence? If the instantaneous magnitude of the lowest pressure is three times more negative than the lowest temporal mean pressure (this is due to turbulence), and if the pressure inside the building is the same as the mean pressure on the leeward wall, then what will be the total wind force on a 2-m by 2-m window in the lowest pressure zone?

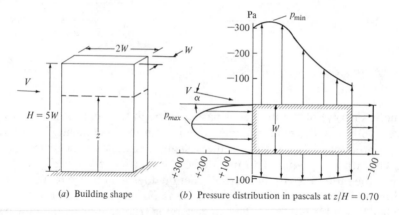

(*a*) Building shape   (*b*) Pressure distribution in pascals at $z/H = 0.70$

Solution   We are assuming the same C_p for the model as for the full-scale prototype building; therefore, we have

$$C_{p_m} = C_{p_p}$$

$$\frac{\Delta p_m}{\rho_m V_m^2/2} = \frac{\Delta p_p}{\rho_p V_p^2/2}$$

$$\frac{p_m - p_0}{\rho_m V_m^2/2} = \frac{p_p - p_0}{\rho_p V_p^2/2}$$

However, our reference pressure p_0 is zero gage, so we can solve for p_p:

$$p_p = \frac{\rho_p}{\rho_m} \frac{V_p^2}{V_m^2} p_m$$

Therefore we obtain

$$p_{p_{max}} = \frac{1.25}{1.20}\left(\frac{50}{20}\right)^2 p_{m_{max}}$$

$$= \frac{1.25}{1.20} \times 6.25 \times 250$$

$$= 1.63 \text{ kPa}$$

In a similar manner $p_{p_{min}} = -1.95$ kPa and the pressure on the leeward wall (and inside the building) is

$$p_{p_{inside}} = -0.58 \text{ kPa}$$

The force on the window will be the product of the difference of pressure across the window and the area. Here the lowest outside pressure is assumed to be $3 \times p_{min}$ so we obtain

$$F_{\text{on window}} = (p_{\text{inside}} - p_{\text{outside}}) \times A$$
$$= \{-0.58 \text{ kPa} - [3 \times (-1.95 \text{ kPa})]\} \times 4 \text{ m}^2$$
$$= 21.0 \text{ kN} \qquad \blacktriangleleft$$

Before leaving the section on non-free-surface flows, it should be noted that Mach-number effects (compressibility) usually become significant for Mach numbers exceeding 0.3. It is not always possible to tell which parameter, Mach number or Reynolds number, will be the most significant in a certain situation. It depends a great deal on what information the engineer is seeking as to which similitude parameter is chosen. If the engineer is interested in the viscous motion of fluid near a wall in shock-free supersonic flow, then the Reynolds number would be selected as the significant similitude parameter. However, if the shock-wave pattern over a body were of interest, then the Mach number would obviously be selected as the similitude parameter. The Mach number and its significance is discussed in more detail in Chapter 12.

8-10   FREE-SURFACE MODEL STUDIES

Spillway models

The major influence, besides the spillway geometry itself, on the flow of water over a spillway is the action of gravity; hence, the Froude-similarity criteria is used for such model studies. It can be appreciated that for large spillways with depths of water in the order of 3 or 4 m and velocities in the order of 10 m/s or more the Reynolds numbers will be very large. At high values of the Reynolds number the relative viscous forces are often independent of the Reynolds number, as already noted in the foregoing section (Sec. 8-9). However, if the reduced-scale model is made too small, the viscous forces as well as the surface-tension forces will have a larger relative effect on the flow in the model than in the prototype. Therefore, in practice, spillway models are made large enough so that the viscous effects have about the same relative effect in the model as in the prototype (the viscous effects are nearly independent of the Reynolds number). Then the Froude number is the significant similarity parameter. Most model spillways are made at least 1 m high, and for precise studies, such as calibration of individual spillway bays, it is not uncommon to design and construct models with 2- or 3-m-high spillway sections. Figures 8-6 and 8-7 show a comprehensive model and spillway model for Hells Canyon Dam.

EXAMPLE 8-8   A 1:49 scale model of a proposed dam is used to predict prototype flow conditions. If the design flood discharge over the spillway is 15,000 m³/s, what water flow rate should be established in the model to simulate this flow? If a velocity of 1.2 m/s is measured at a point in the model, what is the velocity at a corresponding point in the prototype?

Solution   The Froude model law will be used; therefore,

$$F_m = F_p$$

$$\frac{V_m}{\sqrt{g_m L_m}} = \frac{V_p}{\sqrt{g_p L_p}}$$

However,
$$g_m = g_p$$

Thus we have
$$\frac{V_m}{V_p} = \sqrt{\frac{L_m}{L_p}}$$

Now multiply both sides of the equation above by the area ratio A_m/A_p:

$$\frac{V_m}{V_p}\frac{A_m}{A_p} = \frac{A_m}{A_p}\sqrt{\frac{L_m}{L_p}}$$

The left-hand side of the equation above is the discharge ratio and $A_m/A_p = L_m^2/L_p^2$; hence, we obtain

$$\frac{Q_m}{Q_p} = \left(\frac{L_m}{L_p}\right)^{5/2}$$

$$Q_m = Q_p \left(\frac{1}{49}\right)^{5/2}$$

$$= 15,000 \, \frac{1}{16,800} = 0.89 \text{ m}^3/\text{s}$$

FIGURE 8-6   Comprehensive model for Hells Canyon Dam. Tests made at the Albrook Hydraulic Laboratory, Washington State University.

FIGURE 8-7   Spillway model for Hells Canyon Dam. Tests made at the Albrook Hydraulic Laboratory, Washington State University.

From the fourth equation in this example we have

$$\frac{V_m}{V_p} = \sqrt{\frac{L_m}{L_p}}$$

Consequently, $\quad V_p = V_m \sqrt{\frac{L_p}{L_m}} = 1.2 \times 7 = 8.4 \text{ m/s}$ ◀

At the given point in the prototype, we would have a velocity of 8.4 m/s.

Ship model tests

The resistance that the propulsion system of the ship must overcome is the sum of both the wave resistance and the surface resistance of the hull.

FIGURE 8-8 Ship-model test at the Naval Ship Research and Development Center.

The wave resistance is a free-surface, or Froude-number, phenomenon, and the hull resistance is a viscous, or Reynolds-number, phenomenon. Because both the wave and viscous effects contribute significantly to the overall resistance, it would appear that both the Froude and Reynolds similarity laws should be used. However, it is impossible to satisfy both laws if the model liquid is water (the only practical test liquid), because the Reynolds-model law dictates a higher velocity for the model than the prototype [equal to $V_p(L_p/L_m)$], whereas the Froude-model law dictates a lower velocity for the model [equal to $V_p(\sqrt{L_m}/\sqrt{L_p})$]. To circumvent such a dilemma, the procedure is to model for the phenomenon that is the most difficult to predict analytically and to account for the other resistance by analytical means. Since the wave resistance is the most difficult problem, the model is operated according to the Froude-model law and the hull resistance is accounted for analytically.

To illustrate how the test results and the analytical solutions for surface resistance are merged to yield design data, the necessary sequential steps are noted below:

1. Model tests are first made according to the Froude-model law and the total model resistance is measured. This will be equal to the model-wave resistance plus the surface resistance of the hull.
2. The model-surface resistance is then estimated by analytical calculations.
3. The model-surface resistance calculated in step 2 above is subtracted from the total model resistance of step 1 to yield the model-wave resistance.
4. By using the Froude-model law on the model-wave resistance, the prototype wave resistance is calculated.
5. The prototype surface resistance of the hull is then estimated by analytical means.
6. The sum of the prototype wave resistance from step 4 and the prototype surface resistance from step 5 yields the total prototype ship resistance.

8-11   CLOSURE

In this chapter we have considered the role of dimensional analysis in the derivation of dimensionless numbers that are used throughout the field of fluid mechanics. We have also seen how functional relationships between dimensionless numbers can be determined experimentally for certain flow processes. In some cases, the functional relationships are published as curves or formulas for use by engineers. In other cases, the specific tests are simply model studies used to assist in the design of a given unique structure. If the student has acquired a basic understanding of the development of these dimensionless numbers and how they are used, he or she will have gained a much better appreciation and understanding of the entire field of fluid mechanics.

PROBLEMS

8-1 Determine which of the following equations are dimensionally homogeneous:

(a)
$$Q = \tfrac{2}{3}CL\sqrt{2g}\,H^{3/2}$$

   where Q = discharge
            C = a pure number
            L = length
            g = acceleration due to gravity
            H = head

(b)
$$V = \frac{1.49}{n}\,R^{2/3}S^{1/2}$$

   where V = velocity
            n = a length to the one-sixth power
            R = length
            S = slope

(c)
$$h_f = f\frac{L}{D}\frac{V^2}{2g}$$

   where h_f = head loss
            f = a dimensionless resistance coefficient
            L = length
            D = diameter
            V = velocity
            g = acceleration due to gravity

(d)
$$D = \frac{0.074}{\text{Re}^{0.2}}\frac{Bx\rho V^2}{2}$$

   where D = drag force
            $\text{Re} = Vx/\nu$
            V = velocity
            x = length
            B = width
            ρ = mass density

8-2 Determine the dimensions of the following variables and combinations of variables in terms of the length, force, and time system of units; and in terms of the length, mass, and time system of units (*Hint:* Convert F to M by Newton's second law.)

(*a*) T (torque)

(*b*) $\rho V^2/2$, where V is velocity and ρ is mass density

(*c*) $\sqrt{\tau/\rho}$, where τ is shear stress

(*d*) Q/ND^3, where Q is discharge, D is diameter, and N is angular speed of a pump.

8-3 The maximum rise of a liquid in a small capillary tube is a function of the diameter of the tube, the surface tension, and the specific weight of the liquid. What are the significant dimensionless parameters for the problem?

8-4 It is known that the displacement of a particle in rectilinear motion is a function of U_0 (its initial velocity), time t, and magnitude of constant acceleration a. Determine by dimensional analysis the dimensionless parameters that apply to this kinematic problem. Then with the use of the formula from basic mechanics verify the validity of the parameters.

8-5 Consider steady-viscous flow through a small horizontal tube. For this type of flow the pressure gradient along the tube $\Delta p/\Delta \ell$ should be a function of the viscosity μ, the mean velocity V, and the diameter of tube D. By dimensional analysis derive a formula relating these variables.

8-6 It is known that the pressure developed by a centrifugal pump Δp is a function of the diameter of the impeller D, the speed of rotation n, discharge Q, and the fluid density ρ. By dimensional analysis determine the dimensionless parameters relating these variables.

8-7 The velocity V of ripples on a liquid surface is a function of the ripple length L, density ρ, and surface tension σ of the liquid. By dimensional analysis derive an expression for V.

8-8 By dimensional analysis develop the relationship between the pressure gradient in the radial direction (dp/dr) for a rotating tank of liquid (vertical axis of rotation) and the significant variables ρ, ω, and r where ρ is mass density, ω is angular velocity, and r is the radial distance from the axis of rotation.

8-9 A smooth circular plate is positioned a distance S away from a smooth boundary as shown in the figure. Oil with viscosity μ fills the space between the plate and boundary. Then it should be obvious that the torque required to rotate the plate as shown will be a function of μ, ω, S, and D, where ω is the angular velocity in radians per second. Determine the dimensionless parameters that would be involved if you were to correlate results by experimental means only.

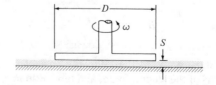

8-10 A general study is to be made of the height of rise of liquid in a capillary tube as a function of time after the start of a test. Other significant variables include surface tension, mass density, specific weight, viscosity, and diameter of the tube. Determine the dimensionless parameters that apply to the problem.

8-11 It takes a certain length of time for the liquid level in a tank, diameter D, to drop from position h_1 to position h_2 as the tank is being drained through an orifice of diameter d at the bottom. Determine the dimensionless parameters that apply to this problem. Assume that the fluid is nonviscous.

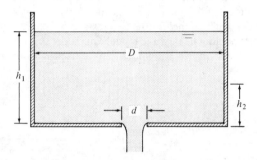

8-12 A spherical balloon that is to be used in air at 60°F (15.5°C) is tested by towing a 1:3 scale model in a lake. If the model is 1 ft (30.5 cm) in diameter and a drag of 20 lbf (89.0 N) is measured when the model is being towed in deep water at 5 ft/sec (1.52 m/s), what drag in pounds force and newtons can be expected for the prototype in air under dynamically similar conditions? Assume that the water temperature is 60°F (15.5°C).

8-13 An airplane travels in air ($p = 100$ kPa, $T = 10°C$) at 200 m/s. If a 1:5 scale model of the plane is tested in a wind tunnel at 25°C, what must be the density of the air in the tunnel so that both the Reynolds- and Mach-similarity criteria are satisfied? For E_v take the adiabatic relation as given in Chapter 2.

8-14 Water at 60°F flows at 7.0 ft/sec in a 6-in. smooth pipe. What must be the velocity of air (standard atm. pressure) at 80°F in a 3-in. smooth pipe for the two flows to be dynamically similar?

8-15 Water at 20°C flows at 3 m/s in a 20-cm smooth pipe. What must be the velocity of air (standard atm. pressure) at 30°C in a 10-cm smooth pipe for the two flows to be dynamically similar?

8-16 Using the Reynolds criterion, a 1:1 scale model of a torpedo is tested in a wind tunnel. If the velocity of the torpedo in water is 7 m/s (23.0 ft/sec), what should be the air velocity (standard atm. pressure) in the wind tunnel? Temperature for both tests is 10°C (50°F).

8-17 A discharge meter to be used in a 40-cm pipeline carrying oil ($\nu = 10^{-5}$ m²/s) is to be calibrated by means of a model (1:4 scale) carrying water ($T = 20°C$ and standard atm. pressure). If the model is operated with a velocity of 1 m/s and a meter coefficient is determined, at what velocity in the prototype may we be certain that the prototype meter coefficient will have the same magnitude as in the model?

8-18 Water at 10°C flowing through a 10-cm-diameter rough pipe is to be simulated by air (20°C) flowing through the same pipe. If the velocity of water is 1.5 m/s, what will the air velocity have to be to achieve dynamic similarity? Assume the absolute air pressure in the pipe to be 150 kPa.

8-19 The "noisemaker" B is towed behind the minesweeper A to set off enemy acoustic mines such as at C. The drag force of the "noisemaker" is to be studied in a water tunnel at a $\frac{1}{4}$ scale ratio (model is $\frac{1}{4}$ the size of the full scale). Then if the full-scale towing speed is 3 m/s, what should be the water velocity in the water tunnel for the two tests to be exactly similar? What will be the prototype drag force if the model drag force is found to be 868 N? Assume sea water at the same temperature is used in both the full-scale and model tests.

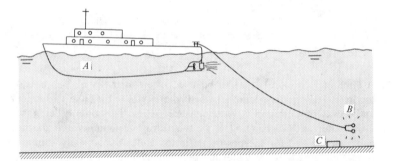

8-20 Oil with a kinematic viscosity of 5×10^{-5} ft²/sec is pumped through a 12-in. smooth pipe at 10 ft/sec. What velocity should water have at 80°F in a 2-in. smooth pipe to be dynamically similar?

8-21 Oil with a kinematic viscosity of 4×10^{-6} m²/s flows through a 20-cm-diameter pipe at 3 m/s. What velocity should water have at 20°C in a 5-cm-diameter pipe to be dynamically similar?

8-22 A large venturi meter is calibrated by means of a 1:10 scale model using the prototype liquid. What is the discharge ratio, Q_m/Q_p, for dynamic similarity?

8-23 The moment acting upon a submarine rudder is studied by a 1:60 scale model. If the test is made in a water tunnel and if the moment measured on the model is 4 ft-lbf when the fresh-water speed in the tunnel is 40 ft/sec, what are the corresponding moment and speed for the prototype? Assume that the prototype operates in sea water. Assume $T = 50°F$ for both the fresh water and sea water.

8-24 The moment acting upon a submarine rudder is studied by a 1:60 scale model. If the test is made in a water tunnel and if the moment measured on the model is 2 m · N when the fresh-water speed in the tunnel is 10 m/s, what are the corresponding moment and speed for the prototype? Assume the prototype operates in sea water. Assume $T = 10°C$ for both the fresh water and the sea water.

8-25 Determine the relationship between the kinematic viscosity ratio ν_m/ν_p and the scale ratio if both the Reynolds- and Froude-similarity criteria are to be satisfied in a given model test.

8-26 The scale ratio between a model dam and its prototype is $1:25$. In the model test the velocity of flow near the crest of the spillway was measured to be 2 m/s (6.56 ft/sec). What is the corresponding prototype velocity?

8-27 A seaplane model is built at a $1:10$ scale. To simulate takeoff conditions at 100 km/h, what should be the corresponding model speed?

8-28 If the scale ratio between a model spillway and its prototype is $1:25$, what velocity and discharge ratio would prevail between model and prototype? If the prototype discharge is 3,000 m³/s (106,000 cfs), what is the model discharge?

8-29 A model spillway is constructed at a scale of $1:100$. For a model velocity of 2 m/s (6.56 ft/sec), what is the corresponding prototype velocity?

8-30 A $1:25$ scale model of a spillway is tested in a laboratory. If the model velocity and discharge are 7.87 ft/sec and 3.53 cfs, respectively, what are the corresponding values for the prototype?

8-31 Flow around a bridge pier is studied by model at $1:10$ scale. When the velocity in the model is 0.9 m/s (2.95 ft/sec), the standing wave at the pier nose is observed to be 2.5 cm (0.08 ft) in height. What are the corresponding values of velocity and wave height in the prototype?

8-32 A $1:25$ scale model of a spillway is tested. If the discharge in the model is 0.1 m³/s, to what prototype discharge does this correspond? If it takes 1 min for a particle to float from one point to another in the model, how long would it take a similar particle to traverse the corresponding path in the prototype?

8-33 The maximum wave force on a $1:36$ model sea wall was found to be 80 N. For a corresponding wave in the full-scale wall what full-scale force would you expect? Assume fresh water is used in the model study. Assume $T = 10°C$ for both model and prototype water.

8-34 A hydraulic model, $1:10$ scale, is built to simulate the flow conditions of a spillway of a dam. For a particular run, the waves downstream were observed to be 10 cm high. Similar waves on the full-scale dam operating under the same conditions would be how high? If the wave period in the model is 1 s, what would be the wave period in the prototype?

8-35 It is desired to build a $1:20$ scale model of a spillway. If the prototype has a capacity of 140 m³/s (4,940 cfs), what must be the water discharge in the model to ensure dynamic similarity? If the total force on part of the model is found to be 22 N (5 lbf), to what prototype forces does this correspond?

8-36 A newly designed dam is to be modeled in the laboratory. The prime objective of the general model study is to determine the adequacy of the spillway design and to observe the water velocities, elevations, and pressures at critical points of the structure. The reach of the river to be modeled is 1,200 m long, the width of the dam (also the maximum width of the reservoir upstream) is to be 300 m and the maximum flood discharge to be modeled is 5,000 m³/s. If the maximum laboratory discharge is limited to 0.90 m³/s and the floor space available for the model construction is 50 m long × 20 m wide, determine the largest feasible scale ratio (model/prototype) for such a study.

8-37 The wave resistance of a 1 : 36 scale model of a ship is 2 lbf at a model speed of 4 ft/sec. What are the corresponding velocity and wave resistance of the prototype?

8-38 The wave resistance of a 1 : 36 scale model of a ship is 10 N at a model speed of 1.5 m/s. What are the corresponding velocity and wave resistance of the prototype?

8-39 A 1 : 10 scale of an automobile is tested in a pressurized wind tunnel. The test is to simulate the automobile traveling at 80 km/h in air at atmospheric pressure and 25°C. The wind tunnel operates with air at 25°C. At what pressure in the test section must the tunnel operate to have the same Mach and Reynolds number? The speed of sound in air at 25°C is 345 m/s.

8-40 If the tunnel in Prob. 8-39 were to operate at atmospheric pressure and 25°C, what speed would be needed to achieve the same Reynolds number of the prototype? At this speed, would you conclude that Mach-number effects were important?

8-41 An important parameter in rarefied gas dynamics is the ratio M/Re. If this ratio exceeds unity, the flow is rarefied. A spherical satellite 2 ft in diameter reenters the earth's atmosphere at 24,000 mph, where the pressure and temperature of the air are 22 psfa and $-67°F$. Can the flow around the satellite be classified as rarefied? The dynamic viscosity at $-67°F$ is 3.0×10^{-7} lbf-sec/ft² and the speed of sound is 975 ft/sec.

8-42 The critical Weber number for breakup of a liquid droplet is 6.0 based on the droplet diameter. The surface tension of heptane is 0.02 N/m. If a spray of heptane is atomized by discharging into air at atmospheric pressure and 100°C at 30 m/s, what is the expected size of the droplets?

8-43 Water is sprayed from a nozzle at 10 m/s into air at atmospheric pressure and 20°C. Estimate the size of the droplets produced if the Weber number for breakup is 6.0 based on the droplet diameter.

8-44 A 1 : 250 scale model of a high-rise office building is tested in a wind tunnel to estimate the pressures and forces on the full-scale structure. The wind-tunnel air speed is 20 m/s at 20°C and the full-scale structure is expected to withstand winds of 160 km/h (10°C). If the extreme values of the pressure coefficient are found to be 1.0, -2.7, and -0.8 on the windward wall, side wall, and leeward wall, respectively, of the model then what corresponding full-scale pressures could be expected for the design wind? *Note:* $C_p = \Delta p / \frac{1}{2} \rho V_0^2$, where $p = p_0$, and p_0 and V_0 are the pressure and velocity in the air approaching the structure. If the lateral wind force (wind force on building normal to wind direction) was measured to be 20 N in the model, then what lateral force might be expected in the prototype in the 160 km/h wind?

REFERENCES

1. Allen, J. *Scale Models in Hydraulic Engineering.* Longmans, Green & Co., Ltd., London, 1952.
2. Bain, D. C., Baker, P. J., and Rowat, M. J. *Wind Tunnels, An Aid to Engineering Structure Design.* British Hydromechanics Research Association, Cranfield, England, 1971.
3. Buckingham, E., "Model Experiments and the Forms of Empirical Equations," *Trans. ASME,* 37, (1915), 263.
4. Freeman, John R. (ed.). *Hydraulic Laboratory Practice.* American Society of Mechanical Engineers (1929).
5. Hickox, G. H. "Hydraulic Models," in C. V. Davis (ed.), *Handbook of Applied Hydraulics,* 2nd ed. McGraw-Hill Book Company, New York, 1952.
6. Ipsen, D. C. *Units, Dimensions and Dimensionless Numbers.* McGraw-Hill Book Company, New York, 1960.
7. Langhaar, Henry L. *Dimensional Analysis and Theory of Models.* John Wiley & Sons, Inc., New York, 1951.
8. "Manual of Engineering Practice No. 25." *Hydraulic Models,* American Society of Civil Engineers (1942).
9. Potthoff, J. "Luftwiderstand und Auftrieb Moderner Kraftfahrzeuge." Paper No. 12, *Proc. 1st Symp. Road Vehicle Aerodyn.,* London (1969).
10. Roberson, J. A., and Crowe, C. T. "Pressure Distribution on Model Buildings at Small Angles of Attack in Turbulent Flow." *Proc. 3rd U.S. Natl. Conf. on Wind Engineering Research,* University of Florida, Gainesville (1978).
11. Warnock, J. E. "Hydraulic Similitude," in H. Rouse (ed.), *Engineering Hydraulics.* John Wiley & Sons, Inc., New York, 1950, Chap. II.

The high speed attained by surfers is owing to the fact that "planing" of the board virtually eliminates wave drag, thus leaving only surface resistance, which is relatively small.

9 SURFACE RESISTANCE

A FLUID MEDIUM THROUGH WHICH BODIES such as aircraft or ships move exerts a resistance to motion on the bodies; this resistance is called *drag*. Aeronautical engineers and naval architects are vitally interested in the drag of the airplane or ship because the success or failure of the craft is directly related to its resistance to motion. If the drag is too large, the craft may be an economic failure because of the excessive costs (initial and operational) of the propulsion system. The drag of a body results from two types of forces applied to the body: shear forces and pressure forces. The drag, or surface, resistance from shearing forces is often called *skin-friction drag*, whereas the drag resulting from pressure forces is called *form drag*. In this chapter we shall concentrate on the mechanics of surface-resistance or skin-friction drag. In Chapter 11, we will cover form drag.

9-1   INTRODUCTION

In general, the shear stress on a smooth plane surface is variable over the surface; hence the total shear force in a given direction is obtained by integrating the component of shear stress in that direction over the total area of the surface. The shear stress on a smooth plane is a direct function of the velocity gradient next to the plane, as given by Eq. (2-5). Therefore, it is seen that any problem involving the shear stress is also involved with the flow pattern in the vicinity of the surface. The layer of fluid near the surface that has undergone a change in velocity because of the shear stress at the surface is called the *boundary layer*, and the general area of study of the flow pattern in the boundary layer as well as the associated shear stress at the boundary is called *boundary-layer theory*.

We will first consider the simplest cases of surface resistance resulting from uniform laminar flow in which the velocity gradient and also the shear stress are constant; then we will consider a laminar boundary layer that develops from the leading edge of a smooth flat plate; and finally we will consider the characteristics of a turbulent boundary layer.

9-2   SURFACE RESISTANCE WITH UNIFORM LAMINAR FLOW

Uniform flow produced by relative movement of parallel plates

Consider the motion of a plate as shown in Fig. 9-1, in which the top plate causes the fluid in contact with it to have the same velocity as the plate. If

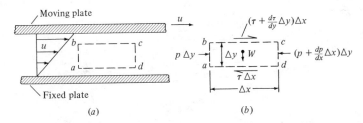

FIGURE 9-1 Flow between fixed plate and moving plate.

the pressure gradient along the plate (x direction) is zero, then the flow produced is called *Couette flow* after a French scientist, M. Couette, who pioneered work on the analysis of shear flows between parallel plates and rotating cylinders. We now consider fluid within the control volume $abcd$ having unit width normal to the paper, Fig. 9-1, and apply the momentum equation in the x direction to obtain

$$p\,\Delta y - \left(p + \frac{dp}{dx}\Delta x\right)\Delta y - \tau\,\Delta x + \left(\tau + \frac{d\tau}{dy}\Delta y\right)\Delta x = \int_{cs} u\rho\mathbf{u}\cdot d\mathbf{A} \quad (9\text{-}1)$$

Because the flow is uniform due to the parallel boundaries, the right-hand side of Eq. (9-1) is zero. Because $dp/dx = 0$, the equation reduces to $d\tau/dy = 0$, which means $\tau = $ constant. Because $\tau = \mu\,du/dy$, it is concluded that the velocity gradient is also constant, as shown in Fig. 9-1. In other words, the velocity varies linearly from zero at the fixed boundary to maximum at the moving surface.

EXAMPLE 9-1   If the fluid between the plates of Fig. 9-1 is SAE 30 lubricating oil, $T = 38°C$, if the plates are spaced 0.3 mm apart, and if the upper plate is moved at a velocity of 1.0 m/s, what is the surface resistance for 1.0 m² of the upper plate?

Solution $\mu = 1.0 \times 10^{-1}$ N $\cdot$ s/m² obtained from Appendix (Table A-4). Then

$$\tau = \mu\frac{du}{dy} = \mu\frac{\Delta u}{\Delta y}$$

$$= (1.0 \times 10^{-1}\text{ N}\cdot\text{s/m}^2)(1.0\text{ m/s})/(3 \times 10^{-4}\text{ m})$$

$$= 333\text{ N/m}^2$$

$$F_s = \tau A = 333\text{ N/m}^2 \times 1.0\text{ m}^2 = 333\text{ N} \qquad \blacktriangleleft$$

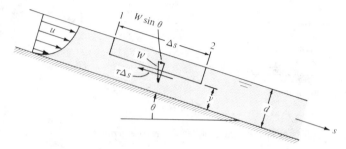

FIGURE 9-2   Free-surface flow down an inclined plane.

Uniform liquid flow over an inclined plane

Consider the case shown in Fig. 9-2, where the plane is inclined at an angle θ to the horizontal and the flow is assumed to be uniform, that is, the depth and velocity do not change with distance along the conduit. Here we apply the momentum equation to an element of fluid, the top of which is the liquid surface. The element has unit width normal to the paper. The pressure is hydrostatic across any section normal to the plane because we are assuming uniform flow. In addition, the depth is constant so that the pressure forces on the end sections of the element will cancel. Furthermore, the shear stress on the liquid surface at the liquid-air interface is assumed to be negligible; therefore, the momentum equation reduces to

$$W \sin \theta - \tau \, \Delta s = 0$$

Consequently, $\qquad \gamma(d - y)\Delta s \sin \theta = \tau \, \Delta s$

$$\tau = \gamma \sin \theta (d - y)$$

This shows that the shear stress across the section varies linearly from zero at the liquid surface to maximum at the plate, which means that physically the shear stress is proportional to the weight of the element and the weight of the element increases directly with the thickness of the element $(d - y)$. Experiments have shown that if the Reynolds number based upon the depth of flow, $\mathrm{Re} = Vd/\nu$ is less than 500, one can expect laminar flow in an open channel such as this. For laminar flow we can substitute $\mu \, du/dy$ for τ and then solve for the velocity distribution:

$$\mu \frac{du}{dy} = \gamma \sin \theta (d - y)$$

$$\frac{du}{dy} = \frac{\gamma}{\mu} \sin \theta (d - y)$$

Upon integrating, the equation above becomes

$$u = \frac{-\gamma}{\mu} \sin \theta \frac{(d - y)^2}{2} + C$$

To evaluate the constant of integration, we observe that $u = 0$ when $y = 0$; therefore,

$$C = \frac{\gamma}{\mu} \sin \theta \frac{d^2}{2}$$

The velocity distribution is then given by

$$u = \frac{\gamma}{2\mu} (\sin \theta) y (2d - y) = \frac{g \sin \theta}{2\nu} y(2d - y) \tag{9-2}$$

We now obtain the discharge per unit width by integrating the velocity u over the depth of flow:

$$q = \int_0^d u \, dy = \frac{\gamma}{2\mu} \sin \theta \left[dy^2 - \frac{y^3}{3} \right]_0^d$$

$$= \frac{1}{3} \frac{\gamma}{\mu} d^3 \sin \theta \tag{9-3}$$

The average velocity is now obtained by dividing Eq. (9-3) by the cross-sectional area d:

$$V = \frac{q}{d} = \frac{1}{3} \frac{\gamma}{\mu} d^2 \sin \theta$$

which reduces to

$$V = \frac{gd^2}{3\nu} \sin \theta \tag{9-4a}$$

The slope, $S_0 = \tan \theta$, is approximately equal to $\sin \theta$ for small slopes, thus Eq. (9-4a) can be given as

$$V = \frac{gS_0 d^2}{3\nu} \tag{9-4b}$$

EXAMPLE 9-2   Crude oil, $\nu = 9.3 \times 10^{-5}$ m²/s, sp. gr. = 0.92, flows over a flat plate that has a slope of $S_0 = 0.02$. If the depth of flow is 6 mm, what is the maximum velocity and what is the flow rate per meter of width of plate? Also determine the Reynolds number for this flow.

Solution   First assume that the flow is laminar. Since $S_0 \approx \sin \theta$,

$$u = \frac{gS_0}{2\nu} y[2d - y]$$

Therefore, the maximum velocity will occur when y is maximum ($y = d$) or

$$u_{max} = \frac{(9.81 \text{ m/s}^2)(0.02)(0.006)^2 \text{ m}^2}{2 \times 9.3 \times 10^{-5} \text{ m}^2/\text{s}}$$

$$= 0.038 \text{ m/s} \qquad \blacktriangleleft$$

The discharge per meter of width is given by Eq. (9-3), and for a small slope we replace $\sin \theta$ by S_0:

$$q = \frac{1}{3}\frac{\gamma}{\mu} S_0 d^3 = \frac{1}{3}\frac{\gamma}{\rho\nu} S_0 d^3 = \frac{1}{3}\frac{g}{\nu} S_0 d^3$$

Then
$$q = \frac{1}{3}\frac{(9.81 \text{ m/s}^2)}{(9.3 \times 10^{-5} \text{ m}^2/\text{s})}(0.02)(0.006)^3 \text{ m}^3$$

$$= 1.52 \times 10^{-4} \text{ m}^2/\text{s} \qquad \blacktriangleleft$$

Now let us check the Reynolds number to see if our original assumption of laminar flow (Re < 500) is correct.

$$\text{Re} = \frac{Vd}{\nu} = \frac{q}{\nu} = \frac{1.52 \times 10^{-4} \text{ m}^2/\text{s}}{9.3 \times 10^{-5} \text{ m}^2/\text{s}}$$

$$= 1.63 \qquad \blacktriangleleft$$

The Reynolds number is less than 500; therefore, our assumption of laminar flow and the use of Eqs. (9-3) and (9-4a) were valid.

Flow between parallel plates with a pressure gradient along the plates

Here we consider the general two-dimensional case where two plates are inclined to the horizontal at an angle θ and a pressure gradient exists along the length of the plates. The schematic arrangement for these conditions is shown in Fig. 9-3. When the momentum equation is applied to the ele-

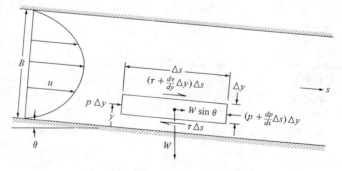

FIGURE 9-3   Flow between parallel boundaries with a pressure gradient.

ment of fluid (having dimensions $\Delta s \times \Delta y \times$ unity) shown in Fig. 9-3, we arrive at the following equation:

$$\sum F_s = 0$$

$$-\frac{dp}{ds} \Delta s\, \Delta y + \frac{d\tau}{dy} \Delta s\, \Delta y + W \sin \theta = 0$$

Here $\sin \theta = -dz/ds$ and $W = \gamma \Delta s\, \Delta y$. Then, upon making these substitutions and dividing by $\Delta s\, \Delta y$, we have

$$\frac{d\tau}{dy} = \frac{d}{ds} (p + \gamma z)$$

or

$$\frac{d\tau}{dy} = \gamma \frac{d}{ds} \left(\frac{p}{\gamma} + z \right)$$

$$\frac{d\tau}{dy} = \gamma \frac{dh}{ds} \tag{9-5}$$

However,

$$\tau = \mu \frac{du}{dy}$$

so we get

$$\frac{d}{dy} \left(\mu \frac{du}{dy} \right) = \gamma \frac{dh}{ds}$$

Here μ is a constant so we can write the above equation as

$$\frac{d^2 u}{dy^2} = \frac{\gamma}{\mu} \frac{dh}{ds}$$

Upon integrating this equation twice we have

$$u = \frac{\gamma}{\mu} \frac{dh}{ds} \frac{y^2}{2} + C_1 y + C_2$$

The constants of integration C_1 and C_2 can be evaluated from the boundary conditions $u = 0$ at $y = 0$ and $y = B$. Thus we obtain

$$C_2 = 0 \quad \text{and} \quad C_1 = -\frac{\gamma}{2\mu} \frac{dh}{ds} \frac{B}{2}$$

so

$$u = -\frac{\gamma}{2\mu} (By - y^2) \frac{dh}{ds} \tag{9-6a}$$

where

$$h = p/\gamma + z$$

As in the case of the unconfined laminar-liquid flow over a plane surface, the velocity distribution is parabolic but the maximum velocity occurs midway between the two plates. It can also be shown by integrating the velocity over the section and dividing by the section area that the mean velocity V is two-thirds of u_{max}. Note furthermore that flow is the result of a change of the piezometric head, not just a change of p or z alone. Experiments reveal that if the Reynolds number VB/ν is less than 1,000 the flow will be laminar.

EXAMPLE 9-3   Oil having a specific gravity of 0.8 and a viscosity of 2×10^{-2} N · m/s^2 flows downward between two vertical smooth plates spaced 10 mm apart. If the discharge per meter of width is 0.01 m^3/s, what is the pressure gradient dp/ds for this flow?

Solution   First check to see if the flow will be laminar or turbulent:

$$\text{Re} = \frac{VB}{\nu} = \frac{VB\rho}{\mu} = \frac{q\rho}{\mu}$$

$$= \frac{(0.01 \text{ m}^2/\text{s}) \times 800 \text{ N} \cdot \text{s}^2/\text{m}^4}{0.02 \text{ N} \cdot \text{s}/\text{m}^2}$$

$$= 400$$

The Reynolds number is less than 1,000; therefore, it is laminar. Then the velocity distribution is given by Eq. (9–6a):

$$u = -\frac{\gamma}{2\mu} (By - y^2) \frac{dh}{ds}$$

and for this example $B = 0.01$ m and

$$\nu = \mu/\rho = \frac{2 \times 10^{-2} \text{ N} \cdot \text{m/s}^2}{0.8 \times 1,000 \text{ N} \cdot \text{s}^2/\text{m}^4}$$

$$= 2.5 \times 10^{-5} \text{ m}^2/\text{s}^2$$

Then we have the u_{max}, where $du/dy = 0$ at $y = B/2$

$$u_{max} = \frac{-0.80 \times 9,810 \text{ N/m}^3}{2 \times 2 \times 10^{-2} \text{ N} \cdot \text{s/m}^2} \left(\frac{0.01^2}{2} \text{ m}^2 - \frac{0.01^2}{4} \text{ m}^2 \right) \frac{dh}{ds}$$

$$= -4.905 \, dh/ds \qquad\qquad (9\text{-}6b)$$

But $\qquad\qquad V = \tfrac{2}{3}u_{max} \qquad$ and $\qquad q = VB$

Therefore $\qquad\qquad q = \tfrac{2}{3}u_{max}B$

or $\qquad\qquad u_{max} = \tfrac{3}{2}q/B$

Here B is the plate spacing and $q = 0.01$ m^3/s

so $\qquad\qquad u_{max} = \tfrac{3}{2} \times 0.01/0.01 = 1.50 \text{ m/s} \qquad\qquad (9\text{-}6c)$

Now we can solve for dh/ds from Eqs. (9-6b) and (9-6c) to obtain

$$dh/ds = -0.306$$

However,
$$\frac{dh}{ds} = \frac{d}{ds}\left(\frac{p}{\gamma} + z\right)$$

Therefore
$$\frac{d}{ds}\left(\frac{p}{\gamma} + z\right) = -0.306$$

Because the plates are vertically oriented and s is positive downward, $dz/ds = -1$.

Thus
$$\frac{d(p/\gamma)}{ds} = 1 - 0.306$$

or
$$\frac{dp}{ds} = (0.8 \times 9{,}810 \text{ N/m}^3) \times 0.694$$

$$= 5446 \text{ N/m}^2 \text{ per meter} \qquad \blacktriangleleft$$

In other words, the pressure is increasing downward at a rate of 5.45 kPa per meter of length of plate.

9-3 QUALITATIVE DESCRIPTION OF THE BOUNDARY LAYER

Flow pattern in a boundary layer

As noted before (Sec. 9-1) the *boundary layer* is the region next to a boundary of an object in which the fluid has had its velocity diminished because of the shearing resistance created by the boundary. Outside the boundary layer the velocity is essentially the same as if an ideal (non-viscous) fluid were flowing past the object. In Sec. 9-2 we considered liquid flow over a plane surface and flow between parallel plates. In a narrow sense, this is boundary-layer flow, that is, uniform flow for a fully developed laminar boundary layer. Now we will look at the boundary layer that is still developing. That is, we will look at boundary layers that grow in thickness and have significant changes of velocity with distance along the boundary. To visualize the flow pattern associated with the boundary layer, we will qualitatively analyze the interaction between the fluid and the surface of a flat plate as a fluid (for example, air) passes by the plate. Figure 9-4 illustrates such a plate. Fluid passes over the top of the plate and underneath the plate, so two boundary layers are depicted in Fig. 9-4 (one on top and one on the bottom of the plate). In Fig. 9-4 the fluid has a constant velocity U_0 before it reaches the vicinity of the plate. However,

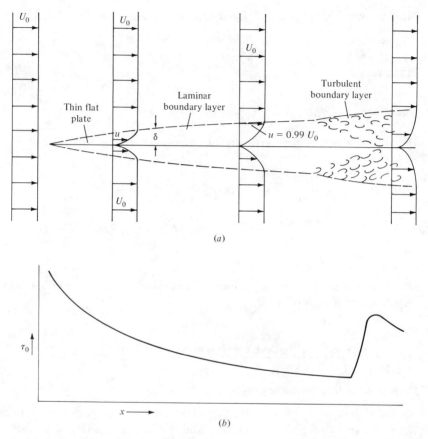

FIGURE 9-4   Boundary-layer development and shear-stress distribution along a flat plate. (*a*) Flow pattern in boundary layers above and below a thin plate. (*b*) Shear-stress distribution on either side of plate.

the fluid touching the plate has zero velocity (the velocity of the plate), because of the no-slip condition characterizing continuum flows. Therefore, a velocity gradient must exist between the fluid in the free stream and the fluid next to the plate. Consistent with this gradient will be a shear stress at the plate surface. As the fluid particles next to the plate pass the leading edge of the plate, a retarding force (from the shear stress) begins to act on them. As these particles progress downstream they will continue to be subjected to shear stress from the plate; consequently, they will continue to decelerate. In addition, these particles (because of their lower velocity) will retard other adjacent particles farther out from the plate. Thus the boundary layer becomes thicker or "grows" in the downstream direction. The broken line in Fig. 9-4 identifies the outer limit of the boundary layer.

Because the boundary layer becomes thicker, the velocity gradient becomes less steep as one proceeds along the plate, because the change of velocity between the free stream and the wall remains the same, U_0, whereas the distance over which the change occurs becomes greater in the downstream direction. This does not, however, imply that the velocity distribution is linear in the boundary layer.

Thickening of the laminar boundary layer continues smoothly in the downstream direction until the thickness is so great that the flow becomes unstable and the boundary layer becomes turbulent. In the turbulent boundary layer, eddies mix higher-velocity fluid into the region close to the wall so that the velocity gradient, du/dy, at the wall now becomes greater than that at the wall in the laminar boundary layer just upstream of the transition point.

Shear-stress distribution along the boundary

In the foregoing section it was noted that it is the shearing force of the plate that decelerates the fluid to produce the boundary layer, but no mention was made of the way the shear stress changes along the boundary. Because the shear stress is given by $\tau = \mu \, du/dy$, it is now easy to visualize that the shear stress must be relatively large near the leading edge of the plate where the velocity gradient is steep; and it will become smaller as the velocity gradient becomes smaller in the downstream direction. However, where the boundary layer becomes turbulent, the shear stress at the boundary again becomes larger to be consistent with the greater velocity gradient next to the wall in the turbulent boundary layer. Figure 9-4*b* depicts the distribution of shear stress we have just discussed. These qualitative aspects of the boundary layer will serve as a foundation for the quantitative relations presented in the next section.

9-4 QUANTITATIVE RELATIONS FOR THE LAMINAR BOUNDARY LAYER

Boundary-layer equations

In 1904 Prandtl (6) first stated the essence of the boundary layer hypothesis, which is that viscous effects are concentrated in a thin layer of fluid (the boundary layer) next to solid boundaries. Along with his discussion of the qualitative aspects of the boundary layer, he also simplified the general equations of motion of a fluid (Navier-Stokes equations) for application to the boundary layer. Then in 1908, Blasius, one of Prandtl's students, obtained a solution for the flow in a laminar boundary layer. The solution was for the case of zero pressure gradient along the plate,

$dp/dx = 0$, and one of his key assumptions was that the shape of the non-dimensional velocity distribution did not vary from section to section along the plate. That is, he assumed that a plot of the relative velocity u/U_0 versus the relative distance from the boundary, y/δ, would be the same at each section. Here δ is the thickness of the boundary layer defined as the distance from the boundary to the point in the fluid when the velocity is 99% of the free-stream velocity. With this assumption and with Prandtl's equations of motion for boundary layers, he obtained a solution for the relative velocity distribution shown in Fig. 9-5. In this plot, x is the distance from the leading edge of the plate, and Re_x is the Reynolds number based upon the free-stream velocity and the length along the plate ($\text{Re}_x = U_0 x/\nu$). In Fig. 9-5 the outer limit of the boundary layer ($u/U_0 = 0.99$) occurs at approximately $y\text{Re}_x^{1/2}/x = 5$. Since $y = \delta$ at this point, we have a relationship for the thickness of the boundary layer:

$$\frac{\delta}{x} \text{Re}_x^{1/2} = 5$$

or

$$\delta = \frac{5x}{\text{Re}_x^{1/2}} \qquad (9\text{-}7)$$

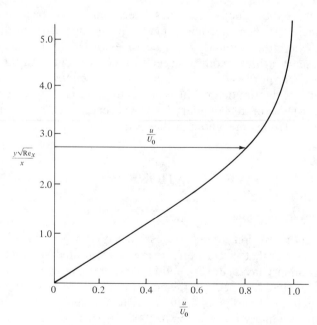

FIGURE 9-5   Velocity distribution in laminar boundary layer. [After Blasius (1).]

We can also obtain from Fig. 9-5 the inverse of the slope of the curve at the boundary, which is equal to 0.332, or

$$\frac{d(u/U_0)}{d[(y/x)\mathrm{Re}_x^{1/2}]}\bigg|_{y=0} = 0.332$$

However, at any given section, x, Re_x, and U_0 will be constants; therefore, we express the velocity gradient at the boundary as follows:

$$\frac{du}{dy}\bigg|_{y=0} = 0.332\,\frac{U_0}{x}\,\mathrm{Re}_x^{1/2} \tag{9-8}$$

$$\frac{du}{dy}\bigg|_{y=0} = 0.332\,\frac{U_0}{x}\left(\frac{U_0 x}{\nu}\right)^{1/2}$$

$$\frac{du}{dy}\bigg|_{y=0} = 0.332\,\frac{U_0^{3/2}}{x^{1/2}\nu^{1/2}} \tag{9-9}$$

Equation (9-9) shows, indeed, that the velocity gradient decreases as the distance along the boundary, x, increases.

Shear stress

The shear stress at the boundary is obtained by multiplying the velocity gradient at the wall, Eq. (9-8), by the absolute viscosity:

$$\tau_0 = 0.332\,\mu\,\frac{U_0}{x}\,\mathrm{Re}_x^{1/2} \tag{9-10}$$

Equation (9-10) is used to obtain the local shear stress at any given section (any given value of x) for the laminar boundary layer as shown in the following example.

EXAMPLE 9-4   Crude oil at 70°F ($\nu = 10^{-4}$ ft^2/sec, sp. gr. $= 0.86$) with a free-steam velocity of 10 ft/sec flows past a thin flat plate that is 4 ft wide and 6 ft long in a direction parallel to the flow. Determine and plot the boundary-layer thickness and the shear-stress distribution along the plate.

Solution

$$\mathrm{Re}_x = \frac{U_0 x}{\nu} = \frac{10x}{10^{-4}} = 10^5 x$$

$$\mathrm{Re}_x^{1/2} = 3.16(10^2 x^{1/2})$$

The shear stress is given by

$$\tau_0 = 0.332\mu\,\frac{U_0}{x}\,\mathrm{Re}_x^{1/2}$$

where

$$\mu = \rho\nu = 1.94 \times 0.86 \times 10^{-4}$$

$$= 1.67 \times 10^{-4}\ \mathrm{lbf\text{-}sec/ft^2}$$

Then

$$\tau_0 = 0.332(1.67 \times 10^{-4})\,\frac{10}{x}\,(3.16)(10^2 x^{1/2})$$

$$= \frac{0.175}{x^{1/2}}\ \mathrm{psf}$$

The thickness of the boundary layer is

$$\delta = \frac{5x}{\mathrm{Re}_x^{1/2}} = \frac{5x}{3.16(10^2 x^{1/2})} = 1.58(10^{-2}x^{1/2})\mathrm{ft}$$

$$= 1.58(12)(10^{-2}x^{1/2}) = 0.190x^{1/2}\ \mathrm{in.}$$

The results for Example 9-4 are plotted in the figure below and are listed in Table 9-1.

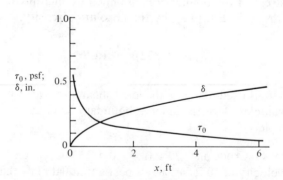

TABLE 9-1   RESULTS—δ AND τ_0 FOR DIFFERENT VALUES OF x

	$x = 0.1$ ft	$x = 1.0$ ft	$x = 2$ ft	$x = 4$ ft	$x = 6$ ft
$x^{1/2}$	0.316	1.00	1.414	2.00	2.45
τ_0, psf	0.552	0.174	0.123	0.087	0.071
δ, ft	0.005	0.016	0.022	0.031	0.039
δ, in.	0.060	0.189	0.270	0.380	0.460

Shearing resistance for a surface of given size

Because the shear stress at the boundary, τ_0, varies along the plate, it is necessary to integrate this stress over the entire surface to obtain the total shearing force on the surface. In equation form this is

$$F_s = \int_0^L \tau_0 B \, dx \tag{9-11}$$

where F_s = surface resistance produced by viscous stresses on one side of the plate

     B = width of the plate

     L = length

When Eq. (9-10) is substituted into Eq. (9-11) we get

$$F_s = \int_0^L 0.332 \, B\mu \, \frac{U_0 U_0^{1/2} x^{1/2}}{x \nu^{1/2}} \, dx$$

$$= 0.664 \, B\mu U_0 \frac{U_0^{1/2} L^{1/2}}{\nu^{1/2}}$$

$$= 0.664 \, B\mu U_0 \mathrm{Re}_L^{1/2} \tag{9-12}$$

In Eq. (9-12) Re_L is the Reynolds number based upon the approach velocity and the length of the plate.

Shear-stress coefficients

It is convenient to express the shear stress at the boundary, τ_0, and the total shearing force F_s in terms of dimensionless resistance coefficients and the dynamic pressure of the free stream, $\rho U_0^2/2$. The coefficients c_f and C_f are defined as follows:

$$c_f = \frac{\tau_0}{\rho U_0^2/2} \tag{9-13}$$

and
$$C_f = \frac{F_s}{BL\rho U_0^2/2} \tag{9-14}$$

Combining Eq. (9-10) with Eq. (9-13) and Eq. (9-12) with Eq. (9-14) produces the relationship between these coefficients and the corresponding Reynolds numbers for each case:

$$c_f = \frac{0.664}{\mathrm{Re}_x^{1/2}} \tag{9-15}$$

$$C_f = \frac{1.33}{\mathrm{Re}_L^{1/2}} \tag{9-16}$$

EXAMPLE 9-5   For the conditions of Example 9-4, determine the resistance of one side of the plate.

Solution

$$F_s = \frac{C_f B L \rho U_0^2}{2}$$

Here

$$C_f = \frac{1.33}{Re_L^{1/2}} = \frac{1.33}{(3.16)(10^2)(6^{1/2})} = 0.0017$$

Then

$$F_s = 0.0017(4)(6)(0.86)(1.94)\left(\frac{10^2}{2}\right)$$

$$= 3.40 \text{ lbf} \quad \blacktriangleleft$$

SI units   For Example 9-4 the significant variables in SI units are $\rho = 860$ kg/m³, $U_0 = 3.05$ m/s, $B = 1.22$ m, and $L = 1.88$ m. Then

$$F_s = 0.0017(1.22 \text{ m})(1.88 \text{ m})(860 \text{ kg/m}^3)\left(\frac{305^2 \text{ m}^2/\text{s}^2}{2}\right)$$

$$= 15.6 \text{ kg} \cdot \text{m/s}^2 = 15.6 \text{ N} \quad \blacktriangleleft$$

Experiment versus theory for the laminar boundary layer

Experimental evidence indicates that the Blasius solution is valid except very near to the leading edge of the plate. In the vicinity of the leading edge, an error results because of certain simplifying assumptions; however, the discrepancy is not of significance for most engineering problems. For very thin, smooth plates the laminar boundary layer can be expected to change to a turbulent boundary layer at a Reynolds number of approximately 500,000; however, if the approach flow is turbulent and/or if the plate is rough, the laminar boundary layer can be expected to become turbulent at a smaller Reynolds number.

9-5   QUANTITATIVE RELATIONS FOR THE TURBULENT BOUNDARY LAYER

Velocity distribution in the turbulent boundary layer along a smooth wall

The turbulent boundary layer is more complex than the laminar boundary layer because the former has three zones of flow that require different equations for the velocity distribution rather than the single relationship of the laminar case. Figure 9-6 shows a greatly enlarged portion of a turbulent boundary layer in which the three different zones of flow are identified. Each of these zones will be discussed separately.

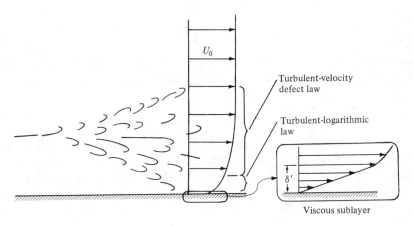

FIGURE 9-6   Flow pattern in turbulent boundary layer.

The zone immediately adjacent to the wall is a layer of fluid, which because of the damping effect of the wall, remains relatively smooth even though most of the flow in the boundary layer is turbulent. This very thin layer is called the *viscous sublayer,* and the velocity distribution in this layer is related to the shear stress and viscosity by Newton's viscosity law, which was introduced in Chapter 2. For convenience, it is repeated here as follows: $\tau = \mu \, du/dy$. In the viscous sublayer, τ is virtually constant and equal to the shear stress at the wall τ_0. Thus $du/dy = \tau_0/\mu$, which upon integration yields

$$u = \frac{\tau_0 \, y}{\mu} \tag{9-17}$$

If we multiply and divide the right side of Eq. (9-17) by ρ, we obtain the following:

$$u = \frac{\tau_0/\rho}{\mu/\rho} \, y$$

$$\frac{u}{\sqrt{\tau_0/\rho}} = \frac{\sqrt{\tau_0/\rho}}{\nu} \, y \tag{9-18}$$

The combination of variables $\sqrt{\tau_0/\rho}$ recurs again and again in derivations involving boundary-layer theory; therefore, it has been given the special name *shear velocity.* The name is appropriate because the variable τ_0 relates to the shear stress and the units of the combination are of velocity. The shear velocity has also been given a special symbol—u_*.[1] Thus, by definition.

---

[1] Also note that u_* is sometimes called *friction velocity.*

$$u_* = \sqrt{\frac{\tau_0}{\rho}} \qquad (9\text{-}19)$$

Now, when we substitute u_* for $\sqrt{\tau_0/\rho}$ in Eq. (9-18), we have the customary form for expressing the velocity distribution in the viscous sublayer:

$$\frac{u}{u_*} = \frac{y}{\nu/u_*} \qquad (9\text{-}20)$$

Equation (9-20) shows that the relative velocity (velocity relative to the shear velocity) in the viscous sublayer is equal to a dimensionless distance from the wall. Experimental results show that the viscous sublayer occurs only in the film of fluid for which yu_*/ν is less than approximately 5. Consequently, the thickness of the viscous sublayer, identified by δ', is given as

$$\delta' = \frac{5\nu}{u_*} \qquad (9\text{-}21)$$

It then follows that the thickness of the viscous sublayer will be small for flows in which the shear stress is large, because u_* will also be large for these cases. Furthermore, the viscous sublayer will become larger along the wall in the direction of flow because the shear stress decreases along the wall in the downstream direction.

The flow zone outside the viscous sublayer is turbulent; therefore, a completely different type of flow is involved. In fact, the turbulence alters the flow regime so much that the shear stress as given by $\tau = \mu \, du/dy$ is not significant. What happens is that the mixing action of turbulence causes small fluid masses to be swept back and forth in a direction transverse to the mean flow direction. Thus, as a small mass of fluid is swept from a low-velocity zone next to the sublayer into a higher-velocity zone farther out in the stream, the mass has a retarding effect on the higher-velocity stream. This mass of fluid through an exchange of momentum creates the effect of a retarding shear stress applied to a higher velocity stream much like the "conveyor belt" analogy introduced on page 18. Similarly a small mass of fluid that originates farther out in the boundary layer in a high-velocity flow zone and is swept into a region of low velocity has an effect on the low-velocity fluid much like shear stress augmenting the flow velocity. In other words, the mass of fluid with relatively higher momentum will tend to accelerate the lower-velocity fluid in the region into which it moves. Although the process described above is primarily a momentum-exchange phenomenon, it has the same effect as a shear stress applied to the fluid; thus in turbulent flow these "stresses" are termed *apparent shear stresses,* or *Reynolds stresses* after the British

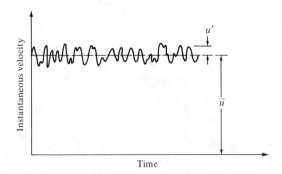

FIGURE 9-7   Velocity fluctuations in turbulent flow.

scientist-engineer who first did extensive research in turbulent flow in the late 1800s.

The mixing action of turbulence causes the velocities at a given point in a flow to fluctuate with time. If one were to place a velocity-sensing device, such as a hot-wire anemometer, in a turbulent flow, one would measure a fluctuating velocity as illustrated in Fig. 9-7. It is convenient to think of the velocity as composed of two parts: a mean value, $\bar{u}$, plus a fluctuating part, u'. The fluctuating parts of the velocity are responsible for the mixing action and the momentum exchange, which manifests itself as an apparent shear stress as noted above. In fact, the apparent shear stress is related to the fluctuating parts of the velocity by

$$\tau_{\text{app}} = -\rho\,\overline{u'v'} \tag{9-22}$$

where u' and v' refer to the x and y components of the velocity fluctuations and the bar over these terms denotes the product of $u'v'$ averaged over a period of time. The expression for apparent shear stress is not very useful in this form, so Prandtl developed a theory to relate the apparent shear stress to the temporal mean velocity distribution.

Prandtl's mixing-length theory

Prandtl hypothesized that a mass of fluid that moves across the boundary layer a certain distance ℓ (this ℓ is called a Prandtl's mixing length) will experience a longitudinal velocity change in accordance with the mean velocity gradient. That is, in Fig. 9-8, if a mass of fluid moves outward from the wall a distance ℓ, then it will increase in velocity an amount $u' = \ell\,du/dy$. In addition, Prandtl assumed that the magnitudes of u' and v' were approximately equal ($u' \approx v'$), which would seem to be a reasonable assumption because they both arise from the same sets of eddies.

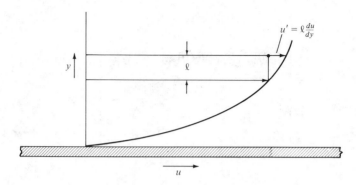

FIGURE 9-8   Concept of mixing length.

When these assumptions are incorporated in Eq. (9-22) and y is measured outward from the wall, we obtain

$$\tau_{\text{app}} = \rho \ell^2 \left(\frac{du}{dy}\right)^2 \tag{9-23}$$

A more general form of Eq. (9-23) is

$$\tau_{\text{app}} = \rho \ell^2 \left|\frac{du}{dy}\right| \frac{du}{dy}$$

however, application of this form is not needed in this text.

The theory leading to Eq. (9-23) is called Prandtl's mixing-length theory and it is used extensively in analyses involving turbulent flow. The advantage of Eq. (9-23) over Eq. (9-22) is that Eq. (9-23) is in terms of the temporal mean velocity distribution and can thus be integrated to obtain formulas for the velocity variation. In Eq. (9-23), ℓ is the mixing length, which was shown by Prandtl to be essentially proportional to the distance from the wall for the region close to the wall: $\ell = \kappa y$. If we consider the velocity distribution in a boundary layer where du/dy is positive, such as is shown in Fig. 9-6, and if we substitute κy for ℓ, then Eq. (9-23) reduces to

$$\tau_{\text{app}} = \rho \kappa^2 y^2 \left(\frac{du}{dy}\right)^2$$

For the zone of flow near the boundary, it is assumed that the shear stress is uniform and approximately equal to the shear stress at the wall; thus, the above equation becomes

$$\tau_0 = \rho \kappa^2 y^2 \left(\frac{du}{dy}\right)^2 \tag{9-24}$$

Taking the square root of each side of Eq. (9-24) and rearranging yields

$$du = \frac{\sqrt{\tau_0/\rho}}{\kappa} \frac{dy}{y}$$

Integrating the equation above and substituting u_* for $\sqrt{\tau_0/\rho}$ then gives

$$\frac{u}{u_*} = \frac{1}{\kappa} \ln y + C \tag{9-25}$$

Experiments on smooth boundaries indicate that the constant of integration C can be given in terms of u_*, ν, and a pure number as

$$C = 5.56 - \frac{1}{\kappa} \ln \frac{\nu}{u_*}$$

When this expression for C is substituted into Eq. (9-25), we have

$$\frac{u}{u_*} = \frac{1}{\kappa} \ln \frac{yu_*}{\nu} + 5.56 \tag{9-26}$$

In Eq. (9-26), κ has sometimes been called the universal turbulence constant, and experiments show this to have a value of approximately 0.40 for the turbulent zone next to the viscous sublayer. Introducing this into Eq. (9-26) and expressing the equation in terms of the logarithm to the base 10, we get

$$\frac{u}{u_*} = 5.75 \log \frac{yu_*}{\nu} + 5.56 \tag{9-27}$$

This logarithmic velocity distribution is valid for values of yu_*/ν from approximately 30 to 500. Thus we have a form of velocity distribution for this zone far different from that for the viscous sublayer. However, the relative velocity is still a function of yu_*/ν; thus, for the range of yu_*/ν from 0 to approximately 500 (Fig. 9-9), the velocity distribution is called the *law of the wall*.

By making a semilogarithmic plot of the velocity distribution in a turbulent boundary layer as shown in Fig. 9-9, it is easy to identify the velocity distribution in the laminar sublayer and in the region where the logarithmic equation applies. However, note that this form of plot accentuates the distance variable yu_*/ν near the wall. So that the student may view this plot in better perspective, the same graph shown in Fig. 9-9 is repeated in Fig. 9-10, except that in the latter, both the relative distance yu_*/ν and the relative velocity are plotted on linear scales. Figure 9-10 properly indicates that the laminar sublayer and the buffer zone, which is defined below, are a small part of the thickness of the turbulent boundary layer.

For $y/\delta > 0.15$ the law of the wall is no longer valid. Therefore, in this

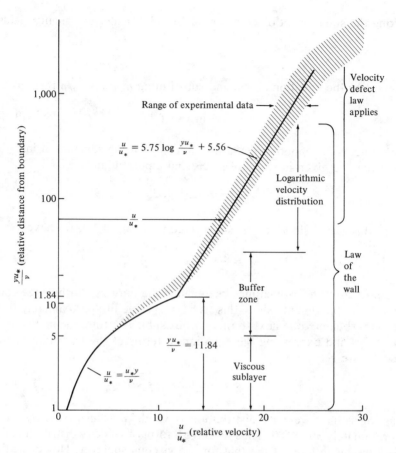

FIGURE 9-9   Velocity distribution in turbulent boundary layer. [Adapted from Schlichting (8) and Daily and Harleman (10).]

outer region we have the third zone given by the *velocity-defect law,* Fig. 9-11. In fact, the zone of applicability of the velocity-defect law not only applies to this outer region, but overlaps well into the logarithmic zone, Fig. 9-9. The velocity-defect law relates the relative defect of velocity $(U_0 - u)/u_*$ to y/δ. Another important point about this law is that it applies to rough as well as smooth surfaces.

The foregoing discussions about the three zones of flow (viscous sublayer, logarithmic velocity distribution, and the outer zone where the velocity-defect law applies) perhaps imply that there is a sharp demarcation between zones. This is definitely not the case, as can be seen in Fig. 9-9, which shows a smooth transition of velocity between the viscous sublayer and zone of logarithmic velocity distribution. The band of

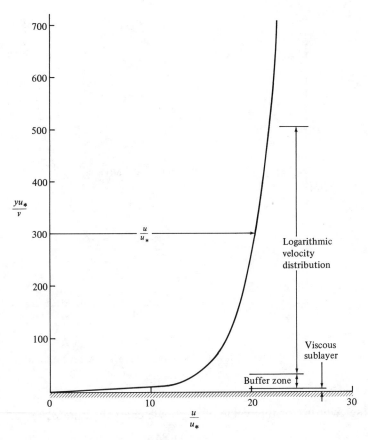

FIGURE 9-10   Velocity distribution in turbulent boundary layers—linear scales.

experimental data shows that there is a range of distance over which neither law applies. This region has been aptly called the *buffer* zone. In some cases it may be desirable to define the velocity in the buffer zone by a separate equation, for which the student is referred to Rouse (7). However, for some boundary-layer calculations, it is convenient to ignore the precise form of velocity distribution in the buffer zone and simply let the distribution be given by the extension of the equations already cited. In this case we refer to the nominal thickness of the viscous sublayer, δ'_N, which is the distance from the boundary to the point where the velocity-distribution curves for the viscous sublayer and the logarithmic velocity distribution intersect. This point of intersection occurs (see Fig. 9-9) at

$$\frac{yu_*}{\nu} = 11.84 \qquad (9\text{-}28)$$

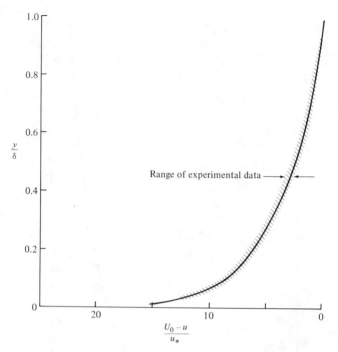

FIGURE 9-11   Velocity-defect law for boundary layers. [After Schlichting (8).]

In other words, for this value of yu_*/ν we can substitute δ_N' for y:

$$\delta_N' = \frac{11.84\nu}{u_*} \tag{9-29}$$

As noted earlier in this section, the actual limit of the viscous sublayer occurs at about $\delta' = 5\nu/u_*$. That is, turbulence affects the velocity distribution in the boundary layer from $\delta' = 5\nu/u_*$ outward; however, the effect is not appreciable up to the point of the intersection of the velocity-distribution curve for zone of logarithmic distribution (see Fig. 9-9). Thus, in derivations and analyses involving turbulent boundary layers it is not uncommon to assume that the velocity distribution in the viscous sublayer extends out to $\delta_N' = 11.84 \, \nu/u_*$.

Power-law formula for velocity distribution

Analyses have shown that for a wide range of Reynolds numbers ($10^5 < \text{Re} < 10^7$) the velocity profile in the turbulent boundary layer is reasonably approximated by the equation

$$\frac{u}{U_0} = \left(\frac{y}{\delta}\right)^{1/7} \tag{9-30}$$

Comparisons with experimental results show that this formula conforms to experimental results very closely over about 90% of the boundary layer ($0.1 < y/\delta < 1$). For the inner 10% of the boundary layer, one must resort to equations for the law of the wall (see Fig. 9-9) to obtain a more precise indication of velocity. Because Eq. (9-30) is valid over the major portion of the boundary layer, it is used to advantage in deriving the overall thickness of the boundary layer as well as other relations for the turbulent boundary layer. These will be considered in the next sections.

Momentum equation applied to the boundary layer

If the form of velocity distribution in the boundary layer is known, then it is possible to derive equations for the shear stress and the thickness of the boundary layer by utilizing the momentum equation. These basic relationships involved with the momentum equation will be introduced in this section. Consider the control volume in Fig. 9-12 for the boundary layer over a flat plate with zero pressure gradient along the plate. Flow comes into the control volume at section 1-1 and through the top of the boundary layer; it leaves the control volume at section 2-2.

When we write the momentum equation for the x direction for the fluid in this control volume, we have

$$\sum F_x = \int_{cs} u\rho \mathbf{V} \cdot d\mathbf{A} \tag{9-31}$$

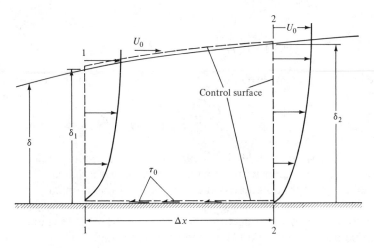

FIGURE 9-12   Control volume applied to boundary layer.

We assume a unit width normal to the paper and designate the mass inflow through the top of the boundary layer as $\dot{m}_\delta$; therefore, Eq. (9-31) becomes

$$-\tau_0\,\Delta x = \int_0^{\delta_2} \rho u_2^2\,dy - \int_0^{\delta_1} \rho u_1^2\,dy - U_0\dot{m}_\delta \qquad (9\text{-}32)$$

The mass flow rate into the control volume from the top of the boundary layer can be expressed as the difference in the mass flow rate past sections 1 and 2, or

$$\dot{m}_\delta = \int_0^{\delta_2} \rho u_2\,dy - \int_0^{\delta_1} \rho u_1\,dy \qquad (9\text{-}33)$$

Now, when Eq. (9-33) is substituted into Eq. (9-32), we obtain

$$-\tau_0\,\Delta x = \int_0^{\delta_2} \rho u_2^2\,dy - \int_0^{\delta_1} \rho u_1^2\,dy -$$
$$U_0\left[\int_0^{\delta_2} \rho u_2\,dy - \int_0^{\delta_1} \rho u_1\,dy\right] \qquad (9\text{-}34)$$

Assuming ρ is constant the equation above can be rearranged to yield the following:

$$-\tau_0\,\Delta x = \rho\int_0^{\delta_2} (u_2^2 - U_0 u_2)\,dy - \rho\int_0^{\delta_1} (u_1^2 - U_0 u_1)\,dy$$

$$= \rho U_0^2\left\{\int_0^{\delta_2}\left[\left(\frac{u_2}{U_0}\right)^2 - \frac{u_2}{U_0}\right]dy - \int_0^{\delta_1}\left[\left(\frac{u_1}{U_0}\right)^2 - \frac{u_1}{U_0}\right]dy\right\}$$

$$= \rho U_0^2\,\Delta\left\{\int_0^{\delta}\left[\left(\frac{u}{U_0}\right)^2 - \frac{u}{U_0}\right]dy\right\}$$

If we let Δx approach zero in the limit, then the equation above reduces to

$$\tau_0 = \rho U_0^2 \frac{d}{dx}\int_0^{\delta} \frac{u}{U_0}\left(1 - \frac{u}{U_0}\right)dy \qquad (9\text{-}35)$$

Equation (9-35) states that the shear stress at the wall is equal to ρU_0^2 times the rate of change with respect to x of an integral that is a function of the velocity distribution across the section. This equation can be used to evaluate the shear stress on a boundary indirectly by measuring the velocity distribution at various sections or, as will be shown in the next section, it can be used to derive other boundary-layer equations.

Thickness of the turbulent boundary layer on a flat plate

Using the power law for velocity distribution, Eq. (9-30), in Eq. (9-35) we have

$$\frac{\tau_0}{\rho} = U_0^2 \frac{d}{dx} \int_0^\delta \left(\frac{y}{\delta}\right)^{1/7} \left[1 - \left(\frac{y}{\delta}\right)^{1/7}\right] dy \qquad (9\text{-}36)$$

When the integration in Eq. (9-36) is performed, the following is obtained:

$$\frac{\tau_0}{\rho} = \frac{7}{72} U_0^2 \frac{d\delta}{dx} \qquad (9\text{-}37)$$

Here we have an equation involving δ; however, it is not very useful because τ_0 and δ are both unknowns. We can get another equation to obtain a solution by referring to the velocity distribution in Fig. 9-9. By fitting a power-law distribution to the curve for $100 < yu_*/\nu < 1,000$ one obtains

$$\frac{u}{u_*} = 8.74 \left(\frac{yu_*}{\nu}\right)^{1/7} \qquad (9\text{-}38)$$

Now, if we consider the outer limit of the boundary layer where $u = U_0$ and $y = \delta$ and if we recall that $u_* = \sqrt{\tau_0/\rho}$ and substitute all of these into Eq. (9-38) and solve for τ_0/ρ, we obtain

$$\frac{\tau_0}{\rho} = 0.0225 \, U_0^2 \left(\frac{\nu}{U_0\delta}\right)^{1/4} \qquad (9\text{-}39)$$

This is valid for smooth surfaces up to a Reynolds number, based upon length of boundary, of about 10^7. When Eqs. (9-37) and (9-39) are combined, the resulting equation is

$$\frac{0.0225 \, \nu^{1/4}}{U_0^{1/4}} = \frac{7}{72} \delta^{1/4} \frac{d\delta}{dx}$$

Then when the variables are separated and the integration is performed, we have

$$\delta^{5/4} = \frac{5}{4} \left[\frac{(0.0225)(72)}{7}\right] \frac{x\nu^{1/4}}{U_0^{1/4}} + C \qquad (9\text{-}40)$$

We are assuming that the boundary layer starts from the leading edge, so we evaluate C by taking $x = 0$ when $\delta = 0$; therefore, $C = 0$. Now when Eq. (9-40) is solved for δ and simplified, we obtain

$$\delta = \frac{0.37x}{\mathrm{Re}_x^{1/5}} \qquad (9\text{-}41)$$

Thus we have the thickness of the turbulent boundary layer as a function of the distance along the boundary and the Reynolds number based upon the distance along the boundary.

Shearing resistance of the turbulent boundary layer on a flat plate

When δ of Eq. (9-41) is substituted back into Eq. (9-39), we can express τ_0 in terms of the Reynolds number based upon the distance along the boundary:

$$\tau_0 = \rho \frac{U_0^2}{2} \frac{0.058}{\mathrm{Re}_x^{1/5}} \tag{9-42}$$

Since $c_f = \tau_0/\frac{1}{2}\rho U_0^2$, we can solve for c_f from Eq. (9-42):

$$c_f = \frac{\tau_0}{\rho U_0^2/2} = \frac{0.058}{\mathrm{Re}_x^{1/5}} \tag{9-43}$$

When τ_0 from Eq. (9-42) is integrated over the area of the boundary, we obtain the overall shearing resistance, which is

$$F_s = \frac{0.072\,BL}{\mathrm{Re}_L^{1/5}} \rho \frac{U_0^2}{2}$$

From the definition of C_f, Eq. (9-14), we then conclude that C_f for a turbulent boundary layer along a smooth plate is $C_f = 0.072/\mathrm{Re}_L^{1/5}$; however, Schlichting (8) indicates that better agreement with experimental results is obtained if the numerical coefficient is given a value of 0.074. Then C_f is

$$C_f = \frac{0.074}{\mathrm{Re}_L^{1/5}} \tag{9-44}$$

Summary of relations for the turbulent boundary layer on a flat plate

To summarize, the equations for the boundary-layer thickness and resistance for the tubulent boundary layer are

Boundary-layer thickness:

$$\delta = \frac{0.37x}{\mathrm{Re}_x^{1/5}}$$

Local shear stress:

$$\tau_0 = c_f \rho \frac{U_0^2}{2}$$

where

$$c_f = \frac{0.058}{\mathrm{Re}_x^{1/5}}$$

Overall shearing resistance:

$$F_s = C_f B L \rho \frac{U_0^2}{2}$$

where
$$C_f = \frac{0.074}{\text{Re}_L^{1/5}} \qquad \text{for Re} < 10^7$$

In the foregoing developments we have applied the integral momentum equation over the boundary layer in order to derive useful equations for local shear stress as well as overall plate resistance. As a point of interest, the integral momentum equation, Eq. (9-31), is also used in a variety of other applications to relate the drag of a body to the difference in pressure distribution and momentum flux between an upstream section and downstream section. Applications of this type include the determination of the drag of an airfoil section, the analyses of boundary layers over rough surfaces, the analysis of boundary layers over evaporating surfaces [see Crowe (2)], and the prediction of points of separation on curved surfaces.

EXAMPLE 9-6 Air at a temperature of 20°C (68°F) and with a free-stream velocity of 30 m/s (98 ft/sec) flows past a smooth, thin plate, which is 3 m wide (9.8 ft) by 6 m long (19.7 ft), in the direction of flow. Assuming that the boundary layer is forced to be turbulent from the leading edge, determine the shear stress, the thickness of the viscous sublayer, and the thickness of the boundary layer 5 m (16.4 ft) downstream of the leading edge.

Solution First compute Re_x at a distance 5 m from the leading edge:

$$\text{Re}_x = U_0 \frac{x}{\nu} = \frac{(30 \text{ m/s})(5 \text{ m})}{1.49(10^{-5})\text{m}^2/\text{s}} = 10^7$$

$$\text{Re}_x^{1/5} = (10^7)^{1/5} = 25$$

Compute τ_0, where $\tau_0 = c_f \rho U_0^2/2$. Here

$$c_f = 0.058 \text{ Re}_x^{-1/5} = \frac{0.058}{25}$$

Also
$$\rho = 1.20 \text{ kg/m}^3 \text{ (0.00232 slugs/ft}^3) \qquad \text{(from Appendix)}$$

Then
$$\tau_0 = \frac{0.058}{25} (1.20 \text{ kg/m}^3) \frac{30^2}{2} \text{ m}^2/\text{s}^2$$

$$= 1.25 \text{ kg/m} \cdot \text{s}^2 = 1.25 \text{ N/m}^2 \qquad \blacktriangleleft$$

Traditional units

$$\tau_0 = \frac{0.058}{25} (0.00232) \frac{98^2}{2} = 0.0259 \text{ lbf/ft}^2 \qquad \blacktriangleleft$$

Now compute $u_* = \sqrt{\tau_0/\rho}$ and the thickness of the boundary layer and viscous sublayer.

$$u_* = \left(\frac{\tau_0}{\rho}\right)^{1/2} = \left(\frac{1.25 \text{ N/m}^2}{1.20 \text{ kg/m}^3}\right)^{1/2} = 1.02 \text{ m/s}$$

Traditional units

$$u_* = \left(\frac{0.0259 \text{ lbf/ft}^2}{0.00232 \text{ slugs/ft}^3}\right)^{1/2} = 3.34 \text{ ft/sec}$$

The thickness of the viscous sublayer is given by

$$\delta' = \frac{5\nu}{u_*}$$

but $\nu = 1.49 \times 10^{-5} \text{ m}^2/\text{s}$, so

$$\delta' = \frac{5(1.49)(10^{-5}) \text{ m}^2/\text{s}}{1.02 \text{ m/s}} = 7.30 \times 10^{-5} \text{ m}$$

$$= 0.07 \text{ mm} \qquad \blacktriangleleft$$

Traditional units

$$\delta' = \frac{5(1.61)(10^{-4}) \text{ ft}^2/\text{sec}}{3.33 \text{ ft/sec}}$$

$$= 2.42 \times 10^{-4} \text{ ft} \qquad \blacktriangleleft$$

Compute the thickness of the boundary layer:

$$\delta = 0.37 \frac{x}{\text{Re}_x^{1/5}} = \frac{(0.37)(5) \text{ m}}{25} = 0.074 \text{ m} = 74 \text{ mm} \qquad \blacktriangleleft$$

Traditional units

$$\delta = \frac{(0.37)(16.4) \text{ ft}}{25} = 0.24 \text{ ft} = 2.9 \text{ in.} \qquad \blacktriangleleft$$

The foregoing equations for the completely turbulent boundary layer on a smooth boundary are valid up to a Reynolds number of 10^7. For higher Reynolds numbers, the 1/7 power law is not precise and more refined analyses are required. See Schlichting (8) for the local shear stress and C_f at these higher Reynolds numbers. One formula for C_f at these high Reynolds numbers is given as

$$C_f = \frac{0.455}{(\log_{10}\text{Re}_L)^{2.58}} - \frac{1,700}{\text{Re}_L} \quad \text{for Re} > 10^7 \qquad (9\text{-}45)$$

A plot of the average-shear-stress coefficient C_f is shown in Fig. 9-13. Here the curve marked "completely turbulent boundary layer" is for the

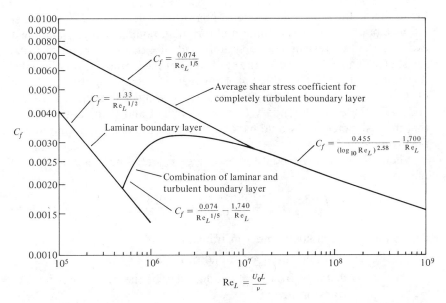

FIGURE 9-13   Average-shear-stress coefficients. [After Schlichting (8).]

case where, by artificial roughening at the upstream edge of the boundary, the boundary layer is forced to be turbulent from the outset. It is of interest to note that marine engineers utilize this technique on ship models to produce a boundary layer that can be predicted more precisely than the combination laminar and turbulent boundary layer.

To this point we have derived formulas for the velocity distribution and shearing resistance of the laminar and turbulent boundary layers. Now we shall discuss the existence of laminar and turbulent boundary layers together on a smooth, flat plate and how one determines the total resistance when they both occur on the same plate. As noted in Sec. 9-1, the laminar boundary layer first develops on the upstream end of the plate. Then, as this layer grows in thickness, it becomes unstable and turbulence sets in, and thereafter a turbulent boundary layer develops over the remainder of the plate. The onset of turbulence depends to a certain extent on the degree of smoothness of the plate and on the degree of turbulence. However, when the approach flow is nonturbulent, the transition or critical region on a smooth plate occurs at a Reynolds number, Re_x, of about 500,000. Then, when the turbulent boundary layer develops downstream of the laminar layer, we wonder whether the turbulent boundary layer will have the characteristics of one with the origin at the leading edge of the plate (same as the origin for the laminar layer) or those of one with the origin at the downstream end of the laminar layer. Experiments reveal that the former is the most valid model. Thus, when calculating the overall resistance of a plate with a laminar and turbulent boundary layer, one must evaluate the two resistances separately and sum them to obtain the total resistance. The calculation of the resistance of the laminar part is straightforward, as given in Example 9-4. However, in calculating the resistance of the turbulent part of the boundary layer, one must compute the resistance as if the entire boundary layer is turbulent and then subtract from that the resistance that would have occurred on the plate up to the transition zone.

When the foregoing model for resistance is stated mathematically, we have

$$F_s = \left(\frac{1.33}{\mathrm{Re}_{cr}^{1/2}} B x_{cr} + \frac{0.074}{\mathrm{Re}_L^{1/5}} BL - \frac{0.074}{\mathrm{Re}_{cr}^{1/5}} B x_{cr} \right) \rho \frac{U_0^2}{2} \qquad (9\text{-}46)$$

where Re_{cr} = Reynolds number at transition

Re_L = Reynolds number at end of plate

x_{cr} = distance from the leading edge of the plate to critical or transition zone

Then if we define the average resistance coefficient as $F_s = C_f BL\rho U_0^2/2$, we can solve for C_f utilizing Eq. (9-46) to yield

$$C_f = \frac{1.33}{\mathrm{Re}_{cr}^{1/2}} \frac{x_{cr}}{L} + \frac{0.074}{\mathrm{Re}_L^{1/5}} - \frac{0.074}{\mathrm{Re}_{cr}^{1/5}} \frac{x_{cr}}{L}$$

Here $x_{cr}/L = \mathrm{Re}_{cr}/\mathrm{Re}_L$; therefore, we get

$$C_f = \frac{1.33}{\mathrm{Re}_{cr}^{1/2}} \frac{\mathrm{Re}_{cr}}{\mathrm{Re}_L} + \frac{0.074}{\mathrm{Re}_L^{1/5}} - \frac{0.074}{\mathrm{Re}_{cr}^{1/5}} \frac{\mathrm{Re}_{cr}}{\mathrm{Re}_L}$$

or

$$C_f = \frac{0.074}{\mathrm{Re}_L^{1/5}} - \frac{\mathrm{Re}_{cr}}{\mathrm{Re}_L} \left(\frac{0.074}{\mathrm{Re}_{cr}^{1/5}} - \frac{1.33}{\mathrm{Re}_{cr}^{1/2}} \right)$$

Then for $\mathrm{Re}_{cr} = 500,000$ we have

$$C_f = \frac{0.074}{\mathrm{Re}_L^{1/5}} - \frac{1,740}{\mathrm{Re}_L} \qquad (9\text{-}47)$$

When Eq. (9-47) is applied for various values of Re_L from 5×10^5 to 10^7, we obtain the laminar-turbulent curve for C_f as shown in Fig. 9-13.

Even though the equations in this chapter have been developed for flat plates, they are useful for engineering estimates for some surfaces that are not truly flat plates. For example, the skin-friction drag of the submerged part of the hull of a ship can be estimated with Eq. (9-45).

EXAMPLE 9-7   Assume that a boundary layer over a smooth, flat plate is first laminar and then becomes turbulent at a critical Reynolds number of

5×10^5. If we have a plate that is 3 m long by 1 m wide and if air, 20°C, at normal atmospheric pressure flows past this plate with a velocity of 30 m/s, what will be the total shearing resistance of one side, the resistance due to the turbulent and laminar part of the boundary layer, and the average resistance coefficient C_f for the plate?

Solution The total resistance is $F_s = C_f B L \rho U_0^2 / 2$, where from Eq. (9-47) we get C_f:

$$C_f = \frac{0.074}{\mathrm{Re}_L^{1/5}} - \frac{1,740}{\mathrm{Re}_L}$$

In addition, $\mathrm{Re}_L = UL/\nu$, or

$$\mathrm{Re}_L = \frac{30 \text{ m/s} \times 3 \text{ m}}{(1.49)(10^{-5}) \text{ m}^2/\text{s}} = 6.04 \times 10^6$$

Then solving Eq. (9-47), we have

$$C_f = 0.00326 - 0.000290 = 0.00297 \qquad \blacktriangleleft$$

The total resistance is now calculated:

$$F_s = C_f B L \rho \frac{U_0^2}{2}$$

$$F_s = 0.00297 \times 1 \times 3 \times 1.2 \times \frac{30^2}{2} = 4.81 \text{ N}$$

Then x_{cr} is determined:

$$\frac{U x_{cr}}{\nu} = 500,000$$

or $$x_{cr} = \frac{500,000 \times 1.49 \times 10^{-5}}{30} = 0.248 \text{ m}$$

Thus the laminar resistance will be

$$F_{s,\mathrm{lam}} = \frac{1.33}{(5 \times 10^5)^{1/2}} \, 1 \times 0.248 \times 1.2 \times \frac{30^2}{2} = 0.252 \text{ N} \qquad \blacktriangleleft$$

Then $\quad F_{s,\mathrm{turb}} = 4.81 \text{ N} - 0.25 \text{ N} = 4.56 \text{ N} \qquad \blacktriangleleft$

EXAMPLE 9-8 Determine the total drag of the plate given in Example 9-6.

Solution The shearing resistance of one side will be given as $F_s = C_f B L \rho U_0^2 / 2$; therefore, the total drag will be twice this for two sides of the plate:

$$F_s = \frac{2 C_f B L \rho U_0^2}{2}$$

Since C_f is a function of Re_L, we compute that as

$$\mathrm{Re}_L = U_0 \frac{L}{\nu} = \frac{(30 \text{ m/s})(6 \text{ m})}{(1.49)(10^{-5} \text{ m}^2/\text{s})}$$

$$= 1.21(10^7)$$

From Fig. 9-13, $C_f = 0.0028$. Then

$$F_s = 0.0028(3 \text{ m})(6 \text{ m})(1.20 \text{ N} \cdot \text{s}^2/\text{m}^4)(30^2 \text{ m}^2/\text{s}^2)$$

$$= 54.4 \text{ N} \qquad \blacktriangleleft$$

Traditional units

$$F_s = (0.0028)(9.8)(19.2)(0.00232)(98^2) = 11.7 \text{ lbf} \qquad \blacktriangleleft$$

PROBLEMS

9-1 A cube weighing 200 N and having dimensions of 30 cm on a side is allowed to slide down an inclined surface on which there is a film of oil having a viscosity of 10^{-2} N · s/m². What is the terminal velocity of the block if the oil has a thickness of 0.1 mm?

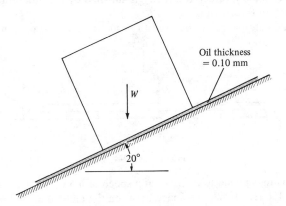

9-2 A 3-ft by 3-ft board weighing 40 lbf slides down an inclined ramp with a velocity of 0.5 fps. The board is separated from the ramp by a layer of oil 0.02 in. thick. Neglecting the edge effects of the board, calculate the approximate dynamic viscosity μ of the oil.

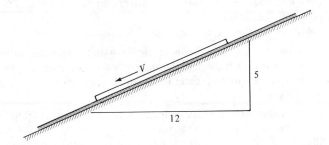

PROBLEMS 9-2, 9-3

9-3 A 1-m by 1-m board weighing 15 N slides down an inclined ramp with a velocity of 12 cm/s. The board is separated from the ramp by a layer of oil 0.5 mm thick. Neglecting the edge effects of the board, calculate the approximate dynamic viscosity μ of the oil.

9-4 A flat plate is pulled to the right at a speed of 30 cm/s. Oil with a viscosity of 3 N · s/m² fills the space between the plate and solid boundary. The plate size is 1 m long $(L = 1$ m) by 30 cm wide and the spacing between the plate and boundary is 2.0 mm.

(a) Mathematically express the velocity in terms of the coordinate system shown.

(b) By mathematical means determine whether this flow is rotational or irrotational.

(c) Determine whether continuity is satisfied by the differential form of the continuity equation.

(d) Calculate the force required to produce this plate motion.

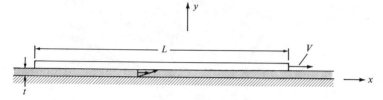

PROBLEMS 9-4, 9-5

9-5   A flat plate is pulled to the right at a speed of 0.5 ft/sec. Oil with a viscosity of 0.06 lbf-sec/ft² fills the space between the plate and solid boundary. The plate size is 3 ft long $(L = 3$ ft) by 1 ft wide and the spacing between the plate and boundary is 0.008 ft.

(a) Mathematically express the velocity in terms of the coordinate system shown.

(b) By mathematical means determine whether this flow is rotational or irrotational.

(c) Determine whether continuity is satisfied by the differential form of the continuity equation.

(d) Calculate the force required to produce this plate motion.

9-6   A tube and a wire positioned concentrically with the tube are submerged in oil. If the wire is drawn through the tube at a constant rate, will the viscous shear stress on the wire be greater than, equal to, or less than the shear stress on the tube wall?

9-7   A circular horizontal disk 12 in. in diameter has a clearance of 0.001 ft from a horizontal plate. What torque is required to rotate the disk at an angular velocity of 180 rpm when the clearance space contains oil $(\mu = 0.10$ lbf-sec/ft²)?

9-8   A circular horizontal disk 20 cm in diameter has a clearance of 2.0 mm from a horizontal plate. What torque is required to rotate the disk at an angular speed of 100 rad/s when the clearance space contains oil $(\mu = 6$ N · s/m²)?

9-9   A movable cone fits inside a stationary conical depression as shown. If a torque is applied to the cone, the cone will rotate at a certain speed depending upon the angles θ and β, the radius r_0, and the viscosity μ of the liquid. Derive an equation for the torque in terms of the other variables including only the viscous resistance. Assume that θ is very small.

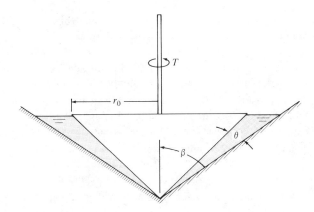

9-10 An important application of surface resistance is found in lubrication theory. Consider a shaft that turns inside a stationary cylinder, with a lubricating fluid used in the annular region. By considering a ring of fluid of radius r and width Δr and realizing that under steady-state operation the net torque on this ring is zero, show that $d(r^2\tau)/dr = 0$, where τ is the viscous shear stress. For a fluid that has a tangential component of velocity only, the shear stress is related to the velocity by $\tau = \mu r d(v/r)/dr$. Show that the torque per unit length acting on the inner cylinder is given by $T = 4\pi\mu\omega r_s^2/(1 - r_s^2/r_0^2)$, where ω is the angular velocity of the shaft.

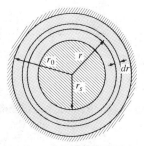

9-11 Using the equation developed in Prob. 9-10, find the power necessary to rotate a 2-cm shaft at 400 rad/s if the inside diameter of the casing is 2.2 cm, the bearing is 4 cm long, and SAE 30 oil at 38°C is the lubricating fluid.

9-12 The analysis developed in Prob. 9-10 applies to a device used to measure the viscosity of a fluid. By applying a known torque to the inner cylinder and measuring the angular velocity achieved, the viscosity of the fluid can be calculated. Assume you have a 4-cm inner cylinder and a 4.5 cm outer cylinder. The cylinders are 10 cm long. When a force of 0.6 N is applied to the tangent of the inner cylinder, it rotates at 30 rpm. Calculate the viscosity of the fluid.

9-13 If a thin film of oil ($\nu = 10^{-3}$ m²/s) that is 3.5 mm thick flows down a surface inclined at 30° to the horizontal, what will be the maximum velocity of flow?

9-14 If a thin film of oil ($\nu = 10^{-2}$ ft^2/sec) that is 0.015 in. thick flows down a surface inclined at 30° to the horizontal, what will be the maximum velocity of flow?

9-15 What is the depth and discharge per unit width of oil (SAE 30 at 100°F) that flows down a 45° incline at a Reynolds number of 200?

9-16 Rain falls on a 15-ft by 40-ft smooth roof at a rate of 0.4 in./hr and the roof has a slope, in the 15-ft direction, of 10°. Estimate the depth and average velocity of flow at the lower end of the roof assuming a temperature of 50°F.

9-17 Rain falls on a 6-m by 12-m smooth roof at a rate of 1 cm/h and the roof has a slope, in the 6-m direction, of 10°. Estimate the depth and average velocity of flow at the lower end of the roof assuming a temperature of 10°C.

9-18 Show that the mean velocity for laminar flow over a plate is $2u_{\max}/3$.

9-19 The upper plate shown is moving to the right with a velocity of 2 ft/sec and the lower plate is free to move laterally under the action of the viscous forces applied to it. After steady-state conditions have been established, what velocity will the lower plate have for the given conditions? Assume that the area of oil contact is the same for the upper plate, each side of the lower plate, and fixed boundary. $t_1 = 0.02$ ft, $\mu_1 = 0.004$ lbf-sec/ft^2, $t_2 = 0.01$ ft, and $\mu_2 = 0.001$ lbf-sec/ft^2.

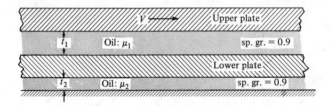

PROBLEMS 9-19, 9-20

9-20 The upper plate shown is moving to the right with a velocity of 80 cm/s and the lower plate is free to move laterally under the action of the viscous forces applied to it. After steady-state conditions have been established, what velocity will the lower plate have for the given conditions? Assume that the area of oil contact is the same for the upper plate, each side of the lower plate, and the fixed boundary. $t_1 = 2$ mm, $\mu_1 = 10^{-1}$ N · s/m^2, $t_2 = 1$ mm and $\mu_2 = 4 \times 10^{-2}$ N · s/m^2.

9-21 Two horizontal parallel plates are spaced 2 mm apart. If the pressure decreases at a rate of 1.20 kPa/m in the horizontal x direction in the fluid between the plates, what is the maximum fluid velocity in the x direction? The fluid has a dynamic viscosity of 10^{-1} N · s/m^2 and a specific gravity of 0.80. What is the magnitude of the shearing force on the upper plate if it is 2 m long (in direction of flow) and 1.5 m wide?

9-22 Two horizontal parallel plates are spaced 0.01 ft apart. If the pressure decreases at a rate of 12 psf/ft in the horizontal x direction in the fluid between the plates, what is the maximum fluid velocity in the x direction? The fluid has a dynamic viscosity of 10^{-3} lbf-sec/ft² and a specific gravity of 0.80.

9-23 Two vertical parallel plates are spaced 0.01 ft apart. If the pressure decreases at a rate of 8 psf/ft in the vertical z direction in the fluid between the plates, what is the maximum fluid velocity in the z direction? The fluid has a viscosity of 10^{-3} lbf-sec/ft² and a specific gravity of 0.80.

9-24 Two vertical parallel plates are spaced 2 mm apart. If the pressure decreases at a rate of 1.20 kPa/m in the positive z direction (vertically upward) in the fluid between the plates, what is the maximum fluid velocity in the z direction? The fluid has a viscosity of 10^{-1} N · s/m² and a specific gravity of 0.84.

9-25 Two vertical parallel plates are spaced 2 mm apart. If the pressure decreases at a rate of 9 kPa/m in the positive z direction (vertically upward) in the fluid between the plates, what is the maximum fluid velocity in the z direction? The fluid has a viscosity of 10^{-1} N · s/m² and a specific gravity of 0.85.

9-26 Two vertical parallel plates are spaced 0.01 ft apart. If the pressure decreases at a rate of 60 psf/ft in the vertical z direction in the fluid between the plates, what is the maximum fluid velocity in the z direction? The fluid has a viscosity of 10^{-3} lbf-sec/ft² and specific gravity of 0.80.

9-27 Two parallel plates are spaced 0.10 in. apart and motor oil (SAE 30) with a temperature of 100°F flows at a rate of 0.0083 cfs per foot of width between the plates. What is the pressure gradient in the direction of flow if the plates are inclined at 60° with the horizontal and if the flow is downward between the plates?

9-28 Two parallel plates are spaced 2 mm apart and oil ($\mu = 10^{-1}$ N · s/m², sp. gr. = 0.80) flows at a rate of 24×10^{-4} m³/s per meter of width between the plates. What is the pressure gradient in the direction of flow if the plates are inclined at 60° with the horizontal and if the flow is downward between the plates?

9-29 One type of bearing that can be used to support very large structures is shown below. Here fluid under pressure is forced from the bearing midpoint (slot A) to the exterior zone B. Thus a pressure distribution occurs as shown. For this bearing, which is 30 cm wide, what discharge of oil from slot A per meter of length of bearing is required to support a 50-kN load per meter of bearing length with a clearance space, t, between the floor and bearing surface of 0.60 mm? Assume an oil viscosity of 10^{-1} N · s/m². How much oil per hour would have to be pumped per meter of bearing length for the given conditions?

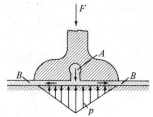

9-30 A thin plate 5 ft long and 3 ft wide is submerged and held stationary in a stream of water $(T = 60°F)$ that has a velocity of 6 ft/sec. What is the thickness of the boundary layer on the plate for $Re_x = 500,000$ (assume the boundary layer is still laminar) and at what distance downstream of the leading edge does this Reynolds number occur? What is the shear stress on the plate at this point?

9-31 A thin plate 2 m long and 1 m wide is submerged and held stationary in a stream of water $(T = 10°C)$ that has a velocity of 2 m/s. What is the thickness of the boundary layer on the plate for $Re_x = 500,000$ (assume the boundary layer is still laminar) and at what distance downstream of the leading edge does this Reynolds number occur? What is the shear stress on the plate at this point?

9-32 For the conditions of Prob. 9-31, what is the shearing resistance on one side of the plate for the part of the plate that has a Reynolds number, Re_x, less than 500,000? What is the ratio of the laminar shearing force to the total shearing force on the plate?

9-33 For this hypothetical boundary layer on the flat plate, what is the skin friction drag on the top side per meter of width and what is the shear stress on the plate at the downstream end? Given constants: $\rho = 1.2$ kg/m^3 and $\mu = 1.8 \times 10^{-5}$ N $\cdot$ s/m^2.

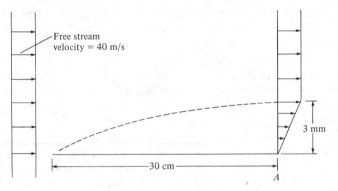

PROBLEMS 9-33, 9-34

9-34 Because of the reduction of velocity associated with the boundary layer, the streamlines outside the boundary layer are shifted away from the boundary. This amount of displacement of streamlines is defined as the displacement thickness δ^*. For the boundary layer at section A of Prob. 9-33, what is the magnitude of the displacement thickness?

9-35 Verify Eqs. (9-15) and (9-16).

9-36 Starting with Eq. (9-36), perform the integration and simplify to obtain Eq. (9-37).

9-37 You want to use the integral technique to determine the thickness of a laminar boundary layer. Assume that the velocity profile can be approximated by $u/U_0 = (y/\delta)^{1/2}$. Experimental data show that $\tau_0 = 1.66\ U\mu/\delta$. Use Eq. (9-35) and the above relations to obtain a differential equation for δ and solve for $\delta = f(x)$. Compare your results with Blasius' equation, Eq. (9-7).

9-38 Starting with Eq. (9-38), carry out the steps leading to Eq. (9-39).

9-39 Oil ($\nu = 10^{-3}$ ft²/sec) flows past a thin plate as shown. If the free-stream velocity is 20 ft/sec, what is the velocity 4 ft downstream from the leading edge and 0.050 ft away from the plate?

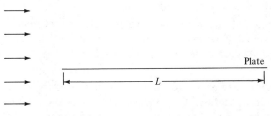

PROBLEMS 9-39, 9-40

9-40 Oil ($\nu = 10^{-4}$ m²/s) flows past a thin plate as shown. If the free-stream velocity is 8 m/s, what is the velocity 2.0 m downstream from the leading edge and 15 mm away from the plate?

9-41 In Prob. 9-40, what is the ratio of the shearing force on the upstream half of the plate to shearing force on the downstream half if $L = 2$ m?

9-42 Estimate the power required to pull the sign if it is towed at 30 m/s and if it is assumed that the sign has the same resistance characteristics as a flat plate. Assume standard atmospheric pressure and a temperature of 10°C.

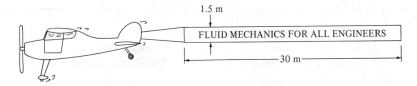

9-43 A thin (3 mm thick) plastic panel is lowered from a ship to a construction site on the ocean floor. The plastic panel weighs 200 N in air and is lowered at a rate of 2 m/s. Assuming that the panel remains vertically oriented, calculate the tension in the cable.

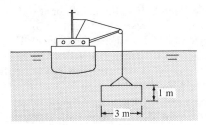

9-44 A javelin is approximately 265 cm long with an average diameter of approximately 25 mm, and it weighs 8.0 N. With a straight throw (javelin oriented parallel to line of flight so relative air speed is parallel to javelin) what will be the air drag at a speed of 30 m/s? What will be the javelin's deceleration (parallel to line of flight) at its trajectory azimuth for the 30-m/s speed? What will be the corresponding drag and accelerations if it is thrown into a 5-m/s wind and then thrown with a 5-m/s tailwind? Estimate the maximum distance of the throw if the thrower releases the javelin with a speed of 32 m/s. Assume air temperature = 20°C.

9-45 A motor boat pulls a long, smooth water-soaked log (1 m in diameter by 50 m long) at a speed of 1.5 m/s. Assuming total submergence, estimate the force required to overcome the surface resistance of the log. Assume a water temperature of 10°C.

9-46 Modern high-speed passenger trains are streamlined to reduce surface resistance. The cross section of a passenger car of one such train is shown. For a train 150 m long, estimate the surface resistance for a speed of 100 km/h and 200 km/h. What power is required for just the surface resistance at these speeds? Assume $T = 10$°C.

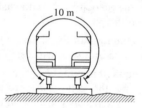

9-47 A flat plate 1.5 m long and 1.5 m wide is towed in water at 20°C in the direction of its length at a speed of 20 cm/s. Determine the resistance of the plate and the boundary-layer thickness at the aft end of the plate.

9-48 A 500-ft long ship steams at a rate of 35 fps through still water ($T = 50$°F). If the submerged surface area of the ship is 50,000 ft², what is the surface resistance?

9-49 A river barge that is 70 m long and 13 m wide has a flat bottom; therefore, its resistance is similar to one side of a flat plate. If the barge is towed at a speed of 4 m/s through still water ($T = 10$°C), what towing force F is required to overcome the viscous resistance and what is the boundary-layer thickness at midlength?

9-50 A supertanker has length, breadth and draught (fully loaded) dimensions of 325 m, 48 m, and 19 m, respectively. In open seas the tanker normally operates at a speed of 15 kts (1 kt = 0.515 m/s). For these conditions estimate (assuming flat-plate boundary layer conditions are approximated) the skin friction drag of such a ship steaming in 10°C water. What power is required to overcome the skin friction drag?

9-51 A ship is designed to be 250 m long, beam = 30 m, and draft = 12 m. The surface area of the ship below the water line is 8,800 m². A 1:30 scale model of the ship is tested and found to have a total drag of 38.0 N when being towed at a speed of 1.45 m/s. Using the methods outlined in Sec. 8-10, answer the following questions:

(a) To what speed does the 1.45 m/s correspond in the prototype?

(b) What is the model skin friction drag and wave drag?

(c) What would the ship drag be in salt water corresponding to the model-test conditions in fresh water?

Assume model tests are made in fresh water (20°C) and prototype conditions are sea water (10°C).

9-52 A 4 m-long hydroplane skims across a very calm lake ($T = 20°C$) at a speed of 20 m/s. For this condition, what will be the minimum shear stress along the smooth bottom?

9-53 Estimate the power required to overcome the surface resistance of a water skier if he is towed at 30 mph and each ski is 4 ft by 6 in. Assume the water temperature = 60°F.

9-54 If the wetted area of a 200-ft ship is 8,500 ft², approximately how great is the surface drag when the ship is traveling at a speed of 15 knots (1 knot = 1.69 fps)? What is the thickness of the boundary layer at the stern?

9-55 If the wetted area of an 80-m ship is 1,500 m², approximately how great is the surface drag when the ship is traveling at a speed of 10 m/s? What is the thickness of the boundary layer at the stern? Assume $T = 10°C$.

9-56 An outboard racing boat "planes" at 50 mph over water at 60°F. The hull has an average width of 5 ft and a length of 8 ft. Estimate the power required to overcome its surface resistance.

9-57 A racing boat "planes" at 30 m/s over 20°C water. The hull in contact with water has an average width of 2.0 m and a length of 3 m. Estimate the power required to overcome its surface resistance.

REFERENCES

1. Blasius, H. "Grenzschichten in Flüssigkeiten mit kleiner Reibung." *Z. Mat. Physik.*, (1908) p. 1. Summarized in Durand (4).

2. Crowe, C. T., Nicholls, J. A., and Morrison, R. B. "Drag Coefficients of Inert and Burning Particles Accelerating in Gas Streams," in *Ninth International Symposium on Combustion*. Academic Press, Inc., New York, 1963, pp. 395–406.

3. Daily, J. W., and Harleman, D. J. R. *Fluid Dynamics*. Addison-Wesley Publishing Company, Inc., Reading, Mass., 1966.

4. Durand, William Frederick (ed.). *Aerodynamic Theory*. Dover Publications, Inc., New York, 1963.

5. Kline, S. J., Morkovin, M. V., Sarron, G., and Cockrell, D. J. (eds.). "Procedures of Computation of Turbulent Boundary Layers." Thermosciences Divi-

sion, Department of Mechanical Engineers, Stanford University, Palo Alto, Calif., 1968.

6. Prandtl, L. "Über Flussigkeitsbewegung bei sehr kleiner Reibung." *Verhandlungen des III. Internationalen Mathematiker-Kongresses.* Leipzig (1905).

7. Rouse, H., et al. *Advanced Mechanics of Fluids.* John Wiley & Sons, Inc., New York, 1959.

8. Schlichting, H. *Boundary Layer Theory.* McGraw-Hill Book Company, New York, 1968.

9. Todd, F. H. "Viscous Resistance of Ships," in *Advances in Hydroscience,* vol. 3. Academic Press, Inc., New York, 1966.

10. White, Frank M. *Viscous Fluid Flow.* McGraw-Hill Book Company, New York, 1974.

Pipes, pipes, pipes . . .

10 FLOW IN CONDUITS

W<small>HEN ONE CONSIDERS THE CONVENIENCES AND NECESSITIES</small> of our everyday life, it is truly amazing to note the role played by conduits. For example, all the water that we use in our homes is pumped through pipes so that it will be available when and where we want it. In addition, virtually all this water leaves our homes as dilute wastes through sewers, another type of conduit. In addition to domestic use, the consumption of water by industry is enormous, whether it be in the processing of agricultural products or in the manufacturing of steel or paper, to cite examples. All the water used in these manufacturing processes is transported by means of piping systems; the petroleum industry in the United States alone transports approximately 20 million barrels of liquid petroleum per day in addition to the billions of cubic feet of gas transported by pipeline.

In the foregoing examples it is the transportation of the fluid that is the primary objective. However, there are numerous applications in which flow is a necessary but secondary part of the process. For example, heating and ventilating systems as well as electric generating stations utilize conduit flow to circulate fluids to transport energy from one location to another. Piping systems are also used extensively for controlling the operation of machinery.

Thus it is seen that the application of flow in conduits cuts across all fields of engineering; consequently, every engineer should understand the basic fluid mechanics involved with such flow. In this chapter we shall introduce the fundamental theory of flow in conduits as well as basic design procedures.

10-1 SHEAR STRESS DISTRIBUTION ACROSS A PIPE SECTION

The velocity distribution in a pipe is closely linked to the shear-stress distribution; hence, it is important to understand the latter. To determine the shear-stress distribution, we start with the equation of equilibrium applied to a cylindrical element of fluid that is oriented coaxially with the pipe, Fig. 10-1. For the conditions shown in Fig. 10-1, it is assumed that the flow is uniform (streamlines are straight and parallel); therefore, the pressure across any section of the pipe will be hydrostatically distributed. Thus the pressure force acting on an end face of the fluid element will be the product of the pressure at the center of the element (also at the center of the pipe) and the area of the face of the element. With steady-uniform flow, equilibrium between the pressure, gravity, and shearing forces

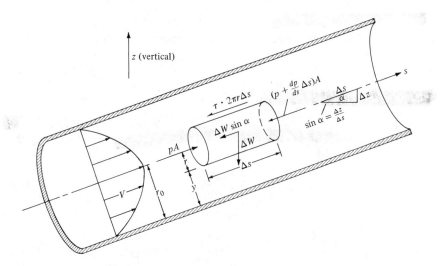

FIGURE 10-1   Variation of shear stress in a pipe.

acting on the fluid will prevail; consequently, the equilibrium equation yields the following:

$$\sum F_s = 0$$

$$pA - \left(p + \frac{dp}{ds}\,\Delta s\right)A - \Delta W \sin \alpha - \tau(2\pi r)\Delta s = 0 \qquad (10\text{-}1)$$

In Eq. (10-1) $\Delta W = \gamma A\,\Delta s$ and $\sin\alpha = dz/ds$; therefore, Eq. (10-1) reduces to

$$-\frac{dp}{ds}\,\Delta s\,A - \gamma A\,\Delta s\,\frac{dz}{ds} - \tau(2\pi r)\Delta s = 0 \qquad (10\text{-}2)$$

Then, in Eq. (10-2) when we divide through by $\Delta s\,A$ and simplify, we obtain

$$\tau = \frac{r}{2}\left[-\frac{d}{ds}(p + \gamma z)\right] \qquad (10\text{-}3)$$

Since the gradient itself, $d/ds(p + \gamma z)$, is negative (see Sec. 7-5) and constant across the section for uniform flow,[1] it follows that $-d/ds(p + \gamma z)$ will be positive and constant across the pipe section. Thus τ in Eq. (10-3) will be zero at the center and increase linearly to a

[1] $p + \gamma z$ is constant across the section because the streamlines are straight and parallel in uniform flow and for this condition there will be no acceleration of the fluid normal to the streamline; thus, hydrostatic conditions prevail across the flow section. For a hydrostatic condition, $p/\gamma + z =$ constant or $p + \gamma z =$ constant, as shown in Chapter 3.

maximum at the pipe wall. Equation (10-3) will be used in the following section to derive the velocity distribution for laminar flow.

10-2   LAMINAR FLOW IN PIPES

We determine how the velocity varies across the pipe by substituting for τ in Eq. (10-3) its equivalent $\mu \, dV/dy$ and integrating. First, making the substitution, we have

$$\mu \frac{dV}{dy} = \frac{r}{2}\left[-\frac{d}{ds}(p + \gamma z)\right] \qquad (10\text{-}4)$$

Noting that $dV/dy = -dV/dr$, Eq. (10-4) becomes

$$\frac{dV}{dr} = -\frac{r}{2\mu}\left[-\frac{d}{ds}(p + \gamma z)\right] \qquad (10\text{-}5)$$

When we separate variables and integrate across the section, we obtain

$$V = -\frac{r^2}{4\mu}\left[-\frac{d}{ds}(p + \gamma z)\right] + C \qquad (10\text{-}6)$$

We can evaluate the constant of integration in Eq. (10-6) by noting that when $r = r_0$ the velocity $V = 0$. Therefore, the constant of integration is given by $C = (r_0^2/4\mu)[-d/ds(p + \gamma z)]$, and Eq. (10-6) then becomes

$$V = \frac{r_0^2 - r^2}{4\mu}\left[-\frac{d}{ds}(p + \gamma z)\right] \qquad (10\text{-}7)$$

which indicates that the velocity distribution for laminar flow in a pipe is parabolic across the section with the maximum velocity at the center of the pipe. Figure 10-2 shows the variation of the magnitude of the shear stress and velocity in the pipe.

EXAMPLE 10-1   Oil (sp. gr. = 0.90; viscosity = $5 \times 10^{-1}\,\text{N}\cdot\text{s/m}^2$) flows steadily in a 3-cm pipe. The pipe is vertical and the pressure at elevation 100 m is 200 kPa. If the pressure at elevation 85 m is 250 kPa, is the flow direction up or down? What is the velocity at the center of the pipe and 6 mm from the center, assuming that the flow is laminar?

Solution   First determine the rate of change of $p + \gamma z$. Taking s in the z direction,

$$\frac{d}{ds}(p + \gamma z) = \frac{(p_{100} + \gamma z_{100}) - (p_{85} + \gamma z_{85})}{15}$$

$$= \frac{[200 \times 10^3 + 8,830(100)] - [250 \times 10^3 + 8,830(85)]}{15}$$

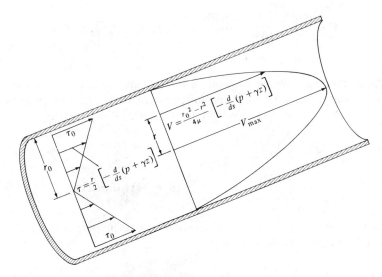

FIGURE 10-2   Distribution of shear stress and velocity for laminar flow in a pipe.

$$= \frac{(1.083 \times 10^6 - 1.00 \times 10^6)\text{N}/\text{m}^2}{15 \text{ m}} = 5.53 \text{ kN}/\text{m}^3$$

The quantity $p + \gamma z$ is not constant with elevation—it increases upward (decreases downward); therefore, the direction of flow is downward. This can be seen by substituting $d(p + \gamma z)/ds = 5.53 \text{ kN}/\text{m}^3$ into Eq. (10-7). When this is done, V is negative for all values of r in the flow. When $r = 0$ (center of pipe), the velocity will be maximum. Thus

$$V_{\text{center}} = V_{\text{max}} = \frac{r_0^2}{4\mu} (-5.53 \text{ kN}/\text{m}^3)$$

$$= \frac{0.015^2 \text{ m}^2}{4(5 \times 10^{-1} \text{ N} \cdot \text{s}/\text{m}^2)} (-5.53 \times 10^3 \text{ N}/\text{m}^3)$$

$$= -0.622 \text{ m}/\text{s} \qquad \blacktriangleleft$$

At first it may seem strange that the velocity is in a direction opposite to the direction of decreasing pressure; however, it may not seem so peculiar if one realizes that in this example the pipe is vertical, so that the gravitational force as well as pressure helps to establish the flow. What counts when flow is other than in the horizontal direction is how the combination $p + \gamma z$ changes with s. If $p + \gamma z$ is constant, then we have the equation of hydrostatics and no flow occurs; however, if $p + \gamma z$ is not constant, flow will occur in the direction of decreasing $p + \gamma z$.

Next determine the velocity at $r = 6 \text{ mm} = 0.006 \text{ m}$. Using Eq. (10-7),

$$V = \frac{0.015^2 \text{ m}^2 - 0.006^2 \text{ m}^2}{(5 \times 10^{-1} \text{ N} \cdot \text{s/m}^2)} (-5.53 \times 10^3 \text{ N/m}^3)$$

$$= -0.522 \text{ m/s} \qquad \blacktriangleleft$$

For many problems we wish to relate the pressure change to the rate of flow or mean velocity $\bar{V}$ in the conduit; therefore, it is necessary to integrate $dQ = V \, dA$ over the cross-sectional area of flow. That is,

$$Q = \int V \, dA$$

$$= \int_0^{r_0} \frac{(r_0^2 - r^2)}{4\mu} \left[-\frac{d}{ds} (p + \gamma z) \right] (2\pi r \, dr) \qquad (10\text{-}8)$$

The factor $\pi[d(p + \gamma z)/ds]/4\mu$ is constant across the pipe section; therefore, upon integration we obtain

$$Q = \frac{\pi}{4\mu} \left[\frac{d}{ds} (p + \gamma z) \right] \frac{(r^2 - r_0^2)^2}{2} \Big|_0^{r_0} \qquad (10\text{-}9)$$

which reduces to

$$Q = \frac{\pi r_0^4}{8\mu} \left[-\frac{d}{ds} (p + \gamma z) \right] \qquad (10\text{-}10)$$

If we divide through by the cross-sectional area of the pipe, we have an expression for the mean velocity:

$$\bar{V} = \frac{r_0^2}{8\mu} \left[-\frac{d}{ds} (p + \gamma z) \right] \qquad (10\text{-}11)$$

By comparing Eqs. (10-11) and (10-7) it can be seen that $\bar{V} = V_{\text{max}}/2$. Also, by substituting $D/2$ for r_0 we have

$$\bar{V} = \frac{D^2}{32\mu} \left[-\frac{d}{ds} (p + \gamma z) \right] \qquad (10\text{-}12)$$

or

$$\frac{d}{ds} (p + \gamma z) = -\frac{32\mu \bar{V}}{D^2} \qquad (10\text{-}13)$$

Upon integrating Eq. (10-13) along the pipe between sections 1 and 2 we obtain

$$p_2 - p_1 + \gamma(z_2 - z_1) = -\frac{32\mu \bar{V}}{D^2} (s_2 - s_1) \qquad (10\text{-}14)$$

Here $s_2 - s_1$ is the length L of pipe between the two sections; therefore, Eq. (10-14) can be rewritten as

$$\frac{p_1}{\gamma} + z_1 = \frac{p_2}{\gamma} + z_2 + \frac{32\mu L \bar{V}}{\gamma D^2} \qquad (10\text{-}15)$$

It can be seen that when the general energy equation for incompressible flow in conduits Eq. (7-26), is reduced to that for uniform flow in a constant-diameter pipe where $V_1 = V_2$, one obtains

$$\frac{p_1}{\gamma} + z_1 = \frac{p_2}{\gamma} + z_2 + h_f \qquad (10\text{-}16)$$

Here h_f is used instead of h_L to signify head loss due to frictional resistance of the pipe. Comparing Eqs. (10-15) and (10-16) it is then seen that the head loss is given by

$$h_f = \frac{32\mu L V}{\gamma D^2} \qquad (10\text{-}17)$$

Here the bar over the V has been omitted to conform to the standard practice of denoting the mean velocity in one-dimensional flow analyses by V without the bar.

10-3 CRITERION FOR LAMINAR OR TURBULENT FLOW IN A PIPE

To predict whether flow will be laminar or turbulent, it is necessary to explore the characteristics of flow in both laminar and turbulent states. Although other scientists before him had sensed the marked physical difference between laminar and turbulent flow, it was Osborne Reynolds (21) who first developed the basic laws of turbulent flow. With his analytical and experimental work he showed that the Reynolds number, already introduced in Chapter 4, was a basic parameter relating to laminar as well as turbulent flow. For example, using an experimental apparatus such as that shown in Fig. 10-3, he found that the onset of turbulence was related to the Reynolds number, $VD\rho/\mu$, in a very interesting way. If the fluid in the upstream reservoir were not completely still or if the pipe had some vibration in it, the flow in the pipe as it was gradually increased from a low rate to higher rates was initially laminar, but then it changed from laminar to turbulent flow at a Reynolds number in the neighborhood of 2,100. However, he found that if the fluid were initially completely motionless and if there were no vibration in the equipment while the flow was increased, it was possible to reach a much higher Reynolds number before the flow became turbulent. He also found when going from high-velocity turbulent flow to low-velocity flow that the change from turbulent flow always occurred at about a Reynolds number of 2,000.

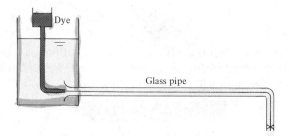

FIGURE 10-3   Apparatus similar to that used by Reynolds to study laminar and turbulent flow.

These experiments of Reynolds indicate that under carefully controlled experiments it is possible to have laminar flow in pipes at Reynolds numbers much higher than 2,000; however, the slightest disturbance will trigger the onset of turbulence at high values of Re. Because most engineering applications involve some vibration or flow disturbances, it is reasonable to expect that pipe flow will be laminar for Reynolds numbers less than 2,000 and turbulent for Reynolds numbers greater than 3,000. When Re is between 2,000 and 3,000, the type of flow is very unpredictable and often changes back and forth between the laminar and turbulent states. Fortunately, however, most engineering applications either are not in this range or are not significantly affected by the unstable flow.

EXAMPLE 10-2   Oil (sp. gr. = 0.85) with a kinematic viscosity of 6×10^{-4} m²/s flows in a 15-cm pipe at a rate of 0.020 m³/s. What is the head loss per 100 m of length of pipe?

Solution   First, we determine whether the flow is laminar or turbulent by checking to see if the Reynolds number is below 2,000 or above 3,000.

$$V = \frac{Q}{A} = \frac{0.020 \text{ m}^3/\text{s}}{(\pi/4)D^2 \text{ m}^2}$$

$$= \frac{0.020 \text{ m}^3/\text{s}}{0.785(0.15^2 \text{ m}^2)} = 1.13 \text{ m/s}$$

Then
$$\text{Re} = \frac{VD}{\nu} = \frac{(1.13 \text{ m/s})(0.15 \text{ m})}{6(10^{-4} \text{ m}^2/\text{s})}$$

$$= 283$$

Since the Reynolds number is less than 2,000, the flow is laminar. The head loss per 100 m is obtained from Eq. (10-17):

$$h_f = \frac{32\mu L V}{\gamma D^2}$$

Here $\mu/\gamma = \nu/g$; hence,

$$h_f = \frac{32\nu LV}{gD^2}$$

Then $\qquad h_f = \dfrac{32(6)(10^{-4}\text{ m}^2/\text{s})(100\text{ m})(1.13\text{ m/s})}{(9.81\text{ m/s}^2)(0.15^2\text{ m}^2)}$

$$= 9.83\text{ m}$$

The head loss is 9.83 m/100 m of length. ◀

EXAMPLE 10-3 Kerosene (0°C) flows under the action of gravity in the 6-mm-diameter by 100-m-long pipe shown. Determine the rate of flow in the pipe.

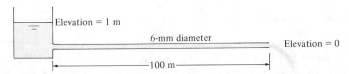

Elevation = 1 m

6-mm diameter

Elevation = 0

100 m

Solution Because the pipe diameter is small and because the head producing flow is also quite small, it is expected that the velocity in the pipe will be small; hence, it will be initially assumed that the flow is laminar and $V^2/2g$ is nil. Then, to solve for the velocity we apply the energy equation to the problem. We write this equation between a section at the upstream water surface to the outlet of the pipe. Thus we have

$$\frac{p_1}{\gamma} + \frac{\alpha_1 V_1^2}{2g} + z_1 = \frac{p_2}{\gamma} + \frac{\alpha_2 V_2^2}{2g} + z_2 + \frac{32\mu LV}{\gamma D^2}$$

With the noted assumption, the equation above reduces to

$$0 + 0 + 1 = 0 + 0 + 0 + \frac{32\mu LV}{\gamma D^2}$$

or $\qquad \dfrac{32\mu LV}{\gamma D^2} = 1$

For 0°C the viscosity from Figs. A-1 and A-2 is

$$\mu = 3.2 \times 10^{-3}\text{ N} \cdot \text{s/m}^2 \qquad \nu = 3.9 \times 10^{-6}\text{ m}^2/\text{s}$$

Then $\qquad V = \dfrac{1 \times \gamma D^2}{32\mu L} = \dfrac{1(814\text{ N/m}^3)(0.006^2\text{ m}^2)}{32(3.2 \times 10^{-3}\text{ N} \cdot \text{s/m}^2)(100\text{ m})}$

$$= 0.00286\text{ m/s} = 2.86\text{ mm/s}$$

Now check Re to see if the flow is laminar and check $V^2/2g$ to see if it is indeed nil:

$$\text{Re} = \frac{VD}{\nu} = \frac{(0.00286 \text{ m/s})(0.006 \text{ m})}{3.9 \times 10^{-6} \text{ m}^2/\text{s}} = 4.40$$

also

$$\frac{V^2}{2g} = \frac{(0.00286 \text{ m/s})^2}{(2)(9.81 \text{ m/s})} = 4.17 \times 10^{-7} \text{ m}$$

Therefore, the flow is laminar and the velocity as determined is valid. The discharge is then calculated as follows:

$$Q = VA = 0.00286 \text{ m/s} \left(\frac{\pi}{4}\right)(0.006 \text{ m})^2 = 8.09 \times 10^{-8} \text{ m}^3/\text{s} \quad \blacktriangleleft$$

10-4   TURBULENT FLOW IN PIPES

Turbulence and its influence in pipe flow

In the preceding section it was pointed out that pipe flow will be turbulent if the Reynolds number is larger than approximately 3,000. However, to say that the flow is turbulent is only a gross description of the flow. A better "feel" for the flow can be obtained if we explore the similarities between turbulent flow in a pipe and flow in a turbulent boundary layer and if we relate the shear stress in the pipe to the level of turbulence. Once we understand these basic physical relationships, we will be better equipped to proceed to the development of equations for the velocity distribution and the resistance to turbulent flow in pipes.

The similarities between turbulent-boundary-layer flow and turbulent-pipe flow are many. In fact, it is valid to think of turbulent-pipe flow as a turbulent boundary layer which has become as thick as the radius of the pipe. With this perspective we then realize that flow in a smooth pipe will have a viscous sublayer just as a flat-plate-boundary layer. In addition, the velocity gradient in the viscous sublayer will be consistent with the shear stress as given by $\tau = \mu \, du/dy$. However, outside the viscous sublayer the viscous shear stress is negligible compared with the resistance resulting from turbulence. We have already referred in Chapter 9 to the *apparent shear stress*, $\tau_{\text{app}} = -\rho \overline{u'v'}$, which involves an exchange of momentum, but its effect is like that of a true shear stress. It is zero at the pipe center and increases to a maximum near the wall, as shown in Fig. 10-4. Here it is seen that the apparent shear stress increases linearly almost to the edge of the pipe. This linear change in τ_{app} is in accordance with Eq. (10-3), which was developed in Sec. 10-1. Near the wall, in the viscous sublayer, τ_{app} reduces to zero because all the shear stress here is in the form of viscous shear stress.

We have shown that there are indeed many analogies between turbulent-boundary-layer flow and turbulent flow in pipes. The primary

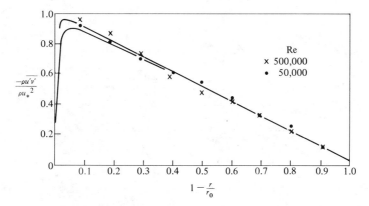

FIGURE 10-4   Apparent shear stress in a pipe. [After Laufer (16).]

difference is that pipe flow is uniform and boundary-layer flow is not. Of course, this difference does not apply near the inlet of the pipe where the flow is also nonuniform.

Velocity distribution and resistance in smooth pipes

In the viscous sublayer and in the turbulent zone near the wall, experiments reveal that the velocity-distribution equations are the same form as those for the turbulent boundary layer. That is for a smooth pipe

$$\frac{u}{u_*} = \frac{u_* y}{\nu} \qquad \text{for} \quad 0 < \frac{y u_*}{\nu} < 5 \tag{10-18}$$

$$\frac{u}{u_*} = 5.75 \log \frac{u_* y}{\nu} + 5.5 \qquad \text{for} \quad 20 < \frac{y u_*}{\nu} \lesssim 10^5 \tag{10-19}$$

Figure 10-5 is a plot of Eqs. (10-18) and (10-19) as well as an indication of the spread of experimental data from various sources. Near the center of the pipe, like flow near the outer limit of the boundary layer, the velocity-defect law is applicable, as shown in Fig. 10-6. Figure 10-6 also includes the range of experimental velocity data obtained from flow in rough conduits. Again, like that for the turbulent boundary layer, a power-law formula is applicable except close to the wall. This law is given as

$$\frac{u}{u_{\max}} = \left(\frac{y}{r_0}\right)^m \tag{10-20}$$

Here y is the distance from the wall and m is an empirically determined quantity. Some references indicate that m has a value of $\frac{1}{7}$ for turbulent

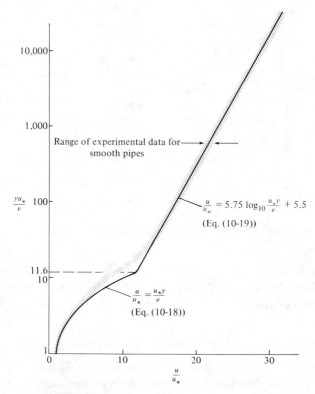

FIGURE 10-5   Velocity distribution for smooth pipes. [After Schlichting (25).]

flow; however, Schlichting (25) shows that m varies from $\frac{1}{6}$ to $\frac{1}{10}$ depending upon the Reynolds number. His values for m are given in Table 10-1.

In Chapter 9 the shear stress on the wall was expressed as

$$\tau_0 = c_f \rho \frac{V_0^2}{2}$$

where c_f was a function of the character of flow (laminar or turbulent) and the Reynolds number. For pipe flow it is customary to express τ_0 in a similar manner; however, we use the mean velocity as the reference velocity, and the coefficient of proportionality is given as $f/4$ instead of c_f. Here f is

TABLE 10-1   EXPONENTS FOR POWER-LAW EQUATION

Re→	4×10^3	2.3×10^4	1.1×10^5	1.1×10^6	3.2×10^6
m→	$\dfrac{1}{6.0}$	$\dfrac{1}{6.6}$	$\dfrac{1}{7.0}$	$\dfrac{1}{8.8}$	$\dfrac{1}{10.0}$

SOURCE: Schlichting (25).

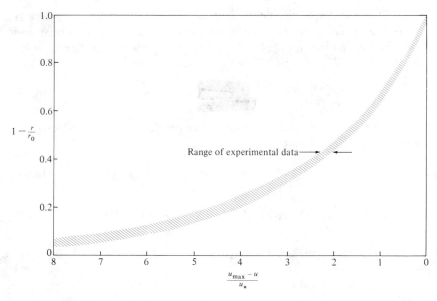

FIGURE 10-6   Velocity-defect law for turbulent flow in smooth and rough pipes. [After Schlichting (25).]

called the resistance coefficient or friction factor of the pipe. Thus we have

$$\tau_0 = \frac{f}{4} \rho \frac{V^2}{2} \tag{10-21}$$

Or, because $\sqrt{\tau_0/\rho} = u_*$, we have

$$\frac{u_*}{V} = \sqrt{\frac{f}{8}}$$

Now by eliminating τ_0 between Eqs. (10-3) ($\tau = \tau_0$ when $r = r_0$) and (10-21) and integrating between two sections along the pipe, we obtain

$$h_1 - h_2 = f \frac{L}{D} \frac{V^2}{2g}$$

$$h_f = f \frac{L}{D} \frac{V^2}{2g} \tag{10-22}$$

where h_f is the head loss created by viscous effects and is equal to the change in piezometric head along the pipe. Equation (10-22) is called the Darcy-Weisbach equation, named after Henry Darcy, a French engineer of the nineteenth century, and Julius Weisbach, a German engineer and scientist of the same era. Weisbach first proposed the use of the nondimensional resistance coefficient and Darcy carried out numerous tests on water pipes. Brief accounts of their works are given by Rouse and Ince (23). For laminar flow it can be easily shown by a simultaneous solution of Eqs. (10-17) and (10-22) that the resistance coefficient is given by

$$f = \frac{64}{\text{Re}} \qquad (10\text{-}23)$$

For turbulent flow analytical and empirical results on *smooth* pipes yield the following approximate relations for f:

$$f = \frac{0.316}{\text{Re}^{1/4}} \qquad \text{for } 3{,}000 < \text{Re} < 100{,}000 \qquad (10\text{-}24)$$

$$\frac{1}{\sqrt{f}} = 2 \log(\text{Re}\sqrt{f}) - 0.8 \qquad \text{Re} > 3{,}000 \qquad (10\text{-}25)$$

Equation (10-24) is attributed to Blasius and Eq. (10-25) was developed by Prandtl.

Velocity distribution and resistance—rough pipes

Numerous tests on flow in rough pipes all show that a semilogarithmic velocity distribution is valid over most of the pipe section (19, 24). This relationship is given in the following form:

$$\frac{u}{u_*} = 5.75 \log \frac{y}{k} + B \qquad (10\text{-}26)$$

Here y is the distance from the rough wall, k is a measure of the height of roughness elements, and B is a function of the character of roughness. That is, B is a function of the type, concentration, and size variation of the roughness. Research by Roberson and Chen (22) shows that B can be analytically determined for artificially roughened boundaries. More recent work by Wright (28), Calhoun (4), and Kumar (15) indicate that the same theory using a numerical approach will yield solutions for B and the coefficient f for natural roughness as found in rock-bedded streams and commercial pipes.

In 1933 Nikuradse (19) carried out a number of tests on the flow in pipes that were roughened with uniform-sized sand grains. From these tests he found that the value for B with this kind of roughness was 8.5. Thus, for his tests Eq. (10-26) becomes

$$\frac{u}{u_*} = 5.75 \log \frac{y}{k_s} + 8.5 \qquad (10\text{-}27)$$

In Nikuradse's tests the distance y was measured from the geometric mean of the wall surface and k_s was the size of the sand grains.

The Nikuradse tests revealed two very important characteristics of rough-pipe flow. First, with low Reynolds numbers and with small-sized sand grains, the flow resistance is virtually the same as for a smooth pipe; and second, for high values of the Reynolds number, the resistance coefficient is solely a function of the relative roughness, k_s/D. These characteristics are shown in Fig. 10-7, where f is plotted as a function of Re for

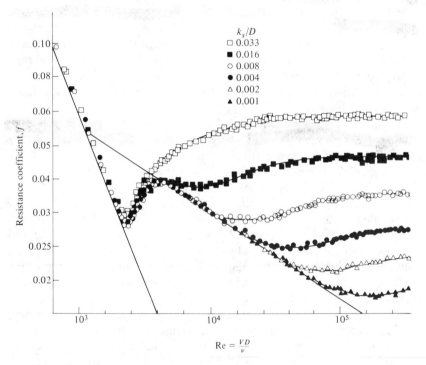

FIGURE 10-7   Resistance coefficient of f versus Re for sand-roughened pipe. [After Nikuradse (19).]

various values of relative roughness, k_s/D. The reason the resistance is like that of a smooth pipe for low values of k_s/D and Re is that for these conditions the roughness elements become submerged in the viscous sublayer and hence have negligible influence on the main flow in the pipe. However, at high values of the Reynolds number, the viscous sublayer is so thin that the roughness elements project into the main stream of flow and the flow resistance is determined by the drag of the individual roughness elements. Hence, for relatively large values of k_s/D and for large Reynolds numbers, the resistance to flow is proportional to V^2; thus f becomes constant for these conditions.

The uniform character of the sand grains used in Nikuradse's tests produces a dip in the f-versus-Re curve (Fig. 10-7) before reaching a constant value of f. However, tests on commercial pipes where the roughness is somewhat random reveal that no such dip occurs. By plotting data for commercial pipe from a number of sources, Moody (18) developed a design chart similar to that shown in Fig. 10-8.

In Fig. 10-8 the variable k_s is the symbol used to denote the *equivalent sand roughness*. That is, a pipe that has the same resistance character-istics at high Re values as a sand-roughened pipe of the same size is said to have a size of roughness equivalent to that of the sand-roughened pipe. Figure 10-9 gives approximate values of k_s and k_s/D for various kinds of pipe. This figure along with Fig. 10-8 is used to solve certain kinds of pipe-flow problems.

In Fig. 10-8 the abscissa (labeled at the bottom) is the Reynolds number, Re, and the ordinate (labeled at the left) is the resistance coeffi-cient f. Each solid curve is for a constant relative roughness, k_s/D, and the values of k_s/D are given on the right at the end of each curve. To find f, given Re and k_s/D, one goes to the right to find the correct relative-roughness curve; then one looks at the bottom of the chart to find the given value of Re and with this value of Re moves vertically upward until the given k_s/D curve is reached. Finally, from this point one moves hori-zontally to the left scale to read the value of f. If the curve for the given value of k_s/D is not plotted in Fig. 10-8, then one simply finds the proper position on the graph by interpolation between curves of k_s/D which bracket the given k_s/D.

For some problems it is convenient to enter Fig. 10-8 using a value of the parameter $\mathrm{Re}f^{1/2}$. This parameter is useful when h_f and k_s/D are known but the velocity, V, is not. Without V the Reynolds number cannot be computed, so f cannot be read by entering the chart with Re and k_s/D. But from $h_f = f(L/D)V^2/2g$ [or $V = (2gh_f/L)^{1/2}(D/f)^{1/2}$] and Re $= VD/\nu$, one can see that Re can be given as

$$\mathrm{Re} = \frac{D^{3/2}}{f^{1/2}} \left(\frac{2gh_f}{L} \right)^{1/2}$$

or, upon multiplying both sides of the above equation by $f^{1/2}$, we get

$$\mathrm{Re}f^{1/2} = \frac{D^{3/2}}{\nu} (2gh_f/L)^{1/2}$$

Thus a value of $\mathrm{Re}f^{1/2}$ can be calculated for this type of flow problem which, in turn, allows a direct determination of f using Fig. 10-8, where curves of constant $\text{Re}f^{1/2}$ are plotted slanting from the upper left to lower right with the values of $\text{Re}f^{1/2}$ for each line given at the top of the chart.

There are basically three types of problems involved with uniform flow in a single pipe. These are

1. Determine the head loss, given the kind and size of pipe along with the flow rate.
2. Determine the flow rate, given the head, kind, and size of pipe.
3. Determine the size of pipe needed to carry the flow, given the kind of pipe, head, and flow rate.

In the first type of problem the Reynolds number and k_s/D are first computed and then f is read from Fig. 10-8, after which the head loss is obtained by the use of Eq. (10-22).

EXAMPLE 10-4   Water, 20°C, flows at a rate of 0.05 m³/s in a 20-cm asphalted cast-iron pipe. What is the head loss per kilometer of pipe?

Solution   First compute the Reynolds number, VD/ν. Here $V = Q/A$. Thus

$$V = \frac{0.05 \text{ m}^3/\text{s}}{(\pi/4)(0.20^2 \text{ m}^2)} = 1.59 \text{ m/s}$$

$$\nu = 1.0 \times 10^{-6} \text{ m}^2/\text{s} \quad \text{(from Appendix A-5)}$$

then $$\text{Re} = \frac{VD}{\nu} = \frac{(1.59 \text{ m/s})(0.20 \text{ m})}{10^{-6} \text{ m}^2/\text{s}}$$

$$= 3.18 \times 10^5$$

From Fig. 10-9, $k_s/D = 0.0007$. Then from Fig. 10-8, using the values obtained for k_s/D and Re, we find $f = 0.019$. Finally the head loss is computed from the Darcy-Weisbach equation:

$$h_f = f\frac{L}{D}\frac{V^2}{2g} = 0.019 \left(\frac{1,000 \text{ m}}{0.20 \text{ m}}\right)\left(\frac{1.59^2 \text{ m}^2/\text{s}^2}{2(9.81 \text{ m/s}^2)}\right)$$

$$= 12.2 \text{ m}$$

The head loss per kilometer is 12.2 m.   ◄

In the second type of problem, k_s/D and the value of $(D^{3/2}/\nu)\sqrt{2gh_f/L}$ are computed so that the top scale can be used to enter the design chart of Fig. 10-8. Then, once f is read from the chart, the velocity from Eq. (10-22) is solved for and the discharge is computed from $Q = VA$.

EXAMPLE 10-5   The head loss per kilometer of 20-cm asphalted cast-iron pipe is 12.2 m. What is the discharge of water?

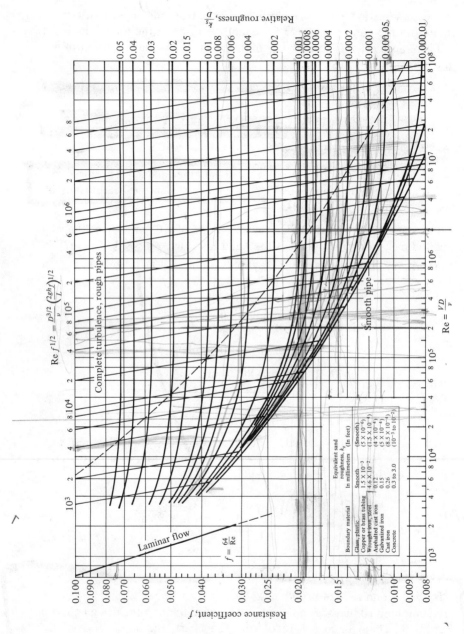

FIGURE 10-8   Resistance coefficient f versus Re. Reprinted with minor variations. [After Moody (18). Reprinted with permission from the A.S.M.E.]

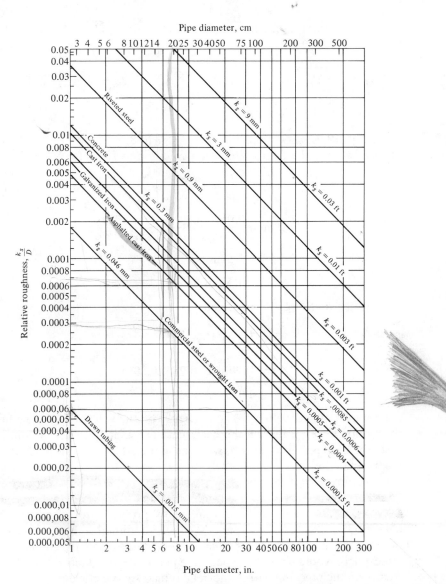

FIGURE 10-9   Relative roughness for various kinds of pipe. [After Moody (18). Reprinted with permission from the A.S.M.E.]

Solution   First compute the parameter $D^{3/2}\sqrt{(2gh_f/L)}/\nu$. Assume $T = 20°C$, so that

$$D^{3/2}\frac{\sqrt{2gh_f/L}}{\nu} = (0.20 \text{ m})^{3/2}\frac{[2(9.81 \text{ m/s}^2)(12.2 \text{ m}/1,000 \text{ m})]^{1/2}}{1.0 \times 10^{-6} \text{ m}^2/s}$$

$$= 4.38 \times 10^4$$

From Fig. 10-9, $k_s/D = 0.0007$. Using Fig. 10-8, we read $f = 0.019$. Now we use this f in the Darcy-Weisbach equation to solve for V:

$$h_f = f\frac{L}{D}\frac{V^2}{2g}$$

$$12.2 \text{ m} = \frac{0.019(1,000 \text{ m})}{0.20 \text{ m}}\frac{V^2}{2(9.81 \text{ m/s}^2)}$$

$$V^2 = 2.52 \text{ m}^2/s^2$$

$$V = 1.59 \text{ m/s}$$

Finally, the discharge is computed:

$$Q = VA = V\frac{\pi}{4}D^2$$

$$= 1.59 \text{ m/s } (0.785)(0.20^2 \text{ m}^2)$$

$$= 0.050 \text{ m}^3/s \qquad \blacktriangleleft$$

Examples 10-4 and 10-5 are good checks on the validity of the methods of solution because therein the basic data are exactly the same—in one case, the head loss is unknown; whereas in the other case, the discharge is unknown.

In the foregoing example, the head loss in the pipe was known; therefore, it was possible to obtain a direct solution by entering Fig. 10-8 with a value of $\text{Re}f^{1/2}$. However, there are many problems for which the discharge Q is desired that cannot be solved directly. For example, a problem in which water flows from a reservoir through a pipe and into the atmosphere would be of this type. Here part of the available head is lost to friction in the pipe and part of the head remains in kinetic energy in the jet as it leaves the pipe. Therefore, at the outset one does not know how much head loss occurs in the pipe itself. To effect a solution one must iterate on f. The energy equation is written and an initial value for f is guessed, then the velocity, V, is solved for. With this value of V a Reynolds number is computed that allows a better value of f to be determined through the use of Fig. 10-8, and so on. This type of solution usually converges quite rapidly because f changes more slowly than Re.

Once f and V have been determined, the discharge can be calculated with the use of the continuity equation.

EXAMPLE 10-6   Determine the discharge of water through the 50-cm steel pipe.

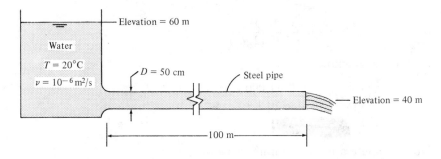

Solution   From the table in Fig. 10-8 the value of $k_s = 4.6 \times 10^{-5}$ m. Therefore, $k_s/D = 9.2 \times 10^{-5}$ m. Now write the energy equation from the reservoir water surface to the free jet at the end of the pipe:

$$\frac{p_1}{\gamma} + \frac{V_1^2}{2g} + z_1 = \frac{p_2}{\gamma} + \frac{V_2^2}{2g} + z_2 + h_L$$

$$0 + 0 + 60 = 0 + \frac{V_2^2}{2g} + 40 + f\frac{L}{D}\frac{V_2^2}{2g}$$

392.4

or
$$V = \left(\frac{2g \times 20}{1 + 200f}\right)^{1/2}$$

First trial: Assume $f = 0.020$; then $V = 8.86$ m/s and Re $= 4.43 \times 10^6$. With Re $= 4.43 \times 10^6$ and $k_s/D = 9.2 \times 10^{-5}$ then $f = 0.012$ (from Fig. 10-8).

Second trial: $V = 10.7$ m/s and Re $= 5.35 \times 10^6$; $f = 0.012$
Thus

$$Q = VA = 10.7 \text{ m/s} \times (\pi/4) \times (0.50)^2 \text{ m}^2 = 2.11 \text{ m}^3/\text{s} \quad \blacktriangleleft$$

In the third type of problem, it is usually best to first assume a value of f and then solve for D, after which a better value of f is computed based upon the first estimate of D. This iterative procedure is continued until a valid solution is obtained. A trial-and-error procedure is necessary because without D one cannot compute k_s/D or Re to enter the Moody diagram.

EXAMPLE 10-7   What size asphalted cast-iron pipe is needed to carry water at a discharge of 3 cfs and with a head loss of 4 ft per 1,000 ft of pipe?

Solution   First assume $f = 0.015$. Then

$$h_f = \frac{fL}{D}\frac{V^2}{2g} = \frac{fL}{D}\frac{Q^2/A^2}{2g} = \frac{fLQ^2}{2g(\pi/4)^2 D^5}$$

or

$$D^5 = \frac{fLQ^2}{0.785^2(2gh_f)}$$

or, for this example,

$$D^5 = \frac{0.015(1,000 \text{ ft})(3 \text{ ft}^3/\text{sec})^2}{0.615(64.4 \text{ ft}/\text{sec}^2)(4 \text{ ft})} = 0.852 \text{ ft}^5$$

$$D = 0.97 \text{ ft}$$

Now compute a more accurate value of f:

$$\frac{k_s}{D} = 0.0004 \qquad V = \frac{Q}{A} = \frac{3 \text{ ft}^3/\text{sec}}{0.785(0.94 \text{ ft}^2)} = 4.07 \text{ ft/sec}$$

Then, $$\text{Re} = \frac{VD}{\nu} = \frac{4.07 \text{ ft/sec } (0.97 \text{ ft})}{1.21(10^{-5} \text{ ft}^2/\text{sec})} = 3.25 \times 10^5$$

From Fig. 10-8, $f = 0.0175$. Now recompute D by applying the ratio of f's to previous calculations for D^5:

$$D^5 = \frac{0.0175}{0.015}(0.85 \text{ ft}^5) = 0.992 \text{ ft}^5$$

$$D = 0.998 \text{ ft}$$

Use a 12-in.-diameter pipe.   ◄

Note   In actual design practice if an odd size of pipe is called for as a result of the design calculation, it is customary to choose the next size larger that is available commercially. By so doing, the cost is less than that for an odd-sized pipe and the pipe will be more than large enough to carry the flow.

10-5   FLOW AT PIPE INLETS AND LOSSES FROM FITTINGS

Introduction

In the preceding sections, formulas were presented which are used to determine the head loss for uniform flow in a pipe; however, pipe systems also include inlets, outlets, bends, and other appurtenances that create additional head losses. Flow separation and the generation of additional turbulence therefrom is usually the cause of these head losses. Therefore, in this section we will consider the flow patterns and resulting head losses for some of these flow transitions.

Flow in pipe inlet

If the inlet to a pipe is well rounded as shown in Fig. 10-10, the boundary layer will develop from the inlet and grow in thickness until it extends to the center of the pipe. Thereafter, the flow in the pipe will be uniform. The

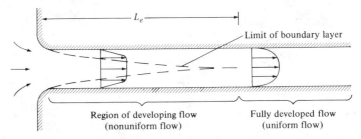

FIGURE 10-10 Flow characteristics at pipe inlet (not to scale).

length L_e of the developing region at the entrance is equal to approximately $0.05D$ Re for laminar flow and approximately $50D$ for turbulent flow. Velocity and pressure distribution for the inlet region of a pipe with turbulent flow are shown in Fig. 10-11. The head loss that is produced by inlets, outlets or fittings is expressed by the following equation:

$$h_L = K \frac{V^2}{2g}$$

(10-28)

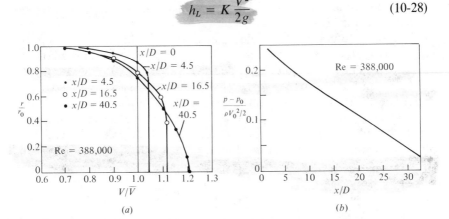

FIGURE 10-11 Velocity and pressure distribution in the inlet region of a pipe (1). (a) Velocity distribution. (b) Pressure distribution.

In Eq. (10-28) V is the mean velocity in the pipe and K is the loss coefficient for the particular fitting that is involved. For example, K for a well-rounded inlet with flow at high values of Re is approximately 0.10; hence, it can be seen that the head loss for such a transition is quite small compared with that for an abrupt pipe outlet for which K is 1.0.

If the pipe inlet is abrupt as in Fig. 10-12, separation will occur just downstream of the entrance; hence, the streamlines converge and then diverge with consequent turbulence and relatively high head loss. The loss coefficient for the abrupt inlet is approximately 0.5.

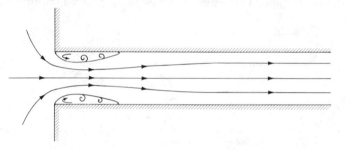

FIGURE 10-12   Flow at sharp-edged inlet.

Flow through an elbow

Although the cross-sectional area of an elbow may not change from section to section, considerable head loss is produced because separation occurs near the inside of the bend and downstream of the midsection. Thus, when the flow leaves the elbow the eddies produced by separation create considerable head loss. The approximate flow pattern for an elbow is shown in Fig. 10-13.

The loss coefficient for an elbow at high Reynolds numbers depends primarily on the shape of the elbow. If it is a very short-radius elbow the loss coefficient will be quite high. For larger radius elbows the coefficient reduces until a minimum value is found at an r/d value of about 4 (see Table 10-2). However, for still larger values of r/d an increase in loss coefficient occurs due to the fact that for larger r/d values the elbow itself is significantly longer than elbows with small r/d values. Thus the greater length creates an additional head loss. The loss coefficients for various types of elbows along with a number of other fittings and flow transitions are given in Table 10-2.[1]

---

[1] Engineering handbooks usually include extensive tables of loss coefficients. For example, references 1, 7, 12, and 27 are particularly useful in this respect.

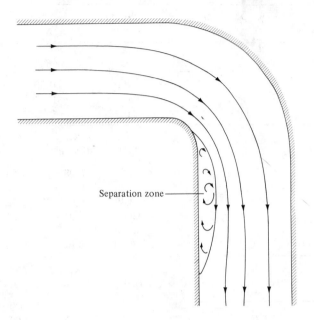

FIGURE 10-13   Flow pattern in an elbow.

EXAMPLE 10-8   If oil ($\nu = 4 \times 10^{-5}$ m²/s, sp. gr. = 0.9) flows from the upper to lower reservoir at a rate of 0.028 m³/s in the 15-cm smooth pipe, what is the elevation of the oil surface in the upper reservoir?

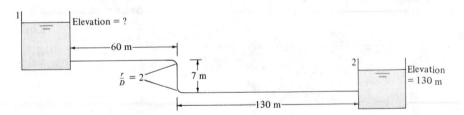

Solution   Apply the energy equation between the surfaces in the upper and lower reservoirs:

$$\frac{p_1}{\gamma} + \frac{V_1^2}{2g} + z_1 = \frac{p_2}{\gamma} + \frac{V_2^2}{2g} + z_2 + \sum h_L$$

$$0 + 0 + z_1 = 0 + 0 + 130 \text{ m} + \frac{fL}{D}\frac{V^2}{2g} + 2K_b\frac{V^2}{2g} + K_e\frac{V^2}{2g} + K_E\frac{V^2}{2g}$$

Here K_b, K_e, and K_E are loss coefficients for bend, entrance and outlet,

TABLE 10-2   LOSS COEFFICIENTS FOR VARIOUS TRANSITIONS AND FITTINGS

Description	Sketch	Additional Data		K		Source
Pipe entrance $h_L = K_e V^2/2g$		r/d 0.0 0.1 >0.2		K_e 0.50 0.12 0.03		(1)
Contraction $h_L = K_c V_2^2/2g$		D_2/D_1 0.0 0.20 0.40 0.60 0.80 0.90	K_C $\theta = 60°$ 0.08 0.08 0.07 0.06 0.05 0.04	K_C $\theta = 180°$ 0.50 0.49 0.42 0.32 0.18 0.10		(1)
Expansion $h_L = K_E V_1^2/2g$		D_1/D_2 0.0 0.20 0.40 0.60 0.80	K_E $\theta = 10°$ 0.13 0.11 0.06 0.03	K_E $\theta = 180°$ 1.00 0.92 0.72 0.42 0.16		(1)
90° miter bend	Vanes	Without vanes		$K_b = 1.1$		(26)
		With vanes		$K_b = 0.2$		(26)
90° smooth bend		r/d 1 2 4 6 8 10		$K_b =$ 0.35 0.19 0.16 0.21 0.28 0.32		(3) and (13)
Threaded pipe fittings	Globe valve—wide open Angle valve—wide open Gate valve—wide open Gate valve—half open Return bend Tee 90° elbow 45° elbow			$K_v = 10.0$ $K_v = 5.0$ $K_v = 0.2$ $K_v = 5.6$ $K_b = 2.2$ $K_t = 1.8$ $K_b = 0.9$ $K_b = 0.4$		(26)

[handwritten annotations: "ASSUME TURBULENT FLOW", "$V_1^2/2g$", "friction factor higher for smaller pipe + vise versa"]

respectively. These have values of 0.19, 0.5, and 1.0. (Table 10-2). To determine f we get Re to enter Fig. 10-8.

$$Re = \frac{VD}{\nu}$$

but

$$V = \frac{Q}{A} = \frac{(0.028 \text{ m}^3/\text{s})}{0.785(0.15 \text{ m})^2} = 1.58 \text{ m/s}$$

Then

$$Re = \frac{1.58 \text{ m/s } (0.15 \text{ m})}{4 \times 10^{-5} \text{ m}^2/\text{s}}$$

$$= 5.93 \times 10^3$$

Now we read f from Fig. 10-8 (smooth pipe curve): $f = 0.035$. Then

$$z_1 = 130 \text{ m} + \frac{V^2}{2g} \left[\frac{0.035(197 \text{ m})}{0.15 \text{ m}} + 2(0.19) + 0.5 + 1 \right]$$

$$= 130 \text{ m} + \left[\frac{(1.58 \text{ m/s})^2}{2(9.81 \text{ m/s}^2)} \right] (46 + 0.38 + 0.5 + 1)$$

$$= 130 \text{ m} + 6.1 \text{ m} = 136.1 \text{ m} \qquad \blacktriangleleft$$

Transition losses and grade lines

In Chapter 7 the effect on the energy and hydraulic grade lines of the pipe head loss and head loss at an abrupt expansion were discussed in some detail. However, no mention was made of head loss due to entrances, bends, and other flow transitions. The primary effect, just like that of the abrupt expansions, is to cause the energy grade line to drop an amount equal to the head loss produced by that transition. Generally this drop will occur over a distance of several diameters downstream of the transition, as seen in Fig. 10-14. The hydraulic grade line also drops sharply immedi-

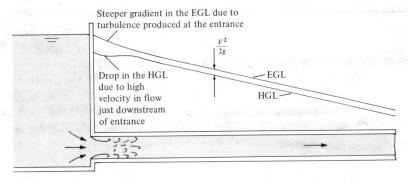

FIGURE 10-14 EGL and HGL at sharp-edged pipe entrance.

ately downstream of the entrance due to the high-velocity flow in the contracted portion of the stream. Then as the turbulent mixing occurs even farther downstream, energy is lost owing to viscous action occurring in the mixing process. Thus the energy grade line at the entrance is steeper than the grade line farther downstream where the flow has become uniform. As the additional energy loss from the transition subsides, the energy grade line takes on the slope created by the head loss of the pipe itself.

Even though many transitions produce grade lines that have interesting local details, such as those noted for the sharp-edged entrance, it is common as a gross indication to simply show abrupt changes in the energy gradeline and to neglect local departures between the EGL and HGL owing to local changes in $V^2/2g$. Thus, in Fig. 10-15 we see a simplified plot of the EGL and HGL for a pipe with several transitions in it.

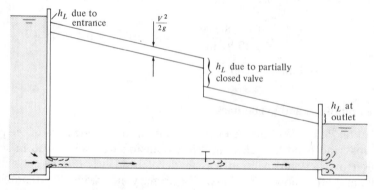

FIGURE 10-15   Head losses in a pipe.

10-6   PIPE SYSTEMS

Simple pump in a pipeline

Up to now, a number of pipe-flow problems have been considered in which the head for producing the flow was explicitly given. Now we shall consider flow in which the head is developed by a pump. However, the head produced by a centrifugal pump is a function of the discharge; hence, a direct solution is usually not immediately available. The solution (that is, the flow rate for a given system) is obtained when the system equation (or curve) of head versus discharge is solved simultaneously with the pump equation (or curve) of head verus discharge. The solution of these two equations (or the point where the two curves intersect) will yield the operating condition for the system. Consider flow of water in the system of Fig. 10-16. When the energy equation is written from the reservoir water surface to the outlet stream we obtain the following equation:

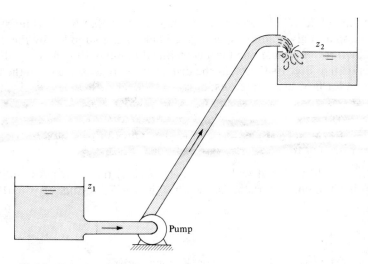

FIGURE 10-16 Pump and pipe combination.

$$\frac{p_1}{\gamma} + \frac{V_1^2}{2g} + z_1 + h_p = \frac{p_2}{\gamma} + \frac{V_2^2}{2g} + z_2 + \sum K_L \frac{V^2}{2g} + \frac{fL}{D}\frac{V^2}{2g}$$

This equation then simplifies to

$$h_p = (z_2 - z_1) + \frac{V^2}{2g}\left(1 + \sum K_L + \frac{fL}{D}\right) \qquad (10\text{-}29)$$

Hence, for any given discharge a certain head h_p must be supplied to maintain that flow. Thus we can construct an h_p-versus-Q curve as shown in Fig. 10-17. Such a curve is called the system curve. Any given centrifu-

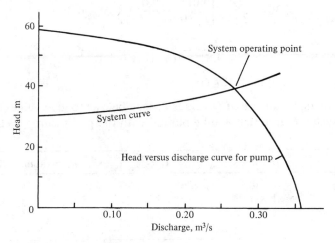

FIGURE 10-17 Pump and system curves.

gal pump will have a head-versus-discharge curve that is characteristic of that pump at a given pump speed. Such curves are supplied by the pump manufacturer; a typical one for a centrifugal pump is shown in Fig. 10-17.

It is seen in Fig. 10-17 that as the discharge increases in a pipe, the head required for flow also increases; however, the head which is produced by the pump decreases as the discharge increases. Consequently, the two curves will intersect, and the operating point is at the point of intersection—that point where the head produced by the pump is just the amount needed to overcome the head loss in the pipe.

EXAMPLE 10-9   What will be the discharge in this water system if the pump has the characteristics shown in Fig. 10-17? Assume $f = 0.015$.

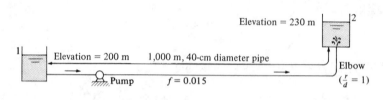

Elevation = 230 m

Elevation = 200 m   1,000 m, 40-cm diameter pipe

Pump $f = 0.015$

Elbow $\left(\frac{r}{d} = 1\right)$

Solution   First write the energy equation from water surface to water surface:

$$\frac{p_1}{\gamma} + \frac{V_1^2}{2g} + z_1 + h_p = \frac{p_2}{\gamma} + \frac{V_2^2}{2g} + z_2 + \sum h_L$$

$$0 + 0 + 200 + h_p = 0 + 0 + 230 + \left(\frac{fL}{D} + K_e + K_b + K_E\right)\frac{V^2}{2g}$$

Here $K_e = 0.5$, $K_b = 0.35$, and $K_E = 1.0$. Hence

$$h_p = 30 + \frac{Q^2}{2gA^2}\left[\frac{0.015(1,000)}{0.40} + 0.5 + 0.35 + 1\right]$$

$$= 30 + \frac{Q^2}{2 \times 9.81 \times [(\pi/4) \times 0.4^2]^2} \quad (39.3)$$

$$= 30 \text{ m} + 127\,Q^2 \text{ m}$$

Now let us make a table of Q versus h_p (see below) to give values to produce a system curve that will be plotted with the pump curve. When the

Q, m³/s	Q^2, m⁶/s²	$127Q^2$	$h_p = 30 \text{ m} + 127Q^2 \text{ m}$
0	0	0	30
0.1	1×10^{-2}	1.3	31.3
0.2	4×10^{-2}	5.1	35.1
0.3	9×10^{-2}	11.4	41.4

system curve is plotted on the same graph as the pump curve, it is seen (Fig. 10-17) that the operating condition occurs at $Q = 0.27$ m³/s. ◄

Pipes in parallel

Consider a pipe that branches into two parallel pipes and then rejoins, as in Fig. 10-18. A problem involving this configuration might be to determine the division of flow in each pipe given the total flow rate. It can be seen that the head loss must be the same in each pipe because the pressure difference is the same. Thus we can write

$$h_{L_1} = h_{L_2}$$

$$f_1 \frac{L_1}{D_1} \frac{V_1^2}{2g} = f_2 \frac{L_2}{D_2} \frac{V_2^2}{2g}$$

Then

$$\left(\frac{V_1}{V_2}\right)^2 = \frac{f_2}{f_1} \frac{L_2}{L_1} \frac{D_1}{D_2}$$

or

$$\frac{V_1}{V_2} = \left(\frac{f_2}{f_1} \frac{L_2}{L_1} \frac{D_1}{D_2}\right)^{1/2}$$

If f_1 and f_2 are known, the division of flow can be easily determined; however, some trial-and-error analysis may be required if f_1 and f_2 are in the range where they are functions of the Reynolds number.

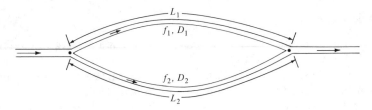

FIGURE 10-18   Flow in parallel pipes.

Pipe networks

The most common pipe networks are the water-distribution systems for municipalities. These systems have one or more sources and numerous loads (one for each household and commercial establishment). The engineer is often engaged to design the original system or to recommend an economical expansion to the network. Such an expansion may involve additional housing or commercial developments or it may simply be an expansion to handle increased loads within the existing area. In any case, the engineer is required to predict pressures throughout the network for various operating conditions; that is, for various combinations of sources and loads. The solution of such a problem must satisfy three basic requirements:

1. Continuity must be satisfied. That is, the flow into a junction of the network must equal the flow out of the junction. This must be satisfied for all junctions.
2. The head loss between any two junctions must be the same regardless of the path in the series of pipes taken to get from one junction point to the other.
3. The flow and head loss must be consistent with the appropriate velocity-head-loss equation.

Only a few years ago such solutions were made by a trial-and-error hand computation, but recent applications using the digital computer have made the older methods obsolete. Even with such advances, however, the engineer charged with the design and/or analysis of such a system must understand the basic fluid mechanics of the system to be able to interpret the results properly and to make good engineering decisions based upon the results. For further information on this subject the student is directed to Refs. 8, 11, and 14.

10-7 TURBULENT FLOW IN NONCIRCULAR CONDUITS

Basic development

Earlier in this chapter (Sec. 10-1), τ_0 was eliminated between Eqs. (10-3) and (10-21) to yield the Darcy-Weisbach equation, Eq. (10-22). It should be noted that Eq. (10-3) was derived by writing the equation of equilibrium in the longitudinal direction for an element of fluid with a circular cross section. If one derives an equation analogous to Eq. (10-3) for flow in a noncircular conduit in which the shear stress acts on the conduit surface with a perimeter P (such as the perimeter of a rectangular conduit) instead of perimeter $2\pi r$, then τ_0 is given by

$$\tau_0 = \frac{A}{P}\left[-\frac{d}{ds}(p + \gamma z)\right] \qquad (10\text{-}30)$$

In Eq. (10-3), to which Eq. (10-30) is analogous, the shear stress τ_0 was everywhere constant around the perimeter of the cylindrical element. In Eq. (10-30) for the noncircular conduit the shear stress will not be constant over the perimeter. However, we still retain Eq. (10-21) to relate τ_0 and V, where τ_0 is now the average shear stress on the boundary. Eliminating τ_0 between Eqs. (10-21) and (10-30) and integrating along the pipe yields

$$h_f = \frac{f}{4}\frac{L}{A/P}\frac{V^2}{2g} \qquad (10\text{-}31)$$

In Eq. (10-31), h_f is the head loss between two points in the conduit, L is the length between the points, and P is the wetted perimeter of the conduit. Thus Eq. (10-31) is the same as the Darcy-Weisbach equation except that D is replaced by $4A/P$. The ratio of the cross-sectional area A to the wetted perimeter P is defined as the *hydraulic radius* R. Experiments have shown we can solve flow problems involving noncircular conduits, such as rectangular ducts, if we apply the same methods and equations that we did in pipes but use $4R$ in place of D. Consequently, the relative roughness is $k_s/4R$ and the Reynolds number is defined as $V(4R)/\nu$.

EXAMPLE 10-10   Air ($T = 20°C$ and $p = 101$ kPa abs) flows at a rate of 2.5 m³/s in a 30 cm × 60 cm commercial steel rectangular duct. What is the pressure drop per 50 m of duct?

Solution   First compute Re and $k_s/4R$:

$$\text{Re} = \frac{V(4R)}{\nu}$$

Here
$$V = \frac{Q}{A} = \frac{2.5 \text{ m}^3/\text{s}}{0.18 \text{ m}^2} = 13.9 \text{ m/s}$$

The hydraulic radius is given by

$$R = \frac{A}{P} = \frac{0.18 \text{ m}^2}{1.8 \text{ m}} = 0.10 \text{ m}$$

$$4R = 4(0.10 \text{ m}) = 0.40 \text{ m}$$

$$\nu = 1.50 \times 10^{-5} \text{ m}^2/\text{s}$$

Hence,
$$\text{Re} = \frac{13.9 \times 0.40}{1.49 \times 10^{-5}}$$

$$= 3.73 \times 10^5$$

$$\frac{k_s}{4R} = \frac{4.6 \times 10^{-5} \text{ m}}{0.40 \text{ m}} = 1.15 \times 10^{-4}$$

Then from Fig. 10-8, $f = 0.015$; thus,

$$h_f = \frac{fL}{4R}\frac{V^2}{2g}$$

or
$$\gamma h_f = \Delta p_f = \frac{fL}{4R}\rho\frac{V^2}{2}$$

$$\rho = 1.2 \text{ kg/m}^3$$

Finally,

$$\Delta p_f \text{ per 50 m} = \frac{0.015 \ (50 \text{ m})}{0.40 \text{ m}} \ (1.2 \text{ N} \cdot \text{s}^2/\text{m}^4) \ \frac{13.9^2}{2} \text{ m}^2/\text{s}^2$$

$$= 217 \text{ Pa} \qquad \blacktriangleleft$$

Uniform flow in open channels

In the case of open-channel flow the same methods are used that were introduced in the preceding section. That is, the Darcy-Weisbach equation with D replaced by $4R$ is used to relate velocity, shape of channel, and depth of flow. It should be noted that the perimeter used for computing the hydraulic radius is the perimeter of the channel that is actually in contact with the flowing fluid. For example, in Fig. 10-19 the hydraulic radius of the channel with rectangular cross section is

$$R = \frac{A}{P} = \frac{By}{B + 2y} \qquad (10\text{-}32)$$

One can see that for very wide shallow channels the hydraulic radius approaches the depth y.

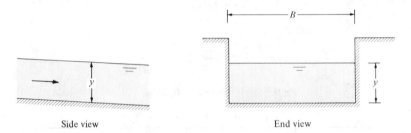

Side view   End view

FIGURE 10-19 Open-channel relations.

EXAMPLE 10-11 Estimate the discharge of water that a 10-ft wide concrete channel can carry if the depth of flow is 6 ft and the slope of the channel is 0.0016.

Solution We will use the Darcy-Weisbach equation:

$$h_f = \frac{fL}{4R} \frac{V^2}{2g}$$

or

$$\frac{h_f}{L} = \frac{f}{4R} \frac{V^2}{2g}$$

but $S = h_f/L = 0.0016$, so we have $V^2/2g = 4RS/f$

or

$$V = \sqrt{\frac{8g}{f} RS}$$

Assume $k_s = 0.005$ ft; then the relative roughness is

$$\frac{k_s}{4R} = \frac{0.005 \text{ ft}}{4(60 \text{ ft}^2/22 \text{ ft})} = \frac{0.005 \text{ ft}}{4(2.74 \text{ ft})}$$

$$= 0.00046$$

Using $k_s/4R = 0.00046$ as a guide and referring to Fig. 10-8, we assume that $f = 0.016$; thus

$$V = \sqrt{\frac{8(32.2 \text{ ft/sec}^2)(2.74 \text{ ft})(0.0016)}{0.016}} = \sqrt{70.6 \text{ ft}^2/\text{sec}^2}$$

$$= 8.40 \text{ ft/sec}$$

Then

$$\text{Re} = V\frac{4R}{\nu} = \frac{8.40 \text{ ft/sec } (10.9 \text{ ft})}{1.2 \ (10^{-5} \text{ ft}^2/\text{sec})}$$

$$= 7.63 \times 10^6$$

Using this new value of Re and with $k_s/4R = 0.00046$, we read f as $f = 0.016$. Our initial guess was good, so now that the velocity is known, we can compute Q:

$$Q = VA$$
$$= 8.40 \text{ ft/sec } (60 \text{ ft}^2) = 504 \text{ cfs} \qquad \blacktriangleleft$$

The Chezy and Mannings equations

Leaders in research and design of open channels have recommended the adoption of the methods involving the Reynolds number and relative roughness already presented (6); however, many engineers continue to use the older, more traditional methods for designing open channels. Because of this current use and for historic reasons, the Chezy and the Mannings equations are presented here. If Eq. (10-31) is rewritten in slightly different form, we have

$$\frac{h_f}{L} = \frac{f}{4R} \frac{V^2}{2g} \qquad (10\text{-}33)$$

For uniform flow, the depth—called *normal depth*—is constant; consequently, h_f/L is the slope S_0 of the channel and Eq. (10-33) can be written as

$$RS_0 = \frac{f}{8g} V^2$$

or
$$V = C\sqrt{RS_0} \qquad (10\text{-}34)$$

where
$$C = \sqrt{8g/f} \qquad (10\text{-}35)$$

Since $Q = VA$, we can express the discharge in a channel as

$$Q = CA\sqrt{RS_0} \qquad (10\text{-}36)$$

This equation is known as the *Chezy equation* after the French engineer of the same name. For practical application the coefficient C must be determined, and the usual design formula for C in the SI system of units is given as

$$C = \frac{R^{1/6}}{n} \qquad (10\text{-}37)$$

When this expression for C is inserted into Eq. (10-36), we obtain a common form of the discharge equation for uniform flow in open channels for SI units:

$$Q = \frac{1.0}{n} AR^{2/3} S_0^{1/2} \qquad (10\text{-}38)$$

In Eq. (10-38), n is a resistance coefficient, called the Mannings n, which has different values for different types of boundary roughness. Table 10-3 gives n for various types of boundary surfaces. The major limitation of this approach is that the viscous effect or relative roughness is not present

TABLE 10-3   TYPICAL VALUES OF THE ROUGHNESS COEFFICIENT n

Lined Canals	n
Cement plaster	0.011
Untreated gunite	0.016
Wood, planed	0.012
Wood, unplaned	0.013
Concrete, troweled	0.012
Concrete, wood forms, unfinished	0.015
Rubble in cement	0.020
Asphalt, smooth	0.013
Asphalt, rough	0.016

Natural Channels	
Gravel beds, straight	0.025
Gravel beds plus large boulders	0.040
Earth, straight, with some grass	0.026
Earth, winding, no vegetation	0.030
Earth, winding	0.050

in the design formula. Hence, application outside the range of normal-size channels carrying water is not recommended.

Mannings equation—Traditional system

In converting from the SI to the traditional ft-lbf-sec system of units, it can be shown that a factor equal to 1.49 must be applied if the same value of n is used in the two systems. Thus in the traditional system the discharge equation using the Manning n is given as

$$Q = \frac{1.49}{n} AR^{2/3} S_0^{1/2}$$

EXAMPLE 10-12   Using the Chezy equation with Mannings n, compute the discharge in the channel described in Example 10-11.

Solution

$$Q = \frac{1.49}{n} AR^{2/3} S_0^{1/2}$$

From Example 10-10,

$$R = \frac{60}{22} = 2.72 \text{ ft} \qquad R^{2/3} = 1.96$$

$$S_0 = 0.0016 \qquad S_0^{1/2} = 0.04$$

$$A = 6(10) = 60 \text{ ft}^2$$

$$n = 0.015 \qquad \text{(Table 10-3)}$$

Then, $Q = \dfrac{1.49}{0.015}(60)(1.96)(0.04) = 467 \text{ cfs}$ ◀

The two results (Examples 10-11 and 10-12) are well within expected engineering accuracy for this type problem.

PROBLEMS

10-1 Liquid in the pipe has a specific weight of $10 \, \text{kN/m}^3$. The acceleration of the liquid is zero. Is the liquid stationary, moving upward, or moving downward in the pipe? If the pipe diameter is 1 cm and the liquid viscosity is $3.125 \times 10^{-3} \, \text{N} \cdot \text{s/m}^2$, what is the magnitude of the mean velocity in the pipe?

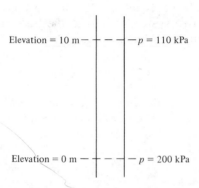

Elevation = 10 m — $p = 110 \, \text{kPa}$

Elevation = 0 m — $p = 200 \, \text{kPa}$

10-2 Liquid flows downward in a 1-cm vertical, smooth pipe with a mean velocity of 1.0 m/s. The liquid has a density of 1,000 kg/m³ and a viscosity of $0.10 \, \text{N} \cdot \text{s/m}^2$. If the pressure at a given section is 300 kPa, then what will be the pressure at a section 10 m below this section?

10-3 Glycerin at a temperature of 30°C flows at a rate of $3 \times 10^{-6} \, \text{m}^3/\text{s}$ through a horizontal 30-mm-diameter tube. What is the pressure drop in pascals per 100 m if the flow is laminar?

10-4 Glycerin at a temperature of 90°F ($\mu = 6 \times 10^{-3} \, \text{lb-sec/ft}^2$) flows at a rate of 10 gpm through a horizontal 1-in.-diameter tube. What is the pressure drop in pounds per square inch per 100 ft if the flow is laminar?

10-5 Oil (sp. gr. = 0.97, $\mu = 10^{-2} \, \text{lb-sec/ft}^2$) is pumped through a 2-in. pipe at the rate of 0.20 cfs. What is the pressure drop per 100 ft of level pipe?

10-6 Oil (sp. gr. = 0.94, $\mu = 0.048 \, \text{N} \cdot \text{s/m}^2$) is pumped through a horizontal 5-cm pipe at the rate of $2.0 \times 10^{-3} \, \text{m}^3/\text{s}$. What is the pressure drop per 100 m of pipe?

10-7 SAE 30 oil at 38°C flows in a vertical 8-mm tube. If the pressure in the oil is constant along the tube, what is the discharge in cubic meters per second and in which direction is flow occurring?

10-8 SAE 30 oil at 100°F flows in a vertical $\frac{3}{8}$-in. tube. If the pressure in the oil is constant along the tube, what is the discharge and in which direction is flow occurring?

10-9 Kerosene (68°F) flows at the rate of 1 cfs in a 10-in.-diameter pipe. Would you expect the flow to be laminar or turbulent?

10-10 Kerosene (20°C) flows at a rate of 0.03 m³/s in a 20-cm pipe. Would you expect the flow to be laminar or turbulent?

10-11 Glycerin ($T = 20°C$) flows in a 20-cm-diameter pipe with a mean velocity of 50 cm/s. Is the flow laminar or turbulent? Plot the velocity distribution across the flow section.

10-12 Glycerin ($T = 68°F$) flows in a 1-ft-diameter pipe with a mean velocity of 2 ft/sec. Is the flow laminar or turbulent? Plot the velocity distribution across the flow section.

10-13 What size steel pipe should be used to carry 0.1 cfs of castor oil at 90°F a distance of 1 mile with an allowable friction loss of 20 psi? Assume specific gravity equals 0.85.

10-14 Glycerin (20°C) flows in a 4-cm wrought-iron tube with a mean velocity of 40 cm/s. Is the flow laminar or turbulent? What is the shear stress at the center of the tube and at the wall? If the tube is vertical and the flow is downward, will the pressure increase or decrease in the direction of flow? At what rate?

10-15 A 4-cm pipe 7 m long is supported vertically over a tank and discharges oil (sp. gr. $= 0.8$, $\mu = 10^{-1} \, N \cdot s/m^2$) at a mean velocity of 3 m/s as shown. If the pipe weighs 200 N, what force is required to support it?

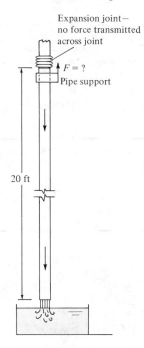

10-16 Velocity measurements are made across a 1-ft pipe. The velocity at the center is found to be 3 fps and the velocity distribution is seen to be parabolic. If the pressure drop is found to be 14 psf per 100 ft of pipe, what is the kinematic viscosity ν of the fluid? Assume that the fluid's specific gravity is 0.90.

10-17 Velocity measurements are made in a 30-cm pipe. The velocity at the center is found to be 1.5 m/s and the velocity distribution is observed to be parabolic. If the pressure drop is found to be 1.8 kPa per 100 m of pipe, what is the kinematic viscosity ν of the fluid? Assume that the fluid's specific gravity is 0.90.

10-18 In a 12-in. smooth pipe, f is 0.017 when oil having a specific gravity of 0.82 flows through it with a mean velocity of 6 ft/sec. If τ_{app} is found to be 0.10 psf at a distance of 1 in. from the wall of the pipe, what is the viscous shear stress on the wall?

10-19 At a Reynolds number of 100,000, what is the nominal thickness of the viscous sublayer in a 6-in. smooth pipe?

10-20 Water is pumped through a heat exchanger consisting of 5-mm-diameter tubes each 5 m long. The velocity in each tube is 10 cm/s. The water temperature increases from 20°C at the entrance to 30°C at the exit. Calculate the pressure difference across the heat exchanger neglecting entrance losses but accounting for the effect of temperature change.

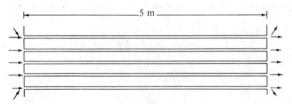

10-21 The velocity of oil (sp. gr. = 0.8) through the 2-in. smooth pipe is 5 ft/s. $L = 30$ ft, $z_1 = 2$ ft, $z_2 = 4$ ft, and the manometer deflection is 4 in. Determine the flow direction, the resistance coefficient f, whether the flow is laminar or turbulent, and the viscosity of the oil.

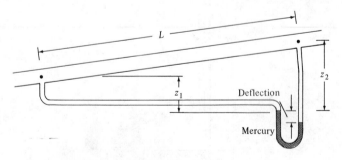

PROBLEMS 10-21, 10-22

10-22 The velocity of oil (sp. gr. = 0.8) through the 5-cm smooth pipe is 1.2 m/s. $L = 10$ m, $z_1 = 1$ m, $z_2 = 2$ m, and the manometer deflection is 10 cm. Determine the flow direction, the resistance coefficient f, whether the flow is laminar or turbulent, and the viscosity of the oil.

10-23 Flow of a liquid in a smooth 3-cm pipe is found to yield a head loss of 1 m per meter of pipe when the mean velocity is 1 m/s. If the rate of flow is doubled would the head loss also be doubled?

10-24 At a Reynolds number of 100,000, what is the nominal thickness of the viscous sublayer in a 12-cm smooth pipe?

10-25 Water (70°F) flows through a 10-in. smooth pipe at the rate of 2 cfs. What is the resistance coefficient f?

10-26 Water (10°C) flows through a 25-cm smooth pipe at a rate of 0.05 m³/s. What is the resistance coefficient f?

10-27 Air flows in a 3-cm smooth tube at a rate of 0.012 m³/s. If $T = 20°C$ and $p = 110$ kPa, what is the pressure drop per meter of length of tube?

10-28 Air flows in a 1-in. smooth tube at a rate of 20 cfm. If $T = 80°F$ and $p = 15$ psia, what is the pressure drop per foot of length of tube?

10-29 Water flows in the pipe shown and the manometer deflects 80 cm. What is f for the pipe if $V = 3$ m/s?

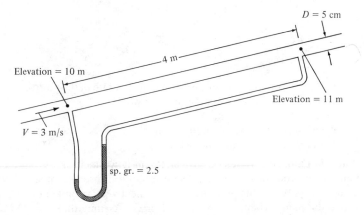

10-30 What is f for the flow of water at 10°C through a 30-cm cast-iron pipe with a mean velocity of 3 m/s? Plot the velocity distribution for this flow.

10-31 What is f for the flow of water at 60°F through a 1-ft-diameter cast-iron pipe with a mean velocity of 6 ft/sec? Also plot the velocity distribution for this flow.

10-32 Water at 20°C flows through a 6-cm smooth brass tube at a rate of 0.003 m³/s. What is f for this flow?

10-33 Water at 60°F is to be pumped at the rate of 600 gpm through 1,000 ft of 6-in. horizontal cast-iron pipe. Determine the head loss and the horsepower required.

10-34 Water at 10°C is to be pumped at a rate of 0.04 m³/s through 1,000 m of 15-cm horizontal cast-iron pipe. Determine the head loss and the power required.

10-35 Irrigation water (20°C) is to be pumped through a 2-km long by 1-m-diameter steel pipe from a river to an irrigation canal at a rate of 2.50 m³/s. The water surface elevation at the pump intake is 100 m and in the canal it is 150 m. What power should be supplied to the pump if the pump efficiency is 82% and the inlet and outlet head losses are nil?

10-36 Referring to the figure, what do you think are at A and C? What do you think is at B? Beyond D complete the physical setup that could yield the EGL and HGL shown. What other information is indirectly revealed by the EGL and HGL?

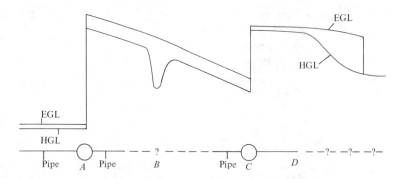

10-37 Water (20°C) flows in a 15-cm cast-iron pipe at a rate of 0.04 m³/s. For these conditions determine or estimate the following.
(a) Shear stress at the wall, τ_0
(b) Shear stress one inch from the wall
(c) Velocity one inch from the wall

10-38 Water is pumped through a vertical 10-cm new steel pipe to an elevated tank on the roof of a building. The pressure on the discharge side of the pump is 1.5 MPa. What pressure can be expected at a point in the pipe 80 m above the pump when the flow is 0.01 m³/s? Assume that $T = 20$°C.

10-39 Suppose that it is possible to prevent turbulence from developing in the flow in a pipe. What would be the ratio of head loss for the laminar flow to that for turbulent when kerosene (sp. gr. = 0.82, $\nu = 2 \times 10^{-6}$ m²/s) is pumped through a 3-cm smooth pipe with an average velocity of 4 m/s?

10-40 In a 1-in. uncoated cast-iron pipe, 0.02 cfs of water flows at 68°F. Determine f from Fig. 10-8.

10-41 Determine the head loss in 1,000 ft of a 6-in.-diameter concrete pipe ($k_s = 0.0002$ ft) carrying 0.75 cfs of fluid. The properties of the fluid are $\nu = 3.33 \times 10^{-2}$ ft²/sec, and $\rho = 1.5$ slugs/ft³.

10-42 Points A and B are 1 km apart along a 15-cm new steel pipe. Point B is 20 m higher than A. With a flow from A to B of 0.03 m³/s of crude oil (sp. gr. = 0.82) at 10°C ($\mu = 10^{-2}$ N · s/m²), what pressure must be maintained at A if the pressure at B is to be 350 kPa?

10-43 Points A and B are 3 km apart along a 60-cm new cast-iron pipe carrying water. Point A is 10 m higher than B. If the pressure at B is 140 kPa greater than at A, determine the direction and rate of flow. $T = 10°C$.

10-44 Points A and B are 3 miles apart along a 24-in. new cast-iron pipe carrying water. Point A is 30 ft higher than B. If the pressure at B is 20 psi greater than at A, determine the direction and amount of flow. $T = 50°F$.

10-45 If a flow of 0.10 m³/s of water is to be maintained, what power must be added to the water by the pump? The pipe is made of steel and it is 15 cm in diameter. Draw the energy and hydraulic grade lines for the system.

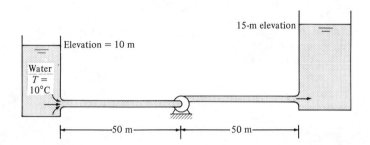

10-46 A fluid with $\nu = 10^{-6}$ m²/s and $\rho = 900$ kg/m³ flows through the 8-cm galvanized-iron pipe. Estimate the flow rate for the conditions shown in the figure.

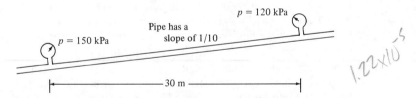

10-47 A new steel pipe 24 in. in diameter and 2 miles long carries water from a reservoir and discharges it into air. If the pipe comes out of the reservoir 10 ft below the water level in the reservoir and the pipe slopes downward from the reservoir on a grade of 2 ft/1,000 ft, determine the discharge.

10-48 What diameter cast-iron pipe is needed to carry water at a rate of 10 cfs between two reservoirs if the reservoirs are 2 miles apart and the elevation difference between the water surfaces in the reservoir is 20 ft?

10-49 A pipeline is to be designed to carry crude oil (sp. gr. = 0.93, $\nu = 10^{-5}$ m²/s) with a discharge of 0.10 m³/s and a head loss per kilometer of 30 m. What diameter steel pipe is needed? What power output from a pump is needed to maintain this flow?

10-50 Determine the discharge of kerosene (20°C) through the 15-cm smooth brass pipe (see page 402) if $h = 3$ m and $L = 40$ m.

10-51 Determine the discharge of kerosene (68°F) through the 6-in. smooth brass pipe (see page 402) if $h = 10$ ft and $L = 80$ ft.

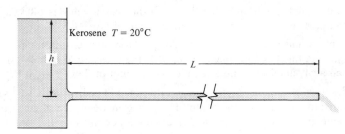

PROBLEMS 10-50, 10-51

10-52 What power must the pump supply to the system to pump the oil from the lower reservoir to the upper reservoir at a rate of 0.20 m³/s? Sketch the HGL and EGL for the system.

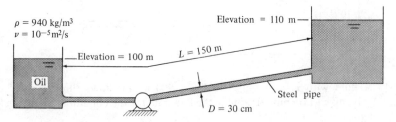

10-53 For a 40-cm pipe the resistance coefficient f was found to be 0.06 when the mean velocity was 3 m/s and the kinematic viscosity was 10^{-5}m²/s. Then if the velocity were doubled, would you expect the head loss per meter of length of pipe to double, to triple, or to quadruple?

10-54 A heat exchanger is being designed as a component of a geothermal power system in which heat is transferred from the geothermal brine to a "clean" fluid in a closed-loop power cycle. The heat exchanger, a shell and the tube type, consists of 100 galvanized iron tubes, 2 cm in diameter and 5 m long, as shown. The temperature of fluid is 200°C, the density 860 kg/m³, and the viscosity 1.35×10^{-4} N · s/m². The total mass flow rate through the exchanger is 50 kg/s.

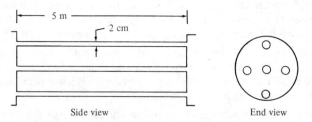

(a) Calculate the power needed to operate the heat exchanger neglecting entrance losses.

(*b*) After continued use, 2 mm of scale develops on the inside surfaces of the tubes. This scale has an equivalent roughness of 0.5 mm. Calculate the power needed under these conditions.

10-55 The heat exchanger shown consists of 2-cm-diameter drawn tubing 20 m long with 19 return bends. Water is pumped through the system at 3×10^{-4} m³/s at 20°C and exits at 80°C. The elevation between the entrance and exit is 0.8 m. Calculate the pump power required to operate the heat exchanger if there is no pressure change between 1 and 2. Use the viscosity corresponding to the average temperature in the heat exchanger.

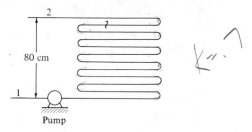

10-56 Gasoline is pumped from the gas tank of an automobile to the carburetor through a 10-ft-long $\frac{1}{4}$-in. fuel line of drawn tubing. The line has five 90° smooth bends with an r/d of 6. The gasoline discharges through a $\frac{1}{32}$in.-diameter jet in the carburetor to a pressure of 14 psia. The pressure in the tank is 14.7 psia. If the pump is 80% efficient, what power must be supplied by the pump if the automobile is accelerating and consuming fuel at the rate of 0.1 gpm? Obtain gasoline properties from Figs. A-2 and A-3.

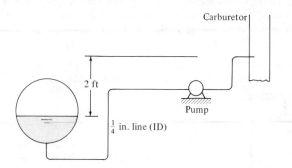

10-57 Find the loss coefficient K_v of the partially closed valve needed to reduce the discharge to 50% of the flow with the valve wide open as shown on page 404.

10-58 The pressure at a water main is 300 kPa gage. What pipe size is needed to carry water from the main at a rate of 0.025 m³/s to a factory that is 140 m from the main? Assume galvanized-steel pipe is to be used and that the pressure required at the factory is 60 kPa gage at a point 10 m above the main connection.

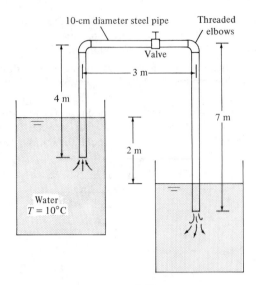

PROBLEM 10-57

10-59 Two reservoirs with a difference in water surface elevation of 11 ft are joined by 45 ft of 1-ft-diameter steel pipe and 30 ft of 6-in. steel pipe in series. The 1-ft line contains 3 bends ($r/D = 1$) and the 6-in. line contains 2 bends ($r/D = 4$). If the 1-ft and 6-in. lines are joined by an abrupt contraction, evaluate the discharge. $T = 60°F$.

10-60 The 2,000-ft of 2-in. galvanized-steel pipe discharges water into the atmosphere. The pipeline has an open globe valve and four threaded elbows. Assuming $f = 0.025$, what is the discharge and what is the pressure at A, the midpoint of the line? $h_1 = 10$ ft and $h_2 = 50$ ft.

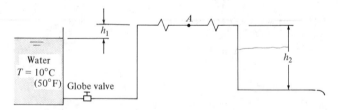

PROBLEMS 10-60, 10-61

10-61 The 1,000-m-long 10-cm galvanized-steel pipe discharges water into the atmosphere. The pipeline has an open globe valve and four threaded elbows. $h_1 = 3$ m and $h_2 = 15$ m. Assuming $f = 0.025$, what is the discharge and what is the pressure at A, the midpoint of the line?

10-62 Oil (sp. gr. = 0.8, $\mu = 3.0 \times 10^{-4}$ lb-sec/ft²) flows in the 1-in. smooth pipe shown. $h_1 = 12$ ft and $h_2 = 1$ ft. Determine the direction of flow and the discharge.

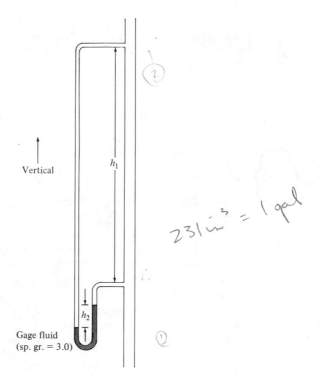

PROBLEMS 10-62, 10-63

10-63 Oil (sp. gr. $= 0.87$, $\mu = 2 \times 10^{-2}$ N · s/m²) flows in the 3-cm smooth pipe. $h_1 = 10$ m and $h_2 = 1$ m. Determine the direction of flow and the discharge.

10-64 The piping system in a typical dwelling is shown below. The exit diameter of the faucet is $\frac{1}{2}$ in. The Darcy-Weisbach friction factor for all the 1-in. pipe is 0.025 and 0.030 for the $\frac{1}{2}$-in. pipe. The main pressure is 40 psig. Determine the time required to fill a 20-gallon bathtub.

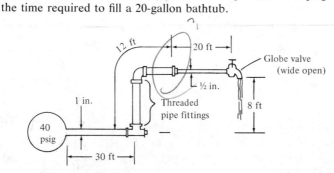

10-65 Water flow is to be maintained at a rate of 0.6 m³/s. The steel pipe is 50 cm in diameter and the water surface elevations in reservoirs ① and ② are

40 m and 50 m, respectively. L_1 and L_2 are 500 m and 700 m, respectively. What power must be supplied by the pump to maintain the flow? Also draw the energy and hydraulic grade lines for the system.

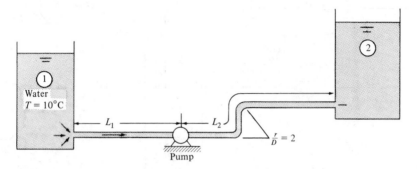

10-66 If the pump for Fig. 10-17 is installed in the system of Prob. 10-65, what will be the rate of discharge of water from reservoir (1) to (2)?

10-67 (a) Determine the discharge of water through the system shown.
   (b) Draw the hydraulic and energy grade line for the system.
   (c) Locate the point of maximum pressure.
   (d) Locate the point of minimum pressure.
   (e) Calculate the maximum and minimum pressures in the system.

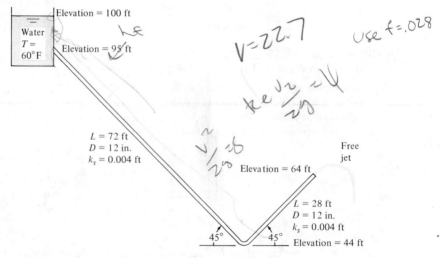

10-68 Water is pumped at a rate of 15 m³/s from the reservoir and out through the pipe, which has a diameter of 1.50 m. What power must be supplied to the water for this discharge?

10-69 Solve for the difference in the elevation of water surfaces. Both pipes

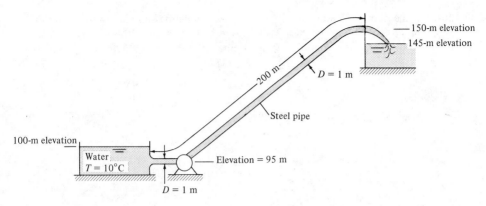

PROBLEM 10-68

shown have an equivalent sand roughness k_s of 0.10 mm and a discharge of 0.1 m³/s. $D_1 = 15$ cm, $L_1 = 50$ m, $D_2 = 30$ cm, and $L_2 = 100$ m.

10-70 Solve for the difference in the elevation of water surfaces. Both pipes shown have an equivalent sand roughness k_s of 0.0002 ft and a discharge of 3 cfs. $D_1 = 6$ in., $L_1 = 100$ ft, $D_2 = 14$ in., and $L_2 = 300$ ft.

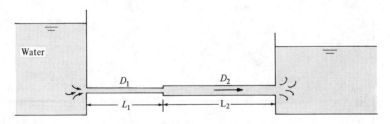

PROBLEMS 10-69, 10-70

10-71 If the discharge through the system shown is 2.0 cfs, what horsepower is the pump supplying to the water? Draw the hydraulic and energy grade lines for the system and determine the water pressure at the midpoint of the long pipe. The four bends have a radius of 12 in. and the 6-in. pipe is smooth.

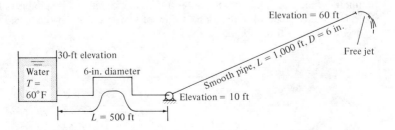

10-72 If the pump efficiency is 70%, what power must be supplied to the pump to pump fuel oil (sp. gr. = 0.94) at a rate of 1 m³/s up to the high reservoir. Assume that the conduit is a steel pipe and $\nu = 5 \times 10^{-5}$ m²/s.

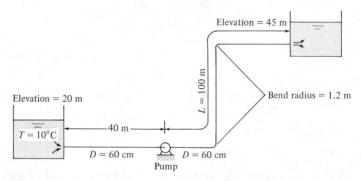

10-73 Determine the elevation of the water surface in the upstream reservoir if the discharge in the system is 0.15 m³/s. Carefully sketch the hydraulic and energy grade lines showing relative magnitudes and slopes. Label $V^2/2g$, p/γ, and z at section A-A.

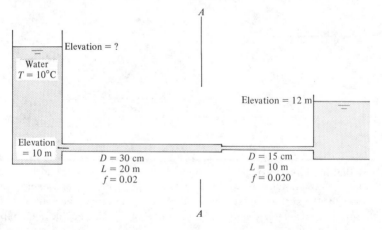

10-74 In the parallel system shown, pipe 1 has a length of 1,000 m and is 50 cm in diameter. Pipe 2 is 1,500 m long and is 40 cm in diameter. The pipe is commercial steel. What is the division of flow of water at 10°C if the total discharge is to be 1.0 m³/s?

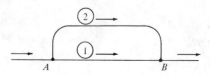

10-75 With a total flow of 14 cfs, determine the division of flow and the loss of head from A to B.

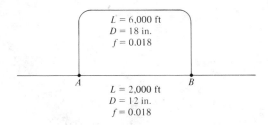

$L = 6,000$ ft
$D = 18$ in.
$f = 0.018$

A B

$L = 2,000$ ft
$D = 12$ in.
$f = 0.018$

10-76 The pipes shown in the system are all concrete ($k_s = 0.01$ ft). With a flow of 20 cfs of water, find the head loss and the division of flow in the pipes from A to B. Assume $f = 0.030$ for all pipes.

$L = 3,000$ ft
$D = 14$ in.

$L = 2,000$ ft $L = 2,000$ ft $L = 4,000$ ft
A $D = 29$ in. $D = 12$ in. $D = 30$ in. B

$L = 3,000$ ft
$D = 16$ in.

10-77 The steel pipe shown carries water from the main pipe A to the reservoir and is 2 in. in diameter by 200 ft long. What must be the pressure in A to provide a flow of 45 gpm?

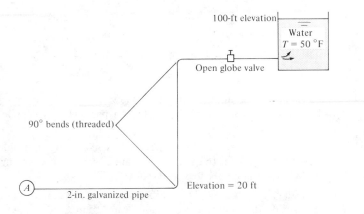

100-ft elevation

Water
$T = 50\,°F$

Open globe valve

90° bends (threaded)

A Elevation = 20 ft
2-in. galvanized pipe

10-78 If the elevation in reservoir B is 100 m, what must be the elevation in reservoir A if a flow of 0.03 m³/s is to occur in the cast-iron pipe? Draw the hydraulic grade line and energy grade line including relative slopes and changes in slope.

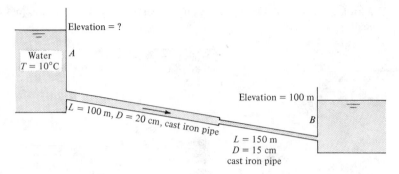

PROBLEM 10-78

10-79 Design a pipe system to supply water flow from the elevated tank to the reservoir at a discharge rate of 2.5 m³/s.

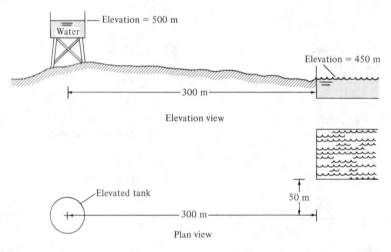

10-80 A pump that has the characteristic curve shown in the accompanying graph is to be installed as shown in the sketch on page 411. What will be the discharge of water in the system?

10-81 A cold-air duct 40 in. by 6 in. in cross section is 300 ft long and made of galvanized iron. This duct is to carry air at a rate of 9,600 cfm at a temperature of 60°F. What is the power loss in the duct?

10-82 A cold-air duct 100 cm by 15 cm in cross section is 100 m long and made of galvanized iron. This duct is to carry air at a rate of 5 m³/s at a temperature of 15°C and at atmospheric pressure. What is the power loss in the duct?

10-83 Estimate the discharge of water ($T = 10°C$) in a long rectangular concrete channel that is 3 m wide and flows 1.5 m deep and which is on a slope of 0.004.

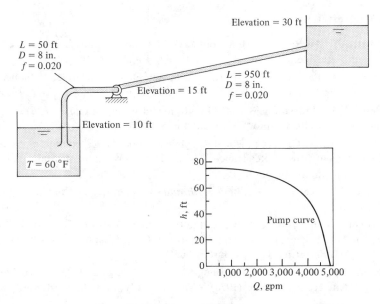

Elevation = 30 ft

L = 50 ft
D = 8 in.
f = 0.020

L = 950 ft
D = 8 in.
f = 0.020

Elevation = 15 ft

Elevation = 10 ft

T = 60 °F

h, ft

Pump curve

Q, gpm

PROBLEM 10-80

10-84 A rectangular concrete channel is 12 ft wide with uniform water flow. If the channel drops 10 ft in a length of 8,000 ft, what is the discharge? Assume that $T = 60°F$. The depth of flow is 4 ft.

10-85 Water flows at a depth of 6 ft in the trapezoidal, concrete-lined channel shown. If the channel slope is 1 ft in 2,000 ft, what is the average velocity and what is the discharge?

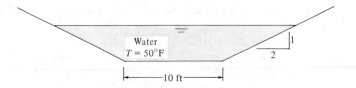

Water
T = 50°F

1

2

10 ft

10-86 A 4-m wide concrete channel on a slope of 0.004 is designed to carry a water ($T = 10°C$) discharge of 25 m³/s. Estimate the uniform flow depth for these conditions.

10-87 A 12-ft-wide rectangular troweled concrete channel with a slope of 10 ft in 8,000 ft is designed for a discharge of 600 cfs. For a water temperature of 40°F estimate the depth of flow.

10-88 Using Eqs. (10-35) and (10-36), estimate the discharge in a rock-bedded stream ($d_{85} = 30$ cm) that has an average depth of 2.21 m, a slope of 0.0037, and a width of 46 m.

REFERENCES

1. ASHRAE. "ASHRAE Handbook, 1977 Fundamentals." Am. Soc. of Heating, Refrigerating and Air Conditioning Engineers, Inc., N.Y. (1977).

2. Barbin, A. R., and Jones, J. B. "Turbulent Flow in the Inlet Region of a Smooth Pipe." *Trans. ASME, Ser. D: J. Basic Eng.*, 85, no. 1, (March 1963).

3. Beij, K. H. "Pressure Losses for Fluid Flow in 90° Pipe Bends." *J. Res. Nat. Bur. Std.*, 21 (1938). Information cited in Ref. 26.

4. Calhoun, Roger G. "A Statistical Roughness Model for Computation of Large Bed-Element Stream Resistance." M.S. Thesis, Washington State University, Pullman, Washington (1975).

5. Chow, Ven Te. *Open Channel Hydraulics.* McGraw-Hill Book Company. New York, 1969.

6. Committee on Hydromechanics of the Hydraulics Division of American Society of Civil Engineers. "Friction Factors in Open Channels." *J. Hydraulics Div., Am. Soc. Civil Eng.*, (March 1963).

7. Crane Co. "Flow of Fluids through Valves, Fittings and Pipe," Technical Paper No. 410, Crane Co. (1978). (Available only through distributor.)

8. Cross, Hardy. "Analysis of Flow in Networks of Conduits or Conductors." *Univ. Illinois Bull.* 286 (November 1936).

9. Daily, James W., and Harleman, Donald R. F. *Fluid Dynamics.* Addison-Wesley Publishing Company, Reading, Mass., 1966.

10. Hamilton, J. B. "The Suppression of Intake Losses by Various Degrees of Rounding." *Univ. Wash. Expt. Sta. Bull.* 51 (1929). Information cited in Ref. 26.

11. Hoag, Lyle N., and Weinberg, Gerald. "Pipeline Network Analysis by Digital Computers." *J. Am. Water Works Assoc.*, 49 (1957).

12. Hydraulic Institute. "Pipe Friction Manual." Hydraulic Institute, 122 E. 42nd St., N.Y. 10017.

13. Idel'chik, I. E. *Handbook of Hydraulic Resistance-Coefficients of Local Resistance and of Friction.* Trans. A. Barouch, Israel Program for Scientific Translations (1966).

14. Jeppson, Roland W. *Analysis of Flow in Pipe Networks.* Ann Arbor Science Publishers, Ann Arbor, Michigan, 1976.

15. Kumar, S. "An Analytical Model for Computation of Rough Pipe Resistance." Ph.D. thesis, Washington State University, Pullman, Washington (1979).

16. Laufer, John. "The Structure of Turbulence in Fully Developed Pipe Flow." *NACA Rept.* 1174 (1954).

17. Marlow, Thomas A., et al. "Improved Design of Fluid Networks With Computers." *J. Hydraulics Div., Am. Soc. Civil Eng.* (July 1966).

18. Moody, Lewis F. "Friction Factors for Pipe Flow." *Trans. ASME* (November 1944), 671.

19. Nikuradse, J. "Strömungsgesetze in rauhen Rohren." *VDI-Forschungsh.,* no. 361 (1933). Also translated in *NACA Tech. Memo* 1292.

20. Prandtl, L. Über die ausgebildete Turbulenz," *Zamm.* 5 (1925), 136–139. Also summarized in Schlicting (25).

21. Reynolds, O. "An Experimental Investigation of the Circumstances which Determine Whether the Motion of Water Shall be Direct or Sinuous and of the Law of Resistance in Parallel Channels." *Phil. Trans. Roy. Soc. London,* 174, part III (1883).

22. Roberson, John A., and Chen, C. K. "Flow in Conduits with Low Roughness Concentration," *J. Hydraulics Div., Am. Soc. Civil Eng.,* 96, no. HY4 (April 1970).

23. Rouse, Hunter, and Ince, Simon. *History of Hydraulics.* Iowa Institute of Hydraulic Research, University of Iowa (1957).

24. Schlichting, Hermann. "Experimentelle Untersuchungen zum Rauhigkietsproblem." *Ingr.-Arch.* 7 (1936), pp. 1–34.

25. Schlichting, Hermann. *Boundary Layer Theory,* 6th ed. McGraw-Hill Book Company, New York, 1968.

26. Streeter, V. L. (ed.). *Handbook of Fluid Dynamics.* McGraw-Hill Book Company, New York, 1961.

27. U.S. Bureau of Reclamation. "Friction Factors for Large Conduits Flowing Full." Engineering Monograph No. 7, U.S. Govt. Printing Office, 1965.

28. Wright, S. J. "A Theory for Prediciting the Resistance to Flow in Conduits With Nonuniform Roughness." M.S. thesis, Washington State University, Pullman, Washington (1973).

Lift by the winds is a lift to the soul.

11 DRAG AND LIFT

A BODY IMMERSED IN A FLOWING FLUID is acted upon by both pressure and viscous forces from the flow. The sum of the forces (pressure, viscous, or both) that acts normal to the free-stream direction is the *lift*, and the sum that acts parallel to the free-stream direction is the *drag*. Buoyant or weight forces may also act on the body; however, lift and drag forces are limited by definition to those forces produced by the dynamic action of the flowing fluid. We will first consider the mathematical formulation for lift and drag in terms of the viscous stresses and pressure, then we will consider each more intensively in subsequent sections.

11-1   BASIC CONSIDERATIONS

Consider the forces acting on the *airfoil* in Fig. 11-1. The vectors normal to the surface of the airfoil are normal forces per unit area, referred to simply as pressure. As shown here, the pressure is referenced to the free-stream pressure. Because the velocity of the flow over the top of the airfoil is greater than the free-stream velocity, the pressure over the top is negative or less than the free-stream pressure. This follows directly from application of the Bernoulli equation. Because the velocity along the underside of the wing is less than the free-stream velocity, the pressure here

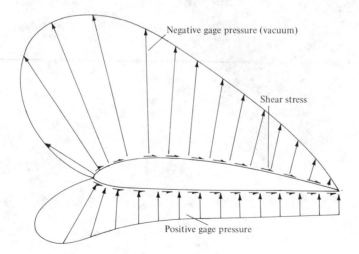

Negative gage pressure (vacuum)

Shear stress

Positive gage pressure

FIGURE 11-1   Pressure and shear stress acting on airfoil.

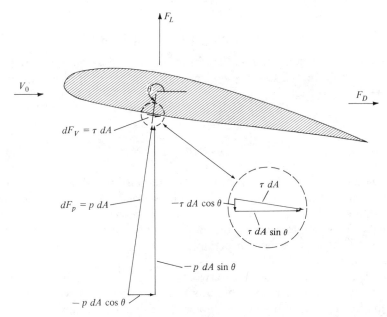

FIGURE 11-2   Pressure and viscous forces on a differential element of area.

is positive or greater than the free-stream pressure. Hence it is seen that both the negative pressure over the top and the positive pressure on the bottom contribute to the lift.

The vectors in Fig. 11-1 that are parallel to the surface of the airfoil represent the shear forces per unit area, referred to simply as shear stress. Except on the front of the airfoil, shear stress acts essentially parallel to the free-stream direction; hence, the shear stress contributes largely to the drag of the airfoil.

The mathematical formulation for lift and drag in terms of the pressure and shear stress can be derived with the aid of Fig. 11-2. Here the pressure and viscous forces acting on a differential area of the surface of the airfoil are shown. The magnitude of the pressure force is $dF_p = p\, dA$ and the magnitude of the viscous force is given by $dF_\tau = \tau\, dA$. However, we are interested in separating the forces into components that are normal and parallel to the free-stream direction to determine lift and drag, respectively. Hence the differential lift force is[1]

$$dF_L = -p\, dA \sin\theta - \tau\, dA \cos\theta$$

---

[1] The sign convention on τ is such that a clockwise sense of $\tau\, dA$ on the surface of the foil signifies a positive sign for τ.

and the differential drag is given by

$$dF_D = -p \, dA \cos \theta + \tau \, dA \sin \theta$$

Then the total lift and drag on the airfoil is obtained by integration of the respective differential forces over the entire surface of the airfoil:

$$F_L = \int (-p \sin \theta - \tau \cos \theta) \, dA \qquad (11\text{-}1)$$

$$F_D = \int (-p \cos \theta + \tau \sin \theta) \, dA \qquad (11\text{-}2)$$

Equations (11-1) and (11-2) are for a two-dimensional flow; i.e., there is no velocity component in the direction normal to the page, so the shear-stress and pressure-force vector lie in the plane of the page. The same basic principle (separation of forces into directions parallel and normal to the free-stream direction) can easily be extended to three-dimensional flows.

In this chapter we shall refer to two-dimensional and three-dimensional bodies. By a two-dimensional body, we mean a body over which the flow is two-dimensional. For example, a very long cylinder which has flow approaching it from a normal direction would be classified as a two-dimensional body because the flow around the ends do not affect the flow pattern and pressure distribution over the central part of the body. However, a short cylinder would be classified as a three-dimensional body because the end effects are significant. For a two-dimensional body, the aerodynamic forces and representative areas are sometimes based on a

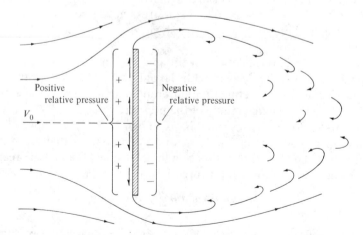

FIGURE 11-3   Flow past a flat plate.

unit length of the body. The two-dimensional body is identified by cross-hatching on the figures indicating the cross section.

Still another classification is the axisymmetric body. Here, if the approach flow is uniform and parallel to the axis of symmetry, then in effect the resulting flow is two-dimensional. That is, for an x-r coordinate system, where x is measured along the axis of symmetry and r is the normal radial distance from the axis, the velocity components exist only in the x and r directions.

Equations (11-1) and (11-2) can be used to evaluate F_L and F_D when the pressure and shear stresses are either obtained analytically or experimentally; however, it is also common to obtain the overall drag and lift by force dynamometer measurements in a wind tunnel. The next section considers the direct application of the pressure and shear-stress variation over the surface of a plate to determine the drag on the plate.

11-2 DRAG OF TWO-DIMENSIONAL BODIES

Drag of a thin plate

To illustrate the relative effect of pressure and viscous forces on drag, we shall consider the drag of a plate first oriented parallel to the flow and then oriented normal to the flow. In the parallel position, the only force acting is viscous shear in the direction of flow; hence, from our considerations of surface resistance in Chapter 9, for both sides of the plate the drag is given as

$$F_D = 2C_f BL\rho \, \frac{V_0^2}{2}$$

When the plate is turned normal to the flow as in Fig. 11-3, both pressure and viscous forces act on the plate. However, the viscous forces act only in the transverse direction and, in addition, are symmetrical about the midpoint of the plate; consequently, the viscous forces do not directly contribute to the lift or drag of the plate. Because the pressure on the plate acts to produce a force only in a direction parallel to the flow, the pressure force contributes totally to the drag of the body; hence, Eq. (11-2) applied to the plate reduces to

$$F_D = \int (-p \cos \theta) \, dA$$

The pressure on the front and rear sides of the plate can be obtained experimentally and are usually given in terms of C_p, as shown in Fig. 11-4 for flow with a relatively high value of Reynolds number (Re $= V_0 B/\nu$).

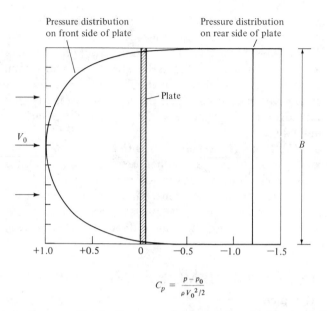

Pressure distribution on front side of plate

Pressure distribution on rear side of plate

Plate

V_0

B

+1.0   +0.5   0   −0.5   −1.0   −1.5

$$C_p = \frac{p - p_0}{\rho V_0{}^2/2}$$

FIGURE 11-4   Pressure distribution on a plate normal to the approach flow.

Since the pressure on the rear side is essentially constant,

$$p = p_0 - 1.2\rho \frac{V_0^2}{2}$$

and since $\theta = 0$, the contribution to drag for the rear side is

$$F_{D,\text{rear}} = -\left(p_0 - 1.2\rho \frac{V_0^2}{2}\right) B\ell$$

$$= -p_0 B\ell + 1.2\rho \frac{V_0^2}{2} B\ell$$

where ℓ is the length of the plate normal to the plane of the paper and, by definition of a two-dimensional body, $\ell >> B$. For the front side, $\theta = \pi$; hence, $\cos \theta = -1$ and the contribution to drag due to pressure on the front side is

$$F_{D,\text{front}} = \int_{-B/2}^{B/2} \left(p_0 + C_p \rho \frac{V_0^2}{2}\right) \ell \, dy$$

$$= p_0 B\ell + \rho \frac{V_0^2}{2} \ell \int_{-B/2}^{B/2} C_p \, dy$$

Then the total drag on the plate is given by

$$F_D = F_{D,\text{front}} + F_{D,\text{rear}}$$

$$= \rho \frac{V_0^2}{2} \ell \left(\int_{-B/2}^{B/2} C_p \, dy + 1.2B \right) \tag{11-3}$$

Evaluation of the first term inside the parentheses on the right side of Eq. (11-3) yields magnitude of approximately $0.80B$. Thus the drag of this plate is given as

$$F_D = \rho \frac{V_0^2}{2} B\ell(0.80 + 1.2) \tag{11-4}$$

At this time we should take note of the pure numbers in the parentheses of Eq. (11-4). The number 0.8 in Eq. (11-4) represents the average pressure coefficient C_p over the front side of the plate. In fact, the sum inside the parentheses $(0.8 + 1.2)$ of this equation reflects the manner in which the pressure is distributed over the front and rear of the body. Because the drag varies directly with the magnitude of this quantity, it has been appropriately defined as the *coefficient of drag* C_D. Thus Eq. (11-4) can be written as

$$F_D = C_D A_p \rho \frac{V_0^2}{2} \tag{11-5}$$

where C_D = coefficient of drag

A_p = projected area of the body

ρ = fluid density

V_0 = free-stream velocity

The projected area A_p is the silhouetted area that would be seen by a person looking at the body from the direction of flow. For example, the projected area of the above plate normal to the flow is $B\ell$ and the projected area of a cylinder with its axis normal to the flow is $D\ell$. In Chapter 8 we saw that C_p was a function of the Reynolds number Re; and because $C_D = f(C_p)$, C_D will be a function of Re. When the drag of bodies is only a result of the shear stress on the body, C_D will still be a function of Re because τ is also a function of Re.

Coefficients of drag for various two-dimensional bodies

We have already seen that C_D can be determined if the pressure and shear-stress distribution around a body are known. The coefficient of drag can also be calculated if the total drag is measured, for example, by means of a force dynamometer in a wind tunnel. Then C_D is calculated by Eq. (11-5) when written as follows:

$$C_D = \frac{F_D}{A_p \rho V_0^2/2} \tag{11-6}$$

Much of the data (C_D versus Re) found in the literature is obtained in this manner.

The coefficient of drag for the flat plate normal to the free stream, along with C_D's for other two-dimensional bodies, for a wide range of Reynolds numbers is given in Fig. 11-5. In general, the total drag of a blunt body is partly due to viscous resistance and partly due to pressure variation. The

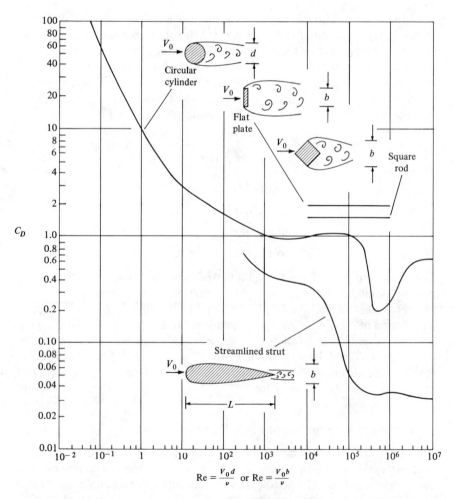

FIGURE 11-5   Drag coefficient versus Reynolds number for two-dimensional bodies. [Data sources: Bullivant (5), Defoe (7), Goett (10), Jacobs (12), Jones (14), Lindsey (18).]

pressure drag is largely a function of the form or shape of the body; hence it is called *form drag*. The viscous drag is often called *skin-friction drag*.

EXAMPLE 11-1   A television transmitting antenna is on top of a pipe, 30 m high (98.4 ft) and 30 cm (11.8 in.) in diameter, which is on top of a tall building. What will be the total drag of the pipe and bending moment at the base of the pipe in a 35 m/s (115 ft/sec) wind at normal atmospheric pressure and a temperature of 20°C (68°F)?

Solution   For the conditions given, the viscosity and density of the air are obtained from the Appendix as

$$\mu = 1.79 \times 10^{-5} \text{ N} \cdot \text{s/m}^2 \qquad (3.75 \times 10^{-7} \text{ lbf-sec/ft}^2)$$
$$\rho = 1.21 \text{ kg/m}^3 \qquad (0.00234 \text{ slugs/ft}^3)$$

Now the Reynolds number is calculated:

$$\text{Re} = \frac{V_0 D \rho}{\mu} = \frac{35 \text{ m/s} \times 0.30 \text{ m} \times 1.21 \text{ kg/m}^3}{(1.79 \times 10^{-5} \text{ N} \cdot \text{s/m}^2)}$$
$$= 7.1 \times 10^5$$

Then, from Fig. 11-5, $C_D = 0.20$. Now we compute the total drag as

$$F_D = \frac{C_D A_p \rho V_0^2}{2}$$
$$= \frac{(0.2)(30 \text{ m})(0.3 \text{ m})(1.21 \text{ kg/m}^3)(35^2 \text{ m}^2/\text{s}^2)}{2}$$
$$= 1,334 \text{ N} \qquad \blacktriangleleft$$

Assuming that the resultant drag force acts midway up the pole, the moment is

$$M = F_D \left(\frac{L}{2}\right) = 1,334 \text{ N} \left(\frac{30}{2}\right) \text{ m}$$
$$= 20,000 \text{ N} \cdot \text{m} \qquad \blacktriangleleft$$

Traditional units

$$F_D = 0.2(98.4 \text{ ft}) \left(\frac{11.8}{12} \text{ ft}\right) (0.00234 \text{ slugs/ft}^3) \left(\frac{115^2 \text{ ft}^2/\text{sec}^2}{2}\right)$$
$$= 299 \text{ lbf} \qquad \blacktriangleleft$$

$$M = 299 \text{ lbf} \left(\frac{98.4}{2} \text{ ft}\right) = 14,700 \text{ ft-lbf} \qquad \blacktriangleleft$$

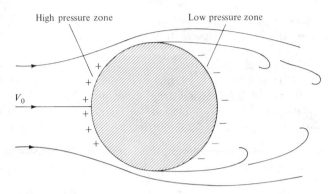

FIGURE 11-6   Flow pattern for $10^3 < \text{Re} < 10^5$.

Discussion of C_D for two-dimensional bodies

At low Reynolds numbers, C_D changes with the Reynolds number. The change is owing to the relative change in viscous resistance, which has already been mentioned in Chapter 8. Beyond $\text{Re} = 10^4$, the flow pattern remains virtually unchanged, thereby producing constant values of C_p over the body. Constancy of C_p at high Reynolds numbers is reflected in the constant value of C_D. This characteristic, the constancy of C_D at high values of Re, is representative of most bodies that have angular form; however, certain bodies with rounded form, such as circular cylinders, show a remarkable decrease in C_D with an increase in Re from about 10^5 to 5×10^5.

This reduction in C_D at a Reynolds number of approximately 10^5 is due to a change in flow pattern triggered by a change in the character of the boundary layer. For Reynolds numbers less than 10^5, the boundary layer is laminar, and separation occurs about midway between the front and rear of the cylinder (Fig. 11-6). Hence, the entire rear half of the cylinder is exposed to a relatively low pressure, which in turn produces a relatively high value for C_D. When the Reynolds number is increased to about 10^5, the boundary layer on the surface of the cylinder becomes turbulent, which causes higher-velocity fluid to be mixed into the region close to the wall of the cylinder. As a consequence of this high velocity, high-momentum fluid in the boundary layer, the flow proceeds farther downstream along the surface of the cylinder against the adverse pressure before separation occurs (Fig. 11-7). Hence the flow pattern causes C_D to be reduced for the following reason: with the turbulent boundary layer, the streamlines downstream of the cylinder midsection diverge somewhat before separation, and hence a decrease in velocity occurs before separation. According to Bernoulli's equation the decrease in velocity produces an increase in pressure at the point of separation over that at the midsection. Thus the pressure at the point of separation and also in the zone of

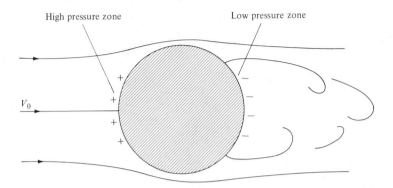

FIGURE 11-7   Flow pattern about a cylinder for Re $> 5 \times 10^5$.

separation is significantly higher than when separation occurs farther upstream. Therefore, the pressure difference between the front and rear surfaces of the cylinder is less at high values of Re, thus yielding a lower drag and lower C_D.

Because the boundary layer is so thin, it is also very sensitive to other conditions. For example, if the surface of the cylinder is slightly roughened upstream of the midsection, the boundary layer will be forced to become turbulent at lower Reynolds numbers than those for a smooth cylinder surface. The same trend can also be produced by creating abnormal turbulence in the approach flow. The effects of roughness are shown in Fig. 11-8 for cylinders that were roughened with sand grains of size k. A small to medium size of roughness ($10^{-3} < k/d < 10^{-2}$) on a cylinder triggers an early onset of reduction of C_D; however, when the relative roughness is quite large ($10^{-2} < k/d$) the characteristic dip in C_D is absent.

11-3 VORTEX SHEDDING FROM CYLINDRICAL BODIES

Figures 11-6 and 11-7 show the average (temporal mean) flow pattern around a cylinder. The phenomenon becomes more complex, however, when we observe the detailed flow pattern as time passes. Observations show that above Re ≈ 50 vortices are formed and shed periodically downstream of the cylinder. Hence, at a given time, the detailed flow pattern might appear as in Fig. 11-9. In this figure a vortex is in the process of formation near the top of the cylinder. Below and to the right of the first vortex is another vortex, which was formed and shed a short period before. Thus the flow process in the wake of a cylinder involves the formation and shedding of vortices alternately from one side and then the other.

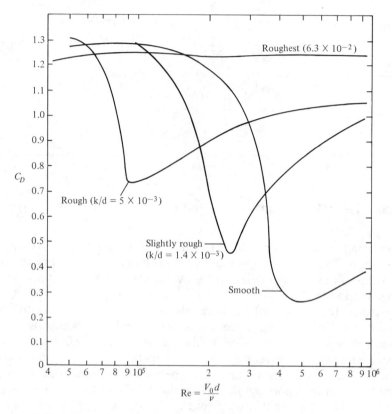

FIGURE 11-8 Effects of roughness on C_D for a cylinder. [After Miller *et al.* (19).]

This phenomenon is of major importance in engineering design because the alternate formation and shedding of vortices also creates a regular change in pressure with consequent periodicity in side thrust on the cylinder. Vortex shedding was the primary cause of failure of the Tacoma Narrows suspension bridge in Washington state in 1940. Another more commonplace effect of vortex shedding is the "singing" of wires in the wind.

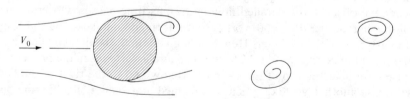

FIGURE 11-9 Vortex formation behind a cylinder.

If the frequency of the vortex shedding is in resonance with the natural frequency of the member which produces it, large amplitudes with consequent large stresses can develop. Experiments show that the frequency of shedding is given in terms of the Strouhal number S, and this in turn is a function of the Reynolds number. Here the Strouhal number is defined as

$$S = \frac{nd}{V_0} \tag{11-7}$$

where n = frequency of shedding of vortices from one side of cylinder, Hz

d = diameter of cylinder

V_0 = free-stream velocity

The relationship between the Strouhal number and the Reynolds number for vortex shedding from a circular cylinder is given in Fig. 11-10. Other cylindrical and two-dimensional bodies also shed vortices; consequently, the engineer should always be alert to vibration problems when designing structures that are exposed to wind or water flow.

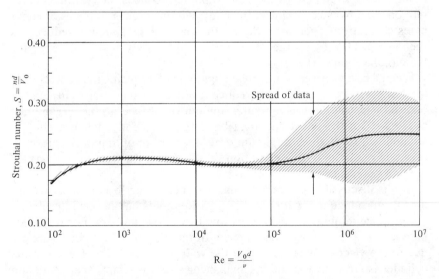

FIGURE 11-10   Strouhal versus Reynolds number for flow past a circular cylinder. [After Jones (14) and Roshko (23).]

EXAMPLE 11-2   For the cylinder and conditions of Example 11-1, at what frequency will the vortices be shed?

Solution   We compute the frequency n from the Strouhal number, which is a function of the Reynolds number given in Fig. 11-10. Thus, with a Reynolds number of 6.08×10^5 from Example 11-1, we read a Strouhal number of 0.23. But

$$S = \frac{nd}{V_0}$$

so

$$n = \frac{SV_0}{d}$$

$$= \frac{0.23 \times 35 \text{ m/s}}{0.30 \text{ m}} = 27 \text{ Hz} \qquad \blacktriangleleft$$

11-4 EFFECT OF STREAMLINING

For Reynolds numbers greater than 10^3, the drag of a cylinder is predominantly due to the pressure variation around the cylinder. The pressure difference between the front and rear sides of the cylinder is the primary cause of drag, and this pressure difference is largely owing to separation; hence, if the separation can be eliminated, the drag will be reduced. That is exactly what streamlining does. Streamlining reduces the extreme curvature on the downstream side of the body and this process reduces or eliminates separation. Therefore, the coefficient of drag is greatly reduced as seen in Fig. 11-5 (C_D for the streamlined shape is only about 10% of C_D for the circular cylinder).

When a body is streamlined by elongating and reducing the curvature of the body, the pressure drag is reduced; however, the viscous drag is increased because there is a greater amount of surface on the streamlined body than on the nonstreamlined body. Consequently, when a body is streamlined to produce minimum drag, there is an optimum condition to be sought. The optimum condition results when the sum of the surface drag and pressure drag is minimum.

In this discussion of streamlining, it it interesting to note that streamlining to produce minimum drag at high Reynolds numbers will probably not produce minimum drag at very low Reynolds numbers. For Reynolds numbers less than unity, the majority of the drag of a cylinder is already due to the viscous shear stress on the wall of the cylinder. Hence, if the cylinder is streamlined, the viscous shear stress is simply magnified and C_D may actually increase for this range of Re where the viscous resistance is predominant.

At high values of the Reynolds number, another advantage of streamlining is that the periodic formation of vortices is inhibited if not eliminated altogether.

EXAMPLE 11-3 Compare the drag of the cylinder of Example 11-1 with the drag of the streamlined shape shown in Fig. 11-5. Assume that they both have the same projected area and that the streamlined shape is oriented for minimum drag.

Solution   Because $F_D = C_D A \rho V_0^2/2$, the drag of the streamlined shape will be

$$F_{DS} = F_{DC} \left(\frac{C_{DS}}{C_{DC}}\right)$$

where F_{DS} = drag of streamlined shape

F_{DC} = drag of cylinder

C_{DS} = coefficient of drag of streamlined shape

C_{DC} = coefficient of drag of cylinder

But $C_{DS} = 0.034$, from Fig. 11-5 with Re = 6.08×10^5; then

$$F_{DS} = F_{DC} \left(\frac{0.034}{0.20}\right)$$

$$= 1,334 \text{ N} \left(\frac{0.034}{0.20}\right)$$

$$F_{DS} = 227 \text{ N} \qquad\blacktriangleleft$$

$$\frac{F_{DS}}{F_{DC}} = 0.17 \qquad\blacktriangleleft$$

11-5   DRAG OF AXISYMMETRIC AND THREE-DIMENSIONAL BODIES

The same principles that applied to the drag of two-dimensional bodies also apply to axisymmetric and three-dimensional bodies. That is, at very low values of the Reynolds number, the coefficient of drag is given by exact equations relating C_D and Re; and at high values of Re, the coefficient of drag becomes constant for angular bodies while rather abrupt changes in C_D occur for rounded bodies. All these characteristics can be seen in Fig. 11-11, where C_D is plotted against Re for several axisymmetric bodies.

For Reynolds numbers less than 0.5, the flow around the sphere is laminar and amenable to analytical solutions. An exact solution by Stokes yielded the following equation, called Stokes' law, for the drag of a sphere:

$$F_D = 3\pi\mu V_0 D \qquad (11\text{-}8)$$

Note that the drag for this laminar-flow condition varies directly with the first power of V_0. This is characteristic of all laminar-flow processes; for

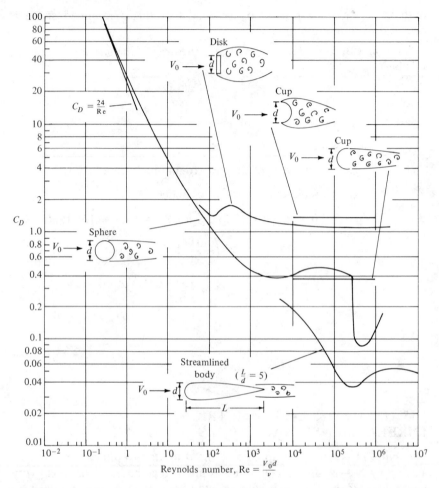

FIGURE 11-11   Coefficient of drag versus Reynolds number for axisymmetric bodies. [Data sources: Abbott (1), Breevoort (4), Freeman (9), and Rouse (24).]

completely turbulent flow, the drag is a function of the velocity to the second power. When Eqs. (11-8) and (11-6) are solved simultaneously, we get the drag coefficient corresponding to Stokes' law:

$$C_D = \frac{24}{\text{Re}} \tag{11-9}$$

Thus, for flow past a sphere, when $\text{Re} \leq 0.5$, one may use the direct relation for C_D as given above. Coefficient of drag data for other axisymmetric and three-dimensional bodies at high Reynolds numbers ($\text{Re} > 10^4$) are given in Table 11-1.

TABLE 11-1   APPROXIMATE C_D VALUES FOR VARIOUS BODIES

Type of Body	Length Ratio	Re	C_D
Rectangular plate	$l/b = 1$	$>10^4$	1.18
	$l/b = 5$	$>10^4$	1.20
	$l/b = 10$	$>10^4$	1.30
	$l/b = 20$	$>10^4$	1.50
	$l/b = \infty$	$>10^4$	1.98
Circular cylinder— axis ∥ to flow	$l/d = 0$ (disk)	$>10^4$	1.17
	$l/d = 0.5$	$>10^4$	1.15
	$l/d = 1$	$>10^4$	0.90
	$l/d = 2$	$>10^4$	0.85
	$l/d = 4$	$>10^4$	0.87
	$l/d = 8$	$>10^4$	0.99
Square rod	∞	$>10^4$	2.00
Square rod	∞	$>10^4$	1.50
Triangular cylinder	∞	$>10^4$	1.39
Semicircular shell	∞	$>10^4$	1.20
Semicircular shell	∞	$>10^4$	2.30
Hemispherical shell		$>10^4$	0.39
Hemispherical shell		$>10^4$	1.40
Cube		$>10^4$	1.10
Cube		$>10^4$	0.81
Cone—60° vertex		$>10^4$	0.49
Parachute		$\approx 3 \times 10^7$	1.20

SOURCES: Brevoort (4), Lindsey (18), Morrison (20), Roberson (22), Rouse (24), Scher (26).

EXAMPLE 11-4   What is the drag of a 12-mm sphere that drops at a rate of 8 cm/s in oil ($\mu = 10^{-1}$ N $\cdot$ s/m^2, sp. gr. $= 0.85$)?

Solution

$$\rho = 0.85 \times 1,000 \text{ kg/m}^3$$

Then $$\text{Re} = \frac{VD\rho}{\mu} = \frac{(0.08 \text{ m/s})(0.012 \text{ m})(850 \text{ kg/m}^3)}{10^{-1} \text{ N} \cdot \text{s/m}^2}$$

$$= 8.16$$

Then from Fig. 11-11

$$C_D = 5.3$$

$$F_D = \frac{C_D A_p \rho V_0^2}{2}$$

Hence, $$F_D = \frac{(5.3)(\pi/4)(0.012^2 \text{ m}^2)(850 \text{ kg/m}^3)(0.08^2 \text{ m}^2/\text{s}^2)}{2}$$

$$= 1.63 \times 10^{-3} \text{ N} \qquad \blacktriangleleft$$

11-6   TERMINAL VELOCITY

The engineer is often required to use drag data in the computation of the *terminal velocity* of a body. When a body is first dropped in the atmosphere or in water, it will accelerate under the action of its weight. Then, as the speed of the body increases, the drag will increase and finally the drag will reach a magnitude such that the sum of all the external forces on the body will be zero; hence, acceleration will cease and the body will have attained its terminal velocity. Thus the terminal velocity is the maximum velocity attained by a falling body. This assumes that the fluid through which it falls has constant properties over the path of descent. The following two examples will illustrate the method of computing terminal velocity of a body, first in air, then in water.

EXAMPLE 11-5   A 50-mm solid plastic sphere (sp. gr. $= 1.3$) is dropped from an airplane. Assuming standard atmospheric conditions ($T = 20°$C, $p_{atm} = 101$ kPa abs), what will be the terminal velocity of the sphere?

Solution   The free body of the sphere is shown on page 433, where F_B is the buoyant force, F_D is the drag, and W is the weight.

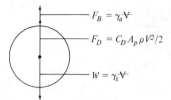

Obviously, the sphere will drop at a high rate of speed; therefore, we make a first guess at C_D by choosing a value from the high range of Re. Assume $C_D = 0.15$ (based upon Fig. 11-11). Then

$$F_D = \frac{C_D A_p \rho V_0^2}{2}$$

or

$$V_0 = \left(\frac{2 F_D}{C_D A_p \rho}\right)^{1/2}$$

Here F_D is the net weight of the sphere, or

$$F_D = (\gamma_s - \gamma_a)\left(\frac{4}{3}\right)\pi r^3 = (\gamma_s - \gamma_a)\left(\frac{4}{3}\right)\pi\left(\frac{D^3}{8}\right) = \frac{1}{6}\pi D^3(\gamma_s - \gamma_a)$$

where

$$\gamma_s = \text{specific weight of the sphere}$$
$$\gamma_a = \text{specific weight of air}$$

and

$$A_p = \frac{\pi}{4}D^2$$

The specific weight of the sphere, γ_s, is the specific gravity of the sphere times γ_{water} or $\gamma_s = 1.3 \times 9.81 \ \text{kN/m}^3 = 12.7 \ \text{kN/m}^3$. Also, $\rho_{\text{air}} = 1.21 \ \text{kg/m}^3$ and $\gamma_a = 1.21 \times 9.81 \ \text{N/m}^3$; thus $\gamma_a << \gamma_s$.

Hence

$$V_0 = \left(\frac{2\gamma_s(1/6)\pi D^3}{C_D(\pi/4)D^2 \ \rho_{\text{air}}}\right)^{1/2} = \left(\frac{\gamma_s(4/3)D}{C_D \ \rho_{\text{air}}}\right)^{1/2}$$

$$V_0 = \left(\frac{(12.7(10^3 \ \text{N/m}^3)(4/3)(0.05 \ \text{m})}{C_D(1.21 \ \text{kg/m}^3)}\right)^{1/2} = \left(\frac{700 \ \text{m}^2/\text{s}^2}{C_D}\right)^{1/2}$$

$$V_0 = \frac{26.4 \ \text{m/s}}{(C_D)^{1/2}}$$

With the first guess at $C_D = 0.15$, we get a velocity of

$$V_0 = \frac{26.4 \ \text{m/s}}{(0.15)^{1/2}} = 68.3 \ \text{m/s}$$

Now we compute Re based upon this velocity of 68.3 m/s so that we can get a better value of C_D:

$$\text{Re} = \frac{VD\rho}{\mu}$$

$$= \frac{68.3 \text{ m/s } (0.05 \text{ m})(1.21 \text{ kg/m}^3)}{1.80(10^{-5} \text{ N} \cdot \text{s/m}^2)}$$

$$= 2.3 \times 10^5$$

Then a better estimate of C_D based upon Re $= 2.3 \times 10^5$ is $C_D = 0.40$. Our first guess for C_D was too low; hence we make another calculation of V_0 based upon the more precise value of C_D:

$$V_0 = \frac{26.4 \text{ m/s}}{(0.40)^{1/2}} = 41.8 \text{ m/s}$$

Another calculation based upon the new velocity of 41.8 m/s yields a C_D of 0.42:

$$V_0 = \frac{26.4 \text{ m/s}}{(0.42)^{1/2}} = 40.8 \text{ m/s}$$

One more computation for Re and then determination of C_D yields Re $= 2.2 \times 10^5$ and $C_D = 0.42$. Therefore, the terminal velocity is $V_0 = 40.8$ m/s. ◄

EXAMPLE 11-6   A 50-mm plastic sphere (sp. gr. = 1.3) is dropped in water. Determine its terminal velocity. Assume $T = 20°C$.

Solution We approach this problem in the same manner as Example 11-5:

$$F_D = (\gamma_s - \gamma_w)(1/6)\pi D^3$$

Then $$V_0 = \left[\frac{(\gamma_s - \gamma_w)(4/3)D}{C_D \rho_w}\right]^{1/2}$$

$$= \left[\frac{(12.7 - 9.79)(10^3 \text{ N/m}^3)(4/3)(0.05 \text{ m})}{C_D(10^3 \text{ kg/m}^3)}\right]^{1/2}$$

$$= \left(\frac{0.194}{C_D}\right)^{1/2} = \frac{0.44}{C_D^{1/2}} \text{ m/s}$$

Now, the velocity of fall will not be as great as in air, so a reasonable first guess at C_D is $C_D = 0.48$. With this value of C_D, the velocity is

$$V_0 = \frac{0.44 \text{ m/s}}{0.48^{1/2}} = 0.63 \text{ m/s}$$

Then $\text{Re} = VD\rho/\mu$, where $\rho = 10^3$ kg/m³ and $\mu = 1.00(10^{-3}$ N · s/m²$)$

$$\text{Re} = \frac{0.63 \text{ m/s}(0.05 \text{ m})(10^3 \text{ kg/m}^3)}{1.00(10^{-3} \text{ N} \cdot \text{s/m}^2)}$$

$$= 3.2 \times 10^4$$

From Fig. 11-11, $C_D = 0.48$; therefore

$$V_0 = 0.63 \text{ m/s} \qquad \blacktriangleleft$$

11-7 EFFECT OF COMPRESSIBILITY ON DRAG

The variation of drag coefficient with Mach number for three axisymmetric bodies is shown on Fig. 11-12. In each case, the drag coefficient increases only slightly with the Mach number at low Mach numbers and then increases sharply as transonic flow ($M \approx 1$) is approached. One notes that the rapid increase in drag coefficient occurs at a higher Mach number, more near unity, if the body is slender with a pointed nose. The drag coefficient reaches a maximum at a Mach number somewhat larger than unity and then decreases as the Mach number is further increased.

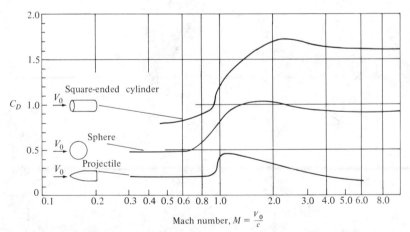

FIGURE 11-12   Drag characteristics of projectile, sphere, and cylinder with compressibility effects. [After Rouse (14).]

The slight increase in drag coefficient with low Mach numbers is attributed to an increase in form drag due to compressibility effects on the pressure distribution. However, as the flow velocity is increased, the maximum velocity on the body finally becomes sonic. The Mach number of the free-stream flow at which sonic flow first appears on the body is called the *critical Mach number*. Further increases in flow velocity result in local regions of supersonic flow ($M > 1$), which leads to wave drag due to shock-wave formation and an appreciable increase in drag coefficient.

The critical Mach number for a sphere is approximately 0.6; and one notes in Fig. 11-12 that the drag coefficient begins to rise sharply at about this Mach number. The critical Mach number for the pointed body is larger and correspondingly, the rise in drag coefficient occurs at a Mach number closer to unity.

The drag-coefficient data for the sphere shown in Fig. 11-12 are for a Reynolds number of the order of 10^4. The sphere-drag data shown in Fig.

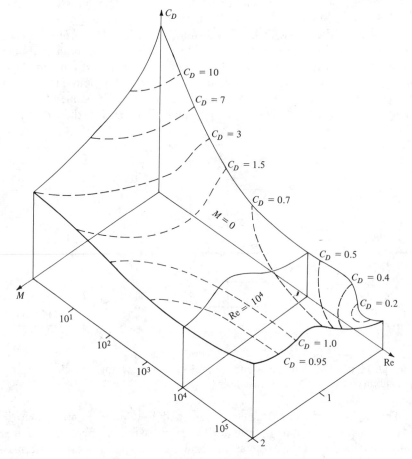

FIGURE 11-13 Contour plot of the drag coefficient of the sphere versus Reynolds and Mach numbers. [After Crowe (6).]

11-11, on the other hand, are for very low Mach numbers. The question then arises about the general variation of the drag coefficient of a sphere with both Mach number and Reynolds number. Information of this nature is often needed to predict the trajectory of a body through the upper atmosphere and sometimes to analyze the flows transporting solid-phase particles.

A contour plot of the drag coefficient of a sphere versus both Reynolds and Mach numbers based on available data (5) is shown in Fig. 11-13. One notices the C_D-versus-Re curve from Fig. 11-11 in the $M = 0$ plane. Correspondingly we see the C_D-versus-M curve from Fig. 11-12 in the $Re = 10^4$ plane. We see, then, that at low Reynolds number, C_D decreases with increasing Mach number, whereas at high Reynolds number the opposite trend is observed. Using this figure the engineer can determine the drag coefficient of a sphere at any Re-M combination. Of course, corresponding C_D contour plots can be generated for any body, provided the data are available.

11-8   LIFT

In Sec. 11-1 it was shown that a differential pressure between the top and bottom of a body will cause a lateral force or lift to be imposed on the body. However, no explanation was given for the cause of the differences in velocity which produces such a pressure distribution. In this section, we will consider circulation—which is the basic cause of lift; then we will consider the lift and drag characteristics of typical airfoils.

Circulation

Consider flow along a closed path such as is shown in Fig. 11-14. Along any differential segment of the path the velocity can be resolved into components that are tangent and normal to the path. Let us signify the tangential component of velocity as V_L. Now, if we integrate $V_L \, dL$ around the curve, the resulting quantity is called circulation, which is identified by the Greek symbol Γ (capital gamma). Hence, we have

$$\Gamma = \oint V_L \, dL \tag{11-10}$$

In applying Eq. (11-10), sign convention dictates that tangential velocity vectors that have a counterclockwise sense around the curve are taken as negative, and velocity vectors that have a clockwise direction have a positive contribution.[1] The circulation for an irrotational vortex is determined

[1] The sign convention is opposite that for the mathematical definition of a line integral.

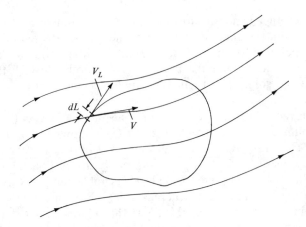

FIGURE 11-14   Concept of circulation.

as follows. The tangential velocity at any radius is C/r, where a positive C means a clockwise rotation. Therefore, if we evaluate the circulation about a curve with radius r, the differential circulation is

$$d\Gamma = V_L \, dL = \frac{C}{r_1} r_1 \, d\theta = C \, d\theta \qquad (11\text{-}11)$$

Then, when we integrate this around the entire circle, we obtain

$$\Gamma = \int_0^{2\pi} C \, d\theta = 2\pi C \qquad (11\text{-}12)$$

One way to physically induce circulation is to rotate a cylinder about its axis. In Fig. 11-15a we see the flow pattern produced by such action. The velocity of the fluid next to the surface of the cylinder is equal to the velocity of the cylinder surface itself because of the nonslip condition that must prevail between the fluid and solid. At some distance from the cylinder, however, the velocity decreases with r, much like that for the irrotational vortex. In the next section we will see how circulation produces lift.

Combination of circulation and uniform flow about a cylinder

If we now superpose the velocity field produced for uniform flow about a cylinder, Fig. 11-15b, onto a velocity field with circulation about a cylinder, Fig. 11-15a, we see that the velocity is reinforced on the top side of the cylinder and reduced on the other side (Fig. 11-15c). We also observe that the stagnation points have both moved toward the low-velocity side of the cylinder. Consistent with the Bernoulli principle (assuming irrotational flow throughout), we find that the pressure on the high-velocity side

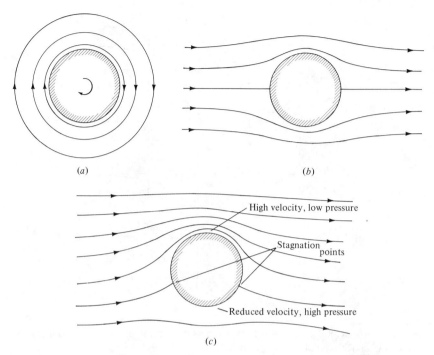

FIGURE 11-15   Ideal flow about a cylinder. (*a*) Circulation. (*b*) Uniform flow. (*c*) Combination of circulation and uniform flow.

will be lower than the pressures on the low-velocity side. Hence, a pressure differential exists that causes a side thrust or lift on the cylinder. According to ideal-flow theory, the lift per unit length of an infinitely long cylinder is given by $F_L/\ell = \rho V_0 \Gamma$, where F_L is the lift on the segment of length ℓ. For this ideal irrotational flow there is no drag on the cylinder; but for the real-flow case, separation and viscous stresses do produce drag, and the same viscous effects will reduce the lift on the cylinder. However, the lift is significant when flow occurs past a rotating body or when a body is translating and rotating through a fluid. Hence we have the reason for the "curve" on a pitched baseball or the "drop" on a Ping-Pong ball that is given a fore spin. This phenomenon of lift produced by rotation of a solid body is called the *Magnus effect* after a nineteenth century German scientist who made early studies of the lift on rotating bodies.

Coefficients of lift and drag for the rotating cylinder with end plates are shown in Fig. 11-16. In this figure, the parameter $r\omega/V_0$ is the ratio of cylinder-surface speed to the free-stream velocity, where r is the radius of the cylinder and ω is the angular speed in radians per second. The corre-

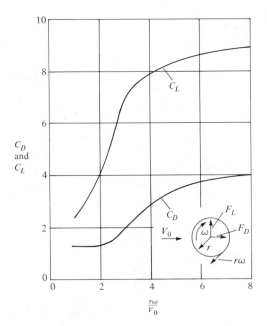

FIGURE 11-16 Coefficients of lift and drag as a function of $r\omega/V_0$ for rotating cylinder. [After Rouse (24).]

sponding curves for the rotating sphere are given in Fig. 11-17. It should be noted that the lift is given as $F_L = C_L A_p \rho V_0^2/2$, where A_p is the projected area.

Lift of an airfoil

Let us first consider the motion of an airfoil through an ideal (nonviscous) fluid. With such a fluid, the flow past an airfoil will be irrotational, as shown in Fig. 11-18a. Here, as for irrotational flow past a cylinder, the lift and drag will be zero. There is a stagnation point on the bottom side near the leading edge and there is also a stagnation point on the top side near the trailing edge of the foil. In the real-flow (viscous-fluid) case, the flow pattern around the front half of the foil is plausible; however, the flow pattern in the region of the trailing edge as shown in Fig. 11-18a cannot occur. A stagnation point on the upper side of the foil indicates that fluid must flow from the lower side around the trailing edge and then toward the stagnation point. Such a flow pattern implies an infinite acceleration of the fluid particles as they turn the corner around the trailing edge of the wing. This is a physical impossibility, and as we have seen in previous

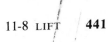

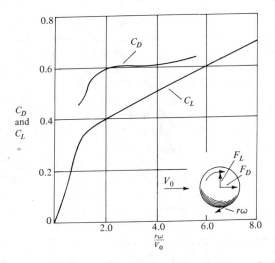

FIGURE 11-17   Lift and drag characteristics of a rotating sphere. [After Barkla (3).]

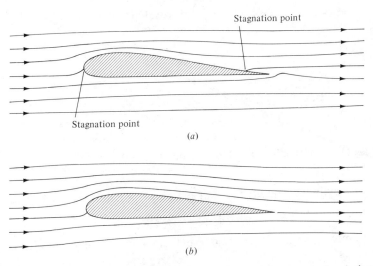

FIGURE 11-18   Flow patterns about an airfoil. (*a*) Ideal flow—no circulation. (*b*) Real flow—circulation.

sections of the text, separation will occur at the sharp edge. As a consequence of the separation, the rear stagnation point moves to the trailing edge. Flow from both the top and bottom sides of the airfoil in the vicinity of the trailing edge then leaves the airfoil smoothly and essentially parallel to these surfaces at the trailing edge (Fig. 11-18b).

To bring theory into line with the physically observed phenomenon, it was hypothesized that a circulation around the airfoil must be induced in just the right amount so that the downstream stagnation point is moved all the way back to the trailing edge of the airfoil, thus allowing the flow to leave the airfoil smoothly at the trailing edge. This is called the Kutta condition (17), named after a pioneer in aerodynamic theory. When analyses are made with this simple assumption concerning the magnitude of the circulation, very good agreement occurs between theory and experiment for the flow pattern and pressure distribution as well as for the lift on a two-dimensional airfoil section (no end effects). Ideal-flow theory then shows that the magnitude of the circulation required to maintain the rear stagnation point at the trailing edge (the Kutta condition) of a symmetric airfoil with a small angle of attack is given by

$$\Gamma = \pi c V_0 \alpha \qquad (11\text{-}13)$$

where Γ = circulation

c = chord length of the airfoil

α = angle of attack of the chord of the airfoil with the free-stream direction (see Fig. 11-19 for a definition sketch)

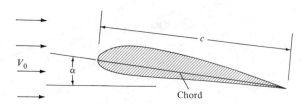

FIGURE 11-19   Definition sketch for airfoil section.

Like the cylinder, the lift per unit length of an infinitely long wing is

$$F_L/\ell = \rho V_0 \Gamma$$

The planform area for the length segment ℓ is ℓc; hence the lift on segment ℓ is

$$F_L = \rho V_0^2 \pi c \ell \alpha \qquad (11\text{-}14)$$

For an airfoil we define the coefficient of lift as

$$C_L = \frac{F_L}{S\rho V_0^2/2} \tag{11-15}$$

where S is the planform area of the wing—that is the area seen from the plan view. Upon combining Eqs. (11-14) and (11-15) and identifying S as the area associated with length segment ℓ, we find that C_L for irrotational flow past a two-dimensional airfoil is given by

$$C_L = 2\pi\alpha \tag{11-16}$$

Equations (11-14) and (11-16) are the theoretical lift equations for an infinitely long airfoil at a small angle of attack. Flow separation near the leading edge of the airfoil will produce deviations (high drag and low lift) from the ideal-flow predictions at high angles of attack; hence, experimental wind-tunnel tests are always made to evaluate the performance of a given type of airfoil section. For example, the experimentally determined lift coefficient versus α for two NACA airfoils are shown in Fig. 11-20. Note in this figure that the coefficient of lift increases with angle of attack, α, to a maximum value and then decreases with further increase in α. This condition, where C_L starts to decrease with a further increase in α, is called *stall*. Stall occurs because of the onset of separation over the top of the airfoil, thus changing the pressure distribution in such a way as to not only decrease lift but also to increase drag. Data for many other airfoil sections are given by Abbott (2).

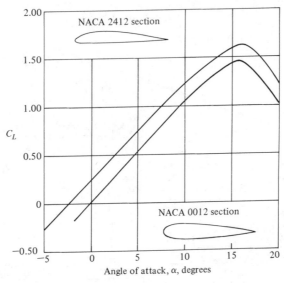

FIGURE 11-20 Lift characteristics of the NACA airfoil sections. [After Abbott (1).]

Airfoils of finite length—Effect on drag and lift

The drag of a two-dimensional foil at a low angle of attack (no end effects) is primarily viscous drag; however, wings of finite length also have an added drag and a reduced lift that is associated with vortices generated at the wing tips. These vortices occur because the high pressure below the wing and the low pressure on top causes fluid to circulate around the end of the wing from the high- to the low-pressure zone as shown in Fig. 11-21. This induced flow has the effect of adding a downward component of velocity w to the approach velocity V_0. Hence, the "effective" free-stream velocity is now at an angle ($\phi \approx w/V_0$) with the direction of the original free-stream velocity, and the resultant force will be tilted back as shown in Fig. 11-22. Thus the effective lift will be smaller than the lift for the infinitely long wing because the effective angle of incidence is now smaller. This resultant force has a component parallel to V_0 that is

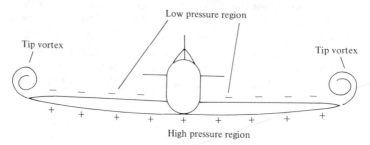

FIGURE 11-21   Formation of tip vortices.

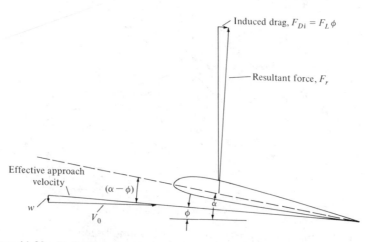

FIGURE 11-22   Definition sketch for induced-drag relations.

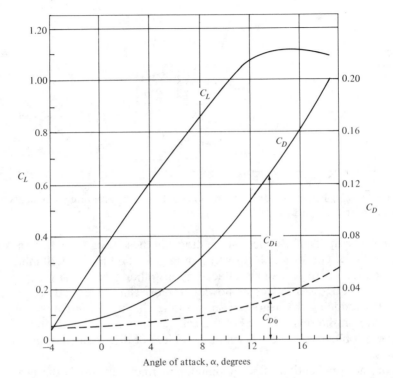

FIGURE 11-24   Lift and drag characteristics of a wing with an aspect ratio of 5. [After Prandtl (21).]

Then

$$11{,}600 = 1.18(15) \left(\frac{1.2}{2}\right) (V_{stall})^2$$

$$V_{stall} = 33.0 \text{ m/s} = 119 \text{ km/h} \qquad \blacktriangleleft$$

Drag and lift on road vehicles

Because of the increased costs of fuel, considerable research is now being done to reduce the drag of automobiles and trucks. One such study was done on a $\frac{3}{8}$-scale model of a typical notchback sedan. Wind-tunnel test results for the sedan are shown in Fig. 11-25. Here the centerline pressure distributions (distribution of C_p) for the conventional sedan are shown by a solid line and for a sedan with a 68-mm rear deck lip by a dashed line. Thus it is seen that the rear deck lip causes the pressure on the rear of the car to increase (C_p is less negative), thereby reducing the drag on the car itself. It also decreases the lift, thereby improving stability effects. Of

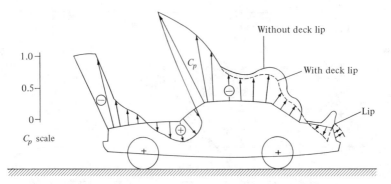

FIGURE 11-25 Effect of rear deck lip on model surface. Pressure coefficients are plotted normal to surface. [After Schenkel (25).]

course, the lip itself produces some drag and these tests show that the optimum lip height for greatest overall drag reduction is about 20 mm.

The drag of trucks can be reduced by installing vanes near the corners of the truck body to deflect the flow of air more sharply around the corner, thereby reducing the degree of separation. This in turn creates a higher pressure on the rear surfaces of the truck, which reduces the drag of the truck. Problem 11-20 addresses this problem of drag reduction by use of vanes.

Another application of aerodynamics to road vehicles is the use of a negative lift vane on racing cars, as shown in Fig. 11-26. Here the negative lift produced by the vane improves stability and traction of the race car.

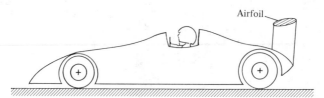

FIGURE 11-26 Race car with negative lift device.

EXAMPLE 11-8 The vane installed on the racing car of Fig. 11-26 is at an angle of attack of 8° and has the characteristics of Fig. 11-23. Estimate the downward thrust (negative lift) and drag from the vane that is 1.5 m long and has a chord length of 250 mm. Assume the racing car travels at a speed of 270 km/h at a track where normal atmospheric pressure prevails and at a temperature of 30°C.

Solution

$$F_L = C_L \ell c \rho V_0^2 / 2$$

where $\quad\quad \ell = 1.5$ m; $\quad c = 0.25$ m; $\quad \ell/c = 6.0$

$\quad\quad\quad\quad\quad\quad \rho = 1.1$ kg/m³

$\quad\quad\quad\quad\quad\quad V_0 = 270$ km/h $= 75$ m/s

$\quad\quad\quad\quad\quad\quad C_L = 0.93 \quad$ (from Fig. 11-23)

Then negative $\quad F_L = 0.93 \times 1.5 \times 0.25 \times 1.1 \times (75)^2/2 = 1{,}079$ N $\quad\blacktriangleleft$

Also $\quad\quad\quad\quad C_D = 0.070 \quad$ (from Fig. 11-23)

Therefore $\quad\quad F_D = (0.070/0.93) \times 1{,}079 = 81.2$ N $\quad\quad\blacktriangleleft$

PROBLEMS

11-1 The hypothetical pressure distribution on a rod of triangular (equilateral) cross section is shown when flow is from left to right. That is, C_p is maximum and equal to $+1.0$ at the leading edge and decreases linearly to zero at the trailing edges. The pressure coefficient on the downstream face is constant with a value of -0.5. Neglecting friction, what is C_D for the rod?

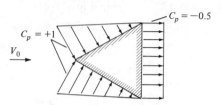

11-2 The pressure distribution on a rod having an equilateral-triangular cross section is as shown when flow is from left to right. What is the coefficient of drag C_D for the rod?

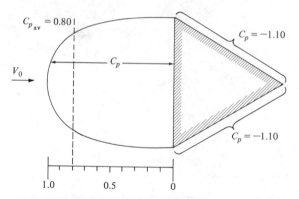

11-3 The pressure distribution on a thick disk (axis parallel to the flow) is as shown. What is the coefficient of drag?

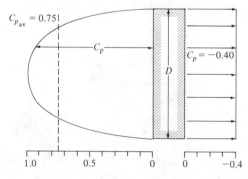

11-4 Estimate the wind force on the outdoor movie screen if the temperature is 0°C and the wind speed is 40 m/s.

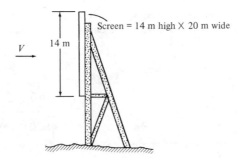

Screen = 14 m high × 20 m wide

V 14 m

11-5 Estimate the ratio of the drag of a large square plate (2 m × 2 m) when towed through water (10°C) in a normal orientation to that when towed edgewise. Assume a towing speed of about 1 m/s.

11-6 If a 1-m-diameter (3.3 ft) disk is submerged and towed behind a boat at a speed of 4 m/s (13.1 ft/sec), what drag will be produced? Assume orientation of the disk so that maximum drag is produced.

11-7 A circular billboard having a diameter of 5 m (16.4 ft) is so mounted as to be freely exposed to the wind. Estimate the total force exerted upon the structure in a wind that has a direction normal to the structure and a speed of 30 m/s (98 ft/sec). Assume $T = 10°C$ and $p = 101$ kPa abs.

11-8 Estimate the moment at ground level on a signpost supporting a 2 by 2 m (6.6 × 6.6 ft) sign if the wind is normal to the surface and has a speed of 35 m/s (115 ft/sec) and the center of the sign is 2 m (6.6 ft) above the ground. Neglect the wind load on the post itself. Assume $T = 10°C$ and $p = 100$ kPa abs.

11-9 Compute the overturning moment exerted by a 30 m/s (98 ft/sec) wind upon a smokestack that has a diameter of 3 m and a height of 90 m. Assume that the air temperature is 20°C and $p_a = 99$ kPa (14.4 psia).

11-10 What is the moment at the bottom of a flag pole 35 m high (115 ft) and 10 cm in diameter in a 25 m/s wind? The atmospheric pressure is 100 kPa (14.5 psia) and the temperature is 20°C (68°F).

11-11 Estimate the additional power required for a truck carrying the rectangular sign over that when not carrying the sign at a speed of 20 m/s (66 ft/sec).

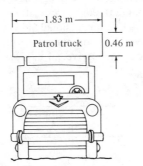

|←——1.83 m——→|

Patrol truck   0.46 m

11-12 Estimate the added power required for the car when the cartop carrier is used compared to without it when the car is driven at 80 km/h in a 20-km/h head wind.

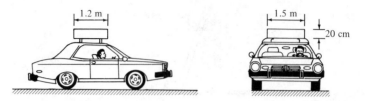

11-13 The resistance to motion of an automobile consists of rolling resistance and aerodynamic drag. The rolling resistance is the product of the weight and coefficient of rolling friction. The weight of an automobile is 3,000 lb and has a frontal area of 20 ft². The drag coefficient is 0.40 and the coefficient of rolling friction is 0.10. Determine the percentage savings in gas mileage in driving at 55 mph instead of 65 mph on a level road. Assume an air temperature of 60°F.

11-14 Estimate the wind force that would act on *you* if you were standing on top of a tower in a 30 m/s (115 ft/sec) wind on a day when the temperature is 20°C (68°F) and the atmospheric pressure is 96 kPa (14 psia).

11-15 Windstorms sometimes blow empty boxcars off their tracks. The dimensions of one type of boxcar is shown. What minimum wind velocity normal to the side of the car would be required to blow the car over?

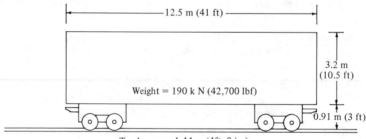

Track gage = 1.44 m (4ft, 8 in.)

11-16 A cylindrical rod of diameter d and length L is rotated about its center in a horizontal plane. Assume the drag force at each section of the rod can be calculated assuming a two-dimensional flow with an oncoming velocity equal to the relative velocity component normal to the rod. Assume C_D is constant along the rod. (*a*) Derive an expression for the average power

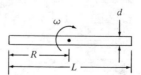

needed to rotate the rod. (*b*) Calculate the power for $\omega = 100$ rad/s, $d = 2$ cm, $L = 1$ m, $\rho = 1.2$ kg/m³, and $C_D = 1.2$.

11-17 A bicyclist coasts down a 5% grade in a 5 m/s headwind. The mass of the bicycle and rider is 80 kg, the projected area is 0.2 m², the drag coefficient is 0.5, and the air density is 1.2 kg/m³. Rolling friction is negligible. Calculate the maximum speed.

11-18 A bicyclist is capable of delivering 100 W of power to the wheels. The rolling resistance of a bicycle is negligible, so the power equals the aerodynamic drag times the velocity. How fast can the bicyclist travel in a 5-m/s head wind if his or her projected area is 0.5 m², the drag coefficient is 0.3, and the air density is 1.2 kg/m³?

11-19 The drag coefficient of a sports car with a convertible roof is 0.3 when the roof is closed and increases to 0.42 when the roof is open. The resistance to motion consists of rolling resistance and aerodynamic drag. The rolling resistance is the product of the coefficient of rolling friction and weight. The sports car has a mass of 800 kg, a frontal area of 4 m² and a coefficient of rolling friction of 0.10. The maximum power delivered to the wheels is 80 kW. Assuming a temperature of 20°C and a level road, determine the maximum possible speed with the roof closed and open.

11-20 One way to reduce drag of a blunt object is to install vanes to suppress the amount of separation. Such a procedure was used on model trucks in a wind tunnel study by Kirsch and Bettes (15). They noted for tests on a van-type truck that without vanes the C_D was 0.78. However, when vanes were installed around the top and side leading edges of the truck body (see the figure) a 25% reduction in C_D was achieved. For a truck with a projected area of 8.36 m² (90 ft²), what reduction in drag force would be effected with installation of vanes when the truck is traveling at 93 km/h (57.8 mph)?

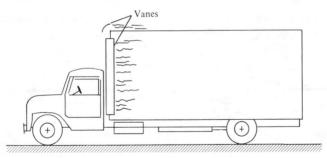

11-21 For the truck of Prob. 11-20 assume that the total resistance is given by $R = F_D + C$, where F_D is the air drag and C is the resistance due to bearing friction, etc. If C is constant at 450 N for the given truck, what fuel savings in percent will be effected by installation of the vanes when the truck travels at 80 km/h and at 100 km/h?

11-22 If the train of Prob. 9-46 has a form drag coefficient (for the locomotive and irregular undercarriage) of 0.80 and a constant bearing resistance of

3,000 N, what will be the total resistance for the train at a speed of 100 km/h and 200 km/h? Assume the projected area is 9 m² and standard atmospheric conditions prevail. What percent of the resistance is due to bearing resistance, form drag and skin-friction drag at the two speeds?

11-23 The two bodies have the same projected area when the flow is from left to right as shown. If the bodies are held in an airstream ($T = 20°C$, $p = 100$ kPa abs) that has a velocity of 30 m/s, which body will have the greatest drag? Briefly explain why. If the bodies are held in a flow of oil ($\nu = 10^{-3}$ m²/s) that has a velocity of 30 mm/s, which body will have the greatest drag? Briefly explain why.

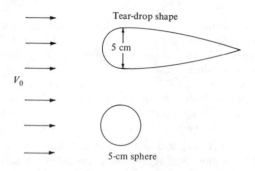

11-24 A paratrooper and parachute weigh 900 N. What rate of descent will they have if the parachute is 7 m in diameter and if the air has a density of 1.20 kg/m³?

11-25 What is the terminal velocity of a ⅜-in.-diameter hailstone in air that has an atmospheric pressure of 14.5 psia and a temperature of 40°F. Assume that the hailstone has a specific weight of 55 lbf/ft³.

11-26 What is the terminal velocity of a 1-cm hailstone in air that has an atmospheric pressure of 96 kPa abs and a temperature of 0°C? Assume that the hailstone has a specific weight of 6 kN/m³.

11-27 If a 30-cm-diameter rock has a specific gravity of 2.5, what is its terminal velocity in water?

11-28 A spherical rock weighs 45 N in air and 25 N in water. Estimate its terminal fall velocity in water (20°C).

11-29 Estimate the terminal velocity of a 0.1-in.-diameter plastic sphere in oil. The oil has a specific gravity of 0.95 and a kinematic viscosity of 10^{-3} ft²/sec. The plastic has a specific gravity of 1.05. Volume of sphere = $\pi D^3/6$.

11-30 Estimate the terminal velocity of a 2.5-mm plastic sphere in oil. The oil has a specific gravity of 0.95 and a kinematic viscosity of 10^{-4} m²/s. The plastic has a specific gravity of 1.2. Volume of sphere = $\pi D^3/6$.

11-31 An 80-cm long by 20-cm diameter cylinder of wood is weighted with lead at one end so that its total weight in air is 260 N. Then, if this cylinder is released at a depth of 100 m in a 200-m deep lake, what will be the terminal velocity of the cylinder? Assume $T = 10°C$.

11-32 If a balloon weighs 0.05 N and if it is inflated with helium to a diameter of 30 cm, what will be its terminal velocity in air (standard atmospheric conditions)?

11-33 If a balloon weighs 0.01 lb and if it is inflated with helium to a diameter of 12 in., estimate its terminal velocity in 60°F air.

11-34 Consider a small air bubble (approximately 4-mm diameter) rising in a very tall column of liquid. Will the bubble accelerate or decelerate as it moves upward in the liquid? Will the drag on the bubble be largely skin-friction drag or form drag? Explain.

11-35 A cylindrical anchor (axis vertical) made of concrete ($\gamma = 15$ kN/m³) is reeled in at a rate of 1.5 m/s by a man in a boat. If the anchor is 30 cm in diameter and 30 cm long, what tension must be applied to the rope to pull it up at this rate? Neglect the weight of the rope.

11-36 Determine the terminal velocity of a ¼-in.-diameter spherical pebble falling in 75°F water. Assume that the specific gravity of the pebble equals 2.94 and the water-kinematic viscosity equals 1×10^{-5} ft²/sec.

11-37 Determine the terminal velocity in water of a 5-in.-diameter ball which weighs 6 lbf in air.

11-38 Determine the terminal velocity in water ($T = 10°C$) of a 15-cm ball which weighs 15 N in air.

11-39 If a 2-m-diameter balloon used for meteorological observations is filled with helium and the balloon itself weighs 3 N, what velocity of ascent will it attain under standard atmospheric conditions?

11-40 If a 4-ft-diameter balloon used for meteorological observations is filled with helium ($\gamma = 0.016$ lbf/ft³) and the balloon itself weighs 1 lbf, what velocity of ascent will it attain under normal atmospheric conditions? Volume $= \pi D^3/6$.

11-41 A 5-cm-diameter sphere of specific gravity 0.20 rises at a terminal velocity of 0.5 cm/s in a fluid with a specific gravity of 0.66. Find the kinematic viscosity of the fluid.

11-42 If Stokes' law is considered valid below a Reynolds number of 0.5, what is the largest raindrop that will fall in accordance with this equation?

11-43 Analyses of pitched baseballs indicate that the C_L of a rotating baseball is approximately three times that shown in Fig. 11-17. This greater C_L is owing to the added circulation caused by the seams of the ball. What is the lift of a ball pitched at a speed of 90 ft/sec and with a spin rate of 35 rps? Also, how much would the ball be deflected from its original path by the time it gets to the plate as a result of the lift force? *Note*: The mound-to-plate distance is 60 ft, the weight of the baseball is 5 oz, and the circumference is 9 in. Assume standard atmospheric conditions.

11-44 A semiautomatic popcorn popper is shown. The unpopped corn is placed in screen S then the fan F blows air past the heating coils C and then past the popcorn. When the corn pops, its projected area increases; thus it is blown up and into a container. Unpopped corn has a mass of about 0.15 grams per kernel, and is approximately 6 mm in diameter. When the corn pops, its average diameter is about 18 mm. For these data what range of air speed in the chamber would make the device operate properly?

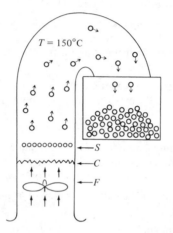

11-45 A boat of the hydrofoil type has a lifting vane with aspect ratio of 4 that has the characteristics shown in Fig. 11-23. If the angle of attack is 4° and the weight of the boat is 5 tons, what foil dimensions are needed to support the boat at a velocity of 60 fps?

11-46 When an airplane is flying in straight and level flight, the lift is equal to the weight. An airplane, which has a mass of 400 kg and a wing area of 10 m², is flying at 50 m/s near sea level (standard atmospheric conditions). Calculate the lift coefficient.

11-47 The landing speed of an airplane is 5 m/s faster than its stalling speed. The lift coefficient at landing speed is 1.2 and the maximum lift coefficient (stall condition) is 1.4. Calculate both the landing and stalling speeds.

11-48 An airplane has a rectangular-planform wing which has an elliptical span-wise lift distribution. The airplane has a mass of 1,000 kg, a wing area of 20 m², and a wingspan of 14 m, and it is flying at 60 m/s at 3,000-m altitude in a standard atmosphere. If the form-drag coefficient is 0.01, calculate the total drag on the wing and the power ($P = F_D V$) necessary to overcome the drag.

11-49 The figure shows the relative pressure distribution for a Göttingen 387-FB lifting vane (16) when the angle of attack is 8°. If such a vane with 20-cm chord were used as a hydrofoil at a depth of 70 cm, at what speed in 10°C fresh water would cavitation begin? Also estimate the lift per unit of length of foil at this speed.

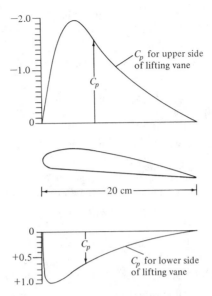

C_p for upper side of lifting vane

C_p

20 cm

C_p

C_p for lower side of lifting vane

11-50 The total drag coefficient for a wing with an elliptical lift distribution is

$$C_D = C_{D_0} + \frac{C_L^2}{\pi \Lambda}$$

where Λ is the aspect ratio. Derive an expression for C_L that corresponds to minimum C_D/C_L (maximum C_L/C_D) and the corresponding C_L/C_D.

11-51 A glider at 1,000-m altitude has a mass of 200 kg and a wing area of 20 m². The glide angle is 1.7° and the air density is 1.2 kg/m³. If the lift coefficient of the glider is 1.0, how many minutes will it take to reach sea level on a calm day?

REFERENCES

1. Abbott, I. H. "The Drag of Two Streamline Bodies as Affected by Protuberances and Appendages." NACA Rept. 451 (1932).

2. Abbott, I. H., and Von Doenhoff, A. E. *Theory of Wing Sections*. Dover Publications, New York, 1949.

3. Barkla, H. M., et al. "The Magnus or Robins Effect on Rotating Spheres." *J. Fluid Mech.*, 47, part 3 (1971).

4. Brevoort, M. J., and Joyner, U. T. "Experimental Investigation of the Robinson-Type Cup Anemometer." *NACA Rept.* 513 (1935).

5. Bullivant, W. K. "Tests of the NACA 0025 and 0035 Airfoils in the Full Scale Wind Tunnel." *NACA Rept.* 708 (1941).

6. Crowe, C. T., et al. "Drag Coefficient for Particles in Rarefied, Low Mach-Number Flows," in *Progress in Heat and Mass Transfer.* Pergamon Press, New York, 1972, vol. 6, pp. 419–431.

7. DeFoe, G. L. "Resistance of Streamline Wires." *NACA Tech. Note* 279 (March 1928).

8. Fage, A., and Warsap, J. H. "The Effects of Turbulence and Surface Roughness on the Drag of a Circular Cylinder." *Rept. Mem.* 1283, *Brit. Aeronaut. Res. Comm.* (October 1929).

9. Freeman, H. B. "Force Measurements on a 1/40-Scale Model of the U.S. Airship 'Akron'." *NACA Rept.* 432 (1932).

10. Goett, H. J., and Bullivant, W. K. "Tests of NACA 0009, 0012, and 0018 Airfoils in the Full Scale Tunnel." *NACA Rept.* 647 (1938).

11. Goldstein, S. *Modern Developments in Fluid Dynamics.* Dover Publications, Inc., New York, 1965.

12. Jacobs, E. N. "The Drag of Streamline Wires." *NACA Tech. Note* 480 (December 1933).

13. Hoerner, S. F. *Fluid Dynamic Drag.* (1958), published directly by author.

14. Jones, G. W., Jr. "Unsteady Lift Forces Generated by Vortex Shedding About a Large, Stationary, and Oscillating Cylinder at High Reynolds Numbers." *Symp. Unsteady Flow, ASME* (1968).

15. Kirsch, J. W., and Bettes, W. H. "Feasibility Study of the S^3 Air Vane and Other Truck Drag Reduction Devices." *Proc. of Conference on Reduction of the Aerodynamic Drag of Trucks, NSF RANN* (October 1974).

16. Knight, Montgomery, and Loeser, Oscar, Jr. *NACA Rept.* No. 288, U.S. Govt. Printing Office (1928).

17. Kuethe, A. M., and Schetzer, J. D. *Foundations of Aerodynamics.* John Wiley & Sons, Inc., 1967.

18. Lindsey, W. F. "Drag of Cylinders of Simple Shapes." *NACA Rept.* 619 (1938).

19. Miller, B. L., Mayberry, J. F., and Salter, I. J. "The Drag of Roughened Cylinders at High Reynolds Numbers." *NPL Rept. MAR Sci* R132 (April 1975).

20. Morrison, R. B. (ed.). *Design Data for Aeronautics and Astronautics.* John Wiley & Sons, Inc., New York, 1962.

21. Prandtl, L. "Applications of Modern Hydrodynamics to Aeronautics." *NACA Rept.* 116 (1921).

22. Roberson, J. A., et al. "Turbulence Effects on Drag of Sharp-Edged Bodies." *J. Hydraulics Div., Am. Soc. Civil Eng.* (July 1972).

23. Roshko, A. "Turbulent Wakes from Vortex Streets." *NACA Rept.* 1191 (1954).

24. Rouse, H. *Elementary Mechanics of Fluids.* John Wiley & Sons, Inc., New York, 1946.

25. Schenkel, Franz K. "The Origins of Drag and Lift Reductions on Automobiles with Front and Rear Spoilers." *Soc. Automotive Engineers, Paper* 770389 (February 1977).

26. Scher, S. H., and Gale, L. J. "Wind Tunnel Investigation of the Opening Characteristics, Drag, and Stability of Several Hemispherical Parachutes." *NACA Tech. Note* 1869, 1949.

27. White, R. G. S. "A Method of Estimating Automobile Drag Coefficients." *SAE Paper* No. 690189, Soc. Automotive Engineers (1969).

Shadowgraph of blunt-nosed cylindrical model moving at Mach 7. The blunt nose and flare at the rear of the model generate shock waves that propagate outward from the model. The "sonic boom" is what we hear when these waves reach ground level.

12 COMPRESSIBLE FLOW

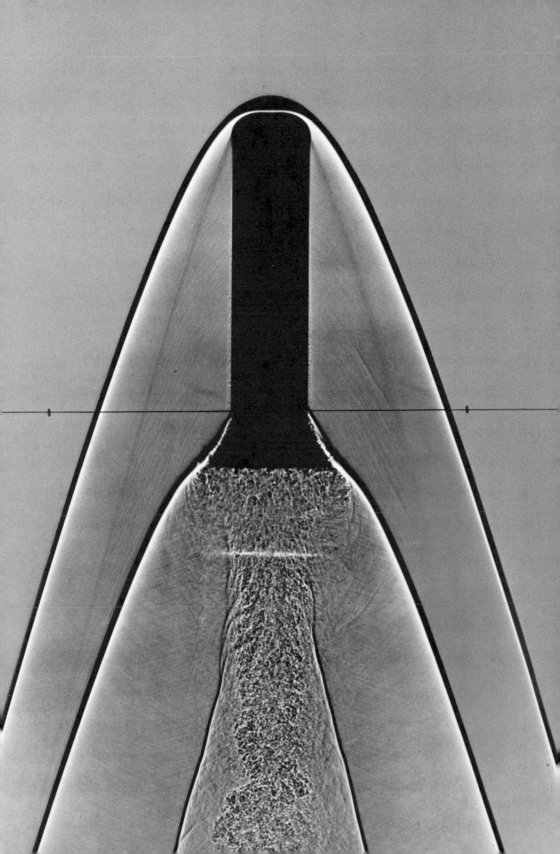

U P TO THIS POINT IN OUR STUDY of fluid flow, we have assumed that density of the fluid is constant. This assumption is well founded for the flow of liquids because very large, atypical pressure changes effect only small density variations. For example, a pressure of 20 MPa will change the density of water by less than 1%. On the other hand, pressure changes normally encountered in gas-flow problems can cause significant density changes. For example, if the same 20 MPa pressure change were applied to air initially at atmospheric pressure, its density would change by 4,370%. A change of this magnitude can no longer be regarded as negligible!

The variables that describe the state of a flowing liquid are velocity, pressure, and temperature. The equations available to solve for the flow variables derive from the conservation of mass, momentum, and energy. If, in addition, density is a variable, one more equation is necessary. This is the equation of state which, for an ideal gas, is

$$p = \rho RT \qquad (12\text{-}1)$$

which was introduced in Chapter 2. The developments in this chapter will be based on ideal-gas relationships.

12-1   WAVE PROPAGATION IN COMPRESSIBLE FLUIDS

The speed of a flowing liquid is typically much less than the speed at which a pressure disturbance is propagated through the liquid. Gas flows, on the other hand, can achieve speeds that are comparable to and even exceed the speed at which pressure disturbances are propagated. In this situation, the propagation speed is an important parameter and must be incorporated into the flow analysis of compressible fluids. We will learn in this section how the speed of an infinitesimal pressure disturbance can be evaluated and what its significance is to flow of a compressible fluid.

Speed of sound

Everyone has had the experience during a thunderstorm of seeing lightning flash and hearing the accompanying thunder an instant later. Obviously the sound was produced by the lightning, so the sound wave must have traveled at a finite speed. Had the air been totally incompressible (if that were possible), the thunder would have been heard simultaneously upon seeing the flash, because all disturbances propagate at infinite speed through incompressible media.[1] It is analogous to striking one end of a bar of incompressible material and recording instantaneously the response at the other end. Actually, all materials are compressible to some degree and propagate disturbances at finite speeds.

The *speed of sound* is defined as the rate at which an *infinitesimal* disturbance (pressure pulse) propagates in a medium with respect to the frame of reference of that medium. Actual sound waves, comprised of finite amplitude pressure disturbances such that the ear can detect them, travel only slightly faster than the "speed of sound." We found in Chapter 6 that the speed at which a pressure wave travels through a fluid depends on the bulk modulus of the fluid and its density. This analysis was done by considering the unsteady flow within a control volume as the wave passed through the control volume. Here we will derive an equation for the speed of sound assuming the control volume moves with the wave, thereby analyzing a steady-flow problem.

Let us consider a small section of a pressure wave as it propagates at velocity c through a medium as depicted in Fig. 12-1. As the wave travels through the gas at pressure p and density ρ, it produces infinitesimal

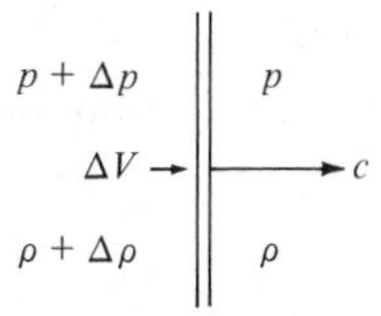

$$p + \Delta p \quad \| \quad p$$
$$\Delta V \rightarrow \| \quad \longrightarrow c$$
$$\rho + \Delta \rho \quad \| \quad \rho$$

FIGURE 12-1   Section of a sound wave.

changes of Δp, $\Delta \rho$, and ΔV. We realize that these changes must be related through the laws of conservation of mass and momentum. Let us draw a control surface around the wave and let the control volume travel with the wave. The velocities, pressures, and densities relative to the control volume (which is assumed to be very thin) are shown in Fig. 12-2. Conservation of mass in a steady flow requires that the net mass flux across the control surface be zero. Thus

$$-\rho c A + (\rho + \Delta \rho)(c - \Delta V)A = 0 \tag{12-2}$$

where A is the cross-sectional area of the control volume. Neglecting

---

[1] Actually, the thunder would have been heard before the lightning was seen, because light also travels at a finite, though very high, speed! However, this would violate one of the basic tenets of relativity theory. No medium can be completely incompressible and propagate disturbances exceeding the speed of light.

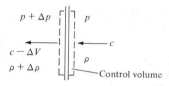

FIGURE 12-2   Flow relative to the sound wave.

products of higher-order terms ($\Delta\rho \, \Delta V$) and dividing by the area reduces the conservation-of-mass equation to

$$-\rho \, \Delta V + c \, \Delta\rho = 0 \qquad (12\text{-}3)$$

The momentum equation for steady flow,

$$\sum \mathbf{F} = \sum \mathbf{V}\rho\mathbf{V} \cdot \mathbf{A} \qquad (12\text{-}4)$$

applied to the control volume containing the pressure wave gives

$$(p + \Delta p)A - pA = (-c)(-\rho Ac) + (-c + \Delta V)\rho Ac \qquad (12\text{-}5)$$

defining the direction to the right as positive. The momentum equation reduces to

$$\Delta p = \rho c \, \Delta V \qquad (12\text{-}6)$$

Substituting the expression for ΔV obtained in Eq. (12-3) into Eq. (12-6) gives

$$c^2 = \frac{\Delta p}{\Delta\rho} \qquad (12\text{-}7)$$

which shows how the speed of propagation is related to the pressure and density change across the wave. We immediately see from this equation that if the flow were ideally incompressible, $\Delta\rho = 0$, the propagation speed would be infinite, which confirms the argument presented earlier.

Equation (12-7) gives us an expression for the speed of a general pressure wave. The sound wave is a special type of pressure wave. By definition, a sound wave produces only infinitesimal changes in pressure and density, so it can be regarded as a reversible process. There is also negligibly small heat transfer, so one can assume the process is *adiabatic*. A reversible, adiabatic process is an *isentropic* process; thus the resulting expression for the speed of sound is

$$c^2 = \frac{\partial p}{\partial\rho}\bigg|_s \qquad (12\text{-}8)$$

This equation is valid for the speed of sound in any substance. However, for many substances the relationship between p and ρ at constant entropy is not very well known.

To reiterate, the speed of sound is the speed at which an infinitesimal pressure disturbance travels through a fluid. Waves of finite strength (finite pressure change across the wave) travel faster than sound waves. Sound speed is the *minimum* speed at which a pressure wave can propagate through a fluid.

We shall now determine the speed of sound in an ideal gas. It can be shown from thermodynamics—Ref. 4, for example—that the following relationship between pressure and density holds for an isentropic process.

$$\frac{p}{\rho^k} = \text{const} \tag{12-9}$$

Here k is the ratio of specific heats, that is, the ratio of specific heat at constant pressure to that at constant volume.

$$k = \frac{c_p}{c_v} \tag{12-10}$$

The values of k for some commonly used gases are given in Table A-2 in the Appendix. Taking the derivative of Eq. (12-9) to obtain $\partial p/\partial \rho \,|_s$ results in

$$\frac{\partial p}{\partial \rho}\bigg|_s = \frac{kp}{\rho} \tag{12-11}$$

However, from the equation of state for an ideal gas,

$$\frac{p}{\rho} = RT$$

so the speed of sound is given by

$$c = \sqrt{kRT} \tag{12-12}$$

Thus we find that the speed of sound in an ideal gas varies as the square root of temperature. Using this equation to predict sound speeds in real gases at standard conditions gives results very near the measured values. Of course, if the state of gas is far removed from ideal conditions (high pressures, low temperatures), then one is cautioned against use of Eq. (12-12).

EXAMPLE 12-1 Calculate the speed of sound in air at 15°C.

Solution From Table A-2 we find that $k = 1.4$ and $R = 287$ J/kg K for air. Using equation (12-12) we calculate

$$c = [(1.4)(287 \text{ J/kg K})(288 \text{ K})]^{1/2}$$
$$= 340 \text{ m/s} \qquad \blacktriangleleft$$

Mach number

We shall now demonstrate in a very simple way the importance of sound speed in a compressible flow. Consider the airfoil traveling at speed V in Fig. 12-3. As this airfoil travels through the fluid, the pressure disturbance generated by the airfoil's motion propagates as a wave at sonic speed ahead of the airfoil. These pressure disturbances travel a considerable distance ahead of the airfoil before being attenuated by the viscosity of the fluid, and they warn the upstream fluid that the airfoil is coming (the Paul Reveres of fluid flow!). The fluid, in turn, responds such that the fluid particles begin to move apart in such a way that there is a smooth flow over the airfoil by the time it arrives. If a pressure disturbance created by the airfoil is essentially attenuated in time Δt, then the fluid at a distance $(c - V)\Delta t$ ahead is alerted to prepare for the airfoil's impending arrival.

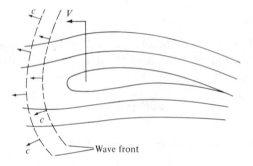

FIGURE 12-3   Sound-wave propagation from an airfoil.

What happens as the speed of the airfoil is increased? Obviously, the relative velocity, $c - V$ is reduced and the upstream fluid has less time to prepare for the airfoil's arrival. The flow field is modified by smaller streamline curvatures and the form drag on the airfoil is increased. If the airfoil speed increases to sound speed or greater, the fluid has no warning whatsoever that the airfoil is coming and cannot prepare for its arrival. Nature, at this point, resolves the problem by creating a shock wave that stands off the leading edge, as shown in Fig. 12-4. As the fluid passes through the shock wave near the leading edge, it is decelerated to a speed less than sonic speed and the fluid has time to divide and flow around the airfoil. Shock waves will be treated later in more detail.

Another approach to appreciating the significance of sound propogation in a compressible fluid is to consider a point source of sound moving in a quiescent fluid, as shown in Fig. 12-5. The sound source is moving at a speed less than the local speed of sound in Fig. 12-5a and faster than the local sound speed in Fig. 12-5b. At time $t = 0$ a sound pulse is generated

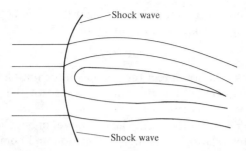

FIGURE 12-4   Bow shock in front of an airfoil.

and propagates radially outward at the local speed of sound. At time t_1 the sound source has moved a distance Vt_1 and the circle representing the sound wave emitted at $t = 0$ has a radius of ct_1. The sound source emits a new sound wave at t_1 that propagates radially outward. At time t_2 the sound source has moved to Vt_2 and the sound waves have moved outward as shown.

When the sound source moves at a speed less than the speed of sound, the sound waves form a family of nonintersecting eccentric circles, as shown in Fig. 12-5a. For an observer stationed at A the frequency of the sound pulses would appear higher than the emitted frequency because the sound source is moving toward the observer. In fact, the observer at A will detect a frequency of

$$f = f_0/(1 - V/c)$$

where f_0 is the emitting frequency of the moving sound source. This change in frequency is known as the *Doppler effect*.

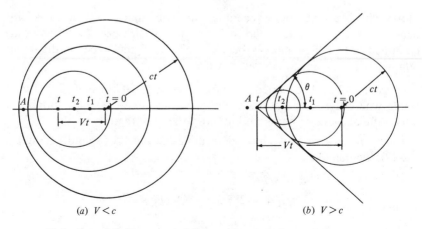

(a) $V < c$ (b) $V > c$

FIGURE 12-5   Sound field generated by moving sound source.

When the sound source moves faster than the local sound speed, the sound waves intersect and form the locus of a cone with a half-angle of

$$\theta = \sin^{-1}(c/V)$$

The observer at A will not detect the sound source until it has passed. In fact, only an observer within the cone is aware of the moving sound source.

In view of the physical arguments given above, one realizes that an important parameter relating to sound propogation and compressibility effects is the ratio V/c. This parameter, already introduced in Chapter 8, was first proposed by Ernst Mach, an Austrian scientist, and bears his name. Thus the Mach number is defined as

$$M = \frac{V}{c} \tag{12-13}$$

The conical wave surface depicted in Fig. 12-5b is known as a *Mach wave* and the conical half-angle as the *Mach angle*.

Besides the heuristic argument presented above for Mach number, we also recall from Chapter 8 that the Mach number is the ratio of the inertial to elastic forces acting on the fluid. If the Mach number is small, the inertial forces are ineffective in compressing the fluid and the fluid can be regarded as incompressible.

Compressible flows are characterized by their Mach-number regimes as follows:

$$M < 1 \qquad \text{Subsonic flow}$$
$$M \approx 1 \qquad \text{Transonic flow}$$
$$M > 1 \qquad \text{Supersonic flow}$$

Flows with Mach number exceeding 5 are sometimes referred to as "hypersonic." Airplanes designed to travel near sonic speeds and faster are equipped with Mach meters because of the significance of the Mach number with respect to aircraft performance.

EXAMPLE 12-1 The Concorde is traveling at 1,400 km/h at 12,000 m altitude where the temperature is $-56°C$. Determine the Mach number at which the airplane is flying and identify the flow regime.

Solution First of all we must determine the speed of sound. Using Eq. (12-12), we calculate

$$c = \sqrt{kRT}$$
$$= [(1.4)(287 \text{ J/kg K})(217 \text{ K})]^{1/2}$$
$$= 295 \text{ m/s}$$

The speed of the airplane in meters per second is

$$V = 1,400 \text{ km/h} \left(\frac{1}{3,600} \text{ h/s}\right) (1,000 \text{ m/km}) = 389 \text{ m/s}$$

The Mach number is therefore

$$M = \frac{389}{295} = 1.32 \qquad \blacktriangleleft$$

so the flow is supersonic.

12-2   MACH-NUMBER RELATIONSHIPS

We are already familiar with Bernoulli's equation and the energy equation and their utility in determining fluid properties along streamlines in liquid flow. In this section, we shall learn how the Mach number is used to determine fluid properties in compressible flows. Let us consider the control volume bounded by two streamlines in a steady compressible flow, as shown in Fig. 12-6. Applying the energy equation, (Eq. 7-14), to this control volume and realizing that the shaft work is zero gives

$$-\dot{m}_1 \left(h_1 + \frac{V_1^2}{2} + gz_1\right) + \dot{m}_2 \left(h_2 + \frac{V_2^2}{2} + gz_2\right) = \dot{Q} \qquad (12\text{-}14)$$

As pointed out in Chapter 7, the elevation terms (z_1 and z_2) can usually be neglected for gaseous flows. If the flow is adiabatic ($\dot{Q} = 0$), the energy equation then reduces to

$$\dot{m}_1 \left(h_1 + \frac{V_1^2}{2}\right) = \dot{m}_2 \left(h_2 + \frac{V_2^2}{2}\right) \qquad (12\text{-}15)$$

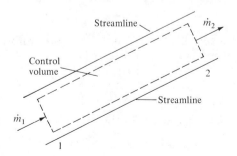

FIGURE 12-6   Control volume enclosed by streamlines.

From continuity, the mass-flow rate is constant, $\dot{m}_1 = \dot{m}_2$, so

$$h_1 + \frac{V_1^2}{2} = h_2 + \frac{V_2^2}{2} \qquad (12\text{-}16)$$

Since positions 1 and 2 are arbitrary points on the same streamline, we say that

$$h + \frac{V^2}{2} = \text{constant along a streamline in an adiabatic flow} \qquad (12\text{-}17)$$

The constant in this expression is called the *total enthalpy* h_t; it is the enthalpy that would arise if the flow velocity were brought to zero in a steady, adiabatic process. Thus the energy equation along a streamline under adiabatic conditions is

$$h + \frac{V^2}{2} = h_t \qquad (12\text{-}18)$$

If h_t is the same for all streamlines, the flow is "homenergic" (or isoenergic).

It is instructive at this point to compare Eq. (12-18) with Bernoulli's equation. Expressing the specific enthalpy as the sum of the specific internal energy and p/ρ, Eq. (12-18) becomes

$$u + \frac{p}{\rho} + \frac{V^2}{2} = \text{const}$$

If the fluid is incompressible and there is no heat transfer, then the specific internal energy is constant and the equation reduces to Bernoulli's equation (excluding the hydrostatic pressure terms).

Temperature

The enthalpy of an ideal gas can be written as

$$h = c_p T \qquad (12\text{-}19)$$

where c_p is the specific heat at constant pressure. Substituting this relation into Eq. (12-18) and dividing by $c_p T$ gives

$$1 + \frac{V^2}{2 c_p T} = \frac{T_t}{T} \qquad (12\text{-}20)$$

where T_t is the total temperature. From thermodynamics (4) we know that

$$c_p - c_v = R \qquad (12\text{-}21)$$

for an ideal gas or $\qquad k - 1 = \dfrac{R}{c_v} = \dfrac{kR}{c_p}$

so
$$c_p = \frac{kR}{k-1} \qquad (12\text{-}22)$$

Substituting this expression for c_p back into Eq. (12-20) and realizing that kRT is the speed of sound squared results in

$$T_t = T\left(1 + \frac{k-1}{2} M^2\right) \qquad (12\text{-}23)$$

The temperature T is called the *static temperature*—the temperature that would be measured by a thermometer moving with the flowing fluid. Total temperature is analogous to total enthalpy in that it is the temperature which would arise if the velocity were brought to zero in a steady, adiabatic process. If the flow is adiabatic, the total temperature is constant along a streamline. If not, the total temperature varies according to the amount of thermal energy transferred.

EXAMPLE 12-3   An aircraft is flying at $M = 1.6$ at an altitude where the atmospheric temperature is $-50°C$. The temperature on the aircraft's surface is approximately the total temperature. Estimate the surface temperature, taking $k = 1.4$.

Solution   This problem can be visualized as the aircraft being stationary and an airstream with a static temperature of $-50°C$ flowing past the aircraft at a Mach number of 1.6. The static temperature in absolute temperature units is

$$T = 223 \text{ K}$$

Using Eq. (12-23) to calculate the total temperature gives

$$T_t = 223[1 + 0.2(1.6)^2]$$
$$= 337 \text{ K or } 64°C \qquad \blacktriangleleft$$

Pressure

If the flow is isentropic, thermodynamics shows that the following relationship for pressure and temperature of an ideal gas between two points on a streamline is valid (4):

$$\frac{p_1}{p_2} = \left(\frac{T_1}{T_2}\right)^{k/(k-1)} \qquad (12\text{-}24)$$

Isentropic flow means that there is no heat transfer, so the total temperature is constant along the streamline. Therefore,

$$T_t = T_1\left(1 + \frac{k-1}{2} M_1^2\right) = T_2\left(1 + \frac{k-1}{2} M_2^2\right) \qquad (12\text{-}25)$$

Solving for the ratio T_1/T_2 and substituting into Eq. (12-24) shows that the pressure variation with the Mach number is given by

$$\frac{p_1}{p_2} = \left\{\frac{1 + [(k-1)/2]M_2^2}{1 + [(k-1)/2]M_1^2}\right\}^{k/(k-1)} \qquad (12\text{-}26)$$

In that the equation of state is used to derive Eq. (12-24), absolute pressures must always be used.

The total pressure in a compressible flow is defined as

$$p_t = p\left(1 + \frac{k-1}{2}M^2\right)^{k/(k-1)} \qquad (12\text{-}27)$$

which is the pressure that would result if the flow were decelerated to zero speed reversibly and adiabatically. Unlike total temperature, total pressure may not be constant along streamlines in adiabatic flows. For example, we will discover that flow through a shock wave, though adiabatic, is not reversible and, therefore, not isentropic. The total pressure variation along a streamline in an adiabatic flow can be obtained by substituting Eqs. (12-27) and (12-25) into Eq. (12-26) to give

$$\frac{p_{t_1}}{p_{t_2}} = \frac{p_1}{p_2}\left\{\frac{1 + [(k-1)/2]M_1^2}{1 + [(k-1)/2]M_2^2}\right\}^{k/(k-1)} = \frac{p_1}{p_2}\left(\frac{T_2}{T_1}\right)^{k/(k-1)} \qquad (12\text{-}28)$$

Unless the flow is, in addition, reversible and Eq. (12-24) is applicable, the total pressures at points 1 and 2 will not be equal. However, if the flow is isentropic, total pressure is constant along streamlines.

Density

Analogous to the total pressure, the total density in a compressible flow is given by

$$\rho_t = \rho\left(1 + \frac{k-1}{2}M^2\right)^{1/(k-1)} \qquad (12\text{-}29)$$

where ρ is the local or static density. If the flow is isentropic, then ρ_t is a constant along streamlines and Eq. (12-29) can be used to determine the variation of gas density with the Mach number.

In literature dealing with compressible flows, one often finds reference to "stagnation" conditions—that is, stagnation temperature and stagnation pressure. By definition, stagnation refers to the conditions that exist at a point in the flow where the velocity is zero, regardless of whether the zero velocity has been achieved by a steady, adiabatic, or reversible process or not. For example, if one were to insert a Pitot tube into a compressible flow, strictly speaking one would measure stagnation pressure, not total pressure, since the deceleration of the flow would not be reversible. In most cases, however, the difference between stagnation and total pressure is negligibly small.

Kinetic pressure

The kinetic pressure, $q = \rho V^2/2$, is often used, as we have seen in Chapter 11, to calculate aerodynamic forces with use of appropriate coefficients. It can also be related to the Mach number. Using the equation of state for an ideal gas to replace ρ gives

$$q = \frac{1}{2} \frac{pV^2}{RT} \tag{12-30}$$

Then using the equation for the speed of sound, Eq. (12-12), results in

$$q = \frac{k}{2} pM^2 \tag{12-31}$$

where p must always be an absolute pressure since it derives from the equation of state.

EXAMPLE 12-3 The drag coefficient for a sphere at a Mach number of 0.7 is 0.95. Determine the drag force on a 10-mm-diameter sphere in air if $p = 101$ kPa.

Solution The drag force on a sphere is

$$F_D = \tfrac{1}{2}\rho V^2 C_D A_p = q C_D A_p$$

where A_p is the projected area. The kinetic pressure is

$$q = \frac{1.4}{2} (101 \text{ kPa})(0.7)^2 = 34.6 \text{ kPa}$$

and the force is calculated to be

$$F = 0.95 \left(34.6 \times 10^3 \, \frac{\text{N}}{\text{m}^2}\right)\left(\frac{\pi}{4}\right) (10^{-2})^2 \text{ m}^2$$

$$= 2.6 \text{ N} \qquad \blacktriangleleft$$

There is one very important fact about compressible flow that must be stressed—Bernoulli's equation is not valid for compressible flows! Let us see what would happen if one decided to measure the Mach number of a high-speed air flow with a Pitot-static tube assuming that Bernoulli's equation is valid. Let us say a total pressure of 180 kPa and a static pressure of 100 kPa were measured. By Bernoulli's equation the kinetic pressure is equal to the difference between the total and static pressures, so

$$\tfrac{1}{2}\rho V^2 = p_t - p$$

or
$$\frac{k}{2}\,pM^2 = p_t - p$$

Solving for the Mach number

$$M = \sqrt{\frac{2}{k}\left(\frac{p_t}{p} - 1\right)}$$

and substituting in the measured values, one obtains

$$M = 1.07$$

Now, what should have been done? The expression relating the total and static pressures in a compressible flow is Eq. (12-27). Solving this equation for the Mach number gives

$$M = \left\{\frac{2}{k-1}\left[\left(\frac{p_t}{p}\right)^{(k-1)/k} - 1\right]\right\}^{1/2} \tag{12-32}$$

and substituting in the measured values yields

$$M = 0.96$$

Thus applying Bernoulli's equation one would have said that the flow was supersonic, whereas the flow was actually subsonic. In the limit of low velocities ($p_t/p \to 1$), Eq. (12-32) reduces to the expression derived from Bernoulli's equation, which is indeed valid for very low ($M \ll 1$) Mach numbers.

It is instructive to see how the pressure coefficient at the stagnation (total pressure) condition varies with Mach number. The pressure coefficient is given by

$$C_p = \frac{p_t - p}{\tfrac{1}{2}\rho V^2}$$

Using Eq. (12-31) for the kinetic pressure allows us to express C_p as a function of Mach number and the ratio of specific heats.

$$C_p = \frac{2}{kM^2}\left[\left(1 + \frac{k-1}{2}M^2\right)^{k/(k-1)} - 1\right]$$

The variation of C_p with Mach number is shown in Fig. 12-7. At a Mach number of zero the pressure coefficient is unity, which corresponds to incompressible flow. The pressure coefficient begins to depart significantly from unity at a Mach number of about 0.3. From this observation we infer that compressibility effects in the flow field are unimportant for Mach numbers less than 0.3.

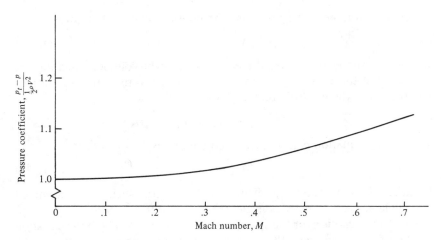

FIGURE 12-7   Variation of pressure coefficient with Mach number.

12-3   NORMAL SHOCK WAVES

Normal shock waves are wave fronts normal to the flow across which a supersonic flow is decelerated to a subsonic flow with an attendant increase in static temperature, pressure, and density. The normal shock wave is analogous to the water hammer introduced in Chapter 6 and somewhat analogous to the hydraulic jump which will be introduced in Chapter 15.

Change in flow properties across a normal shock wave

The most straightforward way to analyze a normal shock wave is to draw a control surface around the wave, as shown in Fig. 12-8, and write down the continuity, momentum, and energy equations. The net mass flux into

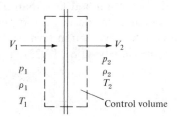

FIGURE 12-8   Control volume enclosing a normal shock wave.

the control volume is zero because the flow is steady. Therefore,

$$-\rho_1 V_1 A + \rho_2 V_2 A = 0 \qquad (12\text{-}33)$$

where A is the cross-sectional area of the control volume. Equating the net pressure forces acting on the control surface to the net efflux of momentum from the control volume gives

$$\rho_1 V_1 A(-V_1 + V_2) = (p_1 - p_2)A \qquad (12\text{-}34)$$

The energy equation can be expressed simply as

$$T_{t_1} = T_{t_2} \qquad (12\text{-}35)$$

because the temperature gradients on the control surface are assumed negligible so heat transfer is neglected (adiabatic).

Using the equation for the speed of sound, Eq. (12-12), and the equation of state for an ideal gas, the continuity equation can be rewritten to include the Mach number as

$$\frac{p_1}{RT_1} M_1 \sqrt{kRT_1} = \frac{p_2}{RT_2} M_2 \sqrt{kRT_2} \qquad (12\text{-}36)$$

The Mach number can be introduced into the momentum equation in the following way:

$$\rho_2 V_2^2 - \rho_1 V_1^2 = p_1 - p_2$$

$$p_1 + \frac{p_1}{RT_1} V_1^2 = p_2 + \frac{p_2}{RT_2} V_2^2$$

$$p_1(1 + kM_1^2) = p_2(1 + kM_2^2) \qquad (12\text{-}37)$$

Rearranging this equation for the static-pressure ratio across the shock wave results in

$$\frac{p_2}{p_1} = \frac{(1 + kM_1^2)}{(1 + kM_2^2)} \qquad (12\text{-}38)$$

As we shall show later, the Mach number of a normal shock wave is always greater than unity upstream and less than unity downstream, so the static pressure always increases across a shock wave.

The energy equation can be rewritten in terms of the temperature and Mach number as done previously (Eq. 12-23), utilizing the fact that $T_{t_2}/T_{t_1} = 1$:

$$\frac{T_2}{T_1} = \frac{\{1 + [(k-1)/2]M_1^2\}}{\{1 + [(k-1)/2]M_2^2\}} \qquad (12\text{-}39)$$

Substituting Eqs. (12-38) and (12-39) into Eq. (12-36) yields the following relationship for the Mach numbers upstream and downstream of a normal shock wave:

$$\frac{M_1}{1 + kM_1^2}\left(1 + \frac{k-1}{2} M_1^2\right)^{1/2} = \frac{M_2}{1 + kM_2^2}\left(1 + \frac{k-1}{2} M_2^2\right)^{1/2} \quad (12\text{-}40)$$

Then, solving this equation for M_2 as a function of M_1, we obtain two solutions. One solution is trivial, $M_1 = M_2$, which corresponds to no shock wave in the control volume. The other solution is

$$M_2^2 = \frac{(k-1)M_1^2 + 2}{2kM_1^2 - (k-1)} \quad (12\text{-}41)$$

Note: Because of the symmetry of Eq. (12-40), Eq. (12-41) can also be used to solve for M_1, given M_2, by simply interchanging the subscripts on the Mach number.

Setting $M_1 = 1$ in Eq. (12-41) results in M_2 also being unity. Equations (12-38) and (12-39) show also that there would be no pressure or temperature increase across such a wave. In fact, the wave corresponding to $M_1 = 1$ is the sound wave across which, by definition, pressure and temperature changes are infinitesimal. Thus the sound wave represents a degenerate normal shock wave.

EXAMPLE 12-5   A normal shock wave occurs in air flowing at a Mach number of 1.5. The static pressure and temperature of the air upstream of the shock wave are 100 kPa abs and 15°C. Determine the Mach number, pressure, and temperature downstream of the shock wave.

Solution   Using Eq. (12-41), the Mach number downstream of the shock wave is thus calculated:

$$M_2^2 = \frac{(0.4)(1.5)^2 + 2}{(2.8)(1.5)^2 - 0.4}$$

$$= 0.49$$

$$M_2 = 0.7 \qquad \blacktriangleleft$$

Equations (12-38) and (12-39) provide the downstream pressure and temperature; namely,

$$p_2 = p_1 \left(\frac{1 + kM_1^2}{1 + kM_2^2}\right)$$

$$= 100 \text{ kPa} \left[\frac{1 + (1.4)(1.5)^2}{1 + (1.4)(0.7)^2}\right]$$

$$= 246 \text{ kPa, abs} \qquad \blacktriangleleft$$

$$T_2 = T_1 \left\{ \frac{1 + [(k - 1)/2]M_1^2}{1 + [(k - 1)/2]M_2^2} \right\}$$

$$= 288 \text{ K} \left[\frac{1 + (0.2)(2.25)}{1 + (0.2)(0.49)} \right]$$

$$= 380 \text{ K or } 107°\text{C} \qquad \blacktriangleleft$$

Note that absolute pressures and temperatures must always be used in carrying out these calculations. The change in flow properties across a shock wave are presented in tabular form in Appendix A-1 for a gas, such as air, for which $k = 1.4$.

A shock wave is an adiabatic process in which no shaft work is done. Thus the total enthalpy and total temperature for ideal gases is unchanged across the wave. The total pressure, however, does change across a shock wave. Using the previous example, the total pressure upstream of the wave is

$$p_{t_1} = p_1 \left(1 + \frac{k - 1}{2} M_1^2 \right)^{k/(k-1)}$$

$$= 100 \text{ kPa}[1 + (0.2)(2.25)]^{3.5}$$

$$= 367 \text{ kPa}$$

The total pressure downstream is

$$p_{t_2} = p_2 \left(1 + \frac{k - 1}{2} M_2^2 \right)^{k/(k-1)}$$

$$= 246 \text{ kPa}[1 + (0.2)(0.49)]^{3.5}$$

$$= 341 \text{ kPa}$$

Thus we see that the total pressure decreases through the wave which, as we will see later, is owing to the fact that the flow through the shock wave is not an isentropic process. Total pressure remains constant along streamlines only in isentropic flow. Tables for the ratio of total pressure across a normal shock wave are provided also in Appendix A-1.

Existence of shock waves only in supersonic flows

Let us look back at Eq. (12-41), which gives the Mach number downstream of a normal shock wave. If one were to substitute a value for M_1 less than unity, one would predict a downstream Mach number larger than unity. For example, if $M_1 = 0.5$ in air, then

$$M_2^2 = \frac{(0.4)(0.5)^2 + 2}{(2.8)(0.5)^2 - 0.4}$$

$$M_2 = 2.64$$

Is it possible to have a shock wave in a subsonic flow across which the Mach number becomes supersonic? We also find that the total pressure would increase across such a wave; i.e.,

$$\frac{p_{t_2}}{p_{t_1}} > 1$$

The existence of such a wave would be a significant scientific discovery!

The only way to determine if such a solution is possible is to invoke the second law of thermodynamics, which states that for any process, the entropy of the universe must remain unchanged or increase.

$$\Delta s_{\text{univ}} \geq 0 \tag{12-42}$$

Because the shock wave is an adiabatic process, there is no change in the entropy of the surroundings, so the entropy of the system must remain unchanged or increase.

$$\Delta s_{\text{sys}} \geq 0 \tag{12-43}$$

The entropy change of an ideal gas between pressures p_1 and p_2 and temperatures T_1 and T_2 is given by Van Wylen (9):

$$\Delta s_{1 \to 2} = c_p \ln \frac{T_2}{T_1} - R \ln \frac{p_2}{p_1} \tag{12-44}$$

Using the relationship between c_p and R, Eq. (12-22), the entropy change can be expressed as

$$\Delta s_{1 \to 2} = R \ln \left[\frac{p_1}{p_2} \left(\frac{T_2}{T_1} \right)^{k/(k-1)} \right] \tag{12-45}$$

One notes that the quantity in the square brackets is simply the total pressure ratio as given by Eq. (12-28), so the entropy change across a shock wave can be rewritten as

$$\Delta s = R \ln \frac{p_{t_1}}{p_{t_2}} \tag{12-46}$$

A shock wave across which the Mach number changes from subsonic to supersonic would give rise to a total pressure ratio less than unity and a corresponding decrease in entropy

$$\Delta s_{\text{sys}} < 0$$

which violates the second law of thermodynamics. Therefore, shock waves can exist only in supersonic flow.

The total pressure ratio approaches unity for sound waves, which conforms with the definition that they are isentropic ($\ln 1 = 0$).

EXAMPLE 12-6   Find the entropy increase across the shock wave considered in Example 12-5.

Solution

$$\Delta s = R \ln \frac{p_{t_1}}{p_{t_2}}$$

$$p_{t_1} = 367 \text{ kPa}$$

$$p_{t_2} = 341 \text{ kPa}$$

$$\Delta s = 287 \text{ J/kg K} \ln \frac{367}{341} = 21 \text{ J/kg K} \quad \blacktriangleleft$$

More examples of shock waves will be given in the next section. We shall conclude this section by qualitatively discussing other features of shock waves.

Besides the normal shock waves studied here, there are oblique shock waves that are inclined with respect to the flow direction. Let us look once again at the shock-wave structure on a blunt body, as depicted qualitatively in Fig. 12-9. The portion of the shock wave immediately in front of the body behaves like a normal shock wave. As the shock wave bends in the free-stream direction, oblique shock waves result. The same relationships derived above for the normal shock waves are valid for the velocity components normal to oblique waves. The oblique shock waves

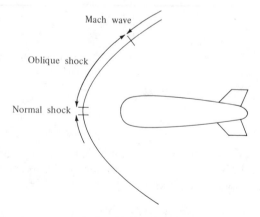

FIGURE 12-9   Shock structure in front of a blunt body.

continue to bend in the downstream direction until the Mach number of the velocity component normal to the wave is unity. Then the oblique shock has degenerated into a so-called Mach wave across which changes in flow properties are infinitesimal.

The familiar sonic booms are the result of weak oblique shock waves that reach ground level. One can appreciate the damage that would ensue from stronger oblique shock waves if aircraft were permitted to travel at supersonic speeds near ground level.

12-4 ISENTROPIC COMPRESSIBLE FLOW THROUGH A DUCT WITH VARYING AREA

We are already familiar with incompressible flow through ducts of varying cross-sectional area, such as the venturi tube. As the flow approaches the throat (smallest area), the velocity increases and the pressure decreases; then as the area again increases, the velocity decreases. The same velocity-area relationship is not always found for compressible flows.

Dependence of the Mach number on area variation

Consider the duct of varying area shown in Fig. 12-10. It is assumed that the flow is isentropic and that the flow properties at each section are uniform. This type of analysis, in which the flow properties are assumed uniform at each section yet in which the cross-sectional area is allowed to vary (nonuniform), is still classified as "one-dimensional."

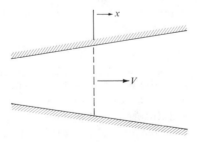

FIGURE 12-10   Variable-area duct.

The mass flow through the duct is given by

$$\dot{m} = \rho A V \tag{12-47}$$

where A is the duct's cross-sectional area. Since the mass flow is constant along the duct, we have

$$\frac{d\dot{m}}{dx} = \frac{d(\rho A V)}{dx} = 0 \tag{12-48}$$

which can be written as[1]

$$\frac{1}{\rho}\frac{d\rho}{dx} + \frac{1}{A}\frac{dA}{dx} + \frac{1}{V}\frac{dV}{dx} = 0 \tag{12-49}$$

The flow is assumed to be inviscid, so Euler's equation for steady flow is applicable:

$$\rho V \frac{dV}{dx} + \frac{dp}{dx} = 0 \tag{12-50}$$

Making use of Eq. (12-8), which relates $dp/d\rho$ to the speed of sound in an isentropic flow, gives

$$\frac{-V}{c^2}\frac{dV}{dx} = \frac{1}{\rho}\frac{d\rho}{dx} \tag{12-51}$$

This equation is now used to eliminate ρ in Eq. (12-49). The result is the following expression:

$$\frac{1}{V}\frac{dV}{dx} = \frac{(1/A)(dA/dx)}{M^2 - 1} \tag{12-52}$$

which, though simple, leads to the following important, far-reaching conclusions.

SUBSONIC FLOW   For subsonic flow, $M^2 - 1$ is negative, which means that a decreasing area leads to an increasing velocity, and correspondingly, an increasing area leads to a decreasing velocity. This velocity-area relationship agrees with our experience relating to flow through pipes with section changes.

SUPERSONIC FLOW   For supersonic flow, $M^2 - 1$ is positive, so a decreasing area leads to a decreasing velocity and an increasing area to an increasing velocity. Thus the velocity at the minimum area of a duct with supersonic compressible flow is a minimum. This is the principle underlying the operation of diffusers on jet engines for supersonic aircraft, as shown in Fig. 12-11. The purpose of the diffuser is to decelerate the flow

---

[1] This step can easily be seen by first taking the logarithm of Eq. (12-48):

$$\ln(\rho A V) = \ln\rho + \ln A + \ln V$$

and then taking the derivative of each term

$$\frac{d}{dx}[\ln(\rho A V)] = 0 = \frac{1}{\rho}\frac{d\rho}{dx} + \frac{1}{A}\frac{dA}{dx} + \frac{1}{V}\frac{dV}{dx}$$

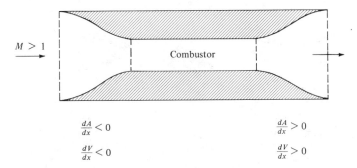

$$\frac{dA}{dx} < 0 \qquad\qquad\qquad \frac{dA}{dx} > 0$$

$$\frac{dV}{dx} < 0 \qquad\qquad\qquad \frac{dV}{dx} > 0$$

FIGURE 12-11   Engine for supersonic aircraft.

so that there is sufficient time for combustion in the chamber. Then the diverging nozzle accelerates the flow again to achieve a larger kinetic energy of the exhaust gases and an increased engine thrust.

TRANSONIC FLOW ($M \approx 1$)   Stations along a duct corresponding to $dA/dx = 0$ represent either a local minimum or local maximum in the duct's cross-sectional area, as illustrated in Fig. 12-12. If at these stations the flow were either subsonic ($M < 1$) or supersonic ($M > 1$), then by Eq. (12-52) $dV/dx = 0$, so the flow velocity would have either a maximum or a minimum value. In particular, if the flow were supersonic through duct a, then the velocity would be a minimum at the throat; if subsonic, a maximum.

Now, what happens if the Mach number is unity? Equation (12-52) tells us that if the Mach number is unity and dA/dx is not equal to zero, the velocity gradient, dV/dx, is infinite—a physically impossible situation. Therefore, dA/dx must be zero where the Mach number is unity to have a finite, physically reasonable velocity gradient.[1]

We can argue one step further here to show that sonic flow can occur only at a minimum area. Consider the nozzle in Fig. 12-12a. If the flow is initially subsonic, the converging duct will accelerate the flow toward a sonic velocity. If the flow is initially supersonic, the converging duct will decelerate the flow toward a sonic velocity. Using this same reasoning, one can prove that sonic flow is impossible in the nozzle depicted in Fig. 12-12b. If the flow is initially supersonic, the diverging duct increases the Mach number even more. If the flow is initially subsonic, the diverging duct decreases the Mach number; thus, sonic flow cannot be achieved at a maximum area. Thus the Mach number in a duct of varying cross-sectional area can be unity only at a local area minimum (throat). This

---

[1] Actually the velocity gradient is indeterminant, because the numerator and denominator are both zero. It can be shown by application of L'Hôpital's rule, however, that the velocity gradient is finite.

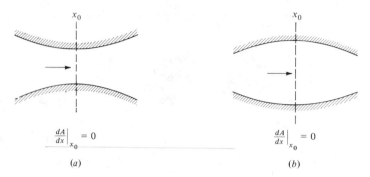

FIGURE 12-12   Nozzle contours for which dA/dx is zero.

does not imply, however, that the Mach number must always be unity at a local area minimum.

Laval nozzle

The Laval nozzle is a duct of varying area which produces supersonic flow. The nozzle is named after its inventor, de Laval (1845–1913), a Swedish engineer. According to the discussion above, the nozzle must consist of a converging section to accelerate the subsonic flow, a throat section for transonic flow, and a diverging section to further accelerate the supersonic flow. Thus the shape of the Laval nozzle is as shown in Fig. 12-13.

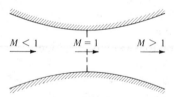

FIGURE 12-13   Laval nozzle.

One very important application of the Laval nozzle is the supersonic wind tunnel, which has been an indispensable tool in the development of supersonic aircraft. Basically the wind tunnel, as illustrated in Fig. 12-14, consists of a high-pressure source of gas, a Laval nozzle to produce supersonic flow, and a test section. The high-pressure source may be from a large pressure tank which is connected to the Laval nozzle through a regulator valve to maintain a constant upstream pressure or from a pumping system, which provides a continuous high-pressure supply.

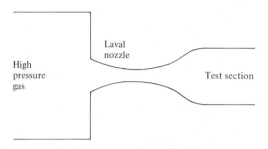

FIGURE 12-14 Wind tunnel.

The equations relating to the compressible flow through a Laval nozzle have already been developed. Since the mass-flow rate is the same at every cross section along the nozzle, we have

$$\rho VA = \text{const}$$

and the constant is usually evaluated corresponding to those conditions that exist when the Mach number is unity. Thus,

$$\rho VA = \rho_* A_* V_* \tag{12-53}$$

where the asterisk signifies conditions where the Mach number is equal to unity. Rearranging Eq. (12-53) gives

$$\frac{A}{A_*} = \frac{\rho_* V_*}{\rho V}$$

However, the velocity is the product of the Mach number and the local speed of sound, therefore,

$$\frac{A}{A_*} = \frac{\rho_*}{\rho} \frac{M_* \sqrt{kRT_*}}{M\sqrt{kRT}} \tag{12-54}$$

By definition $M_* = 1$, so

$$\frac{A}{A_*} = \frac{\rho_*}{\rho} \left(\frac{T_*}{T}\right)^{1/2} \frac{1}{M} \tag{12-55}$$

Because the flow in a Laval nozzle is assumed to be isentropic, the total temperature and total pressure (and total density) are constant throughout the nozzle. From Eq. (12-29), we have

$$\frac{\rho_*}{\rho} = \left\{ \frac{1 + [(k - 1)/2]M^2}{(k + 1)/2} \right\}^{1/(k-1)}$$

and from Eq. (12-25), the temperature ratio is given by

$$\frac{T_*}{T} = \frac{1 + [(k - 1)/2]M^2}{(k + 1)/2}$$

Substituting these expressions into Eq. (12-55) gives the following relationship between area and Mach number in a Laval nozzle:

$$\frac{A}{A_*} = \frac{1}{M}\left\{\frac{1 + [(k - 1)/2]M^2}{(k + 1)/2}\right\}^{(k+1)/2(k-1)} \tag{12-56}$$

This equation is valid, of course, for all Mach numbers—subsonic, transonic, and supersonic. The area ratio, A/A_*, is the ratio of the area at the station where the Mach number is M to the area at the throat (where $M = 1$). Many supersonic wind tunnels are designed to maintain the same test-section area and vary the Mach number by changing the throat area.

EXAMPLE 12-7 Suppose we are designing a supersonic wind tunnel to operate with air at a Mach number of 3. If the throat area is 10 cm², what must the cross-sectional area of the test section be?

Solution Putting $k = 1.4$ for air and $M = 3$ in Eq. (12-56) gives

$$\frac{A}{A_*} = \frac{1}{3}\left[\frac{1 + (0.2)3^2}{1.2}\right]^3$$

$$= 4.23$$

Thus the area of the test section must be 42.3 cm². ◀

One notes by Example 12-7 that it is a straightforward task to calculate the area ratio given the Mach number and ratio of specific heats. However, in practice, one usually knows the area ratio and wishes to determine the Mach number. It is not possible to solve Eq. (12-56) for the Mach number as an explicit function of area ratio. For this reason, compressible flow tables have been developed that allow one to easily obtain the Mach number given the area ratio.

Let us now look again at Table A-1 in the Appendix. This table has been developed for a gas, such as air, for which $k = 1.4$. The symbols that head each column are defined at the beginning of the table. Tables for both subsonic and supersonic flow are provided.

EXAMPLE 12-8 A wind tunnel using air has an area ratio of 10. The absolute total pressure and temperature are 4 MPa and 350 K. Find the Mach number, pressure, and air velocity in the test section.

Solution From the table for supersonic flow we find that the Mach number must be between 3.5 and 4.0. Interpolating between the two points

$$\begin{array}{cc} M & A/A_* \\ 3.5 & 6.79 \\ 4.0 & 10.72 \end{array}$$

gives $M = 3.91$ at $A/A_* = 10.0$.   ◀

The tables can also be used to interpolate for T/T_t and p/p_t or they can be calculated using Eqs. (12-23) and (12-27). The results are

$$\frac{p}{p_t} = 0.00743$$

$$\frac{T}{T_t} = 0.246$$

In the test section,

$$p = 29.7 \text{ kPa}$$
$$T = 86 \text{ K}$$ ◀

The velocity in the test section is obtained from

$$V = Mc$$
$$= M\sqrt{kRT}$$
$$= 727 \text{ m/s}$$ ◀

Mass-flow rate through a Laval nozzle

An important consideration in design of a supersonic wind tunnel is size. A large wind tunnel requires a large mass-flow rate which, in turn, requires a large pumping system for a continuous flow tunnel or a large tank for sufficient run time in an intermittent tunnel. The easiest station at which to calculate the mass-flow rate is the throat, because at this station the Mach number is unity.

$$\dot{m} = \rho_* A_* V_*$$
$$= \rho_* A_* \sqrt{kRT_*}$$

It is more convenient, however, to express the mass flow in terms of total conditions. The local density and static temperature at sonic velocity are related to the total density and temperature by

$$\frac{T_*}{T_t} = \left(\frac{2}{k+1}\right)$$

$$\frac{\rho_*}{\rho_t} = \left(\frac{2}{k+1}\right)^{1/(k-1)}$$

which, when substituted into the equation above, gives

$$\dot{m} = \rho_t \sqrt{kRT_t} \, A_* \left(\frac{2}{k+1}\right)^{(k+1)/2(k-1)} \tag{12-57}$$

Usually the total pressure and temperature are known. Using the equation of state for an ideal gas to eliminate ρ_t, we have

$$\dot{m} = \frac{p_t A_*}{\sqrt{RT_t}} k^{1/2} \left(\frac{2}{k+1}\right)^{(k+1)/2(k-1)} \tag{12-58}$$

For gases with a ratio of specific heats of 1.4,

$$\dot{m} = 0.685 \frac{p_t A_*}{\sqrt{RT_t}} \tag{12-59}$$

and for gases with $k = 1.67$,

$$\dot{m} = 0.727 \frac{p_t A_*}{\sqrt{RT_t}} \tag{12-60}$$

EXAMPLE 12-9   A supersonic wind tunnel with a 15-cm square test section is being designed to operate at a Mach number of 3 using air. The static temperature and pressure in the test section are $-20°C$ and 50 kPa, respectively. Calculate the mass-flow rate.

Solution   From Example 12-6 the area ratio for a Mach 3 wind tunnel is 4.23. Thus the area of the throat must be

$$A_* = \frac{225}{4.23} = 53.2 \text{ cm}^2 = 0.00532 \text{ m}^2$$

The total pressure is obtained from Eq. (12-27).

$$p_t = p \left(1 + \frac{k-1}{2} M^2\right)^{k/(k-1)}$$

$$= 50(36.7) = 1,836 \text{ kPa} = 1.836 \text{ MPa}$$

The total temperature is

$$T_t = T \left(1 + \frac{k-1}{2} M^2\right)$$

$$= 253(2.8) = 708 \text{ K}$$

Finally the mass-flow rate from Eq. (12-59) is

$$\dot{m} = \frac{(0.685)[1.836(10^6 \text{ N/m}^2)](0.00532 \text{ m}^2)}{[(287 \text{ J/kg K})(708 \text{ K})]^{1/2}}$$

$$= 14.8 \text{ kg/s}$$

◀

A pump capable of moving air at this rate against a 1.8-MPa pressure would require over 6,000 kW of power input. Such a system would be large, expensive, and costly to operate.

Classification of nozzle flow by exit conditions

Let us now take a qualitative look at the pressure distribution in a Laval nozzle. Consider the Laval nozzle depicted in Fig. 12-15 with the corresponding pressure and Mach-number distribution plotted beneath it. The pressure at the nozzle entrance is very near the total pressure, because the Mach number is small. As the area decreases toward the throat, the

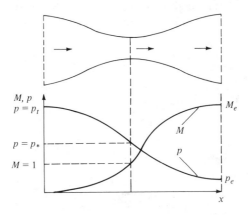

FIGURE 12-15   Static pressure and Mach-number distribution in a Laval nozzle.

Mach number increases and the pressure decreases. The static-to-total-pressure ratio at the throat, which corresponds to sonic conditions, is called the *critical pressure ratio*. It has a value of

$$\frac{p_*}{p_t} = \left(\frac{2}{k+1}\right)^{k/(k-1)}$$

which for air is

$$\frac{p_*}{p_t} = 0.528$$

It is called a critical pressure ratio because to achieve sonic flow with air in a nozzle, it is necessary that the exit pressure be at least less than 0.528 of the total pressure. The pressure continues to decrease until it reaches the exit pressure that corresponds to the nozzle exit-area ratio. Similarly, the Mach number monotonically increases with distance down the nozzle.

Now what happens if the nozzle exit pressure p_e is different from the back pressure (the pressure to which the nozzle exhausts)? If the exit pressure is higher than the back pressure, an expansion wave exists at the nozzle exit, as shown in Fig. 12-16a. These waves, which will not be studied here, effect a turning and further acceleration of the flow to achieve the back pressure. As one watches the exhaust of a rocket motor as it rises through the ever-decreasing pressure of higher altitudes, one can see the plume fan out as the flow turns more to achieve the lower pressure. A nozzle for which the exit pressure is larger than the back pressure is called an *underexpanded* nozzle because the flow could have expanded further.

If the exit pressure is less than the back pressure, shock waves occur. If the exit pressure is only slightly less than the back pressure, then pressure equalization can be obtained by oblique shocks at the nozzle exit, as shown in Fig. 12-16b.

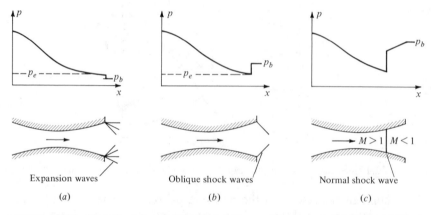

FIGURE 12-16   Nozzle exit conditions. (a) Expansion waves. (b) Oblique shock waves. (c) Normal shock wave.

If, however, the difference between back pressure and exit pressure is larger than can be accommodated by oblique shock waves, a normal shock will occur in the nozzle, as shown in Fig. 12-16c. A pressure jump occurs across the normal shock wave; the flow becomes subsonic and decelerates in the remaining portion of the diverging section in such a way that the exit pressure is equal to the back pressure. As the back pressure is further increased, the shock wave moves toward the throat region until, finally, there is no region of supersonic flow. A nozzle in which the exit pressure corresponding to the exit-area ratio of the nozzle is less than back pressure is called an *overexpanded* nozzle. Any flow that exits from a duct (or pipe) subsonically must always exit at the local back pressure.

A nozzle with supersonic flow in which the exit pressure is equal to the back pressure is *ideally expanded.*

EXAMPLE 12-10   The total pressure in a 4:1 area-ratio nozzle $(A:A_*)$ is 1.3 MPa. Air is flowing through the nozzle. If the back pressure is 100 kPa, is the nozzle overexpanded, ideally expanded, or underexpanded?

Solution   Interpolating between two area-ratio values in the supersonic-flow table (Table A-1),

M	A/A_*
2.90	3.850
3.0	4.235

gives $M = 2.94$ at $A/A_* = 4.0$. The corresponding pressure ratio is

$$\frac{p}{p_t} = 0.0298$$

so
$$p = 38.7 \text{ kPa}$$

Therefore, the nozzle is overexpanded.   ◀

EXAMPLE 12-11   The Laval nozzle shown below has an expansion ratio of 4 (exit area/throat area). Air flows through the nozzle and a normal shock wave occurs at an area ratio of 2. The total pressure upstream of the shock is 1 MPa. Determine the static pressure at the exit.

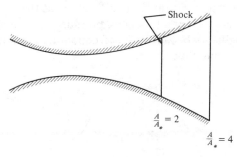

Solution   From the supersonic flow table (Table A-1) we find the Mach number corresponding to an area ratio of 2 is 2.20. From the same entry in the flow table, we find the Mach number downstream of the shock is 0.547 and the ratio of total pressures across the shock wave is

$$\frac{p_{t_2}}{p_{t_1}} = 0.6281$$

Thus the total pressure downstream of the wave is

$$p_{t_2} = 0.6281 \times 1 \text{ MPa} = 628 \text{ kPa}$$

The A/A_* ratio at a Mach number of 0.547 is 1.26. Thus the area ratio that would be needed to generate sonic flow downstream of the shock (if that were to be done) is

$$\frac{A_*}{A_t}\bigg|_{\substack{\text{downstream}\\\text{of shock}}} = \frac{2}{1.26} = 1.59$$

Thus the effective area ratio for the nozzle exit is

$$\frac{A_e}{A_*}\bigg|_{\substack{\text{downstream}\\\text{of shock}}} = \frac{4}{1.59} = 2.52$$

From Table A-1, the subsonic Mach number corresponding to this area ratio is

$$M_e = 0.24$$

The static pressure corresponding to this Mach number and a total pressure of 628 kPa is obtained from Eq. (12-27):

$$p_e = \frac{628 \text{ kPa}}{[1 + (0.2)(0.24)^2]^{3.5}}$$
$$= 603 \text{ kPa} \qquad \blacktriangleleft$$

Mass flow through a truncated nozzle

The truncated nozzle is a Laval nozzle cut off at the throat, as shown in Fig. 12-17. The nozzle exits to a back pressure p_b. This type of nozzle is important to engineers because of its frequent use as a flow-metering device for compressible flows.

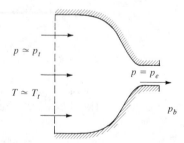

FIGURE 12-17   Truncated nozzle.

To calculate the mass flow, one must first determine whether the flow at the exit is sonic or subsonic. Of course, the flow at the exit could never be supersonic, since the nozzle area does not diverge. First we calculate the value of the critical pressure ratio

$$\frac{p_*}{p_t} = \left(\frac{2}{k+1}\right)^{k/(k-1)}$$

which, for air, is 0.528. We then evaluate the ratio of back pressure to total pressure

$$\frac{p_b}{p_t}$$

and compare it with the critical pressure ratio:

1. If $p_b/p_t < p_*/p_t$, the exit pressure is higher than the back pressure, so the exit flow must be sonic. Pressure equilibration is achieved after exit through a series of expansion waves. The mass flow is then calculated using Eq. (12-58), where A_* is the area at the truncated station.
2. If $p_b/p_t > p_*/p_t$, the flow exits subsonically. If we were to irrationally assume that the flow exited at the speed of sound, then the exit pressure p_* would be less than p_b. There can be no shock waves in a sonic flow (only sound waves) to raise the exit pressure to the back pressure. Therefore, the flow adjusts itself to the back pressure by exiting subsonically.

In case 2, one must first determine the Mach number at the exit by using Eq. (12-32):

$$M_e = \sqrt{\frac{2}{k-1}\left[\left(\frac{p_t}{p_b}\right)^{(k-1)k} - 1\right]}$$

Then using this value for Mach number, the static temperature and speed of sound at the exit are calculated:

$$T_e = \frac{T_t}{\{1 + [(k-1)/2]M_e^2\}}$$

$$c_e = \sqrt{kRT_e}$$

The gas density at the nozzle exit is determined using the exit temperature and back pressure

$$\rho_e = \frac{p_b}{RT_e}$$

Finally, the mass flow is given by

$$\dot{m} = \rho_e A_e M_e c_e$$

where A_e is the area at the truncated section.

EXAMPLE 12-12   Air exhausts through a 3-cm-diameter truncated nozzle from a reservoir at a pressure of 160 kPa and a temperature of 80°C. Calculate the flow rate if the back pressure is 100 kPa.

Solution   First we must determine the nature of the flow at the nozzle exit by evaluating the pressure ratio $p_b/p_t = 100/160 = 0.625$. Because 0.625 is larger than the critical pressure ratio for air (0.528), the flow at the nozzle exit must be subsonic. The Mach number is

$$M_e^2 = \frac{2}{k-1}\left[\left(\frac{p_t}{p_b}\right)^{(k-1)/k} - 1\right]$$

$$M_e = 0.85$$

The exit static temperature is found to be

$$T_e = \frac{T_t}{\{1 + [(k-1)/2]M_e^2\}}$$

$$= 309 \text{ K}$$

Correspondingly the density at the exit is

$$\rho_e = \frac{p_b}{RT_e}$$

$$= \frac{100 \times 10^3 \text{ N/m}^2}{(287 \text{ J/kg K})(309 \text{ K})}$$

$$= 1.13 \text{ kg/m}^3$$

The speed of sound at the exit is

$$c_e = [(1.4)(287 \text{ J/kg K})(309 \text{ K})]^{1/2}$$

$$= 352 \text{ m/s}$$

Finally, we calculate the mass-flow rate to be

$$\dot{m} = (1.13 \text{ kg/m}^3)(0.785)(0.03^2 \text{ m}^2)(0.85)(352 \text{ m/s})$$

$$\dot{m} = 0.239 \text{ kg/s} \qquad \blacktriangleleft$$

Had p_b/p_t been less than 0.528, then Eq. (12-58) would have been used to calculate the mass-flow rate.

12-5   COMPRESSIBLE FLOW IN A PIPE WITH FRICTION

The flow of liquid through a pipe was studied in Chapter 10. The analysis of a compressible flow in a pipe is somewhat more difficult because of the dependence of density and pressure on temperature. The problem is also complicated by the fact that wall friction and heat transfer cannot be simply combined into a single head-loss parameter because of their distinct effect on the Mach-number distribution along the pipe. In most engineering problems, however, the effect of wall friction is the most significant parameter and will be studied here. Two problems will be considered: first, flow in an insulated pipe in which the fluid is treated as an adiabatic system, and second, an isothermal flow that approximates flow in long pipelines.

Adiabatic flow

The conservation-of-mass equation for uniform flow through a constant area duct is

$$\rho V = \text{const}$$

which, expressed in differential form, becomes

$$\frac{dV}{V} + \frac{d\rho}{\rho} = 0 \qquad (12\text{-}61)$$

The conservation-of-energy equation, Eq. (12-17), can be written as

$$h + \frac{V^2}{2} = \text{const}$$

since the flow is adiabatic. Using the relationships between enthalpy and temperature for an ideal gas, Eqs. (12-19) and (12-22), one can rewrite the energy equation in differential form as

$$\frac{kR\,dT}{k-1} + V\,dV = 0 \qquad (12\text{-}62)$$

The conservation-of-momentum equation can be obtained by applying the momentum equation to a control volume of length Δx contained in a pipe, as shown in Fig. 12-18. Equating the forces acting on the system to

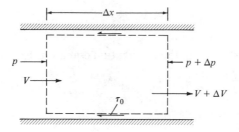

FIGURE 12-18 Control volume in a pipe.

the net efflux of momentum from the control volume results in

$$A[p - (p + \Delta p)] - \tau_0 C \, \Delta x = \rho V A(-V + V + \Delta V) \quad (12\text{-}63)$$

where C is the circumference of the pipe and τ_0 is the shear stress at the wall. Introducing the Darcy-Weisbach friction factor for τ_0, Eq. (10-21),

$$\tau_0 = \frac{f\rho V^2}{8}$$

and simplifying, we can rewrite the momentum equation in differential form as

$$\rho V \, dV + dp + \frac{f\rho V^2 \, dx}{2D} = 0 \quad (12\text{-}64)$$

where D is the pipe diameter.

Mach-number distribution along a pipe

Our goal now is to combine the conservation equations above with the equation of state, Eq. (12-1), to obtain an expression for the Mach-number distribution along a pipe. Dividing each term in Eq. (12-64) by the pressure p, and realizing that

$$\frac{p}{\rho} = RT = \frac{c^2}{k}$$

from Eqs. (12-1) and (12-12), results in

$$kM^2 \frac{dV}{V} + \frac{dp}{p} + \frac{kfM^2 \, dx}{2D} = 0 \quad (12\text{-}65)$$

The equation of state can be written in differential form as

$$\frac{dp}{p} = \frac{d\rho}{\rho} + \frac{dT}{T} \quad (12\text{-}66)$$

Using the continuity equation, Eq. (12-61), to replace ρ by V and the energy equation, Eq. (12-62), to replace T by V yields

$$\frac{dp}{p} = \frac{-dV}{V} - (k - 1)M^2 \frac{dV}{V} \quad (12\text{-}67)$$

which when substituted in the momentum equation, Eq. (12-65), results in

$$(M^2 - 1) \frac{dV}{V} + \frac{kfM^2}{2} \frac{dx}{D} = 0 \quad (12\text{-}68)$$

The Mach number is defined as

$$M = \frac{V}{(kRT)^{1/2}}$$

which can be written in differential form as

$$\frac{dM}{M} = \frac{dV}{V} - \frac{1}{2}\frac{dT}{T} \tag{12-69}$$

Once again using Eq. (12-62) to eliminate T yields

$$\frac{dM}{M} = \frac{dV}{V}\left[1 + \frac{(k-1)M^2}{2}\right] \tag{12-70}$$

Using this equation to eliminate V in Eq. (12-68) results in the following differential equation for the Mach number and distance:

$$\frac{(1-M^2)dM}{M^3\{1+[(k-1)/2]M^2\}} = \frac{kf\,dx}{2D} \tag{12-71}$$

This equation tells us that if the flow is subsonic, $dM/dx > 0$ and the Mach number increases with distance along the pipe. Conversely, if the flow is supersonic, then $dM/dx < 0$ and the Mach number decreases along the pipe. Thus the effect of wall friction is to cause the Mach number always to approach unity. It is impossible for the Mach number of a compressible flow in a pipe to change from subsonic to supersonic. Consequently, the maximum Mach number that an initially subsonic flow can attain is unity, and this can be reached only at the end of the pipe. Shock waves, of course, can occur in the pipe to change an initially supersonic flow to a subsonic flow.

From Chapter 10 we know that the Darcy-Weisbach friction factor f is a function of the Reynolds number and the relative roughness of the pipe. The Reynolds number is constant along the length of the pipe transporting a liquid. The continuity equation requires also that ρV is constant along a pipe transporting a compressible fluid. Temperature, however, may vary by as much as 20% for the subsonic flow of air in a pipe, which would correspond to a viscosity variation of approximately 10%. Referring back to Fig. 10-8 on p. 376, one notes that a 10% change in Reynolds number gives rise to a considerably smaller change in f if the flow is turbulent, which is usually the case. Thus it is reasonable to assume that f is a constant when integrating Eq. (12-71) and equal to the average value, $\bar{f}$, in the pipe.

We are now ready to integrate Eq. (12-71) to determine the variation of the Mach number with distance along the pipe. The left-hand side of the equation can be reduced to a sum of partial fractions to facilitate integration.

$$\left(\frac{1}{M^3} - \frac{k+1}{2M} + \frac{(k+1)(k-1)M}{4\{1+[(k-1)/2]M^2\}}\right)dM = \frac{k\bar{f}\,dx}{2D} \tag{12-72}$$

Integrating each side gives

$$\frac{-1}{2M^2} - \frac{k+1}{2} \ln M + \frac{k+1}{4} \ln \left(1 + \frac{k-1}{2} M^2\right) = \frac{k\bar{f}x}{2D} + C \quad (12\text{-}73)$$

where C is the integration constant. It is convenient to evaluate C by defining x_* as the distance corresponding to a Mach number of unity.

$$C = \frac{-k\bar{f}x_*}{2D} - \frac{1}{2} + \frac{k+1}{4} \ln \left(\frac{k+1}{2}\right) \quad (12\text{-}74)$$

Substituting this expression for C into Eq. (12-73) results in

$$\frac{1 - M^2}{kM^2} + \frac{k+1}{2k} \ln \left[\frac{(k+1)M^2}{2 + (k-1)M^2}\right] = \frac{\bar{f}(x_* - x_M)}{D} \quad (12\text{-}75)$$

where x_M is the distance corresponding to a Mach number of M.

EXAMPLE 12-13   The initial Mach number of the flow of air in a pipe is 0.2. The average friction factor is 0.015. Calculate the distance in pipe diameters required to achieve (a) sonic flow and (b) a Mach number of 0.8 in the pipe.

Solution   Substituting $M = 0.2$ and $k = 1.4$ into Eq. (12-75) gives

$$\frac{\bar{f}(x_* - x_{0.2})}{D} = 14.53$$

Thus the number of pipe diameters to reach sonic flow is

$$\frac{x_* - x_{0.2}}{D} = 969 \qquad \blacktriangleleft$$

Substituting $M = 0.8$ into Eq. (12-75) gives

$$\frac{x_* - x_{0.8}}{D} = 0.07$$

The distance required to increase the Mach number from 0.2 to 0.8 can be obtained by subtraction.

$$\frac{\bar{f}(x_{0.8} - x_{0.2})}{D} = \frac{\bar{f}(x_* - x_{0.2})}{D} - \frac{\bar{f}(x_* - x_{0.8})}{D}$$

$$= 14.53 - 0.07 = 14.46$$

$$\frac{x_{0.8} - x_{0.2}}{D} = 964 \qquad \blacktriangleleft$$

One notes that the Mach number increases very rapidly near the end of the pipe.

Thus we see that it is relatively easy to solve for distance once the Mach number is known. However, it would be more difficult to solve for the Mach number change given a distance along the pipe. For this reason a plot of Mach number versus $\bar{f}(x_* - x_M)/D$ is presented in Fig. 12-19 to handle problems of this type.

EXAMPLE 12-14 Air flows into a 5-cm-diameter commercial steel pipe at 60 m/s. The pressure and temperature of the air are 1 MPa and 100°C. Determine the Mach number at a distance 50 m down the pipe.

Solution First we must evaluate the friction factor f. Referring to Table A-3, in the Appendix, we find that the dynamic viscosity at 100°C is 2.28×10^{-5} N · s/m². Using the ideal-gas equation, we calculate

$$\rho = \frac{p}{RT}$$

$$= \frac{10^6 \text{ N/m}^2}{(287 \text{ J/kg K})(373 \text{ K})}$$

$$= 9.34 \text{ kg/m}^3$$

The Reynolds number, then, is

$$\text{Re} = \frac{(60)(9.34)(0.05)}{2.28(10^{-5})}$$

$$= 1.23(10^6)$$

Referring back to Figs. 10-8 and 10-9, we determine that $k_s/D = 0.001$ and that the corresponding friction factor is $f = 0.0195$.

The speed of sound at the entrance to the pipe is

$$c = (kRT)^{1/2}$$
$$= [1.4(287 \text{ J/kg K})(373 \text{ K})]^{1/2}$$
$$= 386 \text{ m/s}$$

so the initial Mach number is

$$M = \frac{60}{386} = 0.16$$

and reference to Fig. 12-19 shows that

$$\frac{\bar{f}(x_* - x_{0.16})}{D} = 24$$

Now we can write

$$\frac{\bar{f}(x_M - x_{0.16})}{D} = \frac{\bar{f}(x_* - x_{0.16})}{D} - \frac{\bar{f}(x_* - x_M)}{D}$$

where $x_M - x_{0.16}$ is the distance along the pipe. Substituting the above

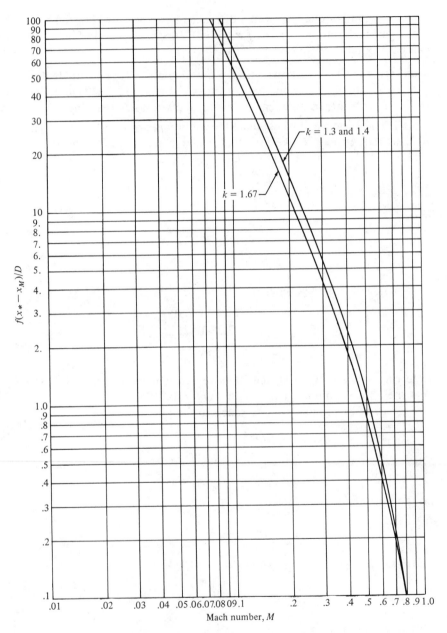

FIGURE 12-19   Variation of $f(x_* - x_M)/D$ with Mach number.

values for distance, diameter, and friction factor into this equation results in

$$\frac{\bar{f}(x_* - x_M)}{D} = 24 - \frac{(50)(0.0195) \text{ m}}{0.05 \text{ m}}$$

$$= 4.5$$

which, from Fig. 12-19, corresponds to a Mach number of

$$M = 0.32 \qquad \blacktriangleleft$$

This calculation was done taking the initial friction factor as the average value. We now ask ourselves how valid this procedure is. The total temperature at the entrance to the tube can be calculated using Eq. (12-23):

$$T_t = (373 \text{ K}) \left[1 + \frac{0.4}{2} (0.16^2) \right]$$

$$= 375 \text{ K}$$

Since it is assumed that the flow is adiabatic, the total temperature does not change along the pipe, so the static temperature 50 m along the pipe is

$$T = \frac{375 \text{ K}}{1 + (0.4/2)(0.32^2)}$$

$$= 367 \text{ K}$$

Thus the temperature changes approximately 8 K, and the viscosity change resulting from this temperature change would be approximately 1%. Therefore, the Reynolds number change is negligible, and the initial friction factor can be used for the average value.

Variation of pressure with distance

The differential equation for pressure as a function of velocity and Mach number is Eq. (12-67),

$$\frac{dp}{p} = \frac{-dV}{V} [1 + (k - 1)M^2]$$

Using Eq. (12-70) we can obtain a differential expression for pressure as a function solely of Mach number, namely,

$$\frac{dp}{p} = \frac{-dM}{M} \left\{ \frac{1 + (k - 1)M^2}{1 + [(k - 1)/2]M^2} \right\} \qquad (12\text{-}76)$$

From this expression we conclude that $dp/dM < 0$, so that pressure decreases with increasing Mach number and increases with decreasing

Mach number. Thus for subsonic flow in a pipe, the pressure decreases with distance, the negative pressure gradient providing the force to overcome the wall shear force and accelerate the fluid.

Carrying out the division of the factors contained in the brackets in Eq. (12-76) and dividing by M gives

$$\frac{dp}{p} = \left\{ -\frac{1}{M} - \frac{[(k-1)/2]M}{1 + [(k-1)/2]M^2} \right\} dM \tag{12-77}$$

Integrating each side, we obtain

$$\ln p = -\ln M - \frac{1}{2} \ln \left(1 + \frac{k-1}{2} M^2 \right) + C \tag{12-78}$$

We can evaluate the constant of integration C by setting $p = p_*$ at $M = 1$, so we have

$$\ln p_* = -\frac{1}{2} \ln \left(\frac{k+1}{2} \right) + C$$

which, when substituted back into Eq. (12-78), gives

$$\frac{p_M}{p_*} = \frac{1}{M} \left[\frac{k+1}{2 + (k-1)M^2} \right]^{1/2} \tag{12-79}$$

where p_M is the pressure at a Mach number M. The variation of M with p_M/p_* is plotted in Fig. 12-20.

EXAMPLE 12-15   Calculate the pressure at the 50-m distance down the pipe in Example 12-14.

Solution   The pressure ratio corresponding to the initial and final Mach numbers can be written as

$$\frac{p_{0.32}}{p_{0.16}} = \frac{p_{0.32}}{p_*} \frac{p_*}{p_{0.16}}$$

Using Fig. 12-20, we find

$$\frac{p_{0.32}}{p_{0.16}} = \frac{3.4}{6.8} = 0.50$$

Therefore the pressure at the 50-m distance is

$$p = 0.50(10^6) \text{ Pa}$$
$$= 500 \text{ kPa} \qquad \blacktriangleleft$$

Let us now consider how the pressure and Mach number vary along a pipe that discharges to the atmosphere as the upstream static pressure is increased. Consider the qualitative static pressure and Mach number-distributions along the pipe shown in Fig. 12-21. The pipe discharges to

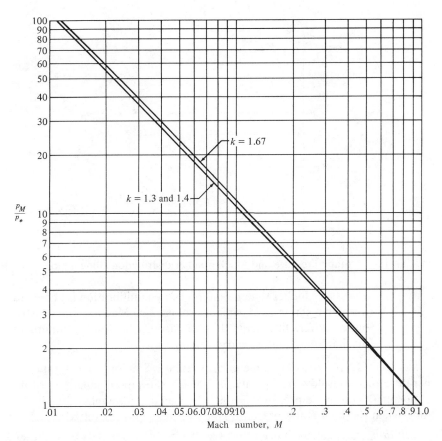

FIGURE 12-20 Variation of f_M/p_* with Mach number for adiabatic viscous flow in constant duct area.

pressure p_0. Case A through D represent a continuous increase in the upstream static pressure:

Case A The static pressure is uniform along the pipe and equal to p_0. Thus, there is no flow in the pipe and the Mach number is everywhere zero.

Case B The static pressure uniformly decreases to p_0 while the Mach number increases along the pipe.

Case C The static pressure decreases more rapidly and causes the flow to accelerate to a Mach number of unity at the exit. In this case $p_0 = p_*$.

Case D The static-pressure distribution is nearly identical to that in case C but shifted to a higher value. Thus the pipe discharges at

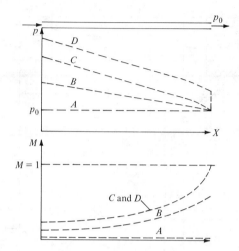

FIGURE 12-21   Static pressure and Mach-number distribution along a pipe.

a pressure higher than p_0 and pressure equilibration is achieved through a series of expansion waves. The Mach-number distribution differs little from that for case C, the flow continuing to discharge at sonic speed.

EXAMPLE 12-16   A 3-cm-diameter brass tube is 8 m long. The total temperature of the airflow in the tube is 300 K. The pipe discharges to the atmosphere where the pressure is 100 kPa. Calculate the mass flow in the tube when the inlet static pressure is (a) 120 kPa and (b) 400 kPa.

Solution   The approach used to solve this problem depends on whether the flow exits subsonically (case B) or sonically (case D). If the flow exits subsonically, then the mass-flow rate must be such that the exit pressure is equal to the atmospheric pressure. On the other hand, if the flow exits sonically, the mass flow is determined by finding the pressure level that gives the correct friction factor in the pipe.

In order to establish which approach is to be used, we first determine the upstream pressure corresponding to case C for which the flow exits sonically at atmospheric pressure. The static temperature and speed of sound at exit for case C are

$$T_e = 300 \text{ K} \frac{2}{k + 1}$$

$$= 250 \text{ K}$$

$$c_e = [(1.4)(287 \text{ J/kg K})(250 \text{ K})]^{1/2}$$

$$= 316 \text{ m/s}$$

The gas density at exit is found to be

$$\rho_e = \frac{100 \times 10^3 \text{ N/m}^2}{(287 \text{ J/kg K})(250 \text{ K})}$$

$$= 1.40 \text{ kg/m}^3$$

Thus, the Reynolds number at exit for case C would have a value of

$$\text{Re} = \frac{(316 \text{ m/s})(1.40 \text{ kg/m}^3)(0.03 \text{ m})}{1.5(10^{-5} \text{ N} \cdot \text{s/m}^2)}$$

$$= 8.85 \times 10^5$$

Referring back to Fig. 10-9, we find that the relative roughness for the 3-cm brass tube is 0.00005. The corresponding Darcy-Weisbach friction factor is found in Fig. 10-8 to be 0.013. Evaluating the parameter $\bar{f}(x_* - x_M)/D$ and entering Fig. 12-19, we find that the initial Mach number would be 0.35. Finally, Fig. 12-20 tells us that the pressure ratio corresponding to this Mach number is 3.1. Thus, for case C, the static pressure 8 m upstream would be 310 kPa. From this we conclude that part a corresponds to case B and part b to case D.

PART a This problem must be solved using an iterative approach. We shall assume an initial Mach number, calculate the exit Mach number as done in Example 12-14, and then determine the exit pressure as was done in Example 12-15. The initial Mach number is varied until the exit pressure matches the atmospheric pressure. The most straightforward approach is to generate the table below, where M_e is the exit Mach number.

M	T, K	Re	$\bar{f}$	M_e	p/p_e	p_e, kPa
0.1	299	5.4×10^4	0.020	0.104	1.04	115
0.2	298	1.08×10^5	0.0185	0.24	1.21	99
0.19	298	1.02×10^5	0.0190	0.23	1.19	101

By interpolation we find that $p_e = 100$ kPa when $M = 0.195$. The temperature, density, and speed of sound at this Mach number are

$$T = 298 \text{ K}$$

$$\rho = 1.40 \text{ kg/m}^3$$

$$c = 346 \text{ m/s}$$

Thus the mass-flow rate is

$$\dot{m} = (0.195)(346 \text{ m/s})(1.40 \text{ kg/m}^3)(0.785)(0.03^2 \text{ m}^2)$$

$$= 0.067 \text{ kg/s} \qquad \blacktriangleleft$$

PART *b*   The way to solve this part is to use the iterative approach again but on the friction factor $\bar{f}$. We begin by assuming an $\bar{f}$, calculating the upstream Mach number and Reynolds number, and finding a new $\bar{f}$ which is used as the assumed value for the next iteration. The solution is found when $\bar{f}$ no longer changes. A check is always made to be sure that p_* exceeds p_0. The iteration using $\bar{f} = 0.02$ as the initial assumed value is demonstrated in the following table.

f	$\bar{f}(x_* - x_M)/D$	M	Re	$\bar{f}$	p/p_*	p/p_0
0.02	5.33	0.3	5.4×10^5	0.0138	3.6	1.11
0.0138	3.67	0.34	6.1×10^5	0.0135	3.25	1.23
0.0135	3.59	0.35	6.3×10^5	0.0135	3.15	1.26

Thus the initial Mach number is 0.35, the exit Mach number is unity, and the exit pressure exceeds the atmospheric pressure. The temperature, density, and speed of sound at $M = 0.35$ are

$$T = 293 \text{ K}$$
$$\rho = 4.77 \text{ kg/m}^3$$
$$c = 342 \text{ m/s}$$

The mass-flow rate is

$$\dot{m} = (0.35)(342 \text{ m/s})(4.77 \text{ kg/m}^3)(0.785)(0.03^2 \text{ m}^2)$$
$$= 0.403 \text{ kg/m}^3 \qquad \blacktriangleleft$$

Isothermal flow

The analysis of isothermal flow in a constant area duct is simplified by the fact that the energy equation is simply

$$T = \text{const}$$

This also means that the speed of sound in the duct is constant.

The momentum equation for flow in the duct is given by Eq. (12-65):

$$kM^2 \frac{dV}{V} + \frac{dp}{p} + \frac{kfM^2 \, dx}{2D} = 0 \qquad (12\text{-}65)$$

Because the temperature is constant, the pressure is proportional to the density so [see Eq. (12-66)]

$$\frac{dp}{p} = \frac{d\rho}{\rho} \qquad (12\text{-}80)$$

Using the continuity equation, Eq. (12-61), to relate ρ and V, gives

$$\frac{d\rho}{\rho} = -\frac{dV}{V} \tag{12-81}$$

which, substituted back into Eq. (12-65), yields

$$\frac{dM}{dx} = \frac{f}{2D}\frac{kM^3}{1 - kM^2} \tag{12-82}$$

One notes that if the Mach number is less than $1/\sqrt{k}$, then the Mach number increases with distance while the opposite trend is noted for Mach numbers exceeding $1/\sqrt{k}$. Thus the Mach number must always approach $1/\sqrt{k}$ as compared to unity for adiabatic flows.

Mach number distribution along duct

The Mach number distribution in the duct is determined by integrating Eq. (12-82) for Mach number as a function of distance. Rewriting Eq. (12-82) as

$$dM\frac{(1 - kM^2)}{kM^3} = \frac{f\,dx}{2D} \tag{12-83}$$

or

$$\frac{dM}{kM^3} - \frac{dM}{M} = \frac{f\,dx}{2D} \tag{12-84}$$

we can integrate each side to obtain

$$\frac{1}{-2kM^2} - \ln M = \frac{fx}{2D} + \text{const} \tag{12-85}$$

Setting x_T as the maximum length at which $M = 1/\sqrt{k}$, the constant of integration becomes

$$-\frac{1}{2} - \ln\left(\frac{1}{\sqrt{k}}\right) - \frac{fx_T}{2D} = \text{const} \tag{12-86}$$

and substituted in Eq. (12-86) gives

$$\frac{f(x_T - x_M)}{D} = \ln(kM^2) + \frac{(1 - kM^2)}{kM^2} \tag{12-87}$$

The variation of $f(x_T - x_M)/D$ with kM^2 is plotted in Fig. 12-22.

Pressure variation

The pressure variation with Mach number is obtained from Eq. (12-81), which can be expressed as

$$\frac{dp}{p} = -\frac{dM}{M} \tag{12-88}$$

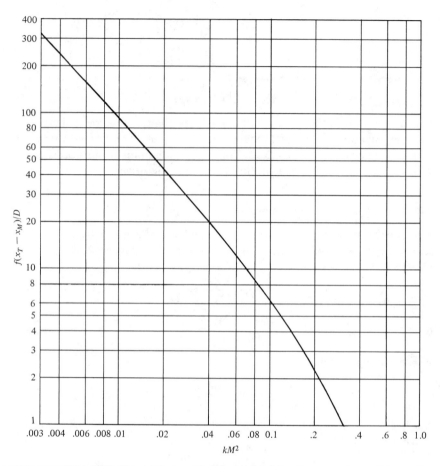

FIGURE 12-22 Variation of $f(x_T - x_M)/D$ with kM^2 for isothermal viscous flow in constant duct area.

and integrated to give

$$\ln p = -\ln M + \text{const} \qquad (12\text{-}89)$$

Letting p_T be the pressure corresponding to the maximum distance, we find the constant of integration and substitute it back into Eq. (12-89) to give

$$\frac{p_M}{p_T} = \frac{1}{\sqrt{k}\, M} \qquad (12\text{-}90)$$

Similar expressions can be found for the variation of density, total pressure and total temperature along the duct.

EXAMPLE 12-17   Methane is to be transported at 15°C in 50-cm commercial steel pipe that is 1 km long. The pressure at the pipe exit is 100 kPa. Determine the maximum flow rate through the pipe and the pressure at the pipe entrance.

Solution   From Table A-2, the value of k for methane is 1.31, the gas constant is 518 J/kg K, and the kinematic viscosity at 15°C is 1.59×10^{-5} m²/s.

The Mach number at the pipe exit is

$$M = \frac{1}{\sqrt{k}}$$

$$= \frac{1}{\sqrt{1.31}}$$

$$= 0.874$$

The corresponding velocity is

$$V = Mc$$
$$= 0.874 \times \sqrt{1.31 \times 518 \times 288}$$
$$= 386 \text{ m/s}$$

The corresponding mass flow, which is the maximum flow rate, is

$$\dot{m} = \rho A V$$

$$= \left(\frac{10^5}{518 \times 288}\right)(0.5)^2 \left(\frac{\pi}{4}\right)(386)$$

$$= 50.8 \text{ kg/s} \qquad \blacktriangleleft$$

The Reynolds number at the exit is

$$\text{Re} = \frac{VD}{\nu}$$

$$= \frac{(386)(0.5)}{1.59 \times 10^{-5}}$$

$$= 1.2 \times 10^7$$

Since ρV is constant along the pipe (continuity equation), and μ is constant (depends primarily on temperature), the Reynolds number is the same along the pipe; therefore, the friction factor does not change. Referring to Figs. 10-8 and 10-9, we find the friction factor is 0.012. Thus, the left-hand side of Eq. 12-87 has the value

$$\frac{f(x_T - x_M)}{D} = \frac{(0.012)(10^3)}{0.5}$$

$$= 24$$

Using Fig. 12-22 we find

$$kM^2 = 0.035$$

The pressure at the entrance is found using Eq. 12-90:

$$p_M = \frac{100 \text{ kPa}}{\sqrt{0.035}} = 535 \text{ kPa} \qquad \blacktriangleleft$$

In this section we have treated the adiabatic and isothermal flow of a viscous compressible gas in a constant area duct. The reader is referred to more specialized texts on compressible flow, such as Ref. 5, to study other types of flows such as the effect of heat addition due to chemical reaction.

PROBLEMS

12-1 Calculate the speed of sound in helium at 20°C.

12-2 Calculate the speed of sound in hydrogen at 68°F.

12-3 How fast (meters per second) will a sound wave travel in methane at 100°C?

12-4 How much faster will a sound wave propogate in helium than in nitrogen if the temperature of both gases is 15°C?

12-5 Determine the equation for the speed of sound in an ideal gas if the sound wave were an isothermal process.

12-6 The relationship between pressure and density with the propogation of a sound wave through a fluid is

$$p - p_0 = E_v \ln(\rho/\rho_0)$$

where p_0 and ρ_0 are reference densities and pressures (constants) and E_v is the bulk modulus of elasticity. Determine the equation for speed of a sound wave in terms of E_v and ρ. Calculate the sound speed for water with $\rho = 1000$ kg/m³ and $E_v = 2.20$ GN/m².

12-7 A Starfighter is flying at a Mach number of 2.0 at 10,000 m where the temperature is -44°C and the pressure is 30.5 kPa.
  (a) How fast is the aircraft traveling in kilometers per hour?
  (b) The total temperature is an estimate of surface temperature on the aircraft. What is the total temperature under these conditions?
  (c) Calculate also the total pressure.
  (d) If the airplane slows down, at what speed (kilometers per hour) will it be traveling to be in the transonic regime?

12-8 An airplane travels at 800 km/h at sea level where the temperature is 15°C. How fast would the airplane be flying at the same Mach number at an altitude where the temperature is -40°C?

12-9 An airplane flies at a Mach number of 0.95 at 10,000-m altitude where the static temperature is -44°C and the pressure is 30 kPa abs. If the lift coefficient of the wing is 0.05, determine the wing loading (lift force/wing area).

12-10 An object is immersed in an air flow with a static pressure of 200 kPa abs, a static temperature of 20°C, and a velocity of 200 m/s. What is the pressure and temperature at the stagnation point?

12-11 An airflow at $M = 0.3$ passes through a conduit with a cross-sectional area of 65 cm². The total absolute pressure is 340 kPa and the total temperature is 10°C. Calculate the mass-flow rate through the conduit.

12-12 Oxygen flows from a reservoir in which the temperature is 200°C and the pressure is 300 kPa abs. Assuming isentropic flow, calculate the velocity, pressure, and temperature where the Mach number is 0.8.

12-13 One problem in creating high Mach-number flows is condensation of the oxygen component in the air when the temperature reaches 50 K. If the temperature of the reservoir is 300 K and the flow is isentropic, at what Mach number will condensation of the oxygen occur?

12-14 Hydrogen flows from a reservoir where the temperature is 20°C and the pressure is 500 kPa abs to a 2-cm-diameter section where the velocity is 300 m/s. Assuming isentropic flow, calculate the temperature, pressure, Mach number, and mass flow rate at the 2-cm-diameter section.

12-15 The total pressure in a Mach 2 wind tunnel operating with air is 800 kPa. A 1-cm-diameter sphere, positioned in the wind tunnel, has a drag coefficient of 0.95. Calculate the drag force on the sphere.

12-16 Using Eq. (12-27), develop an expression for the pressure coefficient at stagnation conditions, that is, $C_p = (p_t - p)/(\frac{1}{2}\rho V^2)$, in terms of Mach number and ratio of specific heats, $C_p = f(k, M)$, and evaluate C_p at $M = 0$, 2, and 4 for $k = 1.4$. What would its value be for incompressible flow?

12-17 For low velocities, the total pressure is only slightly larger than the static pressure. Thus, one can write $p_t/p = 1 + \epsilon$ where ϵ is a small positive number ($\epsilon \ll 1$). Using this approximation, show that Eq. (12-32) reduces to $M = [2(p_t/p - 1)/k]^{1/2}$ as $\epsilon \to 0$ (M $\to$ 0).

12-18 A normal shock wave exists in a 500 m/s stream of nitrogen with a static temperature of -40°C and static pressure of 70 kPa. Calculate the Mach number, pressure, and temperature downstream of the wave and the entropy increase across the wave.

12-19 A normal shock wave exists in a $M = 2$ stream of air with a static temperature and pressure of 45°F and 30 psia, respectively. Calculate the Mach number, pressure, and temperature downstream of the shock wave.

12-20 A Pitot-static tube is used to measure the Mach number on a supersonic aircraft. The tube, because of its bluntness, effects a normal shock wave as shown. The absolute total pressure downstream of the shock wave (p_{t_2}) is 150 kPa. The static pressure of the free stream ahead of the shock wave (p_1) is 50 kPa. Determine the Mach number (M_1) graphically.

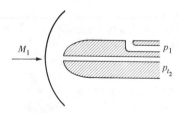

12-21 A shock wave occurs in a methane stream in which the Mach number is 2, the static pressure is 100 kPa abs, and the static temperature is 20°C. Determine the downstream Mach number, static pressure, static temperature, and density.

12-22 The Mach number downstream of a shock wave in helium is 0.7 and the static temperature is 100°C. Calculate the velocity upstream of the wave.

12-23 Show that the lowest Mach number possible downstream of normal shock is

$$M_2 = \sqrt{\frac{k - 1}{2k}}$$

and that the largest density ratio possible is

$$\frac{\rho_2}{\rho_1} = \frac{k + 1}{k - 1}$$

12-24 Show for a weak wave, $(M \simeq 1)$ that the Mach number downstream of the wave is approximated by

$$M_2^2 = 2 - M_1^2$$

(*Hint*: Let $M_1^2 = 1 + \epsilon$ where $\epsilon \ll 1$ and expand Eq. (12-41) in terms of ϵ.)

12-25 A truncated nozzle with an exit area of 5 cm² is used to measure an air-mass flow of 0.25 kg/s. The static temperature of the air at the exit is 20°C and the back pressure is 100 kPa. Determine the total pressure.

12-26 A truncated nozzle with a 12-cm² exit area is supplied from a helium reservoir in which the absolute pressure is first 130 kPa then 350 kPa. The temperature in the reservoir is 28°C and the back pressure is 100 kPa. Calculate the mass-flow rate of helium for the two reservoir pressures.

12-27 A truncated nozzle is used to meter the mass flow of air in a pipe. The area of the nozzle is 3 cm². The total pressure and total temperature measured upstream of the nozzle in the pipe is 200 kPa abs and 20°C. The pressure downstream of the nozzle (back pressure) is 90 kPa abs. Calculate the mass flow rate.

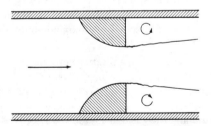

12-28 The truncated nozzle shown in Prob. 12-27 is used to monitor the mass flow rate of methane. The area of the nozzle is 3 cm² and the area of the pipe is 12 cm². The upstream total pressure and total temperature are 150 kPa abs and 30°C. The back pressure is 100 kPa. (*a*) Calculate the mass flow rate of methane. (*b*) Calculate the mass flow rate assuming Bernoulli's equation is valid, the density being the density of the gas at the nozzle exit.

12-29 A sampling probe is used to draw gas samples from a gas stream for analysis. In sampling, it is important that the velocity entering the probe equal the velocity of the gas stream (isokinetic condition). Consider the sampling probe shown below, which has a truncated nozzle inside the probe to control the mass flow rate. The probe has an inlet diameter of 4 mm and a truncated nozzle diameter of 2 mm. The probe is in a hot-air stream with a static temperature of 600°C, a static pressure of 100 kPa abs, and a velocity of 50 m/s. Calculate the pressure in the probe (back pressure) to maintain the isokinetic sampling condition.

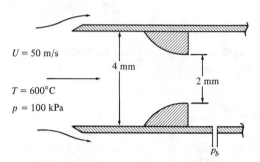

$U = 50$ m/s

4 mm

2 mm

$T = 600°C$

$p = 100$ kPa

p_b

12-30 A wind tunnel is designed to have a Mach number of 3.0, a static pressure of 1.5 psia, and static temperature of $-10°F$ in the test section. Determine the area ratio of the nozzle required and the reservoir conditions that must be maintained if air is to be used.

12-31 A Laval nozzle is to be designed to operate supersonically and expand ideally to an absolute pressure of 30 kPa. If the stagnation pressure in the nozzle is 1 MPa, calculate the nozzle area ratio required. Determine the nozzle throat area for a mass flow of 5 kg/s and stagnation temperature 550 K. Assume that the gas is nitrogen.

12-32 A rocket nozzle, with an area ratio of 5:1, is operating at a total absolute pressure of 1.3 MPa and exhausting to an atmospheric absolute pressure of 35 kPa. Determine if the nozzle is overexpanded, underexpanded, of ideally expanded. Assume that $k = 1.4$.

12-33 A rocket motor operates at an altitude where the atmospheric pressure is 20 kPa. The expansion ratio of the nozzle is 4 (exit area/throat area). The chamber pressure of motor (total pressure) is 1.2 MPa and the chamber temperature (total temperature) is 3,000°C. The ratio of specific heats of the exhaust gas is 1.2 and the gas constant is 400 J/kg K. The throat area of the rocket nozzle is 100 cm². (*a*) Determine the Mach number, density, pressure and velocity at the nozzle exit. (*b*) Determine the mass flow rate. (*c*) Calculate the thrust of the rocket using [see Eq. (6-13)]

$$T = \dot{m}V_e + (p_e - p_0)A_e$$

(*d*) What would the chamber pressure of the rocket have to be to have an ideally expanded nozzle? Calculate the rocket thrust under this condition.

12-34 A rocket motor is being designed to operate at sea level where the pressure is 100 kPa abs. The chamber pressure (total pressure) is 1.5 MPa and the chamber temperature (total temperature) is 3,300 K. The throat area of the nozzle is 10 cm². The ratio of specific heats (k) of the exhaust gas is 1.2 and the gas constant is 400 J/kg K. (*a*) Determine the nozzle expansion ratio to achieve an ideally expanded nozzle and determine the nozzle thrust under these conditions (see Prob. 12-33 for the thrust equation). (*b*) Determine the thrust obtained if the expansion ratio were reduced by 10% to achieve an underexpanded nozzle.

12-35 Air flows through a de Laval nozzle with an expansion ratio of 4. The total pressure of the air entering the nozzle is 200 kPa and the back pressure is 100 kPa. Determine the area ratio at which the shock wave occurs in the expansion section of the nozzle. (*Hint*: This problem can be solved graphically by calculating the exit pressure corresponding to different shock locations and finding the location where the exit pressure is equal to the back pressure.)

12-36 A rocket nozzle has a configuration as shown below. The diameter of the throat is 4 cm and exit diameter is 8 cm. The half-angle of the expansion cone is 15°. Air flows into the nozzle with a total pressure of 250 kPa. The back pressure is 100 kPa. First determine the area ratio at which the shock occurs using an iterative or graphical method and then determine the shock wave's distance in centimeters from the throat.

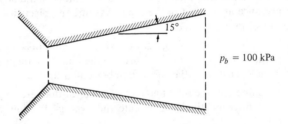

12-37 A normal shock wave occurs in a nozzle at an area ratio of 4:1. Determine the entropy increase if the gas is hydrogen.

12-38 Determine the Mach number and area ratio at which the dynamic pressure is maximized in a Laval nozzle with air. [*Hint*: Express q in terms of p and M and use Eq. (12-27) for p. Differentiate with respect to M and equate to zero.]

12-39 Consider airflow in the varying area channel shown below. Determine the

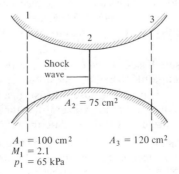

Mach number, static pressure, and stagnation pressure at station 3. Assume isentropic flow except for normal shock waves.

12-40 Determine the value of the atmospheric pressure necessary for the shock to position itself as shown.

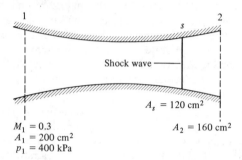

$M_1 = 0.3$
$A_1 = 200 \text{ cm}^2$
$p_1 = 400 \text{ kPa}$

$A_s = 120 \text{ cm}^2$
$A_2 = 160 \text{ cm}^2$

12-41 The velocity of air entering a 1-in.-diameter commercial steel pipe is 150 ft/sec. The static temperature and pressure are 67°F and 30 psia, respectively. Calculate the length of pipe needed to achieve sonic flow. Calculate also the pressure at the end of the pipe. Assume the viscosity of the air is independent of pressure.

12-42 The Mach number of air flowing out the exit of a 3-cm brass tube is 0.7. The total temperature of the airflow is 373 K and the atmospheric pressure is 100 kPa. How far upstream is the Mach number equal to 0.2.

12-43 The Mach numbers at the inlet and exit of a 0.5-in.-diameter pipe 20 ft long are 0.2 and 0.6, respectively. Calculate the average friction factor in the pipe for $k = 1.4$.

12-44 Oxygen flows through a 2.5-cm-diameter wrought-iron pipe 10 m long. It discharges to the atmosphere where the pressure is 100 kPa. If the absolute static pressure at the beginning of the pipe is 300 kPa and the total temperature is 293 K, calculate the mass flow through the pipe.

12-45 Repeat Prob. 12-44, but with a 500-kPa pressure at the beginning of the pipe.

12-46 An engineer is designing a piping system for airflow. The pipe must be 10 m long and fabricated from brass and carry a mass flow of 0.2 kg/s. The total temperature of the flow in the pipe is 373 K and the tolerable pressure loss is 140 kPa. The pipe discharges to a pressure of 100 kPa. Determine the diameter of the pipe.

12-47 A 10-ft-long pressure hose is to be connected to the outlet of the regulator valve on a nitrogen bottle. The hose must deliver 0.06 lbm/sec when the regulator pressure is 45 psia. The hose exhausts to a back pressure of 7 psia. The total temperature is 100°F. Assume that the hose has a roughness equivalent to galvanized iron and calculate the required hose diameter.

12-48 An air blower and pipe system is to be designed to convey agricultural products through a 20-cm steel pipe. The pipe is to be 200 m long and the outlet velocity is to be 50 m/s. If the pipe discharges into the atmosphere (100 kPa, 15°C) what will be the pressure, velocity, and density of the air at the inlet end of the pipe? Assume that the ratio of specific heat for the particle-laden flow is 1.4.

12-49 Methane is pumped into a 15-cm steel pipe at a pressure of 1 MPa and a temperature of 320 K $[\mu = 1.5(10^{-5} \text{ N} \cdot \text{s/m}^2)]$ and with a velocity of 20 m/s. What is the pressure 3,000 m downstream?

12-50 Hydrogen is transported in an underground pipe line. The pipe is 50 m long, 10 cm in diameter, and maintained at a temperature of 15°C. The initial pressure and velocity are 200 kPa and 200 m/s. Determine the pressure drop in the pipe.

12-51 Helium flows in a 5-cm brass tube 100 m long and maintained at a temperature of 15°C. The entrance pressure is 120 kPa and the exit pressure is 100 kPa. Determine the mass flow rate in the pipe. This problem requires an iterative solution.

REFERENCES

1. Chapman, Alan J., and Walker, William F. *Introductory Gas Dynamics.* Holt, Rinehart and Winston, Inc., New York, 1971.

2. Crocco, L. "One-dimensional Treatment of Steady Gas Dynamics." in H. W. Emmons (ed.). *Fundamentals of Gas Dynamics,* vol. 3. Princeton University Press, Princeton, N.J., 1958.

3. Mises, R. von. *Mathematical Theory of Compressible Fluid Flow.* Academic Press, Inc., New York. 1958.

4. Oswatitsch, K. *Gas Dynamics.* Trans. G. Kuerti. Academic Press, Inc., New York, 1956.

5. Owczarek, J. A. *Fundamentals of Gas Dynamics.* McGraw-Hill Book Company, New York, 1971.

6. Pope, Alan. *Aerodynamics of Supersonic Flight.* Pitman Publishing Corporation, New York, 1950.

7. Shapiro, A. H. *The Dynamics and Thermodynamics of Compressible Fluid Flow.* The Ronald Press Company, New York, 1953, 2 vols.

8. Thompson, P. A. *Compressible-Fluid Dynamics.* McGraw-Hill Book Company, New York, 1971.

9. Van Wylen, G. J., and Sonntag, R. E. *Fundamentals of Classical Thermodynamics.* John Wiley & Sons, Inc., New York, 1965.

Flow-measuring instruments such as these used by the U.S. National Weather Service continuously monitor atmospheric flows, playing a major role in modern forecasting procedures.

13 FLOW MEASUREMENTS

THE MOST COMMON FLOW MEASUREMENTS are pressure, rate of flow, and velocity. In Chapter 3 several methods were presented for measurements of pressure and in Chapter 5 the basic theory of the stagnation and Pitot tubes for the measurement of velocity was given. In this chapter we will consider the Pitot, static, and stagnation tubes in more detail, and we will describe other ways of measuring velocity. Finally, several methods will be presented for the measurement of the rate of flow of fluids.

13-1 INSTRUMENTS FOR THE MEASUREMENT OF VELOCITY AND PRESSURE

Stagnation tube

In Chapter 5, where the stagnation tube was first introduced, it was assumed that viscous effects were negligible. This assumption is valid for tubes of normal size when measuring moderately high velocity air or water. However, if the velocity to be measured is very low or if the tube is very small, viscous effects become significant and a correction must be applied to the basic equation. That is, as the Reynolds number decreases, the viscous effects become significant. This influence is shown in Fig. 13-1, where the pressure coefficient $C_p = (p - p_0)/(\rho V_0^2/2)$ is plotted as a function of the Reynolds number.

In Fig. 13-1 it is seen that if the Reynolds number for the circular stagnation tube is greater than 100, the error in measured velocity is less than 1%. Also in Fig. 13-1 are data for stagnation tubes, flattened at the end, which are often used to measure boundary-layer velocities. By flattening the end of the tube, the velocity measurement can be made nearer the boundary than if a circular tube were used. For these flattened tubes, note that the pressure coefficient remains near unity for a Reynolds number as low as 30.

Flow direction with stagnation-type tubes

When the direction of flow is not known, simple pressure-type *yaw* meters can be used to sense the flow direction. Three types of yaw meters are shown in Fig. 13-2. The first two can be used for two-dimensional flow where flow direction in only one plane needs to be found, and the third is used for the determination of flow direction in three dimensions. In all these devices, the tube is turned until the pressure on symmetrically

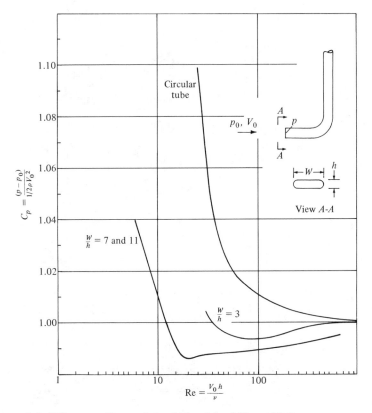

FIGURE 13-1   Viscous effects of C_p. [After Macmillan (12).]

opposite openings is equal. This pressure is sensed by connecting a differential pressure gage or manometer to the openings in the yaw meter. The flow direction is sensed when a null reading is indicated on the differential gage.

Static tube

At times it is desired to measure the static pressure in a flowing fluid. This is accomplished by sensing the pressure at a point along the static tube where the pressure is the same as the free-stream pressure. Such a tube is shown in Fig. 13-3. This is like the Pitot tube except that the stagnation pressure tap is omitted. The placement of the holes along the probe is important for sensing the static pressure because the rounded nose on the tube causes some decrease of pressure along the tube and the downstream stem causes an increase in pressure in front of it. Hence the location for

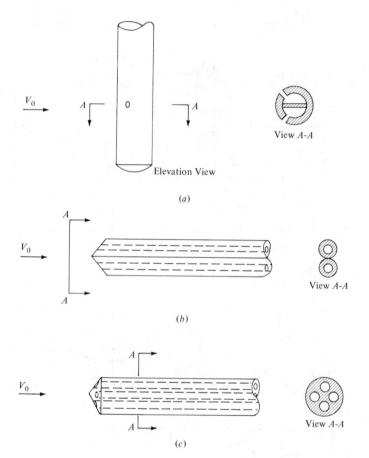

FIGURE 13-2   Various types of yaw meters. (*a*) Cylindrical-tube yaw meter. (*b*) Two-tube yaw meter. (*c*) Three-dimensional yaw meter.

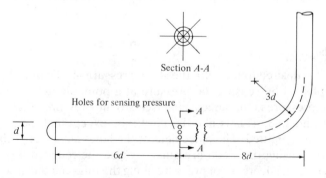

FIGURE 13-3   Static tube.

sensing the static pressure must be at the point where these two effects cancel each other. Experiments reveal that the optimum location is at a point approximately six diameters downstream of the front of the tube and eight diameters upstream from the stem.

The vane or propeller anemometer

The basic vane or propeller type of anemometer has been used in various applications for a number of years to measure the velocity of either gases or liquids. Basically, the type used for airflows, Fig. 13-4, consists of vanes (propeller blades) attached to a rotor that drives a low-friction gear train that, in turn, drives a pointer that indicates feet on a dial. Thus, if the anemometer is held in an airstream for 1 min and the pointer indicates a 300-ft change on the scale, the average airspeed is 300 ft/min.

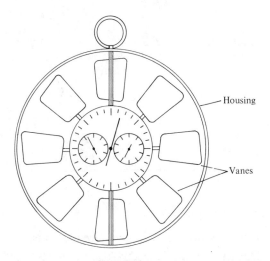

FIGURE 13-4   Vane anemometer for air flow.

Another type of vane anemometer is that used for measuring the velocity of flowing water. The usual commercial vane meter of this type is connected to an electrical circuit in such a manner that an electrical signal is triggered for a given number of revolutions. Thus the frequency of rotation of the vane is obtained, which is directly converted to velocity by a calibration curve.

Vane anemometers are also used inside pipes, where they are calibrated to indicate the flow rate directly in cubic meters per second, or any other desired units of flow rate.

Another common anemometer used in meteorological measurements is an ordinary propeller attached to the forward part of a wind vane, all of which is mounted on a mast. The propeller drives an electromagnetic generator, the voltage from which is proportional to the wind speed.

Cup anemometer

The most common type of cup anemometer is that used by meteorologists to measure wind velocity (Fig. 13-5); however, hydraulic engineers also use a cup-type anemometer to measure the velocity of flow in streams and rivers. Again, the frequency of rotation is directly related to the velocity of flow by appropriate calibration data.

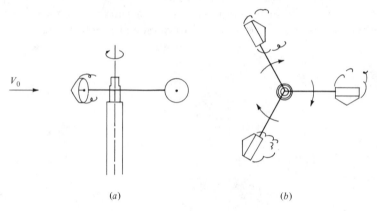

(a)   (b)

FIGURE 13-5   Cup anemometer.

Hot-wire and hot-film anemometers

The velocity-measuring devices we have already considered are suitable for measuring velocity that either is steady or changes slowly with time. However, if we want to measure the velocity fluctuations due to the eddies in turbulent flow, the response of the aforementioned instruments would be too slow to record the rapid changes in velocity. These instruments would also be too large for local velocity measurements, which one might wish to make in a thin boundary layer. The hot-wire anemometer is an instrument that is very sensitive to rapid fluctuations in velocity, and its sensing element is small so that its presence does not seriously disturb the nature of the surrounding flow. Another advantage of the hot-wire anemometer is that it is sensitive to low-velocity flows, a characteristic lacking in the Pitot tube. The main disadvantages of the hot-wire anemometer are its delicate nature (the sensor wire is easily broken) and its relatively high cost (approximately $1,000 and up).

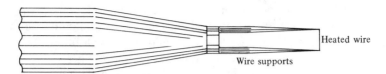

FIGURE 13-6   Probe for hot-wire anemometer (enlarged).

The basic principle of the hot-wire anemometer is described as follows: A very small diameter wire—the sensing element of the hot-wire anemometer—is welded to supports as shown in Fig. 13-6. In operation the wire is either heated by a fixed flow of electric current (the constant-current anemometer) or the wire is maintained at a constant temperature by adjusting the current (the constant-temperature anemometer).

A flow of fluid past the hot wire causes the wire to cool owing to convective heat transfer. In the constant-current anemometer, the cooling of the wire causes its resistance to change, and a corresponding voltage change occurs across the wire. Because the rate of cooling is a function of the speed of flow past the heated wire, the voltage across the wire is correlated with the flow velocity. The more popular type of anemometer, the constant temperature anemometer, operates by varying the current in such a manner as to keep the resistance (and temperature) constant. The flow of current is correlated with the speed of the flow: The higher the speed, the greater the current needed to maintain a constant temperature. Typically, the wires are 1 to 2 mm in length and heated to 150°C. The wires may be 10 μm in diameter and smaller, the time response improving with the smaller wire. The lag of the wire's response to a change in velocity (thermal inertia) can be compensated for more easily in constant-temperature anemometers than constant-current anemometers using modern electronic circuitry. The signal from the hot wire is processed electronically to give the desired information, such as mean velocity or the root-mean-square of the velocity fluctuation.

To illustrate the versatility of these instruments, note that the hot-wire anemometer can measure accurately gas-flow velocities from 30 cm/s to 150 m/s; it can measure fluctuating velocities with frequencies up to 100,000 Hz; and it has been used satisfactorily for both gases and liquids.

The single hot wire mounted normal to the mean-flow direction measures the fluctuating component of velocity in the mean-flow direction. Other probe configurations and electronic circuitry can be used to measure other components of velocity.

For velocity measurements in liquids or dusty gases where wire breakage is a problem, the hot-film anemometer is more suitable. This anemometer consists of a thin conducting metal film (less than 0.1 μm) mounted on a ceramic support, which may be 50 μm in diameter. The hot film operates in the same fashion as the hot wire. Recently the split film has been introduced, which consists of two semicylindrical films mounted on the same cylindrical support and electrically insulated from each other. The split film provides both speed and directional information.

For more detailed information on the hot-wire and hot-film anemometer, see Refs. 1 and 8.

Laser-Doppler anemometer

The laser-Doppler anemometer (LDA) is a relatively new instrument for fluid-velocity measurements. The major advantage of the LDA over the Pitot probe and hot-wire anemometer is that the flow field is not disturbed by the presence of a probe or wire support. A further advantage is that the velocity is measured in a very small flow volume, providing excellent spatial resolution.

There are several different configurations for the LDA depending on the properties and accessibility of the flow. One configuration, known as the dual-beam mode, is shown in Fig. 13-7. The laser beam, which is a highly coherent, monochromatic light source, is first split into two parallel

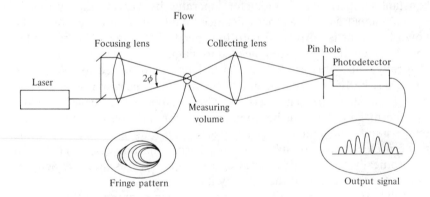

FIGURE 13-7   Dual-beam laser-Doppler anemometer.

beams and then passed through a converging lens. The point where the two beams cross is the measuring volume, which might be best described as an ellipsoid typically 0.3 mm in diameter and 2 mm long, illustrating the excellent spatial resolution achievable. The interference of the two beams generates a series of light and dark fringes in the measuring volume perpendicular to the plane of the two beams. As a particle passes through the fringe pattern, light is scattered and a portion of the scattered light passes through the collecting lens toward the photodetector. A typical signal obtained from the photodetector is shown on the figure.

It can be shown from optics theory that the spacing between the fringes is given by

$$\Delta x = \frac{\lambda}{2 \sin \phi}$$

where λ is the wavelength of the laser beam and ϕ is the half-angle between the crossing beams. By suitable electronic circuitry, the frequency of the signal (f) is measured, so the velocity is given

$$U = \frac{\Delta x}{\Delta t} = \frac{\lambda f}{2 \sin \phi}$$

The operation of the laser-Doppler anemometer depends on the presence of particles in the flow to scatter the light, particles sufficiently small so that they always move at the fluid velocity. Typically, in liquid flows the impurities of the fluid serve as scattering centers. In gaseous flows, however, it is sometimes necessary to "seed" the flow with small particles. Often smoke is used for this seeding.

The laser-Doppler anemometer has specific application in flow measurements requiring fine spatial resolution and minimum disturbance of the flow. Currently, many industrial applications of the LDA are being found to measure velocities of materials such as melted glass, in which it is not possible to mount a mechanical probe. The LDA is also used in biomedical research dealing with the flow of blood.

Marker methods

The basic procedure for the marker method for determining velocity stems from the basic concept of the pathline (see Chapter 4). If identifiable particles are placed in the stream, then by analyzing the motion of these particles one can deduce the velocity of the flow itself. Of course, this requires that the markers follow virtually the same path as the surrounding fluid elements. It means, then, that the marker must have nearly the same density as the fluid or that it must be so small that the relative motion with respect to the fluid is nil. Thus for water flow it is common to use colored droplets from a liquid mixture that has nearly the same density as the water. For example, Macagno (11) used a mixture of n-butyl phthalate and xylene with a bit of white paint to yield a mixture that had the same density as water and could be photographed effectively. It is also possible to use solid particles for markers, such as plastic beads, which have densities near that of the liquid being studied.

Hydrogen bubbles have also been used for markers in water flow. Here electrodes placed in flowing water cause small bubbles to be formed and swept downstream, thus revealing the motion of the fluid. It is required that the wire be very small so that the resulting bubbles do not have a significant rise velocity with respect to the water. By pulsing the current through the electrodes, it is possible to add a time frame to the visualization technique, thus making it a useful tool for velocity measurements. Figure 13-8 shows patches of tiny hydrogen bubbles that were released with a pulsing action from noninsulated segments of a wire from the left of the picture. Flow is from left to right and the necked-down section of the flow passage has higher water velocity; therefore, the patches are longer in that region. Next to the walls the patches of bubbles are shorter, indicating less distance traveled per unit of time. Other details concerning the marker methods of flow visualization are described by Macagno (11).

FIGURE 13-8   Combined-time-streak markers (hydrogen bubbles); flow is from left to right. [After Kline (9).]

13-2   INSTRUMENTS AND PROCEDURES FOR MEASUREMENT OF FLOW RATE

Introduction

The methods of flow measurement, in a broad sense, can be classified as either direct or indirect. Direct methods involve the actual measurement of the quantity of flow (volume or weight) for a given time interval; indirect methods involve the measurement of a pressure change (or some other variable) which in turn is directly related to the rate of flow. Flow through *venturi meters*, *orifices*, and *flow nozzles* are all devices with which one employs indirect methods to measure the rate of flow in closed conduits. *Weirs* are devices with which one uses indirect means to obtain flow rates in open channels. Still another indirect meter is the *electromagnetic* flow meter, which operates on the principle that a voltage is generated when a conductor moves in a magnetic field. All these methods, as well as the *velocity-area integration* of flow measurement, will be discussed in this section.

Direct volume or weight measurements

One of the most accurate methods of obtaining liquid-flow rates is to collect a sample of the flowing fluid over a given period of time t. Then if the sample is weighed, the average weight rate of flow is W/t, where W is the weight of the sample. The volume of a sample can also be measured (usually in a calibrated tank), and from this the average volume rate of flow is given as $\forall/t$, where $\forall$ is the volume of the sample.

Velocity-area-integration method

If the velocity in a pipe is symmetrical, the distribution of the velocity along a radial line can be used to determine the volume rate of flow (discharge) in the pipe. The discharge is obtained by numerically or graphically integrating $V\, dA$ over the cross-sectional area of the pipe. Thus a Pitot tube or hot-wire-anemometer traverse across the flow section provides the primary data from which the discharge is evaluated. One procedure for evaluating this discharge is given in the next paragraph.

From test data of V versus r, one computes $2\pi Vr$ for various values of r; then when $2\pi Vr$ versus r is plotted, the area under the resulting curve, Fig. 13-9, will yield the discharge. This is so because $dQ = V\, dA =$

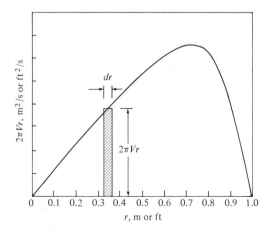

FIGURE 13-9 Graphical integration of $V\, dA$ in a pipe.

$V(2\pi r\ dr)$, which is given by an elemental strip of area in Fig. 13-9. Hence the total area will yield the total discharge.

EXAMPLE 13-1   The data given in the table are for a velocity traverse of airflow in a 100-cm-diameter pipe. What is the volume rate of flow in cubic meters per second?

r, cm	V, m/s
0.00	50.0
5.00	49.5
10.00	49.0
15.00	48.0
20.00	46.5
25.00	45.0
30.00	43.0
35.00	40.5
40.00	37.5
45.00	34.0
47.50	25.0
50.00	0.0

Solution   First make a table as shown of $2\pi Vr$ versus r.

r, cm	$2\pi Vr$, m²/sec
0.00	0.0
5.00	15.5
10.00	30.8
15.00	45.2
20.00	58.4
25.00	70.7
30.00	81.1
35.00	89.1
40.00	94.2
45.00	96.1
47.50	74.6
50.00	0.0

Now plot $2\pi Vr$ versus r. This plot is shown on the next page. The area under the curve is measured to be 29.3 m³/s; hence

$$Q = 29.3 \text{ m}^3/\text{s} \qquad \blacktriangleleft$$

The foregoing procedure involving the velocity-area-integration method is applicable to pipes where the velocity distribution is symmetrical with the axis of the pipe. However, even for flows that are unsymmetrical, it should be obvious that by summing $V \, \Delta A$ over a flow section one can obtain the total flow rate. Such a procedure is commonly used to obtain the discharge in streams and rivers.

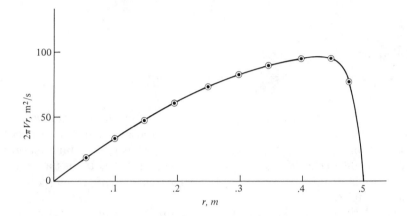

Orifice

A restricted opening through which fluid flows is an *orifice*, and if the geometric characteristics of the orifice plus the properties of the fluid are known, then the orifice can be used to measure flow rates. Consider flow through the sharp-edged pipe orifice shown in Fig. 13-10. It is seen that the streamlines continue to converge a short distance downstream of the plane of the orifice; hence, the minimum-flow area is actually smaller than

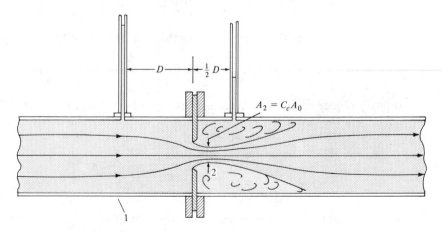

FIGURE 13-10   Flow through a pipe orifice.

the area of the orifice. To relate the minimum-flow area, often called the contracted area of the jet or *vena contracta,* to the area of the orifice A_o, we use the contraction coefficient, which is defined as

$$A_j = C_c A_o$$

$$C_c = \frac{A_j}{A_o}$$

Then, for a circular orifice,

$$C_c = \frac{(\pi/4)d_j^2}{(\pi/4)d^2} = \left(\frac{d_j}{d}\right)^2$$

Because d_j and d_2 are identical, we also have $C_c = (d_2/d)^2$. At low values of the Reynolds number, C_c is a function of the Reynolds number; however, at high values of the Reynolds number, C_c is only a function of the geometry of the orifice. For d/D ratios less than 0.3, C_c has a value of approximately 0.62; however, as d/D is increased to 0.8, C_c increases to a value of 0.72.

The derivation of the discharge equation for the orifice is started by writing the Bernoulli equation between section 1 and section 2 in Fig. 13-10. We then have

$$\frac{p_1}{\gamma} + \frac{V_1^2}{2g} + z_1 = \frac{p_2}{\gamma} + \frac{V_2^2}{2g} + z_2$$

Now, V_1 is eliminated by means of the continuity equation $V_1 A_1 = V_2 A_2$; and solving for V_2 gives

$$V_2 = \left\{ \frac{2g[(p_1/\gamma + z_1) - (p_2/\gamma + z_2)]}{1 - (A_2/A_1)^2} \right\}^{1/2} \tag{13-1a}$$

However, $A_2 = C_c A_o$ and $h = p/\gamma + z$; so that Eq. (13-1a) reduces to

$$V_2 = \sqrt{\frac{2g(h_1 - h_2)}{1 - C_c^2 A_o^2/A_1^2}} \tag{13-1b}$$

Our primary objective is to obtain an expression for discharge in terms of the h_1, h_2, and the geometric characteristics of the orifice. The discharge is given by $V_2 A_2$; hence, when we multiply both sides of Eq. (13-1b) by $A_2 = C_c A_o$ we obtain the desired result:

$$Q = \frac{C_c A_o}{\sqrt{1 - C_c^2 A_o^2/A_1^2}} \sqrt{2g(h_1 - h_2)} \tag{13-2}$$

Equation (13-2) is the discharge equation for the flow of an incompressible inviscid fluid through an orifice. However, this is only valid at relatively high Reynolds numbers. For low and moderate values of the Reynolds number, viscous effects are significant and an additional coefficient must be applied to the discharge equation to relate the ideal to the actual flow. This is called the *coefficient of velocity* C_v; thus for viscous flow in an orifice we have the following discharge equation:

$$Q = \frac{C_v C_c A_o}{\sqrt{1 - C_c^2 A_o^2/A_1^2}} \sqrt{2g(h_1 - h_2)}$$

The product $C_v C_c$ is called the *discharge coefficient* C_d, and the combination $C_v C_c/(1 - C_c^2 A_o^2/A^2)^{1/2}$ is called the *flow coefficient K*. Thus we have $Q = KA_o\sqrt{2g(h_1 - h_2)}$, where

$$K = \frac{C_d}{\sqrt{1 - C_c^2 A_o^2/A_1^2}}$$

If Δh is defined as $h_1 - h_2$, then the final form of the discharge equation for an orifice reduces to

$$Q = KA_o\sqrt{2g\,\Delta h} \tag{13-3}$$

Note that if a differential pressure transducer is connected across the orifice, the transducer will sense a change in pressure that is equivalent to $\gamma\,\Delta h$. Therefore, in this application one simply uses $\Delta p/\gamma$ in place of Δh in Eq. (13-3) and in the parameter at the top of Fig. 13-11. Experimentally determined values of K as a function of d/D and Reynolds number based upon orifice size, $4Q/\pi d\nu$, are given in Fig. 13-11. If Q is given, Re_d is equal to $4Q/\pi d\nu$, K is obtained from Fig. 13-11 (using the vertical lines and the bottom scale), and Δh is then computed from Eq. (13-3). However, we are often confronted with the problem of determining the discharge Q when a certain value of Δh is given. When Q is to be determined we do not have a direct way to obtain K by entering Fig. 13-11 with Re because Re is a function of the flow rate, which is still unknown. Hence, another scale, which does not involve Q, is constructed on the graph of Fig. 13-11. The variables for this scale are obtained in the following manner: because $\mathrm{Re}_d = 4Q/\pi d\nu$ and $Q = K\pi d^2/4\sqrt{2g\,\Delta h}$, we can write Re_d in terms of Δh as

$$\mathrm{Re}_d = K\sqrt{2g\,\Delta h}\,\frac{d}{\nu}$$

or

$$\frac{\mathrm{Re}_d}{K} = \sqrt{2g\,\Delta h}\,\frac{d}{\nu}$$

Thus the slanted lines and the top scale in Fig. 13-11 are used when Δh is known and the flow rate is to be determined.

The literature on orifice flow contains numerous discussions concerning the optimum placement of pressure taps on both the upstream and

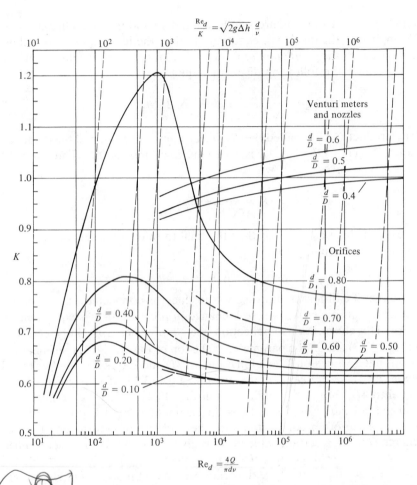

$$\frac{\text{Re}_d}{K} = \sqrt{2g\Delta h}\ \frac{d}{\nu}$$

$$\text{Re}_d = \frac{4Q}{\pi d\nu}$$

FIGURE 13-11 Flow coefficient K and Re_d/K versus the Reynolds number for orifices, nozzles, and venturi meters. [After Johansen (5) and ASME (2).]

downstream side of the orifice. The data given in Fig. 13-11 are for "corner taps." That is, on the upstream side the pressure readings were taken immediately upstream of the plate orifice (at the corner of the orifice plate and the pipe wall) and the downstream tap was at a similar downstream location. However, note that pressure data from flange taps (1 in. upstream and 1 in. downstream) and from the taps shown in Fig. 13-10 all yield virtually the same values for K—the differences are no greater than the deviations involved in reading Fig. 13-11. For more precise values of K with specific types of taps, see the ASME report on fluid meters (1).

EXAMPLE 13-2   A 15-cm orifice is located in a horizontal 24-cm water pipe and a water-mercury manometer is connected to either side of the orifice. When the deflection on the manometer is 25 cm, what is the discharge in the system? Assume that the water temperature is 20°C.

Solution   The discharge is given by Eq. (13-3): $Q = KA_o\sqrt{2g\,\Delta h}$. To either enter Fig. 13-11 or use Eq. (13-3), we will need to first evaluate Δh, the change in piezometric head in meters of fluid that is flowing. This is obtained by applying the equation of hydrostatics to the manometer shown below.

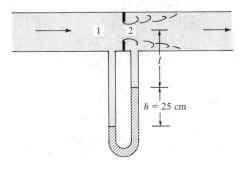

Writing the manometer equation from point 1 to point 2 we have

$$p_1 + \gamma_u l + \gamma_w h - \gamma_{Hg}h - \gamma_w l = p_2$$

Then, $$\frac{p_1 - p_2}{\gamma_w} = \Delta h = \frac{h(\gamma_{Hg} - \gamma_w)}{\gamma_w} = h\left(\frac{\gamma_{Hg}}{\gamma_w} - 1\right)$$

For this example,

$$\Delta h = 0.25 \text{ m} (13.6 - 1)$$
$$\Delta h = 3.15 \text{ m of water}$$

The kinematic viscosity of water at 20°C is 1.0×10^{-6} m²/s; so we now can compute $d\sqrt{2g\,\Delta h}/\nu$, the parameter that is needed to enter Fig. 13-11:

$$\frac{d\sqrt{2g\,\Delta h}}{\nu} = \frac{0.15 \text{ m } \sqrt{2(9.81 \text{ m/s}^2)(3.15 \text{ m})}}{1.0 \times 10^{-6} \text{ m}^2/\text{s}} = 1.2 \times 10^6$$

From Fig. 13-11 with $d/D = 0.625$, we read K to be 0.66 (interpolated). Hence,

$$Q = 0.66\, A_o\sqrt{2g\,\Delta h}$$

$$= 0.66\, \frac{\pi}{4}\, d^2 \sqrt{2(9.81 \text{ m/s}^2)(3.15 \text{ m})}$$

$$= 0.66\, (0.785)(0.15^2 \text{ m}^2)(7.87 \text{ m/s})$$

$$= 0.092 \text{ m}^3/\text{s}$$ ◀

EXAMPLE 13-3   An air-water manometer is connected to either side of an 8-in. orifice in a 12-in. water pipe. If the maximum flow rate is 5 cfs, what is the deflection on the manometer? The water temperature is 60°F.

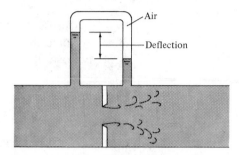

Solution   The Reynolds number $\mathrm{Re}_d = 4Q/\pi d\nu$ is first computed so that we may enter Fig. 13-11 to obtain the flow coefficient K; then K will be used in Eq. (13-3) to compute the deflection Δh.

$$\mathrm{Re}_d = \frac{4Q}{\pi d\nu} = \frac{(4)(5)}{3.14(8/12)(1.22)(10^{-5})}$$

$$= 7.8 \times 10^5$$

From Fig. 13-11 by interpolating between curves of $d/D = 0.6$ and $d/D = 0.8$ for $d/D = 8/12 = 0.667$, we read K to be approximately 0.68. Then from $Q = KA_0\sqrt{2g\,\Delta h}$ we obtain

$$\Delta h = \frac{Q^2}{2gK^2A_0^2} = \frac{25}{64.4(0.68^2)[(\pi/4)(8/12)^2]^2}$$

$$= 6.8 \text{ ft} \qquad \blacktriangleleft$$

The foregoing examples involved the determination of either Q or Δh for a given size of orifice. Another type of problem is the determination of orifice diameter for a given Q and Δh. For this type of problem a trial-and-error procedure is required. Because one knows the approximate value of K, that is guessed first. Then the diameter is solved for, after which a better value of K can be determined, and so on.

Venturi meter

The orifice is a simple and accurate device for the measurement of flow; however, the head loss for the orifice is quite large. It is like an abrupt enlargement in a pipe: $h_L = (V_2 - V_1)^2/2g$. A meter that operates on the same principle as the orifice but with a much smaller head loss is the venturi meter. The lower head loss results from streamlining the flow passage, as shown in Fig. 13-12. Such streamlining eliminates any jet

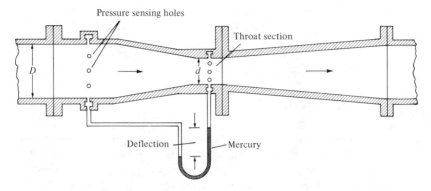

FIGURE 13-12   Typical venturi meter.

contraction beyond the smallest flow section; consequently, the coefficient of contraction has a value of unity and the basic discharge equation for the venturi meter is

$$Q = \frac{A_2 C_d}{\sqrt{1 - (A_2/A_1)^2}} \sqrt{2g(h_1 - h_2)} \tag{13-4}$$

$$= K A_2 \sqrt{2g\, \Delta h} \tag{13-5}$$

Note that the discharge equation for the venturi meter, Eq. (13-5), is the same as for the orifice, Eq. (13-3). However, K for the venturi meter approaches unity at high values of the Reynolds number and small d/D ratios. This trend can be seen in Fig. 13-11, where values of K for the venturi meter are plotted along with similar data for the orifice.

EXAMPLE 13-4   The pressure difference between the taps of a horizontal venturi meter carrying water is 35 kPa. If $d = 20$ cm and $D = 40$ cm, what is the discharge of water at 10°C?

Solution   First compute Δh. No elevation change occurs between the upstream section and the downstream section; therefore Δh is

$$\Delta h = \frac{\Delta p}{\gamma_{\text{water}}} = \frac{35,000 \text{ N/m}^2}{9,810 \text{ N/m}^3} = 3.57 \text{ m of water}$$

From the Appendix we find $\nu = 1.30(10^{-6})$ m²/s. Then we compute

$$\frac{d\sqrt{2g\, \Delta h}}{\nu} = \frac{0.20\sqrt{2(9.81)(3.57)}}{1.30(10^{-6})}$$

$$= 1.29 \times 10^6$$

Then $K = 1.02$

Now we compute the discharge

$$Q = 1.02A_2\sqrt{2g\,\Delta h}$$
$$= 1.02(0.785)(0.20^2)\sqrt{2(9.81)(3.57)}$$
$$= 0.269 \text{ m}^3/\text{s} \qquad \blacktriangleleft$$

Flow nozzles

The plate orifice requires a sharp upstream edge to yield the flow coefficients as given in Fig. 13-11. If the upstream edge becomes rounded because of chemical or mechanical action, the orifice will not be a reliable flow-measuring device (C_d will increase). Hence the flow nozzle, Fig. 13-13, has the advantage that it is less susceptible to wear. The flow nozzle has approximately the same flow coefficients as the venturi meter when the pressure taps are connected as in Fig. 13-13. However, the overall head loss across a flow nozzle will be like that for the sharp-edged orifice. That is, the overall head loss is that of an abrupt expansion.

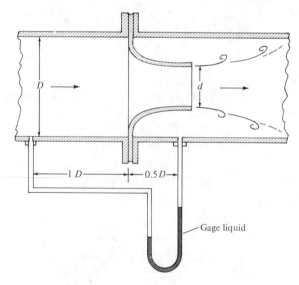

FIGURE 13-13   Typical flow nozzle.

Electromagnetic flow meter

All the flow meters introduced to this point require that some sort of obstruction be placed in the flow. The obstruction may be the rotor of a vane type of meter or the reduced cross section of an orifice or venturi meter. A meter that neither obstructs the flow nor requires pressure taps, which are subject to clogging, is the electromagnetic flow meter. Its basic principle is that a conductor that moves in a magnetic field produces an electromotive force. Hence liquids having a degree of conductivity will generate a voltage between the electrodes as in Fig. 13-14, and this voltage will be proportional to the velocity of flow in the conduit. It is interesting to note that the basic principle of the electromagnetic flow meter was investigated by Faraday in 1832; however, practical use of the principle was not made until approximately a century later, when it was used to measure blood flow. Recently, with the need for a meter to measure the flow of liquid metal in nuclear reactors and with the advent of sophisticated electronic signal detection, the meter has found extensive commercial use.

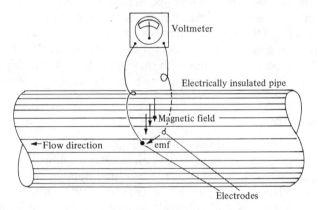

FIGURE 13-14　Electromagnetic flow meter.

The main advantages of the electromagnetic flow meter is that the output signal varies linearly with the flow rate and that the meter causes no resistance to the flow. The major disadvantages are its high cost and its unsuitability for gas flow.

For a summary of the theory and application of the electromagnetic flow meter, the reader is referred to Shercliff (15). This reference also includes a comprehensive bibliography on the subject.

Ultrasonic flowmeter

Another form of nonintrusive flowmeter that is used in diverse applications ranging from blood-flow measurement to open-channel flow is the ultrasonic flowmeter. Basically, there are two different modes of operation for ultrasonic flowmeters. One mode involves measuring the difference in travel time for a sound wave traveling upstream and downstream between two measuring stations. The difference in travel time is proportional to flow velocity. The second mode of operation is based on the Doppler effect. When an ultrasonic beam is projected into an inhomogeneous fluid, some acoustic energy is scattered back to the transmitter at a different frequency (Doppler shift). The measured frequency difference is related directly to the flow velocity.

Turbine flowmeter

The turbine flowmeter consists of a wheel with a set of curved vanes (blades) mounted inside a duct. The volume rate of flow through the meter is related to the rotational speed of the wheel, and this rotational rate is generally measured by a blade passing an electromagnetic pickup mounted in the casing. The meter must be calibrated for the flow conditions of interest. The turbine meter is versatile in that it can be used for either liquids or gases. It has an accuracy of better than 1% over a wide range of flow rates and operates with small head loss. The turbine meter is used extensively in monitoring flow rates in fuel-supply systems.

Vortex flowmeter

The vortex flowmeter consists of a cylinder mounted across the duct, which sheds vortices and gives rise to an oscillatory flow field. By proper design of the cylindrical element, the Strouhal number for vortex shedding will be constant for Reynolds numbers from 10^4 to 10^6. Over this flow range the fluid velocity and volume flow rate are directly proportional to the frequency of oscillation, which can be measured by several different methods. An advantage of this meter is that it has no moving parts (reliability) but it does give rise to a head loss comparable to other obstruction-type meters.

Rotameter

The rotameter consists of a vertical tapered tube through which the fluid flows upward, and inside of which is located the rotor or active element of the meter (Fig. 13-15). Vanes cause the rotor to rotate slowly about the axis of the tube, thus keeping it centered in the tube. Because the velocity is lower at the top of the tube (greater flow section here) than at the bottom, the rotor will seek a neutral position where the drag on the rotor just balances its weight. Thus the rotor will "ride" higher or lower in the tube depending upon the rate of flow. A calibrated scale on the side of the tube then indicates the rate of flow. Although venturi and orifice meters have better accuracy (approximately 1% of full scale) than the rotameter

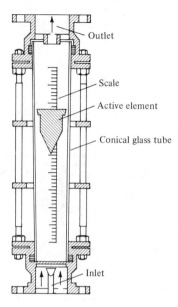

FIGURE 13-15   Rotameter.

(approximately 5% of full scale), the rotameter offers other advantages such as simplicity of design.

Rectangular weir

A *weir* is an obstruction in an open channel over which liquid flows. The discharge over the weir is a function of the weir geometry and of the *head* on the weir. Consider flow over the weir in the rectangular channel shown in Fig. 13-16. The head H on the weir is defined as the vertical distance between the weir crest and the liquid surface taken far enough upstream of the weir so that local free-surface curvature is avoided (see Fig. 13-16).

The basic discharge equation for the weir is derived by integrating $V \, dA = VL \, dh$ over the total head on the weir. Here L is the length of the weir and V is the velocity at any given distance below the free surface h. Neglecting streamline curvature and assuming negligible velocity of approach upstream of the weir, we obtain an expression for V by writing the Bernoulli equation between a point upstream of the weir and a point in the plane of the weir (see Fig. 13-17). This equation is

$$\frac{p_1}{\gamma} + H = (H - h) + \frac{V^2}{2g} \tag{13-6}$$

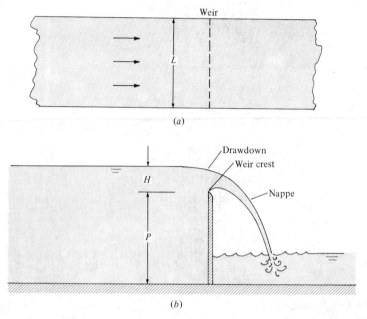

FIGURE 13-16   Definition sketch for a sharp-crested weir. (*a*) Plan view. (*b*) Elevation view.

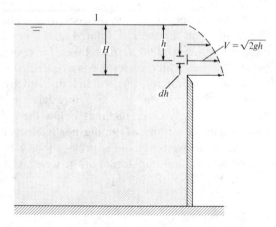

FIGURE 13-17   Theoretical velocity distribution over a weir.

Here the reference elevation is the elevation of the crest of the weir and the reference pressure is atmospheric pressure; therefore, $p_1 = 0$ and Eq. (13-6) reduces to

$$V = \sqrt{2gh}$$

Then $dQ = \sqrt{2gh}\ L\ dh$ and the discharge equation becomes

$$Q = \int_0^H \sqrt{2gh}\ L\ dh$$

$$= \tfrac{2}{3} L \sqrt{2g}\ H^{3/2} \tag{13-7}$$

In the actual case of flow over a weir, the streamlines converge downstream of the plane of the weir and viscous effects are not entirely absent; consequently, a discharge coefficient C_d must be applied to the basic expression on the right side of Eq. (13-7) to bring the theory in line with the actual flow rate. Thus, we have

$$Q = \tfrac{2}{3} C_d \sqrt{2g}\ LH^{3/2}$$

$$= K \sqrt{2g}\ LH^{3/2} \tag{13-8}$$

For low-viscosity liquids, the flow coefficient K is primarily a function of the relative head on the weir, H/P. An empirically determined equation for K adapted from Kindswater (6) is given as

$$K = 0.40 + 0.05 \frac{H}{P} \tag{13-9}$$

This is valid up to an H/P value of 10 as long as the weir is well ventilated so that atmospheric pressure prevails on both the top and bottom of the weir nappe.

EXAMPLE 13-5   The head on a 60-cm-high rectangular weir in a 1.3-m-wide rectangular channel is measured to be 21 cm. What is the discharge of water over the weir?

Solution

$$Q = K\sqrt{2g}\ LH^{3/2} \qquad \text{where } K = 0.40 + 0.05 \frac{H}{P}$$

hence, $K = 0.40 + 0.05 \left(\dfrac{21}{60}\right) = 0.417$

Then $Q = 0.417\sqrt{2(9.81)}\ (1.3)(0.21^{3/2})$

$$= 0.23\ \text{m}^3/\text{s} \qquad \blacktriangleleft$$

When the rectangular weir does not extend the entire distance across the channel, as in Fig. 13-18, additional end contractions occur; therefore,

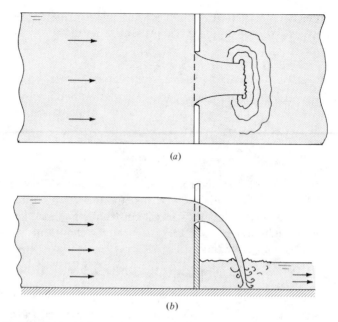

FIGURE 13-18   Rectangular weir with end contractions. (*a*) Plan view. (*b*) Elevation view.

K will be smaller than for the weir without end contractions. The reader is referred to King and Brater (7) for additional information on weir coefficients.

Triangular weir

A definition sketch for the triangular weir is shown in Fig. 13-19. The primary advantage of the triangular weir is that it has a higher degree of accuracy over a much wider range of flow than does the rectangular weir, because the average width of the flow section increases as the head increases.

The basic discharge equation for the triangular weir is derived in the same manner as that for the rectangular weir. The differential discharge $dQ = V\, dA = VL\, dh$ is integrated over the total head on the weir. Thus we have

$$Q = \int_0^H \sqrt{2gh}\, (H - h)\, 2 \tan\left(\frac{\theta}{2}\right) dh$$

which integrates to

$$Q = \frac{8}{15} \sqrt{2g} \tan\left(\frac{\theta}{2}\right) H^{5/2}$$

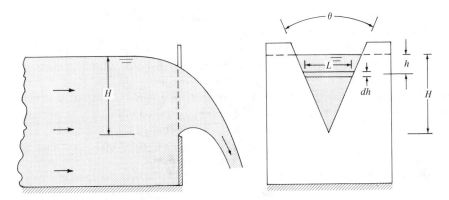

FIGURE 13-19   Definition sketch for triangular weir.

However, a coefficient of discharge must still be used with the basic equation: Hence, we have

$$Q = \frac{8}{15} C_d \sqrt{2g} \tan\left(\frac{\theta}{2}\right) H^{5/2} \qquad (13\text{-}10)$$

Experimental results with water flow over weirs with $\theta = 60°$ and for $H > 2$ cm indicate that C_d has a value of 0.58. Hence, the discharge equation for the triangular weir with these limitations is given by

$$Q = 0.179\sqrt{2g}\ H^{5/2} \qquad (13\text{-}11)$$

EXAMPLE 13-6   The head on a 60° triangular weir is measured to be 43 cm. What is the flow of water over the weir?

Solution   We use the discharge equation $Q = 0.179\sqrt{2g}\ H^{5/2}$. Hence

$$Q = 0.179 \times \sqrt{2 \times 9.81} \times (0.43)^{5/2} = 0.096 \text{ m}^3/\text{s} \qquad \blacktriangleleft$$

13-3   MEASUREMENT IN COMPRESSIBLE FLOW

Many of the measuring techniques used to determine the velocities and flow rates of liquids can be applied to compressible flows. However, since Bernoulli's equation is invalid for compressible flow, alternative relations must be used to correlate velocity and discharge with pressure difference.

Pressure measurements

Static-pressure measurements can be made using conventional static-pressure taps on a surface or a probe. However, if the boundary layer is disturbed by the presence of a shock wave in the vicinity of the pressure tap, the reading may not give the correct static pressure. The effect of the shock wave on the boundary layer is smaller if the boundary layer is turbulent, so an effort is sometimes made to trip the boundary layer and ensure a turbulent boundary layer in the region of the pressure tap.

The stagnation pressure can be measured with a stagnation tube aligned with the local velocity vector. If the flow is supersonic, however, a shock wave will form around the tip of the probe, as shown in Fig. 13-20, and the stagnation pressure measured is that downstream of the shock wave and not of the free stream. The stagnation pressure in the free stream can be calculated using the normal shock relationships provided the free-stream Mach number is known.

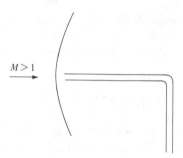

FIGURE 13-20   Total head tube in supersonic flow.

Mach number and velocity measurements

A Pitot-static tube can be used to measure Mach numbers in compressible flows. Taking the measured stagnation pressure as the total pressure, the Mach number in subsonic flows can be calculated from the total-to-static-pressure ratio according to Eq. (12-32), namely

$$M = \left\{ \frac{2}{k-1} \left[\left(\frac{p_t}{p} \right)^{(k-1)/k} - 1 \right] \right\}^{1/2} \qquad (12\text{-}32)$$

It is interesting to note here that one must measure the stagnation and static pressures separately to determine the pressure ratio, whereas only the pressure difference is required to calculate the velocity of a liquid flow.

If the flow is supersonic, then the indicated stagnation pressure is the pressure behind the shock wave standing off the tip of the tube. By taking this pressure as the total pressure downstream of a normal shock wave and the measured static pressure as the static pressure upstream of the shock wave, the Mach number of the free stream (M_1) can be determined from the static-to-total-pressure ratio (p_1/p_{t_2}) according to the expression

$$\frac{p_1}{p_{t_2}} = \frac{\{[2k/(k + 1)]M_1^2 - [(k - 1)/(k + 1)]\}^{1/(k-1)}}{\{[(k + 1)/2]M_1^2\}^{k/(k-1)}}$$ (13-12)

which is called the Rayleigh supersonic Pitot formula. One notes, however, that M_1 is an implicit function of the pressure ratio and must be determined graphically or by some numerical procedure. Many normal shock tables, such as in Ref. 13, have p_1/p_{t_2} tabulated versus M_1, which enables one to find M_1 quite easily by interpolation.

Once the Mach number is determined, more information is needed to evaluate the velocity: namely, the local speed of sound. This can be done by inserting a probe into the flow to measure total temperature and calculating the static temperature using the following equation developed in Chapter 12:

$$T = \frac{T_t}{1 + [(k - 1)/2]M_1^2}$$ (12-23)

The local speed of sound is then determined by

$$c = \sqrt{kRT}$$ (12-12)

and the velocity is calculated from

$$V = M_1 c$$

The hot-wire anemometer can also be used to measure velocity in compressible flows provided it is calibrated to account for Mach-number effects.

Mass-flow measurement

Measuring the flow rate of a compressible fluid using a truncated nozzle has been presented and discussed in some detail in Chapter 12. Basically, the flow nozzle is a truncated nozzle located in a pipe, so the equations developed in Chapter 12 can be used to determine the flow rate through the flow nozzle. Strictly speaking, the flow rate so calculated should be multiplied by the discharge coefficient. For the high Reynolds numbers characteristic of compressible flows, however, the discharge coefficient can be taken as unity. If the flow at the throat of the flow nozzle is sonic, it is conceivable that the complex flow field existing downstream of the nozzle will make the reading from the downstream pressure tap difficult to interpret. That is, there can be no assurance that the measured pressure is the true back pressure. In this case, it is advisable to use a venturi meter because the pressure is measured directly at the throat.

The mass-flow rate of a compressible fluid through a venturi meter can easily be analyzed using the equations developed in Chapter 12. Consider

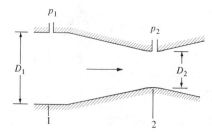

FIGURE 13-21   Venturi meter.

the venturi shown in Fig. 13-21. Writing the energy equation, Eq. (12-16), for the flow of an ideal gas between stations 1 and 2 gives

$$\frac{V_1^2}{2} + \frac{kRT_1}{k-1} = \frac{V_2^2}{2} + \frac{kRT_2}{k-1} \tag{13-13}$$

By conservation of mass, the velocity V_1 can be expressed as

$$V_1 = \frac{\rho_2 A_2 V_2}{\rho_1 A_1}$$

Substituting this result into Eq. (13-13), using the ideal-gas law to eliminate temperature, and solving for V_2 gives

$$V_2 = \left\{ \frac{[2k/(k-1)][(p_1/\rho_1) - (p_2/\rho_2)]}{1 - (\rho_2 A_2/\rho_1 A_1)^2} \right\}^{1/2} \tag{13-14}$$

Assuming that the flow is isentropic,

$$\frac{p_1}{p_2} = \left(\frac{\rho_1}{\rho_2} \right)^k$$

the equation for the velocity at the throat can be rewritten as

$$V_2 = \left\{ \frac{[2k/(k-1)](p_1/\rho_1)[1 - (p_2/p_1)^{(k-1)/k}]}{1 - (p_2/p_1)^{2/k}(D_2/D_1)^4} \right\}^{1/2} \tag{13-15}$$

The mass flow is obtained by multiplying V_2 by $\rho_2 A_2$. This analysis, however, has been based on a one-dimensional flow, and two-dimensional effects can be accounted for by the discharge coefficient C_d. Thus we finally have

$$\dot{m} = C_d \rho_2 A_2 V_2$$

$$= C_d A_2 \left(\frac{p_2}{p_1} \right)^{1/k} \left\{ \frac{[2k/(k-1)]p_1 \rho_1[1 - (p_2/p_1)^{(k-1)/k}]}{1 - (p_2/p_1)^{2/k}(D_2/D_1)^4} \right\}^{1/2} \tag{13-16}$$

This equation is valid for all flow conditions, subsonic or supersonic, provided no shock waves occur between station 1 and 2. It is good design practice to avoid supersonic flows in the venturi meter in order to prevent the formation of shock waves and the attendant total pressure losses. The discharge coefficient can also generally be taken as unity.

EXAMPLE 13-7 Calculate the mass-flow rate of air through a venturi with a 1-cm-diameter throat (D_2) in a 3-cm-diameter pipe (D_1). The upstream static pressure is 150 kPa and the throat pressure is 100 kPa. The static temperature of the air in the pipe is 27°C.

Solution From Table A-2 in the Appendix we find that $k = 1.4$ and $R = 287$ J/kg K. The gas density in the pipe is

$$\rho_1 = \frac{p_1}{RT_1}$$

$$= \frac{150 \times 10^3 \text{ N/m}^2}{(287 \text{ J/kg K})(300 \text{ K})}$$

$$= 1.74 \text{ kg/m}^3$$

Substituting the appropriate values in Eq. (13-16) we find

$$\dot{m} = 1 \times 0.785 \times 10^{-4} \text{ m}^2 \left(\frac{1}{1.5}\right)^{0.714}$$

$$\times \left\{\frac{7 \times 150 \times 10^3 \text{ N/m}^2 \times 1.74 \text{ kg/m}^3[1 - (1/1.5)^{0.286}]}{[1 - (1/1.5)^{1.43}(1/3)^4]}\right\}^{1/2}$$

$$= 0.0264 \text{ kg/s} \qquad \blacktriangleleft$$

The square-edged orifice can also be used to measure the flow rate of compressible fluids. The discharge equation for liquids is multiplied by an empirical factor which accounts for compressibility effects (1). The resulting equation is

$$\dot{m} = YA_0 K\sqrt{2\rho_1(p_1 - p_2)}$$

where K is the flow coefficient defined by Eq. (13-3), A_0 is the orifice area, and Y is called the compressibility factor and given by

$$Y = 1 - \left\{\frac{1}{k}\left(1 - \frac{p_2}{p_1}\right)\left[0.41 + 0.35\left(\frac{A_0}{A_1}\right)^2\right]\right\}$$

One must remember when using the equations above for Mach number and flow rate that absolute pressures, not gage pressures, must always be used.

Shock-wave visualization

When studying the qualitative features of a supersonic flow in a wind tunnel, it is important to be able to locate and identify the shock-wave pattern. Unfortunately, shock waves cannot be seen with the naked eye, so that the application of some type of optical technique is necessary. There are three techniques by which shock waves can be seen: the shadow-graph, the interferometer, and the schlieren. Each technique has its special application related to the type of information needed on density variation. The schlieren technique, however, finds frequent use in shock-wave visualization.

An illustration of the essential features of the schlieren system is given in Fig. 13-22. Light from the source S is collimated by lens L_1 to produce a parallel-light beam. The light then passes through a second lens L_2 and produces an image of the source at plane f. A third lens L_3 focuses the image on the display screen. A sharp edge, usually called the *knife edge*, is positioned at plane f so as to block out a portion of the light.

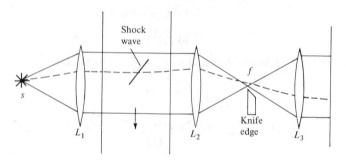

FIGURE 13-22   Schlieren system.

If a shock wave occurs in the test section, the light is refracted by the density change across the wave. As illustrated by the dashed line in Fig. 13-22, the refracted ray will escape the blocking effect of the knife edge and the shock wave will appear as a lighter region on the screen. Of course, if the beam is refracted in the other direction, the knife edge will block out more light and the shock wave will appear as a darker region. The contrast can be increased by intercepting more light with the knife edge.

Interferometry

The interferometer allows one to map contours of constant density and measure the density changes in the flow field. The underlying principle is the phase shift of a light beam on passing through media of different densities. The system now employed almost universally is the Mach-Zender interferometer, shown in Fig. 13-23. Light from a common source is split into two beams as it passes through the first half-silvered mirror. One

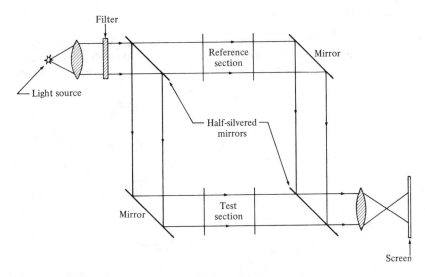

FIGURE 13-23   Schematic diagram of Mach-Zender interferometer.

beam passes through the test section while the other passes through a reference section. The two beams are then recombined and projected onto a screen or photographic plate. If the density in the test section and reference section are the same, there is no phase shift between the two beams and the screen is uniformly bright. However, a change of density in the test section changes the light speed of the test-section beam and a phase shift is generated between the two beams. Upon recombination of the beams, this phase shift gives rise to a series of dark and light bands on the screen. Each band represents a uniform density contour and the change in density across each band can be determined for a given system.

PROBLEMS

13-1   A 2-mm-diameter stagnation tube is used to measure the velocity in a stream of air as shown. What is the air velocity if the deflection on the air-water manometer is 0.80 mm? Air temperature $= 10°C$ and $p = 100$ kPa.

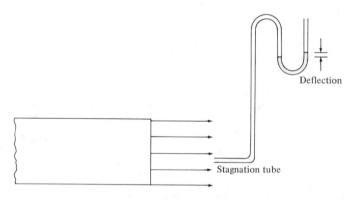

Deflection

Stagnation tube

PROBLEMS 13-1, 13-2

13-2   If the velocity in an airstream ($p_a = 98$ kPa, $T = 10°C$) is 10 m/s, what deflection will be produced on an air-water manometer if the stagnation tube is 2 mm in diameter?

13-3   What would be the error in velocity determination if one used a C_p of 1.00 for a circular stagnation tube instead of the true value of C_p? Assume the measurement is made with a 2-mm-diameter stagnation tube measuring air ($T = 25°C$, $p = 100$ kPa abs) velocity for which the stagnation pressure reading is 5.00 Pa.

13-4   Without exceeding an error of 2%, what is the minimum air velocity that can be obtained using a 2-mm circular stagnation tube if the formula $V = \sqrt{2\,\Delta p_{stag}/\rho} = \sqrt{2\,gh_{stag}}$ is used for computing the velocity. Assume standard atmospheric conditions.

13-5   Without exceeding an error of 2%, what is the minimum water velocity that can be obtained using a 2-mm circular stagnation tube if the formula $V = \sqrt{2\,\Delta p_{stag}/\rho} = \sqrt{2\,gh_{stag}}$ is used for computing the velocity. Assume the water temperature $= 20°C$.

13-6   A Pitot probe used frequently for measuring stack-gas velocities is shown below. The probe consists of two tubes bent away from and toward the flow direction and cut off on a plane normal to the flow direction, as shown. Assume the pressure coefficient is 1.0 at A and -0.4 at B. The probe is inserted in a stack where the temperature is 300°C and the pressure is 100 kPa abs. The gas constant of the stack gases is 410 J/kg K. The probe is connected to a water manometer and a 1-cm deflection is measured. Calculate the stack-gas velocity.

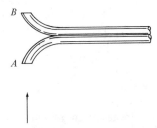

13-7   Water from a pipe is diverted into a tank for 4 min. If the mass of diverted water is measured to be 2,000 kg, what is the discharge in m^3/s? Assume water temperature = 20°C.

13-8   Water from a test apparatus is diverted into a calibrated volumetric tank for 6 min, 12 sec. If the volume of diverted water is measured to be 78 m^3, what is the discharge in cubic meters per second, gallons per minute, and cubic feet per second?

13-9   A velocity traverse in a 24-cm oil pipe yields the following data. What is the discharge, mean velocity, and ratio of maximum to mean velocity? Does the flow appear to be laminar or turbulent?

r, cm	0	1	2	3	4	5	6	7	8	9	10	10.5	11.0	11.5
V, m/s	8.7	8.6	8.4	8.2	7.7	7.2	6.5	5.8	4.9	3.8	2.5	1.9	1.4	0.7

13-10   A velocity traverse inside a 16-in. circular air duct yields the data in the table below. What is the rate of flow in cubic feet per second and cubic feet per minute? What is the ratio of V_{max} to V_{mean}? Does it appear that the flow is laminar or turbulent? If p = 14.3 psia and T = 70°F, what is the mass rate of flow?

y*		0.0	0.1	0.2	0.4	0.6	1.0	1.5	2.0	3.0	4.0	5.0	6.0	7.0	8.0
V, ft/sec	0	72	79	88	93	100	106	110	117	122	126	129	132	135

* Distance from pipe wall, in.

13-11   The asymmetry of the flow in stacks requires that flow velocity be measured at several locations on the cross-flow plane. Consider the cross

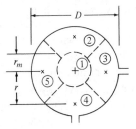

section of the cylindrical stack shown above. The two access holes through which probes can be inserted are separated by 90°. Velocities can be measured at the five points shown (5-point method).

(a) Determine the ratio r_m/D such that the areas of the five measuring segments are equal.

(b) Determine r/D (probe location), which corresponds to the centroid of the segment.

(c) The following data are taken for a 2-m-diameter stack in which the gas temperature is 300°C, the pressure is 110 kPa abs, and the gas constant is 400 J/kg K. The data represent the deflection on a water manometer connected to a conventional Pitot probe located at the measuring stations. Calculate the mass flow rate.

Station	Δh (cm)
1	1.2
2	1.1
3	1.1
4	0.9
5	1.05

13-12 Repeat Prob. 13-11 for the case in which three access holes are separated by 60° and seven measuring points are used. The diameter of the stack is 1.5 m, the gas temperature is 250°C, the pressure is 115 kPa abs, and the gas constant is 420 J/kg K. The following data represent the deflection of a water manometer connected to a conventional Pitot tube at the measuring stations. Calculate the mass flow rate.

Station	Δh (mm)
1	8.2
2	8.6
3	8.2
4	8.9
5	8.0
6	8.5
7	8.4

13-13 Theory and experimental verification indicate that the mean velocity along a vertical line in a wide stream is closely approximated by the velocity at 0.6 depth. Then, if the following velocities at 0.6 depth in a river cross section are measured, what is the discharge in the river?

13-14 For the jet and orifice shown, determine C_v, C_c, and C_d.

13-15 A fluid jet discharging from a 2-cm orifice has a diameter of 1.75 cm at its vena contracta. What is the coefficient of contraction?

13-16 An orifice 6 in. in diameter (d = 6 in.) discharges water from the very large

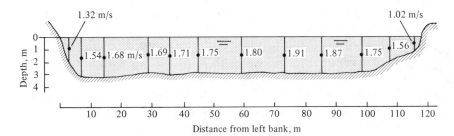

PROBLEM 13-13

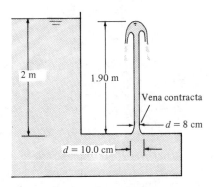

PROBLEM 13-14

tank into the atmosphere. The surface of the water is under a pressure p_0 and $h = 6$ ft. If the discharge rate is 3.6 cfs, what is the pressure p_0?

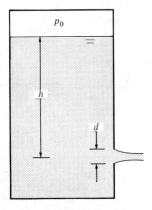

PROBLEMS 13-16, 13-17

13-17 An orifice 20 cm in diameter ($d = 20$ cm) discharges water from a large tank into the atmosphere. The surface of the water is under pressure p_0 and is located 5 m above the center of the orifice ($h = 5$ m). If the discharge rate is 0.40 m³/s, what is the pressure p_0?

13-18 A 6-in. orifice is placed in a 10-in. pipe and a mercury manometer is connected to either side of the orifice. If a flow rate of water (60°F) through this orifice is 3 cfs, what will be the manometer deflection?

13-19 If a liquid (sp. gr. = 0.95, $\nu = 10^{-5}$ m²/s) flows through the pipe and orifice shown, indicate the direction of flow and determine the discharge if the orifice diameter is 3 cm and the pipe diameter is 5 cm.

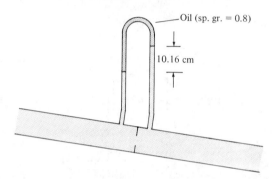

13-20 A 10-cm orifice in the horizontal 30-cm pipe shown is the same size as the orifice in the vertical pipe. The manometers are mercury-water manometers and water ($T = 20°C$) is flowing in the system. The gages are Bourdon-tube gages. The flow, at a rate of 0.1 m³/s, is to the right in the horizontal pipe and therefore downward in the vertical pipe. Is Δp as indicated by gages A

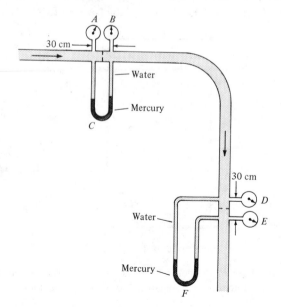

and B the same as Δp indicated by gages D and E? Determine their values. Is the deflection on manometer C the same as the deflection on manometer F? Determine the deflections.

13-21 A 15-cm plate orifice at the end of a 30-cm pipe is enlarged to 20 cm. With the same pressure drop across the orifice, what will be the percentage increase in discharge?

13-22 Water flows through a venturi meter which has a 30-cm throat. The venturi meter is in a 60-cm pipe. What deflection will occur on a mercury-water manometer connected between the upstream and throat sections if the discharge is 0.57 m³/s? Assume $T = 20°C$.

13-23 Determine the rate of flow of water ($T = 10°C$) through the system shown if $h = 2$ m, $D = 10$ cm, and $d = 5$ cm.

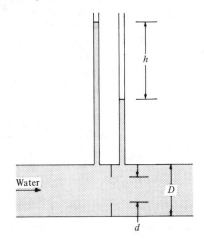

PROBLEMS 13-23, 13-24

13-24 Determine the discharge of water ($T = 60°F$) through the orifice if $h = 5$ ft, $D = 4$ in., and $d = 3$ in.

13-25 If water (20°C) is flowing through this 10-cm orifice, estimate the rate of flow therein.

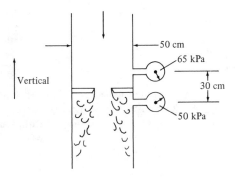

13-26 Determine the size of orifice needed in a 15-cm pipe to measure 0.03 m³/s of water with a deflection of 1 m on a mercury-water manometer.

13-27 What is the discharge of gasoline (sp. gr. = 0.68) in a 10-cm horizontal pipe if the differential pressure across a 6-cm orifice in the pipe is 35 kPa?

13-28 What size of orifice is needed to produce a change in head of 8 m for a discharge of 2 m³/s of water in a 1-m-diameter pipe?

13-29 An orifice is to be designed to have a change in pressure of 50 kPa across the orifice (measured with a differential pressure transducer) for a discharge of 3.0 m³/s of water in a 1.2-m-diameter pipe. What orifice diameter will yield the desired results?

13-30 Semicircular orifices such as this one sometimes used to measure the flow rate of liquids that also transport sediments. The opening at the bottom of the pipe allows free passage of the sediment. Derive a formula for Q as a function of Δp, D, and other relevant variables associated with the problem. Then using that formula and guessing any unknown data, estimate the water discharge through such a semicircular orifice when Δp is read to be 80 kPa and flow is in a 30-cm pipe.

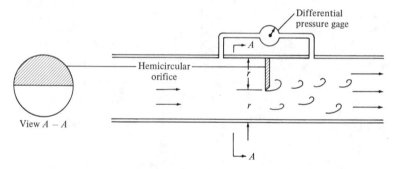

13-31 The pressure differential across this 1-m by 2-m venturi meter is 100 kPa. What is the discharge of water through the venturi meter?

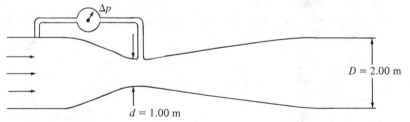

13-32 What throat diameter is needed for a venturi meter in a 200-cm horizontal pipe carrying water with a discharge of 10 m³/s if the differential pressure between the throat and upstream section is to be limited to 200 kPa at this discharge?

13-33 Estimate the rate of flow of water through the venturi meter shown.

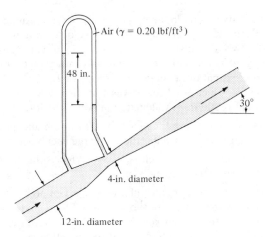

Air ($\gamma = 0.20$ lbf/ft^3)

48 in.

30°

4-in. diameter

12-in. diameter

13-34 The differential pressure gage on the venturi meter shown reads 6.20 psi, $h = 30$ in., $d = 4$ in., and $D = 8$ in. What is the discharge of water in the system? Assume $T = 50°F$.

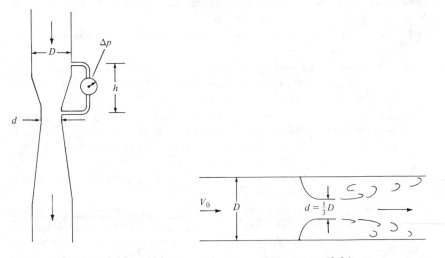

PROBLEMS 13-34, 13-35

PROBLEM 13-36

13-35 The differential gage on the venturi meter reads 50 kPa, $d = 20$ cm, $D = 40$ cm, and $h = 80$ cm. What is the discharge of gasoline (sp. gr. = 0.69) in the system?

13-36 What is the head loss for the flow nozzle shown in terms of $V_0^2/2g$?

13-37 One mode of operation of the acoustic flow meter is to measure the travel times between two stations for the sound wave traveling upstream then downstream with the flow. The wave propogation speed downstream with respect to the measuring stations is $c + V$ where c is the sound speed and V

the flow velocity. Correspondingly, the upstream propogation speed is $c - V$. (a) Derive an expression for the flow velocity in terms of the distance between the two stations, L; the difference in travel times, Δt; and the sound speed. (b) The sound speed is typically much larger than V ($c \gg V$). With this approximation, express V in terms of L, c, and Δt. (c) A 10-ms time difference is measured for waves traveling 20 m in a gas where the speed of sound is 300 m/s. Calculate the flow velocity.

13-38 Water flows over a rectangular weir that is 2 m wide and 30 cm high. If the head on the weir is 15 cm, what is the discharge in cubic meters per second?

13-39 The head on a 60° triangular weir is 40 cm. What is the discharge over the weir in cubic meters per second?

13-40 At one end of a 1-m-wide rectangular tank is a 1-m-high sharp-crested rectangular weir. In the bottom of the tank is a 10-cm sharp-edged orifice. If 0.10 m³/s flows into the tank and leaves the tank both through the orifice and over the weir, what depth of water will be attained in the tank?

13-41 What is the discharge over a 1-m-high rectangular weir in a 2-m-wide channel if the head on the weir is 30 cm?

13-42 What is the discharge over a 2-ft-high rectangular weir in a 6-ft-wide channel if the head on the weir is 1 ft?

13-43 The head on a 60° triangular weir is 1.5 ft. What is the discharge of water over the weir?

13-44 A 3-m-wide rectangular irrigation canal carries water with a discharge of 6 m³/s. What height of rectangular weir installed across the canal will raise the water surface to a level 2 m above the canal floor?

13-45 A Pitot tube is used to measure the Mach number in a compressible subsonic flow of air. The stagnation pressure is 140 kPa and the static pressure is 100 kPa. The total temperature of the flow is 300 K. Determine the Mach number and flow velocity.

13-46 Use the normal shock-wave relationships derived in Chapter 12 to derive the Rayleigh supersonic Pitot formula.

13-47 The static and stagnation pressures measured by a Pitot tube in a supersonic airflow are 54 and 200 kPa, respectively. The total temperature is 350 K. Determine the Mach number and velocity of the free stream.

13-48 A venturi meter is used to measure the flow of helium in a pipe. The pipe is 1 cm in diameter and the throat 0.5 cm. The measured upstream and throat pressures are 120 kPa and 80 kPa, respectively. The static temperature of the helium in the pipe is 17°C. Determine the mass-flow rate.

13-49 The mass-flow rate of methane is measured with a square-edged orifice. The pipe diameter is 2 cm and the orifice diameter is 0.8 cm. The upstream and downstream pressure taps measure 150 kPa and 110 kPa, respectively. The static temperature in the pipe is 300 K. Determine the mass-flow rate.

REFERENCES

1. Bradshaw, P. *An Introduction to Turbulence and Its Measurement*, Pergamon Press, New York, 1971.

2. *Fluid Meters—Their Theory and Applications. ASME* (1959).

3. Grant, H. P., and Kronauer, R. E. "Fundamentals of Hot Wire Anemometry." Symp. *Meas. Unsteady Flow, ASME* (May 1962).

4. Holman, J. P. *Experimental Methods for Engineers.* McGraw-Hill Book Company, New York, 1971.

5. Johansen, F. C. *Proc. Roy. Soc. London, Ser. A,* 125 (1930).

6. Kindswater, Carl E., and Carter, R. W. "Discharge Characteristics of Rectangular Thin-Plate Weirs," *Trans. Am. Soc. Civil Eng.,* 124 (1959) 772–822.

7. King, H. W., and Brater, E. F. *Handbook of Hydraulics.* McGraw-Hill Book Company, New York, 1963.

8. King, L. V. *Phil. Trans. Roy. Soc. London, Ser. A,* 14 (1914), 214.

9. Kline, J. J. "Flow Visualization," in *Illustrated Experiments in Fluid Mechanics, The NCFMF Book of Film Notes.* Educational Development Center, Inc. (1972).

10. Liepmann, H. W., and Roshko, A. *Elements of Gasdynamics.* John Wiley & Sons, Inc., New York, 1957.

11. Macagno, Enzo O. "Flow Visualization in Liquids." *Iowa Inst. Hydraulic Res., Rept.* 114 (1969).

12. Macmillan, F. A. "Viscous Effects on Flattened Pitot Tubes at Low Speeds." *J. Roy. Aeronaut. Soc.,* technical note (December 1954).

13. NACA TR 1135. Equations, Tables, and Charts for Compressible Flow (1953).

14. Ower, E., and Pankhurst, R. C. *The Measurement of Air Flow.* Pergamon Press, New York, 1966.

15. Shercliff, J. A. *Electromagnetic Flow-Measurement.* Cambridge University Press, New York, 1962.

16. Wendt, R. E., Jr. (Ed.). *Flow Measuring Devices.* "Its Measurement and Control in Sciences and Industry," part 2. Instrument Society of America, 1974.

The windmill, a forerunner of modern turbo generators and still used here in Portugal, possesses a unique advantage in its pollution-free operation.

14 TURBOMACHINERY

A LL TURBOMACHINES INCLUDE either a rotating propeller or rotating vanes that change flow velocity and/or pressure within the fluid. In so doing, either the machine does work on the fluid or the fluid does work on the machine. Examples of the former include *pumps, blowers,* and *compressors*, whereas machines that have work done on them are called *turbines*. In this chapter we shall first consider the characteristics of propellers, then we shall see how propellers are incorporated into certain types of pumps in which the flow is essentially parallel to the axis of the pump shaft. These are called *axial-flow* pumps. Next, we shall study *radial-flow* pumps in which the fluid flows radially outward from the shaft axis. Finally we will consider the elementary characteristics and theory of turbines.

14-1   PROPELLER THEORY

Introduction

The design of a propeller is based upon the basic principles of airfoil theory. For example, if we observe a section of the propeller in Fig. 14-1, we will see the analogy between the lifting vane and the airfoil. This propeller is rotating at an angular speed ω and the speed of advance of the airplane and propeller is V_0. Then, if we focus attention on an elemental section of the propeller, Fig. 14-1c, it can be seen that the given section has a velocity that is made of components V_0 and V_t. Here V_t is the tangential velocity, $V_t = r\omega$, resulting from the rotation of the propeller. Now, if the velocity vectors V_0 and V_t are reversed and added, one obtains the velocity of the air relative to the particular propeller section, Fig. 14-1d. Thus we have a flow situation that is directly analogous to the airfoil.

The propeller is designed to produce thrust, and since the greatest contribution to thrust comes from the lift force F_L, the object is to maximize lift and minimize drag F_D. For a given shape of propeller section, the optimum angle of attack can be determined from data such as that given in Fig. 11-23. Because the angle θ, Fig. 14-1d, decreases with an increase in r, for a well-designed propeller the propeller blade must be warped in order to obtain the optimum angle of attack along the length of the blade.

Blade analysis

To analyze the forces on a blade element, we will consider an enlarged view of Fig. 14-1d as shown in Fig. 14-2. By definition, the lift is normal to

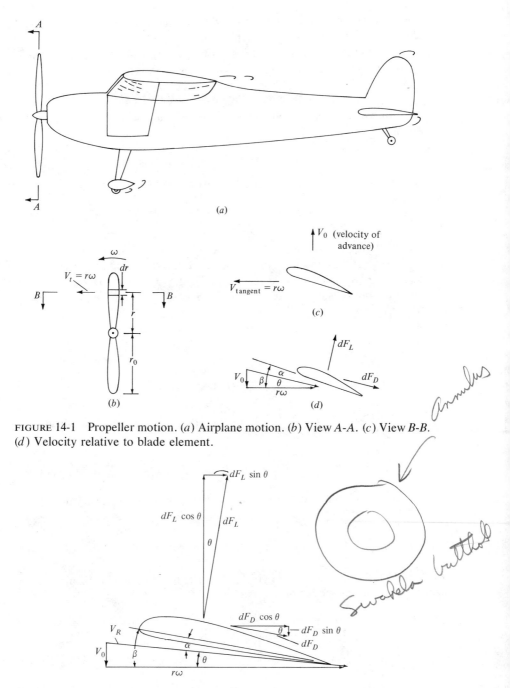

FIGURE 14-1   Propeller motion. (*a*) Airplane motion. (*b*) View *A-A*. (*c*) View *B-B*. (*d*) Velocity relative to blade element.

FIGURE 14-2   Definition sketch for propeller-blade element.

the relative air velocity V_R and the drag is the force acting parallel to the same velocity; thus it is seen that one component of the lift force, $dF_L \cos \theta$, produces a positive thrust component, and a component of the drag, $dF_D \sin \theta$, produces a negative thrust component:

$$dF_{\text{thrust}} = dF_L \cos \theta - dF_D \sin \theta \qquad (14\text{-}1)$$

In a similar manner it can be shown that the tangential force for this blade element is given by

$$dF_{\text{tang}} = dF_L \sin \theta + dF_D \cos \theta \qquad (14\text{-}2)$$

and the torque T opposing rotation of the propeller is given by

$$dT = (dF_L \sin \theta + dF_D \cos \theta)r \qquad (14\text{-}3)$$

From Chapter 11 the lift per unit length for a two-dimensional airfoil is given as $F_L/\ell = \frac{1}{2}C_L c\rho V_0^2$ and the drag per unit length is $F_D/\ell = \frac{1}{2}C_D c\rho V_0^2$. Hence the incremental lift and drag on a radial element dr will be $dF_L = \frac{1}{2}C_L c\rho V_R^2 \, dr$ and $dF_D = \frac{1}{2}C_D c\rho V_R^2 \, dr$, respectively. Then the incremental thrust and torque for the blade element will be, respectively,

$$dF_{\text{thrust}} = \frac{1}{2} (C_L \cos \theta - C_D \sin \theta) \, c\rho V_R^2 \, dr$$

$$dT = \frac{1}{2} (C_L \sin \theta + C_D \cos \theta) \, c\rho V_R^2 r \, dr$$

The total thrust will be obtained by integrating the incremental thrust over the length of the propeller, which is given as follows:

$$F_{\text{thrust}} = \frac{1}{2} N \int_{r_h}^{r_0} (C_L \cos \theta - C_D \sin \theta) \, c\rho V_R^2 \, dr \qquad (14\text{-}4)$$

and in a similar manner, the total torque producing rotation of the propeller is given as

$$T = \frac{1}{2} N \int_{r_h}^{r_0} (C_L \sin \theta + C_D \cos \theta) \, c\rho V_R^2 r \, dr \qquad (14\text{-}5)$$

Note that in Eqs. (14-4) and (14-5), N is the number of blades and r_h is the radius of the propeller hub.

Thrust and power relationships

In Eq. (14-4), ρ is assumed to be constant and $V_R = r\omega/\cos \theta$, so that Eq. (14-4) becomes

$$F_{\text{thrust}} = \frac{1}{2} \rho N \omega^2 \int_{r_h}^{r_0} \left(\frac{C_L}{\cos \theta} - C_D \frac{\tan \theta}{\cos \theta} \right) c r^2 \, dr \qquad (14\text{-}6)$$

$$\frac{F_{\text{thrust}}}{\frac{1}{2} N \rho r_0^2 \omega^2} = \int_{r_h}^{r_0} \left(\frac{C_L}{\cos \theta} - C_D \frac{\tan \theta}{\cos \theta} \right) \left(\frac{r}{r_0} \right)^2 c \, dr \qquad (14\text{-}7)$$

Since $c \, dr$ is the differential area dA of the propeller, Eq. (14-7) is written as follows:

$$\frac{F_{\text{thrust}}}{\frac{1}{2} N \rho r_0^2 \omega^2} = \int_A \left(\frac{C_L}{\cos \theta} - C_D \frac{\tan \theta}{\cos \theta} \right) \left(\frac{r}{r_0} \right)^2 dA \qquad (14\text{-}8)$$

The right side of Eq. (14-8) can be expressed in functional form as

$$\int_A \left(\frac{C_L}{\cos \theta} - C_D \frac{\tan \theta}{\cos \theta} \right) \left(\frac{r}{r_0} \right)^2 dA = C_1 A$$

Here C_1 is a dimensionless coefficient that depends on the shape of the propeller blade, the speed of rotation ω, and the speed of advance. The area A is proportional to D^2, where D is the diameter of the propeller. Then for a particular propeller, Eq. (14-8) can be expressed as

$$F_{\text{thrust}} = \frac{1}{2} \rho r_0^2 \omega^2 C_2 D^2$$

or
$$= C_T \rho n^2 D^4 \qquad (14\text{-}9)$$

where n = speed of rotation, rps (revolutions per second)

D = diameter, meters

C_T = thrust coefficient

The *thrust coefficient* is a function of the geometry of the propeller and of the *advance diameter ratio*, V_0/nD. This ratio in effect establishes the angle of attack for each section of a given propeller. Hence, we have defined the thrust coefficient:

$$C_T = \frac{F_{\text{thrust}}}{\rho D^4 n^2} \qquad (14\text{-}10)$$

where
$$C_T = f\left(\frac{V_0}{nd} \right)$$

The thrust coefficient can be considered as a similarity parameter for propellers in the same way as the lift coefficient C_L was used for wings.[1]

In a manner similar to the development of the thrust coefficient, it can

---

[1] C_T is also a function of Re and M; however, these effects are usually small in comparison with the effect of the advance diameter ratio.

be shown that the torque is given by

$$T = CN\rho r_0^3 \omega^2 D^2$$

However, our main interest in the torque is to provide a means for evaluating the power that must be supplied to the propeller. The power input is $T\omega$; hence, we can express power as

$$P = C_P \rho D^5 n^3$$

or

$$C_P = \frac{P}{\rho D^5 n^3}$$ (14-11)

Here C_P is the *power coefficient* and, like C_T, is a function of the propeller geometry and V_0/nD. The functional relationship between C_P and V_0/nD for a particular propeller is shown along with C_T in Fig. 14-3. Once the relationships between C_T and V_0/nD and between C_P and V_0/nD are found (usually experimentally) for a given propeller, they can be used to predict

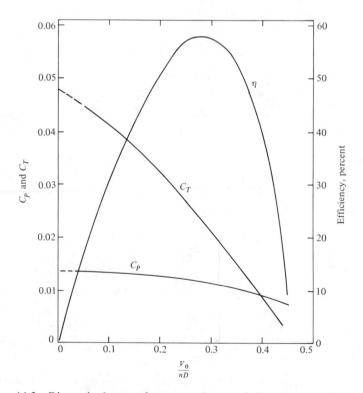

FIGURE 14-3   Dimensionless performance characteristics of a typical propeller. $D = 2.90$ m, $n = 1,400$ *rpm*. [After Weick (15).]

the operation of geometrically similar propellers over a wide range of speed and size. Example 14-1 illustrates such an application.

The efficiency of a propeller is defined as the ratio of the power output, that is, thrust times velocity of advance, to the power input. Hence the efficiency η is given as

$$\eta = \frac{F_T V_0}{P} = \frac{C_T \rho D^4 n^2 V_0}{C_P \rho D^5 n^3}$$

which reduces to $\eta = C_T/C_P$ times advance diameter ratio. Figure 14-3 includes the efficiency for the given propeller.

As a point of interest, the curves of C_T and C_P are obtained from performance characteristics of a given propeller operated with different values of V_0, as shown in Fig. 14-4. Even though the data for the curves are obtained for a given propeller size and a given angular speed, the curves can be applied by similarity principles to the same propeller operating at different angular speeds and to geometrically similar propellers of different size and speed. There will be some change in characteristics with different conditions, but the deviation is usually small. See Ref. 15 for examples of the effect of angular speed on propeller performance.

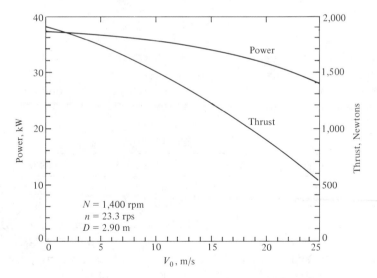

FIGURE 14-4 Power and thrust of a 2.90-m-diameter propeller with a rotational speed of 1,400 rpm. [After Weick (15).]

EXAMPLE 14-1 A propeller having the characteristics shown in Fig. 14-3 is to be used to drive a swamp boat. If the propeller is to have a diameter of 2 m and a rotational speed of $N = 1,200$ rpm, what is the thrust starting from rest? If the boat resistance (air and water) is given by the empirical equation $F_D = 0.003 \rho V_0^2/2$, where V_0 is the boat speed in meters per second, F_D is the drag, and ρ is the mass density of the water, what is the maximum speed of the boat and what power is required to drive the propeller? Assume $\rho_{\text{air}} = 1.20$ kg/m³, or 1.20 N · s²/m⁴.

Solution First evaluate the thrust starting from rest:

$$\frac{V_0}{nD} = \frac{(0 \text{ m/s})}{(20 \text{ rps})(2 \text{ m})} = 0$$

Then, referring to Fig. 14-3, we see that $C_T = 0.048$ for $V_0/nD = 0$. Since $C_T = F_T/\rho_a D^4 n^2$, we can now compute F_T:

$$\begin{aligned} F_T &= C_T \rho D^4 n^2 \\ &= 0.048(1.20 \text{ N} \cdot \text{s}^2/\text{m}^4)(2 \text{ m})^4(20 \text{ rps})^2 \\ &= 368 \text{ N} \end{aligned}$$ ◀

For the second part of the example, it should be recognized that

$$F_T = F_D$$

or

$$C_T \rho_a D^4 n^2 = 0.003 \, \rho_w \frac{V_0^2}{2}$$

Here ρ_w is the mass density of water $= 1,000$ kg/m³, or 1.0 kN · s²/m⁴. To determine the equilibrium condition where $F_T = F_D$, we plot F_T and F_D versus V_0; and where the two curves intersect is the point of equilibrium. The necessary data to construct such curves are included in the table below.

V_0	V_0/nD	C_T	$F_T = C_T \rho_a D^4 n^2$	$F_D = 0.003 \, \rho_w V_0^2/2$
5 m/s	0.125	0.040	307 N	37.5 N
10 m/s	0.250	0.027	207 N	150 N
15 m/s	0.375	0.012	92 N	337 N

Now when F_T and F_D are plotted against V_0 we obtain the graph shown. The curves intersect at $V_0 = 11$ m/s; hence, the boat will attain a speed of 11 m/s. The input power is $P = C_P \rho D^5 n^3$; and since the maximum C_P is 0.014 when $V_0/nD = 0$, we compute the power input as

$$\begin{aligned} P &= 0.014(1.20 \text{ N} \cdot \text{s}^2/\text{m}^4)(2 \text{ m})^5(20 \text{ rps})^3 \\ &= 4300 \text{ m} \cdot \text{N/s} \\ &= 4.30 \text{ kW} \end{aligned}$$ ◀

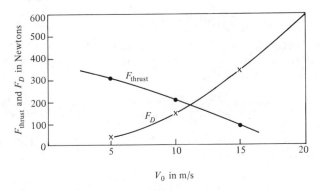

SOLUTION EXAMPLE 14-1

14-2 AXIAL-FLOW PUMPS

Pressure changes

Axial-flow pumps (or blowers) are designed so that the impeller (much like the propeller discussed in Sec. 14-1) is enclosed within a housing (see Fig. 14-5). The action of the propeller applies a force to the fluid that causes a pressure change between the upstream and downstream sections of the pump. Because engineers are usually more interested in the change in pressure head than in the thrust when working with pumps, the thrust coefficient is replaced by the head coefficient and the ratio V_0/nD is replaced by a more useful parameter involving the discharge.

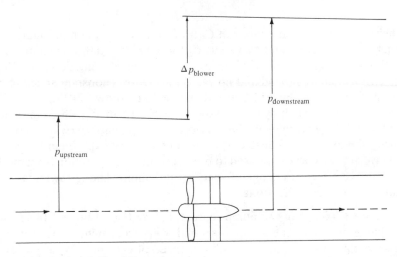

FIGURE 14-5   Axial-flow blower in a duct.

Head and discharge coefficients for pumps

The thrust coefficient is defined as $F_T/\rho D^4 n^2$, and if the same variables are applied to flow in an axial pump, the thrust can be expressed as $F_T = \Delta p A = \gamma \, \Delta H A$ or

$$C_T = \frac{\gamma \Delta H A}{\rho D^4 n^2} = \frac{\pi}{4} \frac{\gamma \Delta H D^2}{\rho D^4 n^2} = \frac{\pi}{4} \frac{g \Delta H}{D^2 n^2} \qquad (14\text{-}12)$$

Now if we define a new parameter, the head coefficient, using the variables of Eq. (14-12), we have

$$C_H = \frac{4}{\pi} C_T = \frac{\Delta H}{D^2 n^2/g} \qquad (4\text{-}13)$$

The independent parameter relating to propeller operations was V_0/nD; however, for pump operation it is convenient to substitute $Q/A = Q/(\pi D^2/4)$ for V_0 and let the numerical factor be absorbed in the functional relationship. Thus the independent parameter for pump similarity studies is Q/nD^3, and this is termed the *discharge coefficient* C_Q. The power coefficient for pumps is exactly like the power coefficient used for propellers. Summarizing, the dimensionless parameters used in similarity analyses of pumps are as follows:

$$C_H = \frac{\Delta H}{D^2 n^2/g} \qquad (14\text{-}14)$$

$$C_P = \frac{P}{\rho D^5 n^3} \qquad (14\text{-}15)$$

$$C_Q = \frac{Q}{nD^3} \qquad (14\text{-}16)$$

where C_H and C_P are functions of C_Q for a given type of pump. Figure 14-6 is set of curves of C_H and C_P versus C_Q for a typical axial-flow pump. Also on this graph is plotted the efficiency of the pump as a function of C_Q. The dimensional curves (head and power versus Q for a constant speed of rotation) from which Fig. 14-6 was developed are shown in Fig. 14-7. Because the curves such as those shown in Fig. 14-6 or Fig. 14-7 characterize the pump performance, they are often called *characteristic curves* or *performance curves*. These curves are obtained by experiment.

Performance curves are used to predict prototype operation from model tests or the effect of change of speed of the pump. The following are examples of these applications.

EXAMPLE 14-2   For the pump represented by Figs. 14-6 and 14-7, what discharge in cubic meters per second will occur when the pump is operating against a 2-m head and at a speed of 600 rpm? What power in kilowatts is required for these conditions?

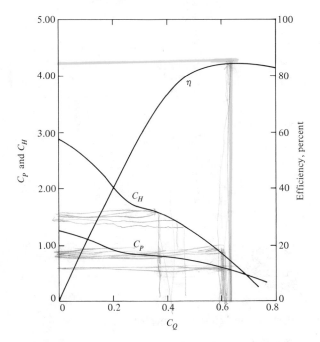

FIGURE 14-6   Dimensionless performance curves for a typical axial-flow pump. [After Stepanoff (13).]

Solution   First compute C_H. Here,

$$D = 35.6 \text{ cm} \qquad \text{and} \qquad n = 10 \text{ rps}$$

Then,
$$C_H = \frac{2 \text{ m}}{(0.356 \text{ m})^2(10^2 \text{ s}^{-2})/(9.81 \text{ m/s}^2)} = 1.55$$

With a value of 1.55 for C_H, a value of 0.40 for C_Q is read from Fig. 14-6. Hence, Q is calculated as follows:

$$C_Q = 0.40 = \frac{Q}{nD^3}$$

or
$$Q = 0.40(10 \text{ s}^{-1})(0.356 \text{ m})^3 = 0.180 \text{ m}^3/\text{s} \qquad \blacktriangleleft$$

From Fig. 14-6 the value of C_P is 0.72 for $C_Q = 0.40$, then

$$
\begin{aligned}
P &= 0.72 \, \rho D^5 n^3 \\
&= 0.72(1.0 \text{ kN} \cdot \text{s}^2/\text{m}^4)(0.356 \text{ m})^5(10 \text{ s}^{-1})^3 \\
&= 4.12 \text{ km} \cdot \text{N/s} = 4.12 \text{ kJ/s} \\
&= 4.12 \text{ kW} \qquad \blacktriangleleft
\end{aligned}
$$

2.54 cm = 1"
.0254 k = 1"

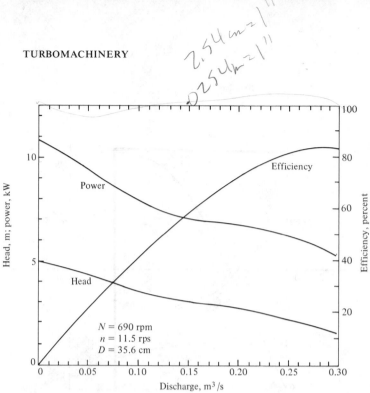

FIGURE 14-7   Performance characteristics of a typical axial-flow pump. [After Stepanoff (13).]

EXAMPLE 14-3   If a 30-cm axial-flow pump having the characteristics shown in Fig. 14-6 is operated at a speed of 800 rpm, what head ΔH will be developed when the water pumping rate is 0.127 m³/s? What power is required for this operation?

Solution   First compute

$$C_Q = \frac{Q}{nD^3}$$

where

$$Q = 0.127 \text{ m}^3/\text{s}$$

$$n = \frac{800}{60} = 13.3 \text{ rps}$$

$$D = 30 \text{ cm}$$

Then,

$$C_Q = \frac{0.127 \text{ m}^3/\text{s}}{13.3 \text{ s}^{-1}(0.30 \text{ m})^3} = 0.354$$

Now enter Fig. 14-6 with a value of $C_Q = 0.354$ and read off a value of 1.70 for C_H and a value of 0.80 for C_P. Then

$$\Delta H = \frac{C_H D^2 n^2}{g} = \frac{1.70(0.30 \text{ m})^2 (13.3 \text{ s}^{-1})^2}{(9.81 \text{ m/s}^2)}$$

$$= 2.76 \text{ m} \qquad \blacktriangleleft$$

and $P = C_P \rho D^5 n^3 = 0.80(1.0 \text{ kN} \cdot \text{s}^2/\text{m}^4)(0.30 \text{ m})^5(13.3 \text{ s}^{-1})^3$

$$= 4.56 \text{ kW} \qquad \blacktriangleleft$$

Range of application of axial-flow machines

In practical applications, axial-flow machines are best suited for relatively low heads and high rates of flow. Hence pumps used for dewatering lowlands, such as those behind dikes, are almost always of the axial-flow type. Water turbines in low head dams (less than 30 m) where the flow rate and power production is large are also generally of the axial type. For larger heads, radial- or mixed-flow machines are more efficient. These are discussed in the next section.

14-3   RADIAL- AND MIXED-FLOW MACHINES

Pumps

In Fig. 14-8 the type of impeller that is used for many radial-flow pumps is shown. Such pumps are also called *centrifugal pumps*. Fluid from the inlet pipe enters the pump through the eye of the impeller and then travels outward between the vanes of the impeller to the edge of the impeller

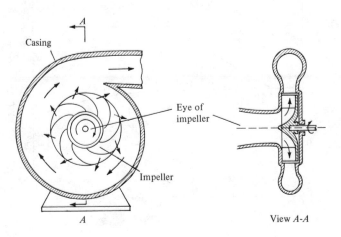

FIGURE 14-8   Centrifugal pump.

where the fluid enters the casing of the pump and is then conducted to the discharge pipe. The principle of the radial-flow pump is different from that of the axial-flow pump in that the change in pressure results in large part by rotary action (pressure increasing outward like that of the rotating tank in Sec. 5-2) produced by the rotating impeller. Additional pressure increase is produced in the radial-flow pump when the high-velocity flow leaving the impeller is reduced in the expanding section of the casing.

Although the basic designs are different for the radial- and axial-flow pumps, it can be shown that the same similarity parameters (C_Q, C_P, and C_H) apply for both types. Thus the methods that have already been discussed for relating size, speed, and discharge in axial-flow machines also apply to the radial-flow machine.

The major practical difference between the axial- and radial-flow pumps so far as the pump user is concerned is the difference in the performance characteristics of the two pumps. In Fig. 14-9 the dimensional performance curves for a typical radial-flow pump operating at a constant speed of rotation are shown; and in Fig. 14-10 the dimensionless performance curves for the same pump are shown. Note that at shutoff flow the power required is less than for flow at maximum efficiency. Normally the motor to drive the pump will be chosen for conditions of maximum pump efficiency. Hence it can be seen that the flow can be throttled between the

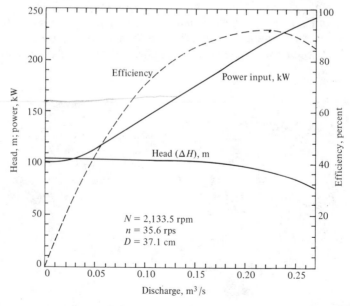

FIGURE 14-9 Typical performance curves for a centrifugal pump, $D - 37.1$ cm. [After Daugherty and Franzini (3).]

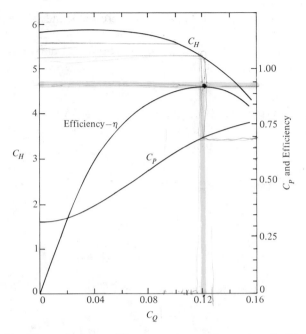

FIGURE 14-10 Dimensionless performance curves for a centrifugal pump, from data given in Figure 14-9. [After Daugherty and Franzini (3).]

limits of shutoff condition and the normal operating condition without any chance of overloading the pump motor. Such is not the case for an axial-flow pump, as seen in Fig. 14-6. In that case, when the pump flow is throttled below maximum efficiency conditions, the required power increases with decreasing flow, thus leading to the possibility of overloading at low-flow conditions. For very large installations, special operating procedures are followed in order to avoid such overloading. For instance, the valve in the bypass from the pump discharge back to the pump inlet can be adjusted to maintain a constant flow through the pump. However, for small-scale applications it is often desirable to have complete flexibility in the flow control without the complexity of special operating procedures. In this latter case, a radial-flow pump offers a distinct advantage.

EXAMPLE 14-4 The pump that has the characteristics given in Fig. 14-9 when operated at 2,133.5 rpm is to be used to pump water at maximum efficiency under a head of 76 m. At what speed should the pump be operated and what will be the discharge for these conditions?

Solution Since the diameter is fixed, the only change that will occur will result from the change in speed (assuming negligible change due to viscous effects). The C_H, C_P, C_Q, and η for this pump operating at maximum efficiency against a ΔH of 76 m are the same as these for operation at maximum efficency with a speed of rotation of 2,133.5 rpm since both operating conditions correspond to the point of maximum efficiency in Fig. 14-10. Thus we can write

$$(C_H)_N = (C_H)_{2,133.5 \, \text{rpm}}$$

Here N refers to the speed of rotation with $\Delta H = 76$ m. The graph of Fig. 14-9 indicates that $\Delta H = 90$ m and $Q = 0.225$ m³/s at maximum efficiency for $N = 2,133.5$ rpm. Thus

$$\frac{76 \text{ m}}{N^2} = \frac{90 \text{ m}}{(2,133.5)^2}$$

$$N^2 = (2,133.5)^2 \, \frac{76}{90}$$

$$N = 2,133.5 \left(\frac{76}{90}\right)^{1/2} = 1,960 \text{ rpm} \qquad \blacktriangleleft$$

Using $(C_Q)_{1,960} = (C_Q)_{2,133.5 \, \text{rpm}}$ and solving for the ratio of discharge, we have

$$\frac{Q_{1,960}}{Q_{2,133.5}} = \frac{1,960}{2,133.5} = 0.919$$

$$Q_{1,960} = 0.207 \text{ m}^3/\text{s} \qquad \blacktriangleleft$$

EXAMPLE 14-5   The pump having the characteristics shown in Figs. 14-9 and 14-10 is a model of a pump that was actually used in one of the pumping plants of the Colorado River Aqueduct (see Ref. 3). For a prototype that is 5.33 times larger than the model and operates at a speed of 400 rpm, what head, discharge, and power would be expected at maximum efficiency?

Solution   From Fig. 14-10 we pick off values of 0.115, 5.35, and 0.69 for C_Q, C_H, and C_P, respectively, for the maximum efficiency condition. Then for $n = (400/60)$ rps and $D = 0.371 \times 5.33 = 1.98$ m, we solve for P, ΔH, and Q:

$$P = C_P \rho D^5 n^3$$

$$= 0.69(1.0 \text{ kN} \cdot \text{s}^2/\text{m}^4)(1.98 \text{ m})^5 \left(\frac{400}{60} \text{ s}^{-1}\right)^3$$

$$= 6,200 \text{ kW} \qquad \blacktriangleleft$$

$$\Delta H = \frac{C_H D^2 n^2}{g} = \frac{5.35(1.98 \text{ m})^2 (400/60 \text{ s}^{-1})^2}{(9.81 \text{ m/s}^2)}$$

$$= 95.0 \text{ m} \qquad \blacktriangleleft$$

$$Q = C_Q n D^3$$

$$= 0.115 \left(\frac{400}{60} \text{ s}^{-1}\right) (1.98 \text{ m})^3$$

$$= 5.95 \text{ m}^3/\text{s} \qquad \blacktriangleleft$$

Compressors

Centrifugal compressors are similar in design to centrifugal pumps. Because the density of the air or gases used is much less than liquid densities, the compressor must turn at much higher speeds than the pump to effect a sizable pressure increase. If the compression process were isentropic and the gases ideal, the power necessary to compress the gas from p_1 to p_2 would be

$$P_{th} = \frac{k}{k-1} Q_1 p_1 \left[\left(\frac{p_2}{p_1}\right)^{(k-1)/k} - 1 \right] \qquad (14\text{-}17)$$

where Q_1 is the volume-flow rate into the compressor. The power calculated using Eq. (14-17) is referred to as the theoretical adiabatic power. The efficiency of a compressor with no water cooling is defined as the ratio of the theoretical adiabatic power to the actual power required at the shaft. Ordinarily the efficiency improves with higher inlet-volume-flow rates, increasing from a typical value of 0.60 at 0.6 m^3/s to 0.74 at 40 m^3/s. Higher efficiencies are obtainable with more expensive design refinements.

EXAMPLE 14-6   Determine the shaft power required to operate a compressor that compresses air at the rate of 1 m^3/s from 100 to 200 kPa. The efficiency of the compressor is 65%.

Solution   We first calculate the theoretical adiabatic power, Eq. (14-17), with $k = 1.4$.

$$P_{th} = \frac{k}{k-1} Q_1 p_1 \left[\left(\frac{p_2}{p_1}\right)^{(k-1)/k} - 1 \right]$$

$$= (3.5)(1 \text{ m}^3/\text{s})(10^5 \text{ N/m}^2)[(2)^{0.286} - 1]$$

$$= 0.767 \times 10^5 \text{ N} \cdot \text{m/s}$$

$$= 76.7 \text{ kW}$$

The shaft power needed is

$$P_{\text{shaft}} = \frac{76.7}{0.65} \text{ kW}$$

$$= 118 \text{ kW} \qquad \blacktriangleleft$$

Cooling is necessary on high-pressure compressors because of the high gas temperatures resulting from the compression process. Cooling can be achieved through the use of water jackets or by intercoolers that cool the gases between stages. The efficiency of water-cooled compressors is based on the power required to compress ideal gases isothermally; namely,

$$P_{th} = p_1 Q_1 \ln \frac{p_2}{p_1} \qquad (14\text{-}18)$$

which is usually called the theoretical isothermal power. The efficiencies of water-cooled compressors are generally lower than the noncooled compressors. If cooled by water jackets the efficiency characteristically ranges between 55 and 60%. The use of intercoolers results in efficiencies from 60 to 65%.

14-4   SPECIFIC SPEED

From the discussion in preceding sections we have seen that a pump's performance is given by the values of its power and head coefficients (C_P and C_H) for a range of values of the discharge coefficient C_Q. It has been noted that certain types of machines are best suited for certain head and discharge ranges. For example, an axial-flow machine is best suited for low heads and high discharges, while a radial-flow machine is best suited for higher heads and lower discharges. The parameter used to pick the type of pump (or turbine) best suited for a given application is specific speed n_s. Specific speed is obtained by combining both C_H and C_Q in such a manner that the diameter D is eliminated:

$$n_s = \frac{C_Q^{1/2}}{C_H^{3/4}} = \frac{(Q/nD^3)^{1/2}}{[\Delta H/(D^2 n^2/g)]^{3/4}} = \frac{n Q^{1/2}}{g^{3/4} \Delta H^{3/4}}$$

Thus specific speed relates different types of pumps without reference to size.

When the actual efficiencies of different types of pumps are plotted against n_s, it is seen that certain types of pumps have higher efficiencies for certain ranges of n_s. In fact, in the range between the completely axial- and radial-flow machine, there is a gradual change in impeller shape to accommodate the particular flow conditions with maximum efficiency (see Fig. 14-11).

It should be noted that the specific speed traditionally used for pumps in the United States is defined as $N_s = NQ^{1/2}/\Delta H^{3/4}$. Here the speed N is in revolutions per minute, Q is in gallons per minute, and ΔH is in feet. This form is not dimensionless; therefore, its values are much larger than those

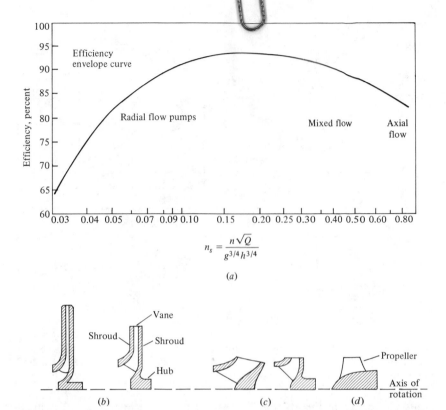

FIGURE 14-11   (*a*) Optimum efficiency and impeller designs versus specific speed n_s. (*b*) Radial-flow impellers. (*c*) Mixed-flow impellers. (*d*) Axial flow.

found for n_s (the conversion factor is 17,200). Thus most texts and references published before the introduction of the SI system of units use the traditional definition for specific speed.

14-5   SUCTION LIMITATIONS OF PUMPS

Because most centrifugal pumps used in a given range of n_s have about the same shape and performance characteristics, it is possible to establish certain general limitations based upon the flow conditions on the suction side of the pump. Such limitations are needed to prevent cavitation, which can cause a loss of efficiency and even structural damage. These limitations are published by the Hydraulic Institute (6) and are given in terms of maximum ΔH versus n_s for different suction lifts or suction heads. For example, a chart for single-suction mixed-flow and axial-flow

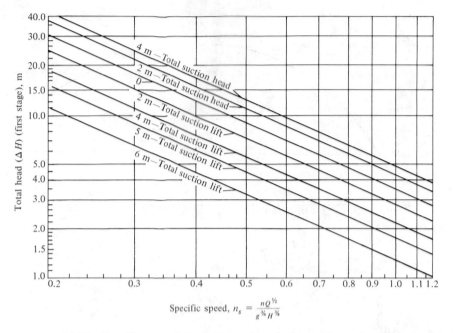

FIGURE 14-12 Specific speed limitations for single-suction mixed-flow and axial-flow pumps (pumping clear water, 30°C at sea level). [Adapted from the Standards of the Hydraulic Institute (6).]

pumps is shown in Fig. 14-12. The use of the chart in Fig. 14-12 is illustrated in Example 14-7.

EXAMPLE 14-7 An axial-flow pump is to be used to lift water from a main irrigation canal to a smaller irrigation canal at higher level. If the total head (elevation difference plus head losses in the pipe) is to be 11 m and if the total suction head is to be 2 m (the impeller is below the water level in the main canal), what is the safe upper limit of specific speed? Would it be safe to operate a pump at a speed of 1,200 rpm and with a discharge of 0.5 m³/s under these conditions?

Solution We enter Fig. 14-12 with a total head of 11 m and a total suction head of 2 m and read a value of n_s of 0.51. Thus the safe upper limit of specific speed is 0.51.

By definition we have

$$n_s = \frac{nQ^{1/2}}{g^{3/4} \Delta H^{3/4}}$$

Hence, for $N = 1,200 \, \text{rpm}$ or $n = 20 \, \text{rps}$, $Q = 0.50 \, \text{m}^3/\text{s}$, and $\Delta H = 11 \, \text{m}$. We compute n_s as follows:

$$n_s = \frac{20(0.50)^{1/2}}{(9.81)^{3/4}(11)^{3/4}} = 0.42$$

The n_s computed here is less than the allowable n_s; consequently, the stated operating conditions are *within the safe range*.   ◄

14-6   TURBINES

Introduction

Most basic theory and similarity parameters used for pumps also apply to turbines. However, there are some differences in physical features as well as terminology. We have also not considered details of the flow through the impellers of radial-flow machines; therefore, these items will now be presented.

The two broad categories of turbines are the *impulse* turbine and the *reaction* turbine. For hydroelectric installations the latter is further subdivided into the Francis type, which is characterized by a radial-flow impeller, and the Kaplan or propeller type, which is an axial-flow machine. We first consider the basic elements of the impulse turbine and then details of the Francis turbine.

Impulse turbine

In the impulse turbine a jet of fluid issuing from a nozzle impinges upon vanes of the turbine wheel or *runner*, thus producing power as the runner rotates (see Fig. 14-13). The primary feature of the impulse turbine with respect to fluid mechanics is the power production as the jet is deflected by the moving vane(s). When the momentum equation is applied to this deflected jet it can be shown [see Daugherty and Franzini (3)] for idealized conditions that the maximum power will be developed when the vane speed is one-half of the initial jet speed. With such conditions the exiting jet speed will be zero—all of the kinetic energy of the jet will have been expended in driving the vane. Thus, if we apply the energy equation, Eq. (7-26), between the incoming jet and the exiting fluid (assuming negligible head loss and nil kinetic energy at exit), we will find that the head given up to the turbine is $h_t = V_j^2/2g$ and the power thus developed will be

$$P = Q\gamma h_t \tag{14-19}$$

where Q = discharge of the incoming jet

γ = specific weight of jet fluid

$h_t = V_j^2/2g$ = velocity head of the jet

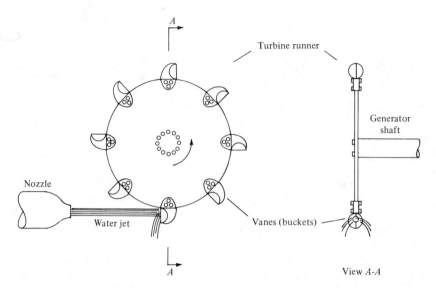

FIGURE 14-13   The impulse turbine.

Thus Eq. (14-19) reduces to

$$P = \rho Q \frac{V_j^2}{2} \qquad (14\text{-}20)$$

To obtain the torque on the turbine shaft, the angular-momentum equation, Eq. (6-27a), is applied to a control volume, as shown in Fig. 14-14. Then for steady flow we have

$$\sum \mathbf{M} = \sum_{cs} (\mathbf{r} \times \mathbf{V})\rho \mathbf{V} \cdot \mathbf{A}$$

or

$$\mathbf{T}_{\text{shaft}} = \sum_{cs} (\mathbf{r} \times \mathbf{V})\rho \mathbf{V} \cdot \mathbf{A} \qquad (14\text{-}21)$$

Now if we consider the angular momentum about the axis of the turbine shaft, we see that $\mathbf{r} \times \mathbf{V}$ for the entering jet will be simply $rV_j\mathbf{k}$ and $\mathbf{V} \cdot \mathbf{A}$ for the entering jet will be $-V_jA_j$ or $-Q$. In addition, we are assuming that the exiting jet has nil angular momentum; hence, the torque acting on the system—that is the torque on the shaft at the section where the control surface passes through the shaft—is given by $\mathbf{T} = -\rho QVr\mathbf{k}$, or the absolute value of the torque will be simply

$$T = \rho Q V_j r \qquad (14\text{-}22)$$

The power thus developed by the turbine will be $T\omega$ or

$$P = \rho Q V_j r \omega \qquad (14\text{-}23)$$

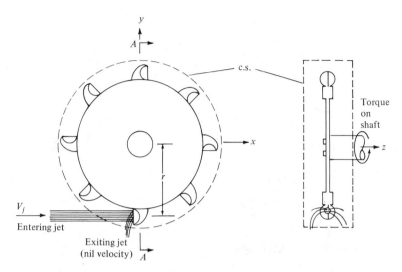

FIGURE 14-14 Control-volume approach for the impulse turbine using the angular-momentum principle.

Furthermore, if the velocity of the turbine vanes are $\frac{1}{2}V_j$ for maximum power, as noted earlier, we have $P = \rho Q V_j^2/2$, the same result as that given by Eq. (14-20).

EXAMPLE 14-8 What power in kilowatts can be developed by the impulse turbine shown if the turbine efficiency is 85%? Assume that the resistance coefficient f of the penstock is 0.015 and the head loss in the nozzle itself is nil. What will be the angular speed of the wheel assuming ideal conditions ($V_j = 2V_{\text{bucket}}$) and what torque will be exerted on the turbine shaft?

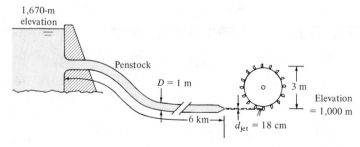

Solution First determine the jet velocity by applying the energy equation from the reservoir to the free jet before it strikes the turbine buckets.

$$\frac{p_1}{\gamma} + \frac{V_1^2}{2g} + z_1 = \frac{p_j}{\gamma} + \frac{V_j^2}{2g} + z_j + h_L$$

where
$$p_1 = 0$$
$$z_1 = 1{,}670 \text{ m}$$
$$V_1^2/2g = 0$$
$$p_j = 0$$
$$z_j = 1{,}000 \text{ m}$$
$$\gamma = 9{,}810 \text{ N/m}^3 \text{ at } 10° \text{ C (assumed)}$$

The penstock water velocity is

$$V_{\text{penstock}} = \frac{V_j A_j}{A_{\text{penstock}}} = 0.0324 \ V_j$$

Then
$$h_L = \frac{fL}{D}\frac{V^2}{2g} = \frac{0.015 \times 6{,}000}{1}(0.0324)^2 \frac{V_j^2}{2g} = 0.094 \frac{V_j^2}{2g}$$

Now solving the energy equation for V_j yields

$$V_j = \left(\frac{2g \times 670}{1.094}\right)^{1/2}$$

$$= 109.6 \text{ m/s}$$

The gross power is

$$P = Q\gamma\frac{V_j^2}{2g} = \frac{\gamma A_j V_j^3}{2g}$$

or
$$P = \frac{9{,}810(\pi/4)(0.18)^2(109.6)^3}{2 \times 9.81}$$

$$= 16{,}760 \text{ kW}$$

Power output of turbine:

$$P = 16{,}760 \times \text{efficiency} = 14{,}245 \text{ kW} \qquad \blacktriangleleft$$

The tangential bucket speed will be $\frac{1}{2}V_j$: therefore,

$$V_{\text{bucket}} = \frac{1}{2} \ 109.6 \text{ m/s} = 54.8 \text{ m/s}$$

or
$$r\omega = 54.8 \text{ m/s}$$

Thus
$$\omega = \frac{54.8 \text{ m/s}}{1.5 \text{ m}} = 36.53 \text{ rad/s}$$

Wheel speed:

$$N = (36.53 \text{ rad/s})\frac{1 \text{ rev}}{2\pi \text{ rad}} \ 60 \text{ s/min} = 349 \text{ rpm} \qquad \blacktriangleleft$$

$$\text{Power} = T\omega$$

Thus $\quad T = \dfrac{\text{power}}{\omega} = 14{,}245 \text{ kW}/36.53 \text{ rad/s} = 390 \text{ kN} \cdot \text{m}$ ◄

Characteristics of the reaction turbine

In contrast to the impulse turbine, where a jet under atmospheric pressure impinges upon only one or two vanes at a time, the flow in a reaction turbine is under pressure and this flow completely fills the chamber in which the impeller is located (see Fig. 14-15). There is a drop in pressure from the outer radius of the impeller, r_1, to the inner radius r_2. This is also different from the impulse turbine, where the pressure is the same for the entering and exiting flow. The original form of the reaction turbine, first extensively tested by J. B. Francis, had a complete radial-flow impeller (Fig. 14-16). That is, the flow passing through the impeller had velocity components only in a plane normal to the axis of the runner. However, more recent impeller designs such as the mixed-flow and axial-flow types are still called *reaction turbines*.

Torque and power relations for the reaction turbine

As we did with the impulse turbine, we will use the angular-momentum equation to develop formulas for the torque and power. The segment of turbine runner shown in Fig. 14-16 depicts the flow conditions that occur for the entire runner. We can see that guide vanes outside the runner itself cause the fluid to have a tangential component of velocity around the entire circumference of the runner. Thus the fluid will have an initial amount of angular momentum with respect to the turbine axis when it approaches the turbine runner. As the fluid passes through the passages of the runner, the runner vanes effect a change in the magnitude and direction of velocity. Thus the angular momentum of the fluid is changed, which produces a torque on the runner. This torque drives the runner which, in turn, generates power.

To quantify the above, we let V_1 and α_1 represent the incoming velocity and angle of the velocity vector with respect to a tangent to the runner, respectively. Similar terms at the inner-runner radius are V_2 and α_2. Applying the angular-momentum equation for steady flow, Eq. 14-21, to the control volume shown in Fig. 14-16 yields

$$
\begin{aligned}
T &= (-r_1 V_1 \cos \alpha_1)\rho(-Q) + (-r_2 V_2 \cos \alpha_2)\rho(+Q) \\
&= \rho Q(r_1 V_1 \cos \alpha_1 - r_2 V_2 \cos \alpha_2)
\end{aligned}
\tag{14-24}
$$

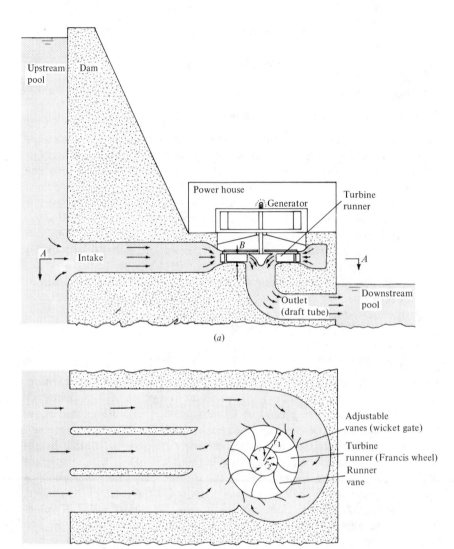

FIGURE 14-15 Schematic view of reaction turbine installation. (*a*) Elevation view. (*b*) Plan view—section *A-A*.

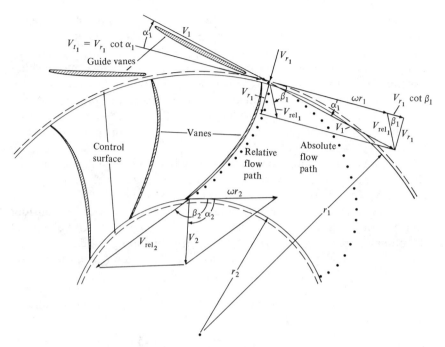

FIGURE 14-16 Velocity diagrams for a Francis-type runner.

The power from this turbine will be $T\omega$, or

$$P = \rho Q\omega(r_1 V_1 \cos \alpha_1 - r_2 V_2 \cos \alpha_2) \tag{14-25}$$

Equation (14-25) shows that the power production is a function of the direction of the flow velocities entering and leaving the impeller, α_1 and α_2.

It is interesting to note that even though the pressure varies within the flow in the reaction turbine, it does not enter into the expressions we have derived using the angular-momentum equation. The reason it does not appear is that the outer and inner control surfaces that we chose are concentric with the axis about which we are evaluating the moments and angular momentum. The pressure forces acting on these surfaces all pass through the given axis; therefore, they do not produce moments about the given axis.

Vane angles

It should be apparent that the head loss in a turbine will be less if the flow enters the runner with a direction tangent to the runner vanes than if the flow approaches the vane with an angle of attack. In the latter case, separation will occur with consequent head loss. Thus vanes of an impeller designed for a given speed and discharge and with fixed guide vanes will have a particular optimum blade angle β_1. However, if the discharge is changed from the original-design condition, the guide vanes and impeller vane angles will not "match" the new flow conditions. Most turbines for hydroelectric installations are made with movable guide vanes on the inlet side to effect a better match at all flows. Thus α_1 is increased or decreased automatically through governor action to accommodate fluctuating power demands on the turbine.

To relate the incoming-flow angle α_1 with the vane angle β_1, we first assume that the flow entering the impeller will be tangent to the blades at the periphery of the impeller. Likewise, the flow leaving the stationary guide vane is assumed to be tangent to the guide vane. Now we will consider both the radial component and the tangential components of velocity at the outer periphery of the wheel ($r = r_1$) in developing the desired equations. We can easily compute the radial velocity, given Q and the geometry of the wheel, by the continuity equation:

$$V_{r_1} = \frac{Q}{2\pi r_1 B} \tag{14-26}$$

where $\qquad\qquad B$ = height of turbine blades.

The tangential (tangent to the outer surface of the runner) velocity of the incoming flow is given as

$$V_{t_1} = V_{r_1} \cot \alpha_1 \tag{14-27}$$

However, in relation to the flow through the runner, this same tangential velocity is equal to the tangential component of relative velocity in the runner, $V_{r_1} \cot \beta_1$, plus the velocity of the runner itself, ωr_1. Thus the tangential velocity when viewed with respect to the runner motion is

$$V_{t_1} = r_1 \omega + V_{r_1} \cot \beta_1 \tag{14-28}$$

Now, upon eliminating V_{t_1} between Eqs. (14-27) and (14-28), we have

$$V_{r_1} \cot \alpha_1 = r_1 \omega + V_{r_1} \cot \beta_1 \tag{14-29}$$

Equation (14-29) can be rearranged to yield

$$\alpha_1 = \text{arccot} \left(\frac{r_1 \omega}{V_{r_1}} + \cot \beta_1 \right) \tag{14-30}$$

EXAMPLE 14-9   A Francis turbine is to be operated at a speed of 600 rpm and with a discharge of 4.0 m³/s. If $r_1 = 0.60$ m, $\beta_1 = 110°$, and the blade height B is 10 cm, what should be the guide vane angle α_1 for a nonseparating flow condition at the runner entrance?

Solution

$$\alpha_1 = \text{arccot}\left(\frac{r_1\omega}{V_{r_1}} + \cot \beta_1\right)$$

where

$r_1\omega = 0.6 \text{ m} \times 600 \text{ rpm} \times 2\pi \text{ rad/rev}$
$\times 1/60 \text{ min/s} = 37.7 \text{ m/s}$

$$V_{r_1} = \frac{Q}{2\pi r_1 B} = \frac{4.00 \text{ m}^3/\text{s}}{2\pi \times 0.6 \text{ m} \times 0.10 \text{ m}} = 10.61 \text{ m/s}$$

$\cot \beta_1 = -0.364$

Then,

$\alpha_1 = \text{arccot}(3.55 - 0.364)$

$= 17.4°$ ◀

Turbine specific speed

Because of the attention focused on the production of power for turbines, the specific speed is defined in terms of power:

$$n_s = \frac{nP^{1/2}}{g^{3/4}\gamma^{1/2}h_t^{5/4}}$$

It should also be noted that large water turbines are innately more efficient than pumps. For additional details of hydropower turbines, the reader is directed to references such as Daugherty and Franzini (3).

Gas turbines

The conventional gas turbine consists of a compressor that pressurizes the air entering the turbine and delivers it to a combustion chamber. The high-temperature high-pressure gases resulting from combustion in the combustion chamber expand through a turbine, which drives the compressor and provides power. The efficiency (power delivered/rate of energy input) of an ideal gas turbine depends on the pressure ratio between the combustion chamber and the intake; the higher the pressure ratio, the higher the efficiency. The reader is directed to other references such as (2) for more detail.

Wind turbines

Extraction of energy from the wind by a wind turbine is discussed frequently as an alternative energy source. In essence, the wind turbine is just the reverse process of introducing energy into an airstream to derive a propulsive force. The wind turbine extracts energy from the wind to produce power. There is one significant difference, however. The theoretical upper limit of efficiency of a propeller supplying energy to an airstream is 100%; that is, it is theoretically possible, neglecting viscous and other effects, to convert all the energy supplied to a propeller into energy of the airstream. The theoretical upper limit of wind-turbine efficiency as given by Glauert (4), on the other hand, is 16/27 or 59.3%. Thus the theoretical maximum power deliverable by a wind turbine with capture area A is

$$P_{max} = \frac{16}{27} \cdot \frac{1}{2} \rho U^3 A$$

where ρ is the air density and U is the wind speed. The capture area is the area swept by the wind turbine viewed from the wind direction. Other factors, such as swirl of the airstream and viscous effects, further reduce the wind turbine's efficiency.

The conventional wind turbine consists of a propeller mounted on a horizontal axis with a vane, or other device, to align the propeller shaft in the wind direction. In recent years considerable effort has been devoted to the assessment of the Savonius and Darrieus wind turbines, both vertical-axis turbines as show in Fig. 14-17. The Savonius rotor consists of two curved blades forming an S-shaped passage for the air flow. The Darrieus turbine consists of two or three airfoils attached to a vertical shaft; thus the unit resembles an egg beater. The advantage of vertical axis turbines is that their operation is independent of wind direction. The Darrieus wind turbine is considered superior in performance but has a disadvantage in that it is not self-starting. Frequently, a Savonius rotor is mounted on the axis of a Darrieus turbine to provide the starting torque. For more information on wind turbines the reader may refer to the proceedings of the Cambridge Wind Energy Systems Symposium (9).

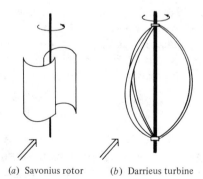

(a) Savonius rotor   (b) Darrieus turbine

FIGURE 14-17

14-7 VISCOUS EFFECTS

In the foregoing sections, similarity parameters have been developed to predict prototype results from model tests, neglecting viscous effects. Such an assumption is not necessarily valid, especially if the model is quite small. To minimize the viscous effects in modeling pumps, the Hydraulic Institute Standards (6) recommend that the size of the model be such that the model impeller is not less than 30 cm in diameter. These same Hydraulic Institute Standards, B-146(e), state that "the model should have complete geometric similarity with the prototype, not only in the pump proper, but also in the intake and discharge conduits."

Even with complete geometric similarity, one can expect the model to be less efficient than the prototype. An empirical formula proposed by Moody that is used for estimating prototype efficiencies of radial- and mixed-flow pumps and turbines from model efficiencies is

$$\frac{1 - e_1}{1 - e} = \left(\frac{D}{D_1}\right)^{1/5} \tag{14-31}$$

Here e_1 is the efficiency of the model and e is the efficiency of the prototype.

EXAMPLE 14-10 A model having an impeller diameter of 45 cm is tested and found to have an efficiency of 85%. If a geometrically similar prototype has an impeller diameter of 1.80 m, estimate its efficiency when it is operating under conditions which are dynamically similar to those in the model test ($C_{Q\,\text{model}} = C_{Q\,\text{prototype}}$).

Solution We apply Eq. (14-31) with the condition that $e_1 = 0.85$ and $D/D_1 = 4$. Then

$$e = 1 - \frac{1 - e_1}{(D/D_1)^{1/5}}$$

$$= 1 - \frac{0.15}{1.32}$$

$$= 1 - 0.11 = 0.89$$

The efficiency of the prototype is estimated to be 89%. ◄

PROBLEMS

14-1   What thrust is obtained from a 3.0-m diameter propeller that has the characteristics given in Fig. 14-3 when the propeller is operated at an angular speed of 1,400 rpm and zero advance velocity? Assume that $\rho = 1.05$ kg/m³.

14-2   What thrust is obtained from the 3-m-diameter propeller that has the characteristics given in Fig. 14-3 when the propeller is operated at an angular speed of 1,400 rpm and an advance velocity of 80 km/h? What power is required to operate the propeller under these conditions? Assume that $\rho = 1.1$ kg/m³.

14-3   An 8-ft-diameter propeller has the characteristics shown in Fig. 14-3. What thrust is produced by the propeller when operating at an angular speed of 1,000 rpm and a forward speed of 30 mph? What power input is required under these operating conditions? If the forward speed is reduced to zero, what is the thrust? Assume that $\rho = 0.0024$ slugs/ft³.

14-4   A 6-ft-diameter propeller like the one for which the characteristics are given in Fig. 14-3 is to be used on a swamp boat and is to operate at maximum efficiency when cruising. If the cruising speed is 30 mph, what should be the angular speed of the propeller?

14-5   For the propeller and conditions described in Prob. 14-4, determine the thrust and power input.

14-6   A 2.0-m-diameter propeller like the one for which the characteristics are given in Fig. 14-3 is to be used on a swamp boat and is to operate at maximum efficiency when cruising. If the cruising speed is 50 km/h, what should be the angular speed of the propeller?

14-7   For the propeller and conditions described in Prob. 14-6, determine the thrust and power input. Assume $\rho = 1.1$ kg/m³.

14-8   A 2.0-m-diameter propeller like the one with the characteristics given in Fig. 14-3 is used on a swamp boat. If the angular speed is 1,000 rpm and if the boat and passengers have a combined mass of 300 kg, estimate the initial acceleration of the boat when starting from rest. Assume $\rho = 1.1$ kg/m³.

14-9   If the tip speed of a propeller is to be kept below 0.9 c, where c is the speed of sound, what is the maximum allowable angular speed of propellers having diameters of 2 m (6.56 ft), 3 m (9.84 ft), and 4 m (13.12 ft)? Take the speed of sound at 335 m/s (1,099 ft/s).

14-10 If the pump having the characteristics shown in Fig. 14-6 has a diameter of 40 cm and is operated at a speed of 1,000 rpm, what will be the discharge when the head is 3 m?

14-11 If a pump that is geometrically similar to the pump having the characteristics of that in Fig. 14-7 is operated at the same speed (690 rpm) but is twice as large, $D = 71.2$ cm, what will be the water discharge and power demand when the head is 10 m?

14-12 For a pump having the characteristics given by Fig. 14-6 or 14-7, what water discharge and head will be produced when operating at maximum efficiency if the pump size is 24 in. and the angular speed is 1,100 rpm? What power is required under these conditions?

14-13 A pump has the characteristics given by Fig. 14-6. What discharge and head will be produced when operating at maximum efficiency if the pump size is 50 cm and the angular speed is 45 rps? What power is required under these conditions when pumping water?

14-14 For a pump having the characteristics of Fig. 14-6, plot the head-discharge curve if the pump is 14 in. in diameter and is operated at a speed of 900 rpm.

14-15 For a pump having the characteristics of Fig. 14-6, plot the head-discharge curve if the pump size is 60 cm and if the speed is 690 rpm.

14-16 The pump used in the system shown has the characteristics given in Fig. 14-7. What discharge will occur under the conditions shown and what power is required?

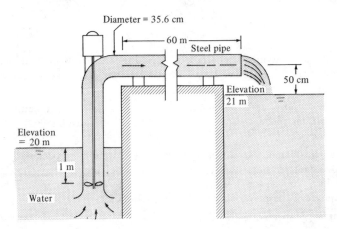

14-17 If the conditions are the same as in Prob. 14-16 except that the speed is increased to 900 rpm, what discharge will occur and what power is required for the operation?

14-18 A pump having the characteristics given in Fig. 14-9 pumps water from a reservoir at elevation 366 m to a reservoir at elevation 450 m through a 36-cm steel pipe. If the pipe is 610 m in length, what will be the discharge through the pipe?

14-19 If a pump having the characteristics as in Fig. 14-9 or 14-10 is operated at a speed of 1,500 rpm, what will be the discharge when the head is 160 ft?

14-20 If the pump having the performance curve shown is operated at a speed of 1,500 rpm, what would be the maximum possible head developed?

14-21 If the pump of Fig. 14-9 is operated at a speed of 30 rps, what would be the shutoff head?

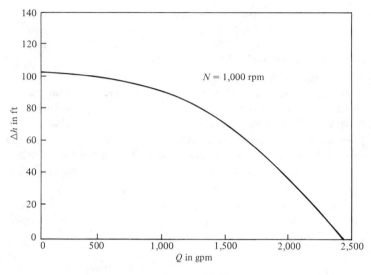

PROBLEM 14-20

14-22 If the pump having the characteristics in Fig. 14-10 is 40 cm in diameter and is operated at a speed of 25 rps, what will be the discharge when the head is 50 m?

14-23 If the pump having the characteristics in Fig. 14-9 is doubled in size but halved in speed, what will be the head and discharge at maximum efficiency?

14-24 A 20-cm-diameter centrifugal pump is used to pump kerosene at a speed of 5,000 rpm. Assume that the pump has the characteristics shown in Fig. 14-10. Calculate the flow rate, pressure rise across the pump, and the power required if the pump operates at maximum efficiency.

14-25 For a pump having the characteristics as shown in Fig. 14-10, plot the head-discharge curve if the pump diameter is 1.52 m and the speed is 500 rpm.

14-26 What is the specific speed for the pump that is operating under the conditions given in Prob. 14-16? Is this a safe operation with respect to the susceptibility to cavitation?

14-27 What type of pump should be used to pump water at a rate of 12 cfs and under a head of 25 ft? Assume that $N = 1,500$ rpm.

14-28 What type of pump should be used to pump water at a rate of 0.30 m³/s and under a head of 8 m for most efficient operation? Assume that $n = 25$ rps.

14-29 What type of pump should be used to pump water at a rate of 0.40 m³/s and under a head of 70 m? Assume that $N = 1,100$ rpm.

14-30 What type of pump should be used to pump water at a rate of 12 cfs and under a head of 600 ft? Assume that $N = 1,100$ rpm.

14-31 An axial-flow pump is to be used to lift water against a head (friction and static) of 5.0 m. If the discharge is to be 0.40 m³/s, what maximum speed in revolutions per minute is allowed if the suction head is 1.5 m?

14-32 You want to pump water at a rate of 1.0 m³/s from the lower to upper reservoir. What type of pump would you suggest using for this operation if the impeller speed is to be 600 rpm?

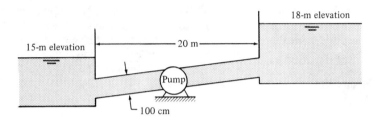

14-33 An axial-flow blower is used for a wind tunnel that has a 60 × 60 cm test section and is capable of airspeeds up to 30 m/s. If the blower is to operate at maximum efficiency at the highest speed and the rotational speed of the blower is 2,000 rpm at this condition, what is the diameter of the blower and the power required? Assume that the blower has the characteristics shown in Fig. 14-6.

14-34 An axial-flow blower is used to air-condition an office building that has a volume of 10^5 m³. It is decided that the air in the building must be completely changed every 15 min. Assume that the blower operates at 600 rpm at maximum efficiency and has the characteristics shown in Fig. 14-6. Calculate the diameter and power requirements for two blowers operating in parallel.

14-35 Methane flowing at the rate of 1 kg/s is to be compressed by a noncooled centrifugal compressor from 100 to 150 kPa. The temperature of the methane entering the compressor is 27°C. The efficiency of the compressor is 65%. Calculate the shaft power necessary to run the compressor.

14-36 A 10-kW (shaft output) motor is available to run a noncooled compressor for carbon dioxide. The pressure is to be increased from 90 to 140 kPa. If the compressor is 60% efficient, calculate the volume flow rate into the compressor.

14-37 A water-cooled centrifugal compressor is used to compress air from 100 to 400 kPa at the rate of 1 kg/s. The temperature of the inlet air is 15°C. The efficiency of the compressor is 50%. Calculate the necessary shaft horsepower.

14-38 A 1-m-diameter by 10-km-long penstock carries water from a reservoir to an impulse turbine. If the turbine is 83% efficient, what power can be produced by the system if the upstream reservoir elevation is 650 m above the turbine jet and the jet diameter is 16.0 cm? Assume that $f = 0.016$ and neglect head losses in the nozzle. What should be the diameter of the turbine wheel if it is to have an angular speed of 360 rpm? Assume ideal conditions for the bucket design ($V_{bucket} = \frac{1}{2}V_j$).

14-39 Consider an idealized bucket on an impulse turbine that turns the water through 180°. Prove that the bucket speed should be one-half the incoming jet speed for maximum power production. *Hint*: Set up the momentum equation to solve for the force on the bucket in terms of the V_j and V_{bucket}; then the power will be given by this force times V_{bucket}. You can use your good mathematical talent to complete the problem.

14-40 Consider a single jet of water striking the bucket of the impulse wheel as shown. Assume ideal conditions for power generation ($V_{bucket} = \frac{1}{2}V_j$ and the jet is turned through 180° of arc). With the foregoing conditions, solve for the jet force on the bucket and then solve for the power thus developed. Note that this power is not the same as that given by Eq. (14-20)! Study the figure shown to resolve this discrepancy.

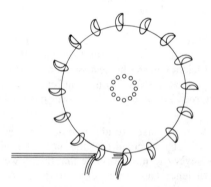

14-41 A Francis turbine is to be operated at a speed of 60 rpm and with a discharge of 4.0 m³/s. If $r_1 = 1.5$ m, $r_2 = 1.20$ m, $B = 30$ cm, $\beta_1 = 85°$, and $\beta_2 = 165°$, what should be α_1 for nonseparating flow to occur through the runner? What power and torque should result with this operation?

14-42 A Francis turbine is to be operated at a speed of 120 rpm and with a discharge of 113 m³/s. If $r_1 = 2.50$ m, $B = 0.90$ m, and $\beta_1 = 45°$, what should be α_1 for nonseparating flow at the runner inlet?

14-43 (a) For a given Francis turbine $\beta_1 = 60°$, $\beta_2 = 90°$, $r_1 = 5$ m, $r_2 = 3$ m, and $B = 1$ m. What should be α_1 for a nonseparating flow condition at the entrance to the runner when the discharge rate is 126 m³/s and $N = 60$ rpm? (b) What is the maximum attainable power with the conditions noted above? (c) If you were to redesign the turbine blades of the runner noted above, what changes would you suggest to increase the power production if the discharge and overall dimensions are to be kept the same?

14-44 Calculate the maximum power derivable from a conventional, 2-m-diameter horizontal axis wind turbine in a 50 km/h wind with a density of 1.2 kg/m³.

14-45 Calculate the minimum possible capture area of a windmill necessary to operate five 100-watt bulbs if the wind velocity is 20 km/h and the density is 1.2 kg/m³.

REFERENCES

1. Church, A. H. *Centrifugal Pumps and Blowers*. John Wiley & Sons, Inc., New York, 1944.

2. Cohen, H., Rogers, G. F. C., and Saravanamuttoo, H. I. H. *Gas Turbine Theory*. John Wiley & Sons, New York, 1972.

3. Daugherty, Robert L., and Franzini, Joseph B. *Fluid Mechanics with Engineering Applications*. McGraw-Hill Book Company, New York, 1957.

4. Glauert, H. "Airplane Propellers." *Aerodynamic Theory*, Vol. IV, ed. W. F. Durand. Dover Publications, Inc., New York, 1963.

5. Hicks, T. G. *Pump Operation and Maintenance*. McGraw-Hill Book Company, New York, 1958.

6. *Hydraulic Institute Standards*, 12th ed., Hydraulic Institute, New York, 1969.

7. Karassick, I. J., and Carter, R. *Centrifugal Pumps*. F. W. Dodge Company, a division of McGraw-Hill, Inc., New York, 1960.

8. Marks, Lionel S. (ed.). *Mechanical Engineers' Handbook*. McGraw-Hill Book Company, New York, 1951.

9. Proceedings of Symposium on Wind Energy Systems. BHRA, Cranfield, England (1976).

10. Sorensen, H. A. *Gas Turbines*. The Ronald Press Company, New York, 1951.

11. Spannhake, W. *Centrifugal Pumps, Turbines, and Propellers*. The Technology Press of the Massachusetts Institute of Technology, Cambridge, Mass., 1934.

12. Spannhake, W. "Problems of Modern Pump and Turbine Design." *Trans. ASME*, 56 (no. 4) (1934), 225.

13. Stepanoff, A. J. *Centrifugal and Axial Flow Pumps*, 2nd ed. John Wiley & Sons, Inc., New York, 1957.

14. Weick, F. E. *Aircraft Propeller Design*. McGraw-Hill Book Company, New York, 1930.

15. Weick, Fred E. "Full Scale Tests on a Thin Metal Propeller at Various Pit Speeds." *NACA Report* 302 (January 1929).

Flow in natural streams, like this one in northern Idaho, is not only pleasing to the eye but is also one of the most challenging phenomena for researchers to analyze.

15 VARIED FLOW IN OPEN CHANNELS

I N CHAPTER 8 IT WAS NOTED that when gravity influences the flow pattern, the Froude number is a significant correlating parameter. To this point in the text, except for dimensional analysis and flow over weirs, all the problems considered were ones for which either the Reynolds number or the Mach number were the significant correlating parameters. In this chapter, we will focus on liquid flow in open channels for which the force of gravity is a very significant variable; and it will be shown that the Froude number is indeed a significant parameter for such flows. Even though the primary emphasis here will be on water flow in open channels, some of the basic principles developed will also have relevance to other gravity-affected phenomena such as the behavior of density currents either in the atmosphere or in lakes and oceans.

15-1 ENERGY RELATIONS IN OPEN CHANNELS

The energy equation applied to open-channel flow

In our consideration of open-channel flow, we will make the simplifying assumption that the channel is rectangular in cross section; and it will be assumed that the kinetic energy factor α is equal to unity. Hence, with these assumptions, the one-dimensional energy equation for open channels (see Fig. 15-1) is

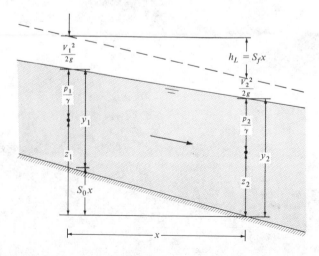

FIGURE 15-1 Definition sketch for flow in open channels.

$$\frac{p_1}{\gamma} + \frac{V_1^2}{2g} + z_1 = \frac{p_2}{\gamma} + \frac{V_2^2}{2g} + z_2 + h_L \qquad (15\text{-}1)$$

The pressure is nearly hydrostatically distributed because the flow is nearly uniform at each section. Thus $p/\gamma + z$ is constant at each section and we see from Fig. 15-1 that the following equalities hold:

$$\frac{p_1}{\gamma} + z_1 = y_1 + S_0 x \qquad \text{and} \qquad \frac{p_2}{\gamma} + z_2 = y_2$$

Here S_0 is the slope of the channel bottom and y is the depth of flow. Hence we can write Eq. (15-1) as

$$y_1 + \frac{V_1^2}{2g} + S_0 x = y_2 + \frac{V_2^2}{2g} + h_L \qquad (15\text{-}2)$$

Now, if we consider the special case where the channel bottom is horizontal ($S_0 = 0$) and the head loss is zero ($h_L = 0$), then Eq. (15-2) becomes

$$y_1 + \frac{V_1^2}{2g} = y_2 + \frac{V_2^2}{2g} \qquad (15\text{-}3)$$

Specific energy

The sum of the depth of flow and the velocity head is defined as the *specific energy*:

$$E = y + \frac{V^2}{2g} \qquad (15\text{-}4)$$

Thus Eq. (15-3) states that the specific energy at section 1 is equal to the specific energy at section 2, or $E_1 = E_2$. The continuity equation between sections 1 and 2 will be

$$y_1 V_1 B_1 = y_2 V_2 B_2 \qquad (15\text{-}5)$$

and $B_1 = B_2 = B$ for a rectangular channel of constant width B; therefore, Eq. (15-5) reduces to

$$y_1 V_1 = y_2 V_2 = q \qquad (15\text{-}6)$$

In Eq. (15-6), q is the discharge per unit of width of channel. Now the velocities V_1 and V_2 can be expressed as a function of q and y:

$$V_1 = \frac{q}{y_1} \qquad (15\text{-}7)$$

$$V_2 = \frac{q}{y_2} \qquad (15\text{-}8)$$

Hence, when Eqs. (15-7) and (15-8) are substituted into Eq. (15-3), one obtains

$$y_1 + \frac{q^2}{2gy_1^2} = y_2 + \frac{q^2}{2gy_2^2} \qquad (15\text{-}9)$$

Thus, for a given discharge (q = constant), it is seen that the magnitude of the specific energy at sections 1 or 2 is solely a function of the depth of flow at each section. Utilizing Eq. (15-9) for the relation between y, q, and E reveals that, in general, the depth may have two values for a given value of specific energy. The relationship is shown in Fig. 15-2, where y is plotted as a function of E for a given discharge. It is seen that for a given

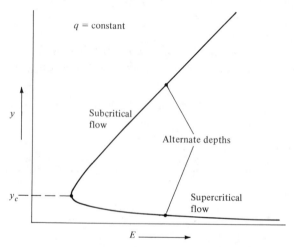

FIGURE 15-2   Depth-versus-specific-energy relation.

value of specific energy, the depth may be either large or small. In a physical sense, this means that for a low depth, the bulk of the energy of flow is in the form of kinetic energy ($q^2/2gy^2$); whereas for the larger depth, most of the energy is in the form of potential energy. Flow under a *sluice gate* (Fig. 15-3) is an example of flow in which two depths occur for a given value of specific energy. The large depth and low kinetic energy occur upstream of the gate; the low depth and large value of kinetic energy occur downstream. The depths as used here are called *alternate depths*. That is, for a given value of E, the large depth is alternate to the smaller depth, or vice versa. Returning to the flow under the sluice gate, we find that if we maintain the same rate of flow but set the gate with a larger opening as in Fig. 15-3b, the upstream depth will drop and the downstream depth will rise. Thus we have different alternate depths and a smaller value of specific energy than before. This is consistent with the diagram in Fig. 15-2.

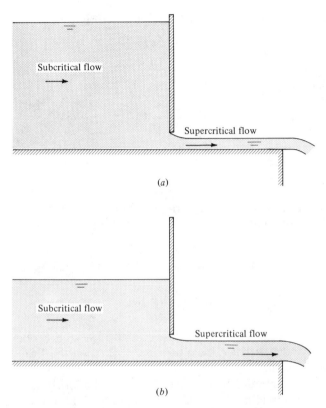

FIGURE 15-3   Flow under a sluice gate.

Finally, it can be seen in Fig. 15-2 that a point will be reached where the specific energy is minimum and only a single depth occurs. At this point the flow is termed *critical*. Thus it is seen that one definition of critical flow is the flow that occurs when the specific energy is minimum for a given discharge. Flow for which the depth is less than critical (velocity is greater than critical) is termed *supercritical flow,* and flow for which the depth is greater than critical (velocity less than critical) is termed *subcritical flow*. Using this terminology, we can see that it is apparent that subcritical flow occurs upstream and supercritical flow occurs downstream of the sluice gate in Fig. 15-3. Other aspects of critical flow will be considered in the next section.

Characteristics of critical flow

We have already seen that critical flow occurs when the specific energy is minimum for a given discharge. This condition may be quantified if we solve for dE/dy from $E = y + q^2/2gy^2$ and set dE/dy equal to zero:

$$\frac{dE}{dy} = 1 - \frac{2q^2}{2gy_c^3} = 0$$

from which we obtain

$$\frac{q^2}{gy_c^3} = 1 \tag{15-10}$$

or

$$y_c = \left(\frac{q^2}{g}\right)^{1/3} \tag{15-11}$$

Thus we now have a means of computing the critical depth y_c.

The relationship between the Froude number and critical flow is given below. We know that $q = Vy$; hence, Eq. (15-10) can be written as

$$\frac{V^2}{gy_c} = 1 \tag{15-12}$$

$$\frac{V}{\sqrt{gy_c}} = 1 \tag{15-13}$$

The left side of Eq. (15-13) will be recognized as the Froude number; hence, we conclude that critical flow prevails when the Froude number has a value of unity.

Critical flow may also be characterized in still another way if we analyze Eq. (15-12) in more detail. Multiplying Eq. (15-12) through by y_c and dividing by 2, we obtain

$$\frac{V^2}{2g} = \frac{y_c}{2} \tag{15-14}$$

Thus it is seen that critical flow exists when the velocity head is equal to one-half the depth.

Finally, if we express E as in Eq. (15-9) we have $E = y + q^2/2gy^2$. Now, if we determine how q varies with y for a constant value of specific energy, it will be seen that critical flow occurs when the discharge is maximum (Fig. 15-4).

Originally the term critical flow probably related to the unstable character of the flow for this condition. If we refer to Fig. 15-2, it may be seen that only a slight change in specific energy will cause the depth to rise or fall a significant amount: this is a very unstable condition. In fact, observations of critical flow in open channels show that the water surface consists of a series of standing waves. Because of the unstable nature of the depth in critical flow, it is usually best to design canals so that normal depth is either well above or well below critical depth. As a point of

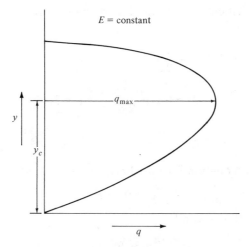

FIGURE 15-4   Variation of q and y with constant specific energy.

interest, the flow in canals and rivers is usually subcritical; however, the flow in steep chutes or over spillways is supercritical.

In this section, various characteristics of critical flow have been explored. The main ones can be summarized as follows:

1. Critical depth is given as $y_c = (q^2/g)^{1/3}$.
2. Critical flow occurs when $F = 1$.
2. Critical flow occurs when the specific energy is minimum for a given discharge.
4. Critical flow occurs when the discharge is maximum for a given specific energy.
5. The velocity head in critical flow is half the depth.

EXAMPLE 15-1   Water flows in a 6-m-wide rectangular channel at a depth of 3 m and has a discharge of 30 m³/s. Compute the Froude number, classify the flow, and determine the alternate depth. Also compute critical depth for this discharge.

Solution   Because $Q = VA = VBy$, we can easily compute the velocity:

$$V = \frac{Q}{A} = \frac{30 \text{ m}^3/\text{s}}{(3 \text{ m})(6 \text{ m})} = 1.67 \text{ m/s}$$

Then

$$F = \frac{V}{(gy)^{1/2}} = \frac{1.67 \text{ m/s}}{[(9.81 \text{ m/s}^2)(3 \text{ m})]^{1/2}}$$

$$= 0.31$$

The flow is subcritical because the Froude number is less than 1. The specific energy

$$E = y + \frac{V^2}{2g}$$

$$= 3m + \frac{1.67^2 \text{ m}^2/\text{s}^2}{2 \times 9.81 \text{ m/s}^2} = 3.14 \text{ m}$$

Also, $$E = y + \frac{q^2}{2gy^2}$$

Solving this last equation for the alternate depth yields

$$y = 0.72 \text{ m} \quad \blacktriangleleft$$

The critical depth is given by

$$y_c = \left(\frac{q^2}{g}\right)^{1/3} = \left(\frac{5.0^2}{9.81}\right)^{1/3} = 1.37 \text{ m} \quad \blacktriangleleft$$

Occurrence of critical depth

Critical flow will occur when a liquid passes over a broad-crested weir (Fig. 15-5a). The principle of the broad-crested weir is illustrated by considering first a closed sluice gate that prevents water from being

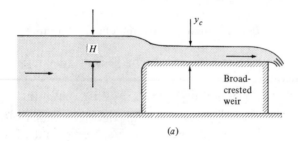

(a)

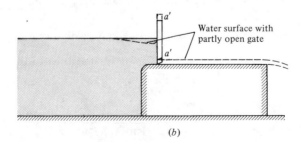

(b)

FIGURE 15-5   Flow over a broad-crested weir. Views (a) and (b).

discharged from the reservoir (Fig. 15-5b). Now, if the gate is opened a small amount (gate position a'-a') the flow upstream of the gate will be subcritical and the flow downstream will be supercritical (like the condition first introduced in Fig. 15-3). Then, as the gate is opened further, a point is finally reached where the depths upstream and downstream are the same. This is the critical condition. At this gate opening and beyond, the gate has no influence on the flow; this is the condition shown in Fig. 15-5a, the broad-crested weir. If the depth of flow over the weir is measured, the rate of flow is easily computed from Eq. (15-10):

$$q = \sqrt{gy_c^3}$$

or

$$Q = L\sqrt{gy_c^3} \tag{15-15}$$

where L is the length of the weir.

Because $\frac{1}{2}y_c = V_c^2/2g$ [from Eq. (15-14)], it is easily shown that $y_c = \frac{2}{3}E$; hence Eq. (15-15) can be rewritten as

$$Q = L\sqrt{g} \left(\frac{2}{3}\right)^{3/2} E_c^{3/2}$$

$$= 0.545\sqrt{g}\, LE_c^{3/2}$$

or, for high weirs where $V_1^2/2g$ is nil,

$$Q = 0.545\sqrt{g}\, LH^{3/2} \tag{15-16}$$

Equations (15-15) and (15-16) are the basic theoretical equations for a broad-crested weir and will yield reasonable discharge approximations; however, for more precise evaluation of the discharge, the reader is directed to references in Chow (2).

The depth also passes through a critical stage in channel flow where the slope changes from a mild one to a steep one. Here a mild slope is defined as a slope for which the normal depth y_n is greater than y_c. Likewise, a steep slope is one for which $y_n < y_c$. Such a condition is shown in Fig. 15-6. Note that y_c is the same for both slopes in the figure because y_c is a function of the discharge only. However, normal depth (uniform flow depth) for the "mild" upstream channel is greater than critical, whereas

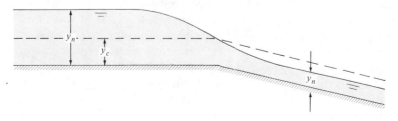

FIGURE 15-6   Critical depth at break in grade.

the normal depth in the "steep" downstream channel is less than critical; hence it is obvious that the depth must pass through a critical stage. Critical depth will occur a very short distance upstream of the intersection of the two channels.

Another place where critical depth occurs is near a free overfall at the end of a channel with a mild slope (Fig. 15-7). Critical depth will occur at a distance of 3 or 4 y_c upstream of the brink. Such occurrences of critical depth (at a break in grade or at a brink) are very useful in the computations of surface profiles because they provide a point for starting surface-profile calculations. The procedure for making such computations will be covered in a later section.

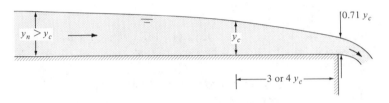

FIGURE 15-7   Critical depth at a free overfall.

Channel transitions

Whenever a channel has its cross-sectional configuration (shape or dimension) changed along its length, the change is termed a *transition*. We will use the basic concepts already presented in the first part of this chapter to show how the flow depth changes when the floor of a rectangular channel is increased in elevation or when the width of the channel is decreased. In these developments we assume negligible energy losses. We will look first at the case where the floor of the channel is raised (an upstep).

Consider the channel shown in Fig. 15-8, where the floor is raised an amount Δz. To help us in evaluating depth changes, we use the diagram of specific energy versus depth, which is similar to Fig. 15-2. This diagram is placed both at the section upstream of the transition and just downstream of the transition. Because the flow per unit width, q, is the same at both sections, the given diagram is valid at both sections. As noted in Fig. 15-8, the depth of flow at section 1 can be either large (subcritical) or small (supercritical) if the specific energy E_1 is greater than that required for critical flow. It can also be seen in Fig. 15-8 that if the flow were initially subcritical, a decrease in depth would occur in the region of the elevated channel bottom. This occurs because the specific energy at this section, E_2, is less than at section 1 by the amount Δz; therefore, the specific-energy diagram indicates that y_2 will be less than y_1. In a similar manner it

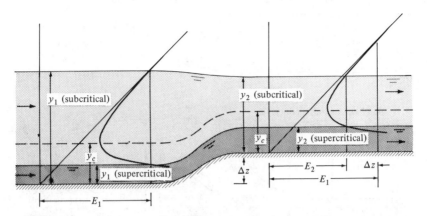

FIGURE 15-8   Changes in depth with change in bottom elevation.

can be seen that if the flow were initially supercritical, the depth as well as the actual water-surface elevation would increase from section 1 to section 2. A further note should be made with respect to possible flow depths with a change in bottom-surface elevation. If the channel bottom at section 2 is at an elevation greater than that just sufficient to establish critical flow at section 2, then there is not enough head at section 1 to cause flow to occur over the rise under steady-flow conditions. What will occur is that the water level upstream will have to rise until it is just sufficient to reestablish steady flow.

When the channel bottom is kept at the same elevation but the channel is decreased in width, then the discharge per unit of width between section 1 and 2 will increase but the specific energy E will remain constant. Thus, when we utilize the diagram of q versus depth for the given specific energy E, we note changes in depth similar to that for a rise in the channel bottom (see Fig. 15-9). For additional details about the design and analysis of more complex transitions refer to Chow (2).

Wave celerity

Wave celerity is the speed at which an infinitesimally small wave travels in a fluid. The following is a derivation of the wave celerity c.

Consider a small solitary wave moving with a velocity c in an otherwise calm body of liquid (Fig. 15-10a). Because the velocity in the liquid changes with time, this is a condition of unsteady flow. However, if we refer all velocities to a reference frame moving with the wave, then the shape of the wave is fixed and the flow is steady. Then the flow is amenable to analysis with the Bernoulli equation. The steady-flow condition is shown in Fig. 15-10b. Now, when the Bernoulli equation is written

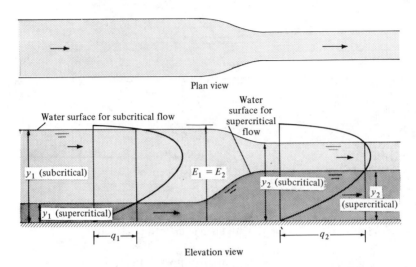

FIGURE 15-9 Changes in depth with change in channel width.

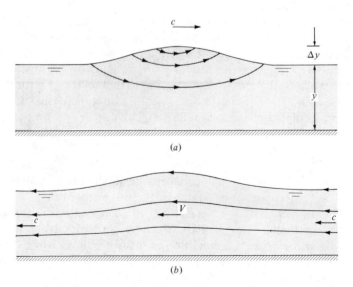

FIGURE 5-10 Solitary wave (exaggerated vertical scale). (*a*) Unsteady. (*b*) Steady.

between a point on the surface of the undisturbed fluid and a point at the wave crest, we have

$$\frac{c^2}{2g} + y = \frac{V^2}{2g} + y + \Delta y \qquad (15\text{-}17)$$

In Eq. (15-17), V is the velocity of the liquid in the section where the crest of the wave is located; and from the continuity equation we have $cy = V(y + \Delta y)$. Hence,

$$V = \frac{cy}{y + \Delta y}$$

and

$$V^2 = \frac{c^2 y^2}{(y + \Delta y)^2} \qquad (15\text{-}18)$$

When Eq. (15-18) is substituted into Eq. (15-17), we obtain

$$\frac{c^2}{2g} + y = \frac{c^2 y^2}{2g[y^2 + 2y \, \Delta y + (\Delta y)^2]} + y + \Delta y \qquad (15\text{-}19)$$

Upon solving Eq. (15-19) for c after discarding terms with $(\Delta y)^2$, since we are assuming an infinitesimally small wave, we obtain

$$c = \sqrt{gy} \qquad (15\text{-}20)$$

It has thus been shown that the speed of a small solitary wave is equal to the square root of the product of the depth and g.

In the previous section, we noted that critical flow occurs when $F = 1$ or when $V/\sqrt{gy} = 1$. In other words, when $V = \sqrt{gy}$ the flow is critical. Thus we see that the critical velocity is exactly the same magnitude as the speed of a small solitary wave in a stationary liquid. This observation can be used to test for critical flow in a channel: if one produces a small wave with a board or by hand, the wave will propagate upstream in subcritical flow, but it will be swept downstream in supercritical flow.

15-2 THE HYDRAULIC JUMP

Introduction

When the flow is supercritical in an upstream section of a channel and when it is then forced to become subcritical in a downstream section (the change in depth can be forced by a sill in the channel or by just the prevailing depth in the stream further downstream), a rather abrupt change in depth usually occurs and considerable energy loss accompanies the process. This flow phenomenon is called the *hydraulic jump* (Fig. 15-11). The hydraulic jump is often considered in the design of open channels and

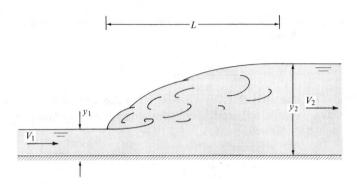

FIGURE 15-11   Definition sketch for the hydraulic jump.

spillways. For example, many spillways are designed so that the jump will occur on the *apron* of the spillway, thereby reducing the downstream velocity so that objectionable scour of the river channel is prevented. If a channel is designed to carry water at supercritical velocities, the designer must be certain that the flow will not become subcritical prematurely; otherwise, overtopping of the channel will undoubtedly occur, with consequent failure of the structure. Because the energy loss in the hydraulic jump is initially not known, the energy equation is not a suitable tool for analysis of the velocity-depth relationships; therefore, the momentum equation is applied to the problem.

Derivation of depth-velocity relationships

Consider flow as shown in Fig. 15-11. Here it is assumed that uniform flow occurs both upstream and downstream of the jump and that the resistance of the channel bottom is negligible. It should also be noted that the derivation is made for a horizontal channel, but experiments show that the results of the derivation will apply to all channels of moderate slope. Our first objective in this derivation is to determine the depth downstream of the jump as a function of the depth and velocity upstream. We start by applying the momentum equation in the x direction to the control volume shown in Fig. 15-12. Hence we have

$$\sum F_x = \sum_{cs} V_x \rho \mathbf{V} \cdot \mathbf{A}$$

The forces are the hydrostatic forces on each end of the system per unit of width ($\gamma y_1^2/2 - \gamma y_2^2/2$); thus, the following is obtained:

$$\gamma \frac{y_1^2}{2} - \gamma \frac{y_2^2}{2} = V_1 \rho(-V_1 y_1) + V_2 \rho(V_2 y_2)$$

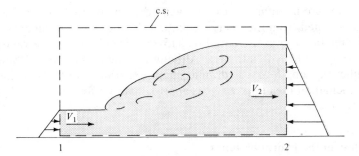

FIGURE 15-12   Control-volume analysis for the hydraulic jump.

The foregoing equation can then be expressed as

$$\frac{\gamma}{2}(y_1^2 - y_2^2) = \frac{\gamma}{g}(V_2^2 y_2 - V_1^2 y_1) \tag{15-21}$$

The continuity equation written between sections 1 and 2 is $V_1 y_1 = V_2 y_2$, which can then be expressed as

$$V_2 = V_1 \frac{y_1}{y_2} \tag{15-22}$$

Now, when the V_2 of Eq. (15-21) is replaced by the V_2 of Eq. (15-22) and simplified, we achieve the following expression:

$$\frac{2V_1^2}{gy_1} = \left(\frac{y_2}{y_1}\right)^2 + \frac{y_2}{y_1} \tag{15-23}$$

The term on the left-hand side of Eq. (15-23) will be recognized as twice F_1^2; hence, Eq. (15-23) is written as

$$\left(\frac{y_2}{y_1}\right)^2 + \frac{y_2}{y_1} - 2F_1^2 = 0 \tag{15-24}$$

By use of the quadratic formula, it is easy to solve for y_2/y_1 in terms of the upstream Froude number. Thus, we obtain

$$\frac{y_2}{y_1} = \frac{1}{2}(\sqrt{1 + 8F_1^2} - 1) \tag{15-25}$$

or

$$y_2 = \frac{y_1}{2}(\sqrt{1 + 8F_1^2} - 1) \tag{15-26}$$

The other solution of Eq. (15-24) gives a negative downstream depth that is not relevant here. Hence we have expressed the downstream depth in terms of the upstream depth and upstream Froude number. In Eqs. (15-25) and (15-26) the depths y_1 and y_2 are said to be *conjugate* or *sequent* (both terms are in common use) to each other in contrast to the alternate depths obtained from the energy equation. Numerous experiments show that the relation represented by Eqs. (15-25) and (15-26) is quite valid over a wide range of Froude numbers. Although no theory has been developed to predict the length of a hydraulic jump, experiments (2) show that the relative length of the jump, L/y_2, is approximately 6 for a range of F_1 from 4 to 18.

Head loss in the hydraulic jump

In addition to dealing with the geometric characteristics of the hydraulic jump, it is often desirable to determine the head loss produced by the hydraulic jump. This is obtained by comparing the specific energy before the jump to that after the jump—the head loss being the difference between the two specific energies. It can be shown that this head loss is given by

$$h_L = \frac{(y_2 - y_1)^3}{4 y_1 y_2} \tag{15-27}$$

For more information on the hydraulic jump, the reader is referred to Chow (2).

EXAMPLE 15-2 Water flows at a depth of 30 cm and with a velocity of 16 m/s in a channel. If a downstream sill causes a hydraulic jump to be formed, what will be the depth and velocity downstream of the jump? What head loss is produced by the jump?

Solution A sketch representing the flow is given above. To solve the problem we must know F_1, which is computed first:

$$F_1 = \frac{V}{\sqrt{g y_1}} = \frac{16}{\sqrt{9.81(0.30)}} = 9.33$$

y_2 is now computed by using Eq. (15-26):

$$y_2 = \frac{0.30}{2} [\sqrt{1 + 8(9.33)^2} - 1] = 3.81 \text{ m} \qquad \blacktriangleleft$$

Then, $$V_2 = \frac{q}{y_2} = \frac{16 \text{ m/s}(0.30 \text{ m})}{3.81 \text{ m}} = 1.26 \text{ m/s} \qquad \blacktriangleleft$$

The head loss is solved by use of Eq. (15-27):

$$h_L = \frac{(3.81 - 0.30)^3}{4(0.30)(3.81)} = 9.46 \text{ m} \qquad \blacktriangleleft$$

We can check the validity of formula (15-27) because the head loss is equal to $E_1 - E_2$, or

$$h_L = \left(0.30 + \frac{16^2}{2 \times 9.81}\right) - \left(3.81 + \frac{1.26^2}{2 \times 9.81}\right)$$

$$= 9.46 \text{ m} \qquad \blacktriangleleft$$

The answers check.

15-3 SURGE OR TIDAL BORE

Tides are generally low enough so that the waves they produce are smooth and nondestructive; however, in some parts of the world the tides are so high that their entry into shallow bays or mouths of rivers cause surges to be formed that may be very hazardous to small boats. A surge is actually a moving hydraulic jump; hence the same analytical methods used for the jump can be used to solve for the speed of the surge. In Fig. 15-13, a surge is shown coming into an otherwise still body of water. As indicated in Fig. 15-13, the flow is unsteady because throughout the surge

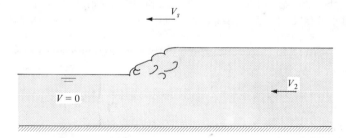

FIGURE 15-13 Surge moving under still water.

itself the velocities are changing with time. However, if all velocities are measured in a coordinate system moving with the surge front, then a steady-flow pattern is obtained. Figure 15-14 shows the steady-flow pattern and control volume. Now our problem is directly analogous to the hydraulic-jump problem. We simply replace V_1 by V_s in Eq. (15-23) to yield

$$\frac{V_s}{\sqrt{g y_1}} = \left[\frac{y_2}{2 y_1}\left(\frac{y_2}{y_1} + 1\right)\right]^{1/2} \qquad (15\text{-}28)$$

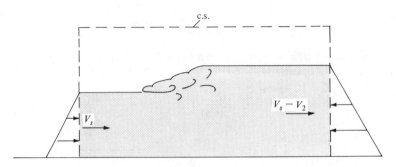

FIGURE 15-14 Control-volume analysis applied to the surge.

15-4 GRADUALLY VARIED FLOW IN OPEN CHANNELS

Basic differential equation for gradually varied flow

There are a number of cases of open-channel flow in which the change in water-surface profile is so gradual that it is possible to integrate the relevant differential equation from one section to another to obtain the desired change in depth. This may be either an analytical integration or, more commonly, a numerical integration. The latter procedure has the advantage that the complete water-surface profile is defined in the process of integration. In Sec. 15-1, the energy equation was written between two sections of a channel x distance apart. If we now consider the energy equation written for a reach Δx in length, we have an equation similar to Eq. (15-2):

$$y_1 + \frac{V_1^2}{2g} + S_0\,\Delta x = y_2 + \frac{V_2^2}{2g} + h_L \qquad (15\text{-}29)$$

The friction slope S_f is defined as the slope of the energy grade line, $\Delta h_f/\Delta L$. Then if we let $\Delta y = y_2 - y_1$, $h_L = S_f\,\Delta x$, and

$$\frac{V_2^2}{2g} - \frac{V_1^2}{2g} = \frac{d}{dx}\left(\frac{V^2}{2g}\right)\Delta x$$

Eq. (15-29) then becomes

$$\Delta y = S_0\Delta x - S_f\Delta x - \frac{d}{dx}\left(\frac{V^2}{2g}\right)\Delta x$$

Dividing through by Δx and taking the limit as Δx approaches zero gives us

$$\frac{dy}{dx} + \frac{d}{dx}\left(\frac{V^2}{2g}\right) = S_0 - S_f \qquad (15\text{-}30)$$

The second term is rewritten as $[d(V^2/2g)/dy]dy/dx$ so that Eq. (15-30) simplifies to

$$\frac{dy}{dx} = \frac{S_0 - S_f}{1 + d/dy\,(V^2/2g)} \qquad (15\text{-}31)$$

To put Eq. (15-31) in a more usable form, we express the denominator in terms of the Froude number. This is accomplished by observing that

$$\frac{d}{dy}\left(\frac{V^2}{2g}\right) = \frac{d}{dy}\left(\frac{q^2}{2gy^2}\right) \qquad (15\text{-}32)$$

After the differentiation on the right side of Eq. (15-32) is performed, Eq. (15-32) then becomes

$$\frac{d}{dy}\left(\frac{V^2}{2g}\right) = \frac{-2q^2}{2gy^3} = -F^2 \qquad (15\text{-}33)$$

Hence, when the expression for $d(V^2/2g)/dy$ is substituted into Eq. (15-31) we obtain

$$\frac{dy}{dx} = \frac{S_0 - S_f}{1 - F^2} \qquad (15\text{-}34)$$

This is the general differential equation for gradually varied flow, used to describe the various types of water-surface profiles that occur in open channels. Note that, in the derivation of the equation, S_0 and S_f were taken as positive when sloping downward in the direction of flow. Also that y is measured from the bottom of the channel; therefore, $dy/dx = 0$ if the slope of the water surface is equal to the slope of the channel bottom, and dy/dx is positive if the water-surface slope is less than the channel slope. With these definitions and limitations in mind, we will now consider the different forms of water-surface profiles.

Basic categories of surface profiles

Water-surface profiles are classified two different ways: according to the slope of the channel (mild, steep, critical, horizontal, or adverse); and according to the actual depth of flow in relation to the critical and normal depths (type 1, 2, or 3).

If the slope is so slight that the normal depth (uniform-flow depth) is greater than critical depth for the given discharge, then the slope of the channel is termed mild; thus, the water surface profile is given an M classification. Similarly if the channel slope is so steep that a normal depth less than critical is produced, then the channel is termed steep and the water-surface profile therein is given an S designation. If the slope is such that the normal depth equals the critical depth, then we have a critical slope, denoted by C. The horizontal and adverse slopes are special categories because normal depth does not exist for such slopes; these are denoted by H and A, respectively. The 1, 2, or 3 designations of water-surface profiles indicate if the actual depth of flow is greater than both normal and critical depths (type 1), between the normal and critical depths (type 2), or less than both normal and critical depths (type 3). For example, consider flow downstream of the sluice gate, Fig. 15-15. Here the discharge and slope are such that the normal depth is greater than the critical depth; therefore, the slope is termed mild. The actual depth of flow shown in Fig. 15-15 is less than either y_c or y_n; hence, a type 3 profile exists. The complete classification of the profile in Fig. 15-15, therefore, is a mild type 3 profile or, simply, an M3 profile. Using these designations it can also be seen that the surface profile upstream of the sluice gate would be a type M1.

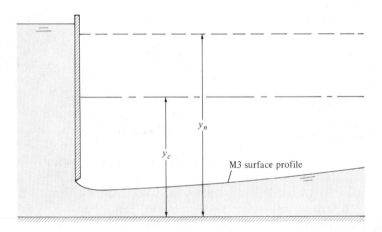

FIGURE 15-15   Surface profile—M3 type.

EXAMPLE 15-3   Classify the water-surface profile for the flow downstream of the sluice gate in Fig. 15-3 if the slope is horizontal and for the profile immediately downstream of the break in grade in Fig. 15-6.

Solution   In Fig. 15-3 the actual depth is less than critical; thus, the profile is a type 3 and the channel is horizontal and the profile is hence designated an H3 profile.   ◀

In Fig. 15-6 the actual depth is greater than normal but less than critical, so it is a type 2 curve. The uniform flow depth (normal depth y_n) is less than critical depth; hence the slope is steep. Therefore the water-surface profile is designated S2.   ◀

With the foregoing introduction to the classification of surface profiles, we can now refer to Eq. (15-34) to describe the shape of the profiles. Again, for example, if we consider the M3 profile, it is known that $F > 1$ because the flow is supercritical ($y < y_c$) and $S_f > S_0$ because the velocity is greater than normal velocity; hence, a head loss greater than that for normal flow must exist. Inserting these relative values into Eq. (15-34), we see that both the numerator and denominator are negative; thus dy/dx must be positive (the depth increases in the direction of flow), and as critical depth is approached, the Froude number approaches unity; hence, the denominator of Eq. (15-34) approaches zero. Therefore, as the depth approaches critical depth, $dy/dx \to \infty$. What actually occurs in cases where the critical depth is approached in supercritical flow is that a hydraulic jump forms and a discontinuity in profile is thereby produced.

Certain general features of profiles shown in Fig. 15-16 will be evident. First, as the depth becomes very great the velocity of flow approaches zero; hence, $F \to 0$ and $S_f \to 0$; and $dy/dx = (S_0 - S_f)/(1 - F^2)$ approaches S_0. In other words, the depth increases at the same rate that the channel bottom drops away from the horizontal. Thus the water surface approaches the horizontal. The curves that tend this way are the M1, S1, and C1 curves. A physical example of the M1 curve is that of the water-surface profile behind a dam, as shown in Fig. 15-17. The second general feature of several of the profiles is that profiles which approach normal depth do so asymptotically. This is shown in the S2, S3, M1, and M2 profiles. Also note in Fig. 15-16 that profiles that approach critical depth are shown by dashed lines. This is done because near critical depth discontinuities develop (hydraulic jump) or the streamlines are very curved (such as near a brink); therefore, the surface profiles cannot be accurately predicted by Eq. 15-35 because this equation is based on one-dimensional flow, which in these regions is invalid.

Quantitative evaluation of the surface profile

An equation that is convenient for quantitative evaluation of the water-surface profile is the energy equation written for a finite reach of channel Δx. We have

$$y_1 + \frac{V_1^2}{2g} + S_0 \, \Delta x = y_2 + \frac{V_2^2}{2g} + S_f \, \Delta x$$

from which we obtain

$$\Delta x(S_f - S_0) = \left(y_1 + \frac{V_1^2}{2g}\right) - \left(y_2 + \frac{V_2^2}{2g}\right)$$

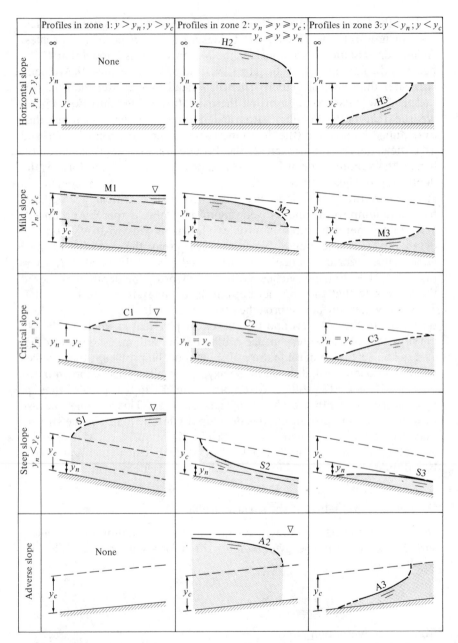

FIGURE 15-16   Classification of flow profiles of gradually vaned flow. [From *Open Channel Hydraulics* by Chow (2), Copyright © 1959, McGraw-Hill Book Company, New York. Used with permission of McGraw-Hill Book Company.]

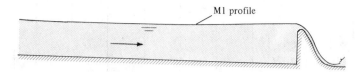

FIGURE 15-17

or $\quad \Delta x = \dfrac{(y_1 + V_1^2/2g) - (y_2 + V_2^2/2g)}{S_f - S_0}$

$$= \dfrac{(y_1 - y_2) + (V_1^2 - V_2^2)/2g}{S_f - S_0} \quad (15\text{-}35)$$

The procedure for evaluation of a profile is to first ascertain the type of profile that applies to the given reach of channel (we use the methods of the preceding section). Then, starting from a known depth, a finite value of Δx is computed for an arbitrarily chosen change in depth. The process of computing Δx, step by step, up (negative Δx) or down (positive Δx) the channel is repeated until the full reach of channel has been covered. Usually small changes of y are taken, so that the friction slope is approximated by the following equation:

$$S_f = \dfrac{h_f}{\Delta x} = \dfrac{fV^2}{8gR} \quad (15\text{-}36)$$

Here V is the mean velocity in the reach and R is the mean hydraulic radius in the reach. That is, $V = (V_1 + V_2)/2$ and $R = (R_1 + R_2)/2$. It is obvious that a numerical approach of this type is ideally suited for solution on the digital computer.

EXAMPLE 15-4 Water discharges from under a sluice gate into a horizontal channel at a rate of 1 m³/s per meter of width as shown. What is the classification of the profile? Quantitatively evaluate the water-surface profile downstream of the gate and determine whether or not it will extend

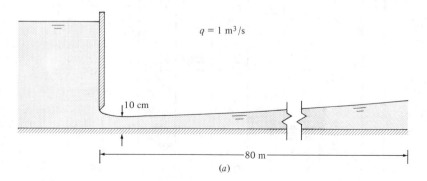

(a)

SOLUTION TO EXAMPLE 15-4

Section number downstream of gate	Depth y, m	Velocity at section V, m/s	Mean velocity in reach $(V_1 + V_2)/2$	V^2	Mean hydraulic radius $R_m = (y + y_2)/2$	$S_f = \dfrac{fV^2_{mean}}{8g\,R_m}$	$\Delta x = \dfrac{(y_1 - y_2) + \dfrac{(V_1^2 - V_2^2)}{2g}}{(S_f - S_o)}$	Distance from gate x
1 (at gate)	0.1	10	...	100	...	...	...	0
	...	...	8.57	73.4	0.12	0.156	15.7	
2	0.14	7.14	...	51.0	...	...	...	15.7
	...	...	6.35	40.3	0.16	0.064	15.3	
3	0.18	5.56	...	30.9	...	...	...	31.0
	...	...	5.05	25.5	0.20	0.032	15.1	
4	0.22	4.54	...	20.6	...	...	...	46.1
	...	...	4.19	17.6	0.24	0.019	13.4	
5	0.26	3.85	...	14.8	...	...	...	59.5
	...	...	3.59	12.9	0.28	0.012	12.4	
6	0.30	3.33	...	11.1	...	...	...	71.9
	...	...	3.13	9.8	0.32	0.008	10.9	
7	0.34	2.94	...	8.6	...	...	...	82.8

all the way to the abrupt drop 80 m downstream. Make the simplifying assumption that the resistance factor f is equal to 0.02 and that the hydraulic radius R is equal to the depth y.

Solution   First determine the critical depth y_c:

$$y_c = (q^2/g)^{1/3} = [(1^2 \text{ m}^4/\text{s}^2)/(9.81 \text{ m/s}^2)]^{1/3}$$
$$= 0.467 \text{ m}$$

The depth of flow from the sluice gate is less than critical depth; hence, the surface profile would be classified as an H3 profile.   ◄

To solve for the depth versus distance along the channel, we apply Eqs. (15-35) and (15-36) using a numerical approach. The computations are shown on page 624. From the numerical results we plot the profile shown, and we also see that the *profile extends to the abrupt drop*.   ◄

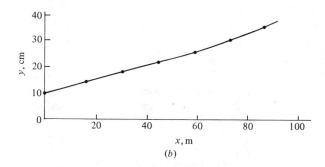

(b)

PROBLEMS

15-1 The water discharge in a 5-m-wide rectangular channel is 15 m³/s. If the depth of water is 1 m, is the flow subcritical or supercritical?

15-2 The discharge in a 5-m-wide rectangular channel is 10 m³/s. If the water velocity is 1.0 m/s, is the flow subcritical or supercritical?

15-3 Water flows at a rate of 10 m³/s in a 3-m wide rectangular channel. Determine the Froude number and the type of flow (subcritical, critical, or supercritical) for depths of 30 cm, 1.0 m, and 2.0 m. What is the critical depth?

15-4 For the discharge and channel of Prob. 15-3, what is the alternate depth to the 30-cm depth? What is the specific energy for these conditions?

15-5 Water flows at critical depth with a velocity of 2 m/s. What is the depth of flow?

15-6 Water flows uniformly at a rate of 9.0 m³/s in a 4.0-m-wide rectangular channel that has a bottom slope of 0.005. If n is 0.014, is the flow therein subcritical or supercritical?

15-7 A small wave is produced in a pond that is 30 cm deep. What is the speed of the wave in the pond?

15-8 A small wave in a pool of water having constant depth travels at a speed of 3.0 m/s. How deep is the water?

15-9 As waves in the ocean approach a sloping beach, they curve so that they are nearly parallel to the beach when they finally break (see figure below). Explain why the waves curve like this.

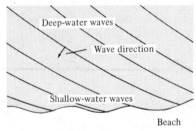

Aerial view of waves

15-10 A rectangular channel is 6 m wide and the discharge of water in it is 18 m³/s. Plot the curve of depth versus specific energy for these conditions. Let specific energy range from E_{min} to $E = 7$ m. What are the alternate and sequent depths to the 30-cm depth?

15-11 A long, 3-m-wide rectangular channel that has a mild slope ends in a free outfall. If the water depth at the brink is 0.250 m, what is the discharge in the channel?

15-12 A 15-ft-wide rectangular channel that has a mild slope ends in a free outfall. If the water depth at the brink is 1.20 ft, what is the discharge in the channel?

15-13 What discharge of water will occur over a high, broad-crested weir that is 20 ft long if the head on the weir is 1.5 ft.?

15-14 What discharge of water will occur over a high, broad-crested weir that is 10 m long if the head on the weir is 60 cm?

15-15 The crest of a high, broad-crested weir has an elevation of 100.00 m. If the weir is 20 m long and the discharge of water over the weir is 50 m³/s, what is the water surface elevation in the reservoir upstream?

15-16 The crest of a high, broad-crested weir has an elevation of 300.00 ft. If the weir is 50 ft long and the discharge of water over the weir is 1,500 cfs, what is the water surface elevation in the reservoir upstream?

15-17 Water flows with a velocity of 3 m/s and at a depth of 3 m in a rectangular channel. What is the change in depth and in water surface elevation produced by a gradual upward change in bottom elevation (upstep) of 30 cm? What would be the depth and elevation changes if there were a gradual downstep of 30 cm? What is the maximum size of upstep before upstream depth changes would result?

15-18 Water flows with a velocity of 2 m/s and at a depth of 3 m in a rectangular channel. What is the change in depth and in water surface elevation produced by a gradual upward change in bottom elevation (upstep) of 60 cm? What would be the depth and elevation changes if there were a gradual downstep of 15 cm? What is the maximum size of upstep before upstream depth changes would result?

15-19 Water flows with a velocity of 3 m/s in a 3-m-wide rectangular channel at a depth of 3 m. What is the change in depth and in water surface elevation that occurs when a gradual contraction in the channel to a width of 2.6 m takes place? Determine the greatest contraction allowable without altering the upstream conditions as specified.

15-20 Water flows with a velocity of 5 m/s in a 3-m-wide rectangular channel at a depth of 60 cm. What is the change in depth that occurs when a gradual contraction in the channel to a width of 2.4 m takes place? Determine the least width (greatest contraction) allowable without altering the upstream conditions as specified.

15-21 Water flows from a reservoir into a steep rectangular channel that is 4 m wide and the reservoir water surface is 3 m above the channel bottom at the channel entrance. What discharge will occur in the channel?

15-22 A spillway as shown has a flow of 3.0 m³/s per meter of width occuring over it. What depth y_2 will exist downstream of the hydraulic jump? Assume negligible energy loss over the spillway.

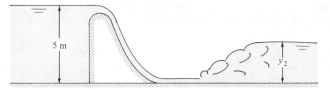

15-23 The flow of water downstream from a sluice gate in a horizontal channel has a depth of 30 cm and a flow rate of 1.8 m³/s per meter of width. Could a jump be caused to form downstream of this section? If so, what would be the depth downstream of the jump?

15-24 Water flows in a channel at a depth of 40 cm and with a velocity of 5 m/s. An obstruction causes a hydraulic jump to be formed. What is the depth of flow downstream of the jump?

15-25 A hydraulic jump occurs in a wide rectangular channel. If the depths upstream and downstream are 30 cm and 1.2 m, respectively, what is the discharge per foot of width of channel?

15-26 If the discharge per foot of width in this rectangular channel is 56.7 cfs, what will be the head loss in the jump that is formed if $y_1 = 1$ ft?

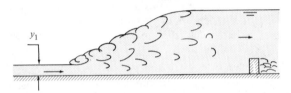

PROBLEMS 15-26, 15-27

15-27 If the discharge per meter of width in this rectangular channel is 30 m³/s, what will be the head loss in the jump that is formed if $y_1 = 1$ m?

15-28 Water flows uniformly at a depth $y_1 = 40$ cm (1.46 ft) in the 10-m (32.8 ft)-wide concrete channel. Estimate the height of the hydraulic jump that will form when a sill is installed to force a hydraulic jump to form.

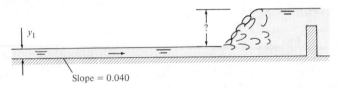

Slope = 0.040

PROBLEMS 15-28, 15-29

15-29 For the derivation of Eq. (15-26) it is assumed that the bottom shearing force is negligible. For the conditions of Prob. 15-28 estimate the magnitude of the shearing force F_s associated with the hydraulic jump and then determine F_s/F_H, where F_H is the net hydrostatic force on the hydraulic jump.

15-30 The normal depth in the channel downstream of the sluice gate is 1 m. What type of surface profile occurs downstream of the sluice gate? Also estimate the shear stress on the smooth bottom at a distance 0.5 m downstream of the sluice gate.

15-31 The partial profile shown is for a rectangular channel that is 3 m wide and has water flowing therein at a rate of 5 m³/s. Sketch in the missing water-surface profile and identify the type(s) of profile(s).

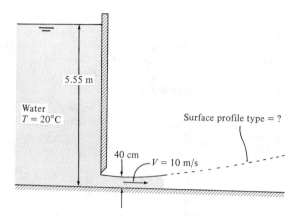

PROBLEM 15-30

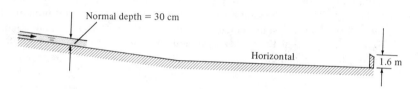

PROBLEM 15-31

15-32 Water flows from under a sluice gate into a horizontal rectangular channel at a rate of 3 m³/s per meter of width. The channel is concrete and the initial depth is 20 cm. Apply Eq. (15-35) to construct the water-surface profile up to a depth of 60 cm. In your solution, compute reaches for adjacent pairs of depths given in the following sequence: d = 20 cm, 30 cm, 40 cm, 50 cm, and 60 cm. Assume that f is constant with a value of 0.02. Plot your results.

15-33 A concrete rectangular channel that has zero slope terminates in a free out-fall. The channel is 4 m wide and carries a discharge of water of 12 m³/s. What is the water depth 300 m upstream from the outfall?

15-34 A 4-m-wide steep rectangular concrete spillway channel is 500 m long; it conveys water from a reservoir and delivers it to a free outfall. The channel entrance is rounded and smooth (negligible head loss at entrance). If the water surface elevation in the reservoir is 2 m above the channel bottom, what will be the discharge in the channel?

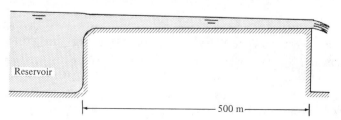

15-35 This concrete rectangular channel is 3.5 m wide and has a bottom slope of 0.001. The channel entrance is rounded and smooth (negligible head loss at the entrance) and the reservoir water surface is 2.5 m above the bed of the channel at the entrance. (a) If the channel is 3,000 m long, estimate the discharge in the channel. (b) If the channel is only 100 m long, tell how you would solve for the discharge in the channel.

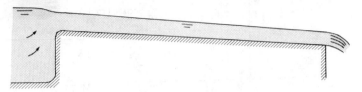

15-36 A 50-m-high dam backs up water in a river valley as shown. During flood flow the discharge per meter of width, q, is equal to 10 m³/s. Making the simplifying assumptions that $R = y$ and $f = 0.030$, determine the water-surface profile from the dam upstream to a depth of 6 m. In your numerical calculation, let the first increment of depth be y_c; use increments of depth of 10 m until a depth of 10 m is reached; and then use 2-m increments until the desired limit is reached.

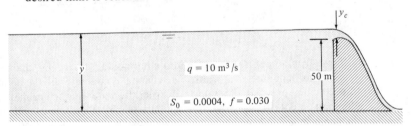

REFERENCES

1. Bakhmeteff, B. A. *Hydraulics of Open Channels*. McGraw-Hill Book Company, New York, 1932.

2. Chow, Ven Te. *Open Channel Hydraulics*. McGraw-Hill Book Company, New York, 1959.

3. Henderson, F. M. *Open Channel Flow*. The Macmillan Company, New York, 1966.

4. Posey, C. J. "Gradually Varied Open Channel Flow," in H. Rouse (ed.), *Engineering Hydraulics*. John Wiley & Sons, Inc., New York, 1950, Chap. IX.

5. Rajaratnam, N. "Hydraulic Jumps," in Chow, V. T. (ed.), *Advances in Hydroscience*, vol. 4. Academic Press, Inc., New York, 1967.

6. Woodward, S. M., and Posey, C. J. *Steady Flow in Open Channels*. John Wiley & Sons, Inc., New York, 1941.

APPENDIX

CONVERSION FACTORS FROM TRADITIONAL TO SI UNITS

Multiply number of	by	to obtain
ft	0.3048	m
mile/hr	0.447	m/s
ft³/sec	0.02832	m³/s
gpm	.00006309	m³/s
in.	25.40	mm
miles	1.609	km
slugs	14.59	kg
lbm	0.4536	kg
slug/ft³	515.4	kg/m³
lbm/ft³	16.02	kg/m³
lbf	4.448	N
lbf/ft²	47.88	N/m²
lbf/in.²	6,895.	N/m²
lbf/ft³	157.1	N/m³
ft-lbf	1.356	J
hp	745.7	W
Btu	1,055.	J

NOMENCLATURE AND DIMENSIONS

Symbol	Dimensions	Description
A	L^2	Area
A_j	L^2	Jet area
A_o	L^2	Orifice area
A_*	L^2	Nozzle area at $M = 1$
a	L/T^2	Acceleration
B	L	Linear measure
B	$\ldots$	Extensive property
b	L	Linear measure
°C	θ	Temperature, centigrade
C_c	$\ldots$	Coefficient of contraction
C_D	$\ldots$	Coefficient of drag

NOMENCLATURE AND DIMENSIONS (CONTINUED)

Symbol	Dimensions	Description
C_d	...	Coefficient of discharge
C_f	...	Shear-force coefficient
C_H	...	Head coefficient
C_L	...	Coefficient of lift
C_P	...	Power coefficient
C_p	...	Pressure coefficient
C_Q	...	Discharge coefficient
C_T	...	Thrust coefficient
C_v	...	Coefficient of velocity
c	...	Centi, multiple $= 10^{-2}$
c	L/T	Speed of sound
c_f	...	Local shear-stress coefficient
c_p	$L^2/T^2\theta$	Specific heat at constant pressure
c_v	$L^2/T^2\theta$	Specific heat at constant volume
D	L	Diameter
d	L	Diameter
d	L	Depth
E	LF	Energy
E	L	Specific energy
E_v	F/L^2	Elasticity, bulk
e	L^2/T^2	Energy per unit mass
F	...	Froude number
F	F	Force
$°F$	θ	Temperature, Fahrenheit
F_D	F	Drag force
F_L	F	Lift force
f	...	Resistance coefficient
G	...	Giga, multiple $= 10^9$
g	L/T^2	Acceleration due to gravity
g_c	...	Proportionality factor
H	L	Head
h	L	Head
h	L	Piezometric head
h	L^2/T^2	Enthalpy
h_f	L	Friction head loss in pipe
h_L	L	Head loss
h_p	L	Head supplied by pump
h_t	L	Head given up to turbine
I	L^4	Area moment of inertia, centroidal
$\mathbf{i}$	...	Unit vector in x direction
J	FL	Joule, unit of work
$\mathbf{j}$	...	Unit vector in y direction
K	...	Flow coefficient
K	θ	Temperature, Kelvin
k	...	Ratio of specific heats

Symbol	Dimensions	Description
k	...	Kilo, multiple $= 10^3$
k	...	Unit vector in z direction
k_s	L	Equivalent sand-roughness size
L	L	Linear measure
l	L	Linear measure
M	...	Mach number
M	FL	Moment
M	FT^2/L	Mass
M	...	Mega, multiple $= 10^6$
m	L	Meter
m	...	Milli, multiple $= 10^{-3}$
$\dot{m}$	FT/L	Mass rate of flow
N	F	Newton, unit of force
N	T^{-1}	Rotational speed
N_s	$L^{3/4}/T^{3/2}$	Specific speed
n	T^{-1}	Frequency in Hertz
n	...	Mannings roughness coefficient
n	T^{-1}	Rotational speed
n_s	...	Specific speed
Pa	F/L^2	Pascal, unit of pressure
p_*	F/L^2	Pressure at $M = 1$
p	F/L^2	Pressure
p_t	F/L^2	Total pressure
p_v	F/L^2	Vapor pressure
Q	L^3/T	Discharge
Q	LF	Heat transferred
q	L^2/T	Discharge per unit width
q	F/L^2	Kinetic pressure
R	L	Hydraulic radius
R	F	Reaction or resultant force
R	L^2/KT^2	Gas constant
°R	θ	Temperature, Rankine
Re	...	Reynolds number
r	L	Linear measure in radial direction
S	L^2	Planform area
S	...	Strouhal number
S_0	...	Channel slope
S	$L^2/T^2\theta$	Entropy
s	T	Time, second
s	L	Linear measure
T	LF	Tourque
T	θ	Temperature
T_t	θ	Total temperature
T_*	θ	Temperature at $M = 1$
t	T	Time

NOMENCLATURE AND DIMENSIONS (CONTINUED)

Symbol	Dimensions	Description
U_0	L/T	Free-stream velocity
u	L/T	Velocity component, x direction
u	L^2/T^2	Internal energy per unit of mass
u_*	L/T	Shear velocity $= \sqrt{\tau_0/\rho}$
u'	L/T	Velocity fluctuation in x direction
V	L/T	Velocity
V_0	L/T	Free-stream velocity
$\forall$	L^3	Volume
v	L/T	Velocity component, y direction
v'	L/T	Velocity fluctuation in y direction
W	LF	Work
W	$\ldots$	Weber number
W	F	Weight
W	FL/T	Watt, unit of power
w	L/T	Velocity component, z direction
x	L	Linear measure
y	L	Linear measure
z	L	Linear measure

GREEK LETTERS

α	$\ldots$	Angular measure
α	$\ldots$	Lapse rate
α	$\ldots$	Kinetic energy coefficient
α	$\ldots$	Angle of attack
β	$\ldots$	Angular measure
β	$\ldots$	Momentum coefficient
β	$\ldots$	Intensive property
Γ	L^2/T	Circulation
γ	F/L^3	Specific weight
Δ	$\ldots$	Increment
δ	L	Boundary-layer thickness
δ'	L	Laminar sublayer thickness
δ'_N	L	Nominal laminar sublayer thickness
η	$\ldots$	Efficiency
θ	$\ldots$	Angular measure
κ	$\ldots$	Turbulence constant
μ	FT/L^2	Viscosity, dynamic
μ	$\ldots$	Micro, multiple $= 10^{-6}$
τ	F/L^2	Shear stress
ν	L^2/T	Kinematic viscosity
π	$\ldots$	3.14
ρ	FT^2/L^4	Mass density

NOMENCLATURE AND DIMENSIONS (CONTINUED)

Symbol	Dimensions	Description
ρ_*	FT^2/L^4	Density at $M = 1$
ρ_t	FT^2/L^4	Total density
Ω	T^{-1}	Vorticity
ω	T^{-1}	Angular speed
σ	F/L	Surface tension

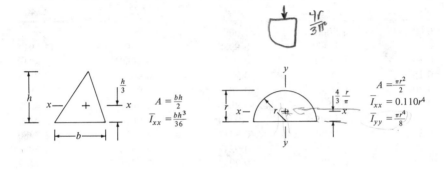

For the triangle:

$$A = \frac{bh}{2}$$

$$\bar{I}_{xx} = \frac{bh^3}{36}$$

For the semicircle:

$$A = \frac{\pi r^2}{2}$$

$$\bar{I}_{xx} = 0.110r^4$$

$$\bar{I}_{yy} = \frac{\pi r^4}{8}$$

For the rectangle:

$$A = bh$$

$$\bar{I}_{xx} = \frac{bh^3}{12}$$

For the circle:

$$A = \pi r^2$$

$$\bar{I}_{xx} = \frac{\pi r^4}{4}$$

For the hexagon:

$$A = 2.5981L^2$$

$$\bar{I}_x = 0.5127L^4$$

For the ellipse:

$$A = \pi ab$$

$$\bar{I}_{xx} = \frac{\pi a^3 b}{4}$$

FIGURE A-1   Centroids and moments of inertia of plane areas.

Volume and Area Formulas: $A_{\text{circle}} = \pi r^2 = \pi D^2/4$

$A_{\text{sphere surface}} = \pi D^2$

$\forall_{\text{sphere}} = \frac{1}{6}\pi D^3$

M or M_1 = local Mach number or Mach number upstream of a normal shock wave; p/p_t = ratio of static pressure to total pressure; ρ/ρ_t = ratio of static density to total density; T/T_t = ratio of static temperature to total temperature; A/A_* = ratio of local cross-sectional area of an isentropic stream tube to cross-sectional area at the point where $M = 1$; M_2 = Mach number downstream of a normal shock wave; p_2/p_1 = static-pressure ratio across a normal shock; T_2/T_1 = static-temperature ratio across a normal shock wave, p_{t_2}/p_{t_1} = total pressure ratio across normal shock wave.

	Subsonic Flow			
M	p/p_t	ρ/ρ_t	T/T_t	A/A_*
0.00	1.0000	1.0000	1.0000	∞
0.05	0.9983	0.9988	0.9995	11.5914
0.10	0.9930	0.9950	0.9980	5.8218
0.15	0.9844	0.9888	0.9955	3.9103
0.20	0.9725	0.9803	0.9921	2.9630
0.25	0.9575	0.9694	0.9877	2.4027
0.30	0.9395	0.9564	0.9823	2.0351
0.35	0.9188	0.9413	0.9761	1.7780
0.40	0.8956	0.9243	0.9690	1.5901
0.45	0.8703	0.9055	0.9611	1.4487
0.50	0.8430	0.8852	0.9524	1.3398
0.52	0.8317	0.8766	0.9487	1.3034
0.54	0.8201	0.8679	0.9449	1.2703
0.56	0.8082	0.8589	0.9410	1.2403
0.58	0.7962	0.8498	0.9370	1.2130
0.60	0.7840	0.8405	0.9328	1.1882
0.62	0.7716	0.8310	0.9286	1.1657
0.64	0.7591	0.8213	0.9243	1.1452
0.66	0.7465	0.8115	0.9199	1.1265
0.68	0.7338	0.8016	0.9153	1.1097
0.70	0.7209	0.7916	0.9107	1.0944
0.72	0.7080	0.7814	0.9061	1.0806
0.74	0.6951	0.7712	0.9013	1.0681
0.76	0.6821	0.7609	0.8964	1.0570
0.78	0.6691	0.7505	0.8915	1.0471
0.80	0.6560	0.7400	0.8865	1.0382
0.82	0.6430	0.7295	0.8815	1.0305
0.84	0.6300	0.7189	0.8763	1.0237
0.86	0.6170	0.7083	0.8711	1.0179
0.88	0.6041	0.6977	0.8659	1.0129
0.90	0.5913	0.6870	0.8606	1.0089
0.92	0.5785	0.6764	0.8552	1.0056
0.94	0.5658	0.6658	0.8498	1.0031
0.96	0.5532	0.6551	0.8444	1.0014
0.98	0.5407	0.6445	0.8389	1.0003
1.00	0.5283	0.6339	0.8333	1.0000

	Supersonic Flow				Normal Shock Wave			
M_1	p/p_t	ρ/ρ_t	T/T_t	A/A_*	M_2	p_2/p_1	T_2/T_1	p_{t_2}/p_{t_1}
1.00	0.5283	0.6339	0.8333	1.000	1.000	1.000	1.000	1.0000
1.01	0.5221	0.6287	0.8306	1.000	0.9901	1.023	1.007	0.9999
1.02	0.5160	0.6234	0.8278	1.000	0.9805	1.047	1.013	0.9999
1.03	0.5099	0.6181	0.8250	1.001	0.9712	1.071	1.020	0.9999
1.04	0.5039	0.6129	0.8222	1.001	0.9620	1.095	1.026	0.9999
1.05	0.4979	0.6077	0.8193	1.002	0.9531	1.120	1.033	0.9998
1.06	0.4919	0.6024	0.8165	1.003	0.9444	1.144	1.039	0.9997
1.07	0.4860	0.5972	0.8137	1.004	0.9360	1.169	1.046	0.9996
1.08	0.4800	0.5920	0.8108	1.005	0.9277	1.194	1.052	0.9994
1.09	0.4742	0.5869	0.8080	1.006	0.9196	1.219	1.059	0.9992
1.10	0.4684	0.5817	0.8052	1.008	0.9118	1.245	1.065	0.9989
1.11	0.4626	0.5766	0.8023	1.010	0.9041	1.271	1.071	0.9986
1.12	0.4568	0.5714	0.7994	1.011	0.8966	1.297	1.078	0.9982
1.13	0.4511	0.5663	0.7966	1.013	0.8892	1.323	1.084	0.9978
1.14	0.4455	0.5612	0.7937	1.015	0.8820	1.350	1.090	0.9973
1.15	0.4398	0.5562	0.7908	1.017	0.8750	1.376	1.097	0.9967
1.16	0.4343	0.5511	0.7879	1.020	0.8682	1.403	1.103	0.9961
1.17	0.4287	0.5461	0.7851	1.022	0.8615	1.430	1.109	0.9953
1.18	0.4232	0.5411	0.7822	1.025	0.8549	1.458	1.115	0.9946
1.19	0.4178	0.5361	0.7793	1.026	0.8485	1.485	1.122	0.9937
1.20	0.4124	0.5311	0.7764	1.030	0.8422	1.513	1.128	0.9928
1.21	0.4070	0.5262	0.7735	1.033	0.8360	1.541	1.134	0.9918
1.22	0.4017	0.5213	0.7706	1.037	0.8300	1.570	1.141	0.9907
1.23	0.3964	0.5164	0.7677	1.040	0.8241	1.598	1.147	0.9896
1.24	0.3912	0.5115	0.7648	1.043	0.8183	1.627	1.153	0.9884
1.25	0.3861	0.5067	0.7619	1.047	0.8126	1.656	1.159	0.9871
1.30	0.3609	0.4829	0.7474	1.066	0.7860	1.805	1.191	0.9794
1.35	0.3370	0.4598	0.7329	1.089	0.7618	1.960	1.223	0.9697
1.40	0.3142	0.4374	0.7184	1.115	0.7397	2.120	1.255	0.9582
1.45	0.2927	0.4158	0.7040	1.144	0.7196	2.286	1.287	0.9448
1.50	0.2724	0.3950	0.6897	1.176	0.7011	2.458	1.320	0.9278
1.55	0.2533	0.3750	0.6754	1.212	0.6841	2.636	1.354	0.9132
1.60	0.2353	0.3557	0.6614	1.250	0.6684	2.820	1.388	0.8952
1.65	0.2184	0.3373	0.6475	1.292	0.6540	3.010	1.423	0.8760
1.70	0.2026	0.3197	0.6337	1.338	0.6405	3.205	1.458	0.8557
1.75	0.1878	0.3029	0.6202	1.386	0.6281	3.406	1.495	0.8346
1.80	0.1740	0.2868	0.6068	1.439	0.6165	3.613	1.532	0.8127
1.85	0.1612	0.2715	0.5936	1.495	0.6057	3.826	1.569	0.7902
1.90	0.1492	0.2570	0.5807	1.555	0.5956	4.045	1.608	0.7674
1.95	0.1381	0.2432	0.5680	1.619	0.5862	4.270	1.647	0.7442
2.00	0.1278	0.2300	0.5556	1.688	0.5774	4.500	1.688	0.7209
2.10	0.1094	0.2058	0.5313	1.837	0.5613	4.978	1.770	0.6742
2.20	0.9352^{-1}†	0.1841	0.5081	2.005	0.5471	5.480	1.857	0.6281
2.30	0.7997^{-1}	0.1646	0.4859	2.193	0.5344	6.005	1.947	0.5833
2.40	0.6840^{-1}	0.1472	0.4647	2.403	0.5231	6.553	2.040	0.5401

TABLE A-1 COMPRESSIBLE FLOW TABLES FOR
AN IDEAL GAS WITH $k = 1.4$ (CONTINUED)

	Supersonic Flow				Normal Shock Wave			
M_1	p/p_t	ρ/ρ_t	T/T_t	A/A_*	M_2	p_2/p_1	T_2/T_1	p_{t_2}/p_{t_1}
2.50	0.5853^{-1}	0.1317	0.4444	2.637	0.5130	7.125	2.138	0.4990
2.60	0.5012^{-1}	0.1179	0.4252	2.896	0.5039	7.720	2.238	0.4601
2.70	0.4295^{-1}	0.1056	0.4068	3.183	0.4956	8.338	2.343	0.4236
2.80	0.3685^{-1}	0.9463^{-1}	0.3894	3.500	0.4882	8.980	2.451	0.3895
2.90	0.3165^{-1}	0.8489^{-1}	0.3729	3.850	0.4814	9.645	2.563	0.3577
3.00	0.2722^{-1}	0.7623^{-1}	0.3571	4.235	0.4752	10.33	2.679	0.3283
3.50	0.1311^{-1}	0.4523^{-1}	0.2899	6.790	0.4512	14.13	3.315	0.2129
4.00	0.6586^{-2}	0.2766^{-1}	0.2381	10.72	0.4350	18.50	4.047	0.1388
4.50	0.3455^{-2}	0.1745^{-1}	0.1980	16.56	0.4236	23.46	4.875	0.9170^{-1}
5.00	0.1890^{-2}	0.1134^{-1}	0.1667	25.00	0.4152	29.00	5.800	0.6172^{-1}
5.50	0.1075^{-2}	0.7578^{-2}	0.1418	36.87	0.4090	35.13	6.822	0.4236^{-1}
6.00	0.6334^{-2}	0.5194^{-2}	0.1220	53.18	0.4042	41.83	7.941	0.2965^{-1}
6.50	0.3855^{-2}	0.3643^{-2}	0.1058	75.13	0.4004	49.13	9.156	0.2115^{-1}
7.00	0.2416^{-3}	0.2609^{-2}	0.9259^{-1}	104.1	0.3974	57.00	10.47	0.1535^{-1}
7.50	0.1554^{-3}	0.1904^{-2}	0.8163^{-1}	141.8	0.3949	65.46	11.88	0.1133^{-1}
8.00	0.1024^{-3}	0.1414^{-2}	0.7246^{-1}	190.1	0.3929	74.50	13.39	0.8488^{-2}
8.50	0.6898^{-4}	0.1066^{-2}	0.6472^{-1}	251.1	0.3912	84.13	14.99	0.6449^{-2}
9.00	0.4739^{-4}	0.8150^{-3}	0.5814^{-1}	327.2	0.3898	94.33	16.69	0.4964^{-2}
9.50	0.3314^{-4}	0.6313^{-3}	0.5249^{-1}	421.1	0.3886	105.1	18.49	0.3866^{-2}
10.00	0.2356^{-4}	0.4948^{-3}	0.4762^{-1}	535.9	0.3876	116.5	20.39	0.3045^{-2}

† x^{-n} means $x \cdot 10^{-n}$

SOURCE: Abridged with permission from R. E. Bolz and G. L. Tuve, *The Handbook of Tables for Applied Engineering Sciences,* CRC Press, Inc., Cleveland, 1973. Copyright 1973 by The Chemical Rubber Co., CRC Press, Inc.

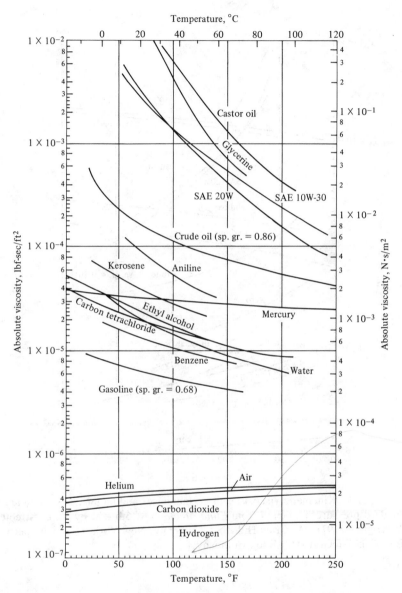

FIGURE A-2 Absolute viscosities of certain gases and liquids. [Adapted from *Fluid Mechanics*, 5th ed., by V. L. Streeter. Copyright © 1971, McGraw-Hill Book Company, New York. Used with permission of the McGraw-Hill Book Company.]

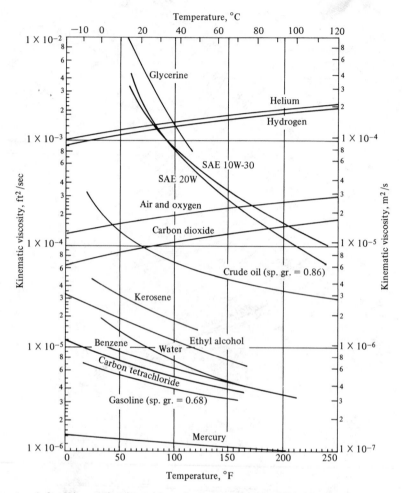

FIGURE A-3   Kinematic viscosities of certain gases and liquids. The gases are at standard pressure. [Adapted from *Fluid Mechanics,* 5th ed., by V. L. Streeter. Copyright © 1971, Mc-Graw Hill Book Company, New York. Used with permission of the McGraw-Hill Book Company.]

TABLE A-2 PHYSICAL PROPERTIES OF GASES AT STANDARD ATMOSPHERIC PRESSURE AND 15°C (59°F)

Gas	Density, kg/m³ (slug/ft³)	Kinematic viscosity, m²/s (ft²/sec)	R Gas constant, J/kg K (ft-lbf/slug-°R)	c_p $\dfrac{\text{J}}{\text{kg K}}$ $\left(\dfrac{\text{Btu}}{\text{lbm-°R}}\right)$	$k = \dfrac{c_p}{c_v}$
Air	1.22 (0.00237)	1.46×10^{-5} (1.58×10^{-4})	287 (1,716)	1,004 (0.240)	1.40
Carbon dioxide	1.85 (0.0036)	7.84×10^{-5} (8.48×10^{-4})	189 (1,130)	841 (0.201)	1.30
Helium	0.169 (0.00033)	1.14×10^{-4} (1.22×10^{-3})	2,077 (12,419)	5,187 (1.24)	1.66
Hydrogen	0.0851 (0.00017)	1.01×10^{-4} (1.09×10^{-3})	4,127 (24,677)	14,223 (3.40)	1.41
Methane (natural gas)	0.678 (0.0013)	1.59×10^{-5} (1.72×10^{-4})	518 (3,098)	2,208 (0.528)	1.31
Nitrogen	1.18 (0.0023)	1.45×10^{-5} (1.56×10^{-4})	297 (1,776)	1,041 (0.249)	1.40
Oxygen	1.35 (0.0026)	1.50×10^{-5} (1.61×10^{-4})	260 (1,555)	916 (0.219)	1.40

SOURCES: V. L. Streeter (ed.), *Handbook of Fluid Dynamics,* McGraw-Hill Book Company, New York, 1961; Also R. E. Bolz and G. L. Tuve, *Handbook of Tables for Applied Engineering Science.* CRC Press, Inc., Cleveland, 1973, *Handbook of Chemistry and Physics,* Chemical Rubber Company, 1951.

TABLE A-3 MECHANICAL PROPERTIES OF AIR AT STANDARD ATMOSPHERIC PRESSURE

Temperature	Density	Specific weight	Dynamic viscosity	Kinematic viscosity
	kg/m^3	N/m^3	$N \cdot s/m^2$	m^2/s
$-20°C$	1.40	13.7	1.61×10^{-5}	1.16×10^{-5}
$-10°C$	1.34	13.2	1.67×10^{-5}	1.24×10^{-5}
$0°C$	1.29	12.7	1.72×10^{-5}	1.33×10^{-5}
$10°C$	1.25	12.2	1.76×10^{-5}	1.41×10^{-5}
$20°C$	1.20	11.8	1.81×10^{-5}	1.51×10^{-5}
$30°C$	1.17	11.4	1.86×10^{-5}	1.60×10^{-5}
$40°C$	1.13	11.1	1.91×10^{-5}	1.69×10^{-5}
$50°C$	1.09	10.7	1.95×10^{-5}	1.79×10^{-5}
$60°C$	1.06	10.4	2.00×10^{-5}	1.89×10^{-5}
$70°C$	1.03	10.1	2.04×10^{-5}	1.99×10^{-5}
$80°C$	1.00	9.81	2.09×10^{-5}	2.09×10^{-5}
$90°C$	0.97	9.54	2.13×10^{-5}	2.19×10^{-5}
$100°C$	0.95	9.28	2.17×10^{-5}	2.29×10^{-5}
$120°C$	0.90	8.82	2.26×10^{-5}	2.51×10^{-5}
$140°C$	0.85	8.38	2.34×10^{-5}	2.74×10^{-5}
$160°C$	0.81	7.99	2.42×10^{-5}	2.97×10^{-5}
$180°C$	0.78	7.65	2.50×10^{-5}	3.20×10^{-5}
$200°C$	0.75	7.32	2.57×10^{-5}	3.44×10^{-5}
	$slugs/ft^3$	lbf/ft^3	$lbf\text{-}sec/ft^2$	ft^2/sec
$0°F$	0.00269	0.0866	3.39×10^{-7}	1.26×10^{-4}
$20°F$	0.00257	0.0828	3.51×10^{-7}	1.37×10^{-4}
$40°F$	0.00247	0.0794	3.63×10^{-7}	1.47×10^{-4}
$60°F$	0.00237	0.0764	3.74×10^{-7}	1.58×10^{-4}
$80°F$	0.00228	0.0735	3.85×10^{-7}	1.69×10^{-4}
$100°F$	0.00220	0.0709	3.96×10^{-7}	1.80×10^{-4}
$120°F$	0.00213	0.0685	2.47×10^{-7}	1.16×10^{-4}
$150°F$	0.00202	0.0651	4.23×10^{-7}	2.09×10^{-4}
$200°F$	0.00187	0.0601	4.48×10^{-7}	2.40×10^{-4}
$300°F$	0.00162	0.0522	4.96×10^{-7}	3.05×10^{-4}
$400°F$	0.00143	0.0462	5.40×10^{-7}	3.77×10^{-4}

SOURCE: Reprinted with permission from R. E. Bolz and G. L. Tuve, *Handbook of Tables for Applied Engineering Science,* CRC Press, Inc., Cleveland, 1973. Copyright 1973 by The Chemical Rubber Co., CRC Press, Inc.

TABLE A-4   APPROXIMATE PHYSICAL PROPERTIES OF COMMON LIQUIDS AT ATMOSPHERIC PRESSURE

Liquid and temperature	Density kg/m³ (slugs/ft³)	Specific gravity (sp. gr.) water at 4°C is ref.	Specific weight, N/m³ (lbf/ft³)	Dynamic viscosity, N·s/m² (lbf-sec/ft²)	Kinematic viscosity, m²/s (ft²/sec)	Surface tension, N/m* (lbf/ft)
Ethyl alcohol[3][1] 20°C (68°F)	799 (1.55)	0.79	7,850 (50.0)	1.2×10^{-3} (2.5×10^{-5})	1.5×10^{-6} (1.6×10^{-5})	2.2×10^{-2} (1.5×10^{-3})
Carbon tetrachloride[3] 20°C (68°F)	1,590 (3.09)	1.59	15,600 (99.5)	9.6×10^{-4} (2.0×10^{-5})	6.0×10^{-7} (6.5×10^{-6})	2.6×10^{-2} (1.8×10^{-3})
Glycerine[3] 20°C (68°F)	1,260 (2.45)	1.26	12,300 (78.5)	6.2×10^{-1} (1.3×10^{-2})	5.1×10^{-4} (5.3×10^{-3})	6.3×10^{-2} (4.3×10^{-3})
Kerosene[2][1] 20°C (68°F)	814 (1.58)	0.81	8,010 (51)	1.9×10^{-3} (4×10^{-5})	2.37×10^{-6} (2.55×10^{-5})	2.9×10^{-2} (2.0×10^{-3})
Mercury[3][1] 20°C (68°F)	13,550 (26.3)	13.55	133,000 (847)	1.5×10^{-3} (3.2×10^{-5})	1.2×10^{-7} (1.3×10^{-6})	4.8×10^{-1} (3.3×10^{-2})
Sea water 10°C at 3.3% salinity	1,026 (1.99)	1.03	10,070 (64.1)	1.4×10^{-3} (3×10^{-5})	1.4×10^{-6} (1.5×10^{-5})	
Oils—38°C (100°F)						
SAE 10W[4]	870 (1.69)	0.87	8,530 (54.4)	3.6×10^{-2} (7.4×10^{-4})	4.1×10^{-5} (4.4×10^{-4})	
SAE 10W-30[4]	880 (1.71)	0.88	8,630 (55.1)	6.7×10^{-2} (1.4×10^{-3})	7.6×10^{-5} (8.2×10^{-4})	
SAE 30[4]	880 (1.71)	0.88	8,630 (55.1)	1.0×10^{-1} (2.0×10^{-3})	1.1×10^{-4} (1.2×10^{-3})	

* Liquid-air surface tension values.

SOURCES: (1) V. L. Streeter, *Handbook of Fluid Dynamics*, McGraw-Hill Book Company, New York, 1961; (2) V. L. Streeter, *Fluid Mechanics*, 4th ed., McGraw-Hill Book Company, New York, 1966; (3) J. Vennard, *Elementary Fluid Mechanics*, 4th ed., John Wiley & Sons, Inc., New York, 1961; (4) R. E. Bolz and G. L. Tuve, *Handbook of Tables for Applied Engineering Science*, CRC Press, Inc., Cleveland, 1973.

TABLE A-5 APPROXIMATE PHYSICAL PROPERTIES OF WATER* AT
ATMOSPHERIC PRESSURE

Temperature	Density	Specific weight	Dynamic viscosity	Kinematic viscosity	Vapor pressure
	kg/m³	N/m³	N · s/m²	m²/s	N/m² abs.
0°C	1,000	9,810	1.79×10^{-3}	1.79×10^{-6}	611
5°C	1,000	9,810	1.51×10^{-3}	1.51×10^{-6}	872
10°C	1,000	9,810	1.31×10^{-3}	1.31×10^{-6}	1,230
15°C	999	9,800	1.14×10^{-3}	1.14×10^{-6}	1,700
20°C	998	9,790	1.00×10^{-3}	1.00×10^{-6}	2,340
25°C	997	9,781	8.91×10^{-4}	8.94×10^{-7}	3,170
30°C	996	9,771	7.96×10^{-4}	7.99×10^{-7}	4,250
35°C	994	9,751	7.20×10^{-4}	7.24×10^{-7}	5,630
40°C	992	9,732	6.53×10^{-4}	6.58×10^{-7}	7,380
50°C	988	9,693	5.47×10^{-4}	5.54×10^{-7}	12,300
60°C	983	9,643	4.66×10^{-4}	4.74×10^{-7}	20,000
70°C	978	9,594	4.04×10^{-4}	4.13×10^{-7}	31,200
80°C	972	9,535	3.54×10^{-4}	3.64×10^{-7}	47,400
90°C	965	9,467	3.15×10^{-4}	3.26×10^{-7}	70,100
100°C	958	9,398	2.82×10^{-4}	2.94×10^{-7}	101,300
	slugs/ft³	lbf/ft³	lbf-sec/ft²	ft²/sec	psia
40°F	1.94	62.43	3.23×10^{-5}	1.66×10^{-5}	0.122
50°F	1.94	62.40	2.73×10^{-5}	1.41×10^{-5}	0.178
60°F	1.94	62.37	2.36×10^{-5}	1.22×10^{-5}	0.256
70°F	1.94	62.30	2.05×10^{-5}	1.06×10^{-5}	0.363
80°F	1.93	62.22	1.80×10^{-5}	0.930×10^{-5}	0.506
100°F	1.93	62.00	1.42×10^{-5}	0.739×10^{-5}	0.949
120°F	1.92	61.72	1.17×10^{-5}	0.609×10^{-5}	1.69
140°F	1.91	61.38	0.981×10^{-5}	0.514×10^{-5}	2.89
160°F	1.90	61.00	0.838×10^{-5}	0.442×10^{-5}	4.74
180°F	1.88	60.58	0.726×10^{-5}	0.385×10^{-5}	7.51
200°F	1.87	60.12	0.637×10^{-5}	0.341×10^{-5}	11.53
212°F	1.86	59.83	0.593×10^{-5}	0.319×10^{-5}	14.70

* Notes: (1) Bulk modulus E_v of water is approximately 2.2 G P_a (3.2×10^5 psi); (2) Water-air surface tension is approximately 7.3×10^{-2} N/m (5×10^{-3} lbf/ft) from 10°C to 50°C.

SOURCE: Reprinted with permission from R. E. Bolz and G. L. Tuve, *Handbook of Tables for Applied Engineering Science*, CRC Press, Inc., Cleveland, 1973. Copyright 1973 by The Chemical Rubber Co., CRC Press, Inc.

ANSWERS TO SELECTED PROBLEMS

CHAPTER 2

2-2 $\gamma = 46.7$ N/m^3

2-4 $M = 3.28 \times 10^8$ slugs

2-6 $W = 95.3$ lbf $= 424$ N

2-8 M and W are extensive; others intensive

2-10 $\Delta\mu_{air} = +2.40 \times 10^{-6}$ N $\cdot$ s/m^2; $\Delta\mu_{water} = -8.44 \times 10^{-4}$ N $\cdot$ s/m^2

2-12 $u_{max} = 0.25$ ft/s; $\tau_0 = 3 \times 10^{-4}$ lbf/ft^2

2-14 $u_{12} = 0.59$ m/s; $\tau_{12} = 20.8$ N/m^2; $u_0 = 0$; $\tau_0 = 40.0$ N/m^2

2-16 $\mu_A/\mu_w = 0.0018$; $\nu_A/\nu_w = 15.1$

2-18 (a) $\tau_2/\tau_3 = 2/3$; (b) $V = 0.50$ ft/s; (c) $\tau = 5.0$ lbf/ft^2

2-20 $\forall = 999.1$ cm^3

2-22 $\Delta p = 4\sigma/R$; $\Delta p = 146$ N/m^2

2-24 $h = 14.9$ mm

CHAPTER 3

3-2 $p = 99.75$ kPa absolute

3-4 $F = 1,767$ lbf

3-6 $\gamma = 45.6$ N/m^3

3-8 $n = 5$

3-10 sp. gr. $= 1.50$

3-12 $p = 68.53$ kPa

3-14 $p = -45.7$ lbf/ft^2

3-16 $p = 13.88$ kPa

3-18 $p_{max} = 127.5$ kPa; $F_{top} = 98.1$ kN

3-20 $p_A = 129.7$ kPa

3-22 $d = 10.2$ m

3-24 $p = 721$ Pa

3-26 $\alpha = 20.9°$

3-28 $p_A = 51.8$ kPa

3-30 sp. gr. $= 0.667$

3-32 $p_A - p_B = -237$ psf

3-34 $z = 2.31$ ft

3-36 $z = 96.4$ mm

3-38 $p_A = 35.5$ kPa

3-40 $p_A = 4{,}742$ lbf/ft^2
3-42 $p_A - p_B = 48.62$ lbf/ft^2; $h_A - h_B = -2.22$ ft
3-44 $T_{1{,}500} = 95°C$; $T_{3{,}000} = 90°C$
3-46 $p = 46.4$ kPa abs
3-48 $z = 2.91$ km; $T = -2.41°C$
3-50 $p = 12.1$ psia; $\rho = 0.00203$ slugs/ft^3; $T = 40°F$
3-52 $M = 1.77 \times 10^5$ N · m
3-54 $F = 6{,}400$ lbf
3-56 $R_A = 2.114$ MN
3-58 $M = 2.26$ kN · m
3-60 $T = 59{,}920$ ft-lbf
3-62 $F = 36{,}766$ lbf
3-64 $\forall = 0.381$ m^3
3-66 Gate will stay.
3-68 $F_{\text{hyd}} = 1.236$ MN; $P = 323.7$ kN
3-70 $P = 18{,}015$ lbf
3-72 $k = \sqrt{2}$
3-74 $R_A = 0.687\ \gamma W \ell^2$
3-76 $M = 2{,}134$ ft-lbf
3-78 $\gamma_{\text{wood}} = \frac{8}{9}\gamma_{\text{water}}$
3-80 1,046 lbf
3-82 Triangular
3-84 $\gamma = 257.6$ lbf/ft^3
3-86 $\forall = 0.0309$ m^3
3-88 $F = 29{,}671$ N $\downarrow$
3-90 114,750 N tension
3-92 $F = 158.6$ N/cm
3-94 $W = 3{,}011$ N
3-96 $F_{\text{vert}} = 3{,}289$ lbf $\uparrow$; line of action is 1.4 ft to right of pt. A.
3-98 $F_{\text{horiz}} = 2{,}763$ lbf$\leftarrow$; $y_{cp} = 4.667$ ft
3-100 $F = 4{,}295$ lbf; $\theta = 40°2'$ from vertical through center of curvature.
3-102 $F_{\text{bolts}} = 27{,}877$ lbf $\downarrow$
3-104 $F_H = 375$ lbf$\rightarrow$; $F_V = 503$ lbf $\downarrow$; $y_{cp} = 2.80$ ft above water surface
3-106 $1.114 <$ sp. gr. < 1.39
3-108 $GM = 0.40$—stable
3-110 $GM = -0.122$ m—unstable
3-112 1.58 Hz

CHAPTER 4

4-2 Unsteady, nonuniform
4-4 (a) Unsteady, nonuniform; (b) Local and convective
4-8 (a) 2D; (b) 1D; (c) 1D; (d) 2D; (e) 3D; (f) 3D; (g) 2D
4-10 $Q = 5{,}290$ gpm
4-12 $\dot{m} = 0.0336$ kg/s
4-14 $D = 17.8$ mm
4-16 $Q = 33.5$ cfs; $Q = 15{,}046$ gpm

4-18 $Q = 173$ m³/s

4-20 $Q = 3.55 \times 10^{-3}$ m³/s

4-22 $q = 5.25$ m³/s; $V = 3.5$ m/s

4-24 $Q = 139$ cfs; $V = 4.63$ cfs

4-26 $V = 4.90$ ft/sec

4-28 $Q = 57.2$ gpm

4-30 $Q = 0.0236$ m³/s

4-32 $a_{conv.} = V^2 D / 2 h^2$

4-34 $a_{local} = 3.56$ ft/sec²; $a_{conv} = 16.86$ ft/sec²

4-36 $a_{conv} = \dfrac{32}{27} \left(\dfrac{q_0}{t_0} \right)^2 \dfrac{t^2}{B^3}$

4-38 $a_{conv} = -1,355$ m/s²

4-40 (a) $\beta = 1$; $dB_{syst}/dt = 0$; $\Sigma \beta \rho \mathbf{V} \cdot \mathbf{A} = -4$ slugs/sec; $\dfrac{d}{dt} \displaystyle\int_{cv} \beta \rho \, d\forall$

   $= +4$ slugs/sec; (b) $\beta = 1$; $dB_{syst}/dt = 0$; $\Sigma \beta \rho \mathbf{V} \cdot \mathbf{A} = 0$; $\dfrac{d}{dt} \displaystyle\int_{cv} \beta \rho \, d\forall = 0$

4-42 $y = 16.97 \, d$

4-44 $V = 12.5$ m/s; $V = 25$ m/s

4-46 $Q_{15} = 0.180$ m³/s; $Q_{20} = 0.320$ m³/s

4-48 $V = 7.56$ m/s

4-50 $V_{fall} = 0.0108$ ft/sec

4-52 $\dot{m} = 5.43$ slugs/sec; $V_c = 15.3$ ft/sec; sp. gr. $= 0.933$

4-54 $A = 6.52 \times 10^{-9}$ ft²

4-56 $t = 77.0$ s

4-60 Continuity is not satisfied.

4-62 Continuity is satisfied; not irrotational

4-64 $V_e = 2,643$ m/s

4-66 $\Delta p_c = 4.541$ MPa

CHAPTER 5

5-2 $\partial p / \partial s = -11.64$ lbf/ft³

5-4 $p = 470$ kPa

5-6 $a_s = 10.19$ m/s²

5-8 $\partial p / \partial x = -3.23$ lbf/ft³

5-10 $a_{conv} = 885$ ft/sec²; $a_{local} = 20.4$ ft/sec²; $p_A = 10,460$ lbf/ft²

5-12 $\partial p / \partial r = 2,032$ N/m³; $p_A = 99.80$ kPa abs

5-14 $a_x = 7.36$ m/s²

5-16 $d_{max} = 2.08$ m

5-18 $p_c - p_A = 46.6$ kPa; $p_B - p_A = 17.1$ kPa

5-20 $p_B = 531$ lbf/ft²

5-22 $z_2 = 70.8$ m

5-24 $z_2 = 0.609$ m

5-26 $\omega = \sqrt{5g/\ell}$ rad/sec

5-28 $F = 38.7$ N
5-30 $\omega = \sqrt{2g/(3\ell)}$
5-32 $p_A = -34.7$ kPa; $p_B = 6.28$ kPa; $p_A = -101$ kPa; $p_B = 13.2$ kPa
5-34 $(dp/dz)_{-1} = -34.8$ kPa/m; $(dp/dz)_{+1} = 15.2$ kPa/m
5-36 $\Delta p_{max} = 523$ lbf/ft^2
5-38 $p = 97.1$ kPa
5-40 $F = 20.3$ lbf
5-42 $V_2 = 33.3$ ft/sec
5-44 $p_B = -36.8$ kPa
5-46 Cavitation is expected to occur.
5-48 $A_c/A_e = 0.172$
5-50 $p_{min} = 86.7$ lbf/ft^2
5-54 $p_B - p_c = 14.0$ kPa
5-56 $V = 27.2$ m/s
5-58 $V = 7.33$ ft/sec
5-60 $V = 7.84$ m/s
5-62 $V = 91.9$ ft/s
5-64 $p_2 - p_1 = 1.98$ kPa
5-66 $V_0 = 12.8$ m/s
5-68 $V_0 = 44.9$ m/s
5-70 $V_0 = 8.51$ m/s
5-72 $V_{true} = 69.3$ m/s

CHAPTER 6

6-2 $V = 30.9$ ft/sec
6-4 $d = 134.9$ mm
6-6 $F_H = -579$ lbf
6-8 $F_B = 204$ N; $F_A = 4.05$ kN
6-10 $F_x = -4.04$ kN; $F_y = -1.03$ kN
6-12 $F_x = 1.24$ kN; $F_y = -512$ N
6-14 $F = 7.95$ kN
6-16 $a = 319$ ft/sec^2
6-18 $P = 32.35$ hp
6-22 Resistance $= 407$ N
6-24 $h = 21.7$ ft
6-26 $F_x = -1,625$ lbf
6-28 $F_H = 380$ kN; $F_v = 14.8$ kN
6-30 $F_v = 3,198$ lbf
6-32 $F_x = -12,820$ lbf
6-34 $F_x = -1,654$ lbf
6-36 $F_x = -900$ lbf; $F_y = -297$ lbf
6-38 $\mathbf{F} = -2,763\mathbf{i} - 1,144\mathbf{j}$ lbf
6-40 $\mathbf{F} = -4,121\mathbf{i} - 1,707\mathbf{j}$ lbf
6-42 $\mathbf{F} = -3.46\mathbf{i} - 1.46\mathbf{j}$ kN
6-44 $\mathbf{F} = -1,095\mathbf{i} - 262\mathbf{j}$ lbf
6-46 $\mathbf{F} = -3.32\mathbf{i} - 3.41\mathbf{j}$ kN

6-48 $F = 1,111$ lbf
6-50 $F_x = -3,622$ lbf
6-52 $F = 1,270$ lbf
6-54 $\mathbf{F}_H = -1,148\mathbf{i} + 161\mathbf{j}$ lbf
6-56 $F_y = -2,513$ lbf
6-58 $\mathbf{F} = -523\mathbf{i} - 58.9\mathbf{j}$ lbf
6-60 $F_x = -28.6$ lbf; $F_y = -80.8$ lbf
6-62 $\mathbf{F} = -11.0\mathbf{j}$ kN
6-64 $F_x = 120$ lbf/ft
6-66 $F_{\text{bolt}} = 1,979$ N
6-68 $F = 159$ N
6-70 $F = 1.38$ kN
6-72 $F_0 = 827$ N; $F_{10} = 551$ N
6-74 $\dot{m} = 2.12$ slug/sec; $V_{\max} = 150$ ft/sec; drag $= 115$ lbf
6-76 $a = 0.03$ m/s²
6-78 $F = 1,007$ lbf
6-80 $\mathbf{R} = -465\mathbf{j} + 1,534\mathbf{k}$ N, $\mathbf{T} = -16.3\mathbf{j} + 413\mathbf{k}$ N · m
6-82 $\mathbf{R} = -12,135\mathbf{i} + 3,071\mathbf{j}$ N
6-86 $L = 1,112$ m
6-88 $F = 29.7$ kN
6-90 $p_{\max} = 246$ kPa gage; $p_{\min} = -50$ kPa gage
6-92 $M = 551$ kg
6-94 $F = 1.02$ MN
6-98 $P = 1.44$ kW

CHAPTER 7

7-2 Power $= 489$ kW
7-4 Heat loss $= 7.56$ kJ/s
7-6 $\alpha = 1.009$
7-8 (a) $\alpha = 1.0$; $\alpha > 1.0$ for (b), (c), and (d)
7-10 $\alpha = 1.35$
7-12 $\alpha = (m + 1)^3/(3m + 1)$; $\alpha = 1.036$
7-14 $p_A - p_B = 33.4$ kPa
7-18 $p_A = -53.5$ kPa gage; $p_B = -19.6$ kPa gage; turbine
7-20 $p_A = -374$ lbf/ft² gage
7-22 $Q = 0.341$ m³/s; $p_B = 15.9$ kPa gage
7-24 $V = 3.13$ m/s; $p_B = -29.4$ kPa gage
7-26 $p_2 = 1,335$ lbf/ft² gage
7-28 $h = 22.8$ m
7-30 Power $= 13.4$ kW
7-32 $p_2/\gamma = 17.8$ ft
7-34 Power $= 26.3$ hp
7-36 Power $= 6.83$ MW
7-38 Power $= 17.8$ kW
7-40 $h_L = 0.566$ m
7-42 $F = 3,390$ lbf

7-44 $F = 222$ lbf
7-46 $Q = 0.303$ m³/s
7-48 $h_L = 2.52$ ft
7-56 $Q = 6.96$ m³/s
7-58 Power $= 29.7$ hp
7-60 Power $= 3.40$ kW
7-62 $F = 0.372$ N

CHAPTER 8

8-2 (a) $[T] = FL = ML^2/T^2$; (b) $[\rho V^2/2] = F/L^2 = M/LT^2$; (c) $[\sqrt{\tau/\rho}] = L/T$; (d) $[Q/ND^3] =$ dimensionless
8-6 $\Delta p/\rho n^2 D^2 = f(Q/nD^3)$
8-8 $dp/dr = C\rho r^2\omega^2$
8-10 $h/d = f(\sigma t^2/\rho d^3, \gamma t^2/\rho d, \mu t/\rho d^2)$
8-12 $F = 4.10$ lb $= 18.2$ N
8-14 $V = 194$ ft/sec
8-16 $V = 75.3$ m/s
8-18 $V = 12.0$ m/s
8-20 $V = 11.2$ ft/sec
8-22 $Q_m/Q_p = 1/10$
8-24 $M = 141$ N · m; $V = 0.179$ m/s
8-26 $V_p = 10$ m/s
8-28 $V_m/V_p = 1/5$; $Q_m/Q_p = 1/3,125$; $Q_m = 0.96$ m³/s
8-30 $V_p = 39.3$ ft/sec; $Q_p = 11,030$ ft³/sec
8-32 $t_p = 5$ min; $Q_p = 312.5$ m³/s
8-34 $T_{wave} = 3.16$ s; $h_{wave} = 1$ m
8-36 $L_m/L_p = 0.0318$
8-38 $V_p = 9$ m/s; $F_p = 467$ kN
8-40 $V_m = 222$ m/s; $M = 0.644$
8-42 $d = 142$ μm
8-44 $p_{windward\ wall} = 1.235$ kPa; $p_{side} = -3.33$ kPa; $F = 6.43$ MN

CHAPTER 9

9-2 $\mu = 5.7 \times 10^{-3}$ lbf-sec/ft²
9-4 (a) $u = 150\ y$; (b) rotational; (c) satisfies continuity; (d) $F = 135$ N
9-6 Greater on wire
9-8 $T = 47.1$ N · m
9-12 $\mu = 1.59$ N · s/m²
9-14 $U_{max} = 1.26 \times 10^{-3}$ ft/s
9-16 $d = 0.012$ in.
9-20 $V_{lower} = 44.4$ cm/s
9-22 $u_{max} = 0.15$ ft/s
9-24 $V_{max} = 35.2$ mm/s downward
9-26 $V_{max} = 0.126$ ft/s
9-28 $dp/ds = -353$ kPa/m

9-30 $x = 1.02$ ft; $\delta = 0.086$ in.; $\tau_0 = 0.0327$ lbf/ft^2
9-32 $F_s = 1.23$ N; $F_{lam}/F_{turb} = 0.099$
9-34 $\delta^* = \delta/2$
9-40 $u = 6.6$ m/s
9-42 $P = 3.34$ kW
9-44 $F_{s,30} = 0.36$ N, $a = 0.44$ m/s^2; $F_{s,\,headwind} = 0.49$ N; $F_{s,\,tailwind} = 0.25$N
9-46 $F_{100} = 1,300$ N; $P_{100} = 36.2$ kW; $F_{200} = 4,919$ N; $P_{200} = 273$ kW
9-48 $F_s = 90,900$ lbf
9-50 $F = 1.24$ MN; $P = 9.59$ MW
9-52 $c_f = 0.00094$; $\tau_{0_{min}} = 188$ N/m^2
9-54 $F_s = 9,537$ lbf; $\delta = 1.49$ ft
9-56 $P = 66.7$ hp

CHAPTER 10

10-2 $p = 78.1$ kPa
10-4 $\Delta p = 78.4$ psi/100 ft
10-6 $p_1 - p_2 = 62.6$ kPa
10-8 $Q = 6.45 \times 10^{-4}$ ft^3/s downward
10-10 Turbulent
10-12 Laminar
10-14 Laminar, $\tau_{center} = 0$; $\tau_{wall} = 49.6$ N/m^2; $dp/ds = 7.34$ kPa/m
10-16 $\nu = 0.00167$ ft^2/sec
10-18 $\tau_0 = 0.120$ psf
10-20 $\Delta p = 570$ Pa
10-22 $f = 0.108$; laminar; $\mu = 0.0812$ N $\cdot$ s/m^2
10-24 $\delta'_N = 0.293$ mm
10-26 $f = 0.016$
10-28 $\Delta p = 1.2$ psf/ft
10-30 $f = 0.020$
10-32 $f = 0.019$
10-34 $h_L = 41.7$ m; $P = 16.4$ kW
10-38 $p = 703.9$ kPa
10-40 $f = 0.040$
10-42 $p_A = 723$ kPa
10-44 $Q = 8.60$ cfs
10-46 $Q = 4.5 \times 10^{-3}$ m^3/s
10-48 $D = 22.5$ in.; use 24-in. commercial size
10-50 $Q = 0.059$ m^3/s
10-52 $P = 26.7$ kW
10-54 (a) $P = 728$ W; (b) $P = 3.03$ kW
10-56 $P = 5.76 \times 10^{-4}$ hp
10-58 $D = 0.104$ m
10-60 $p = -13.3$ psig
10-62 downward; $Q_{avg} \approx 0.038$ cfs
10-64 $t = 140$ sec
10-66 $Q = 0.33$ m^3/s

10-68 $P = 8.52$ MW
10-70 $z_1 - z_2 = 17.2$ ft
10-72 $P = 386$ kW
10-74 $Q_1 = 0.69$ m³/s; $Q_2 = 0.31$ m³/s
10-76 $h_L = 60$ ft; $Q_{12} = 5.15$ cfs; $Q_{14} = 6.19$ cfs; $Q_{16} = 8.64$ cfs
10-78 $z_1 = 104.2$ m
10-80 $Q = 1,640$ gpm
10-82 $P = 23.5$ kW
10-84 $Q = 344$ cfs
10-86 $d = 1.6$ m
10-88 $Q = 310$ m³/s

CHAPTER 11

11-2 $C_D = 1.9$
11-4 $F_D = 341$ kN
11-6 $F_D = 7.35$ kN
11-8 $M = 7.23$ kN · m
11-10 $M = 21.8$ kN · m
11-12 $\Delta P = 3.86$ kW
11-14 $F_D \approx 340$ N
11-16 $P = 450$ W
11-18 $V = 7.32$ m/s
11-20 $F_{D, \text{reduction}} = 653$ N
11-22 $F_{D,100} = 7.77$ kN; $F_{D,200} = 21.8$ kN
11-24 $V_0 = 18.7$ ft/sec
11-26 $V = 12.7$ m/s
11-28 $V = 5.1$ m/s
11-30 $V = 8.9$ mm/s
11-32 $V = 2.3$ m/s upward
11-36 $V = 2.1$ ft/sec
11-38 $V = 0.74$ m/s upward
11-40 $V = 28$ ft/sec upward
11-42 $D = 0.0024$ in.
11-44 6 m/s $< V <$ 18 m/s
11-46 $C_L = 0.25$
11-48 $F_D = 407$ N; $P = 24.4$ kW

CHAPTER 12

12-2 $c = 4,286$ ft/sec
12-4 $\Delta c = 650.5$ m/s
12-6 $c = \sqrt{E_v/\rho}$; $c = 1,483$ m/s
12-8 $V = 719.3$ km/h
12-10 $p_t = 251.8$ kPa
12-12 $V = 312.5$ m/s; $p = 196.8$ kPa; $T = 419.3$ K
12-14 $T = 289.8$ K; $p = 481.6$ kPa; $M = 0.231$; $\dot{m} = 0.038$ kg/s

12-16 $C_p(0) = 1$; $C_p(2) = 2.43$; $C_p(4) = 13.47$; $C_{p,\text{inc}} = 1$
12-18 $M_2 = 0.665$; $p_2 = 200$ kPa; $T_2 = 325$ K; $\Delta s = 34.1$ J/kgK
12-20 $M = 1.39$
12-22 $V_1 = 1,415$ m/s
12-26 $\dot{m}(130 \text{ kPa}) = 0.120$ kg/s; $\dot{m}(350 \text{ kPa}) = 0.386$ kg/s
12-28 (1) $\dot{m} = 0.0733$ kg/s; (2) $\dot{m} = 0.0794$ kg/s
12-30 $A/A_* = 4.23$; $p_t = 55.1$ psia; $T_t = 800°F$
12-32 Overexpanded
12-34 (a) $A/A_* = 2.97$; $T = 2,035$ N; (b) $T = 2,032$ N
12-36 $A_s/A_* = 3.06$; $x_s = 5.59$ cm
12-38 $M = \sqrt{2}$; $A/A_* = 1.123$
12-40 $p_3 = 339$ kPa
12-42 $L = 29.6$ m
12-44 $\dot{m} = 0.149$ kg/s
12-46 $D = 2.96$ cm
12-48 $p_1 = 122$ kPa; $V_1 = 40.8$ m/s; $\rho_1 = 1.48$ kg/m³
12-50 $\Delta p = 36.4$ kPa

CHAPTER 13

13-2 $\Delta h = 6.15$ mm
13-4 $V = 0.29$ m/s
13-6 $V_0 = 18.1$ m/s
13-8 $Q = 0.210$ m³/s; $Q = 7.40$ cfs; $Q = 3,324$ gpm
13-10 $Q = 9,124$ cfm; $V_{max}/V_{mean} = 1.24$; $\dot{m} = 11.1$ lbm/sec
13-12 (a) $r_m/D = 0.189$; (b) $r_c/D = 0.351$; (c) $m = 16.4$ kg/s
13-14 $C_v = 0.975$; $C_c = 0.64$; $C_D = 0.624$
13-16 $p_0 = 530$ psf
13-18 $h = 8.20$ in.
13-20 $p_A - p_B = 224.9$ kPa; $p_D - p_E = 221.9$ kPa; $h = 1.82$ m
13-22 $h = 26.3$ cm
13-24 $Q = 0.67$ cfs
13-26 $d = 62.6$ mm
13-28 $d_0 = 568$ mm
13-30 $Q = 0.290$ m³/s
13-32 $d = 0.798$ m
13-34 $Q = 2.67$ cfs
13-36 $h_L = 64$ $V_0^2/2g$
13-38 $Q = 0.219$ m³/s
13-40 $d = 1.124$ m
13-42 $Q = 20.5$ cfs
13-44 $P = 1.00$ m
13-48 $\dot{m} = 0.0021$ kg/s

CHAPTER 14

14-2 $F_T = 970$ N; $P = 37.4$ kW

14-4 $N = 1,544$ rpm
14-6 $N = 1,462$ rpm
14-8 $a = 0.782$ m/s^2
14-10 $Q = 0.667$ m^3/s
14-12 $Q = 93.8$ cfs; $\Delta h = 31.3$ ft; $P = 417$ hp
14-16 $Q = 0.21$ m^3/s; $P = 6.7$ kW
14-18 $Q = 0.225$ m^3/s
14-20 $H = 229.5$ ft
14-22 $Q = 0.218$ m^3/s
14-24 $Q = 0.0833$ m^3/s; $\Delta h = 146$ m; $P = 104$ kW
14-26 $n_s = 0.357$; safe operating condition
14-28 $n_s = 0.52$; use mixed-flow pump
14-30 $n_s = 0.039$; use radial-flow pump
14-32 $n_s = 0.76$; use axial-flow pump
14-34 $D = 2.07$ m; $P = 27.6$ kW
14-36 $Q = 0.143$ m^3/s
14-38 $P = 10.31$ MW
14-40 $P = (1/4)\, \rho Q V_j^3$
14-42 $\alpha = 11°28'$
14-44 $P_{max} = 2.99$ kW

CHAPTER 15

15-2 Subcritical
15-4 $y_{alt} = 6.58$ m; $E = 6.59$ m
15-6 Supercritical
15-8 0.92 m
15-10 $y_{alt} = 5.38$ m; $y_{seq} = 2.33$ m
15-12 $Q = 187$ cfs
15-14 $Q = 7.93$ m^3/s
15-16 water surface elevation $= 304.55$ ft.
15-18 $\Delta y = -0.76$ m; $\Delta z_{water\ surface} = -0.16$ m; $\Delta y = +0.17$ m;
$\Delta z_{water\ surface} = +0.02$ m
15-20 $\Delta y = +0.227$ m; $B_{min} = 2.06$ m
15-22 $y = 2.27$ m
15-24 $y_2 = 1.24$ m
15-26 $h_L = 37$ ft
15-28 $y_2 = 1.77$ m
15-30 $S3$; $\tau_0 = 132$ N/m^2
15-34 $Q = 19.2$ m^3/s

INDEX

Absolute pressure, 33
Absolute viscosity
  definition of, 17
  of common fluids, 639
Acceleration, 100, 102
  convective, 101
  local, 101
  normal, 101
  pressure variations due to, 143–148
  uniform, of a tank, 144
Advance diameter ratio, 567
Air, properties of, 642
Airfoil, 416, 440–447
  drag, 418, 444
  flow pattern, 94, 466
  lift, 416, 440–447
  pressure distribution on, 417
Alternate depth, 604
Anemometer
  cup, 524
  hot wire, 524
  laser-Doppler, 526
  vane, 523
Angular momentum, 205
Apparent shear stress, 330, 368
Atmosphere, U.S. standard, 39
Atmospheric pressure variation, 39–40
Automobile, pressure distribution on, 448
Axial flow pumps, 571

Bends
  flow in, 382
  forces on, 191–195
Bernoulli equation
  application of, 160
  derivation of, 149–151
  dimensionless form, 156
  limitations of, 151
Body force, 183

Boundary layer
  buffer zone, 335
  description of, 321
  nominal viscous layer, 336
  power law formula, 337
  resistance coefficient, 327, 340, 343
  Reynolds number, 324, 343
  shear stress coefficients in, 327, 340
  shear stress in, 325
  thickness of, 324, 339
  velocity distribution, 325, 333, 337, 339
  viscous sublayer, 329
Bourdon-tube gage, 45
Bulk modulus of elasticity, 21
Buoyancy, 55–58
  center of, 58

Capillary action, 21
Cavitation, 142, 163
Celerity
  of pressure wave, 204, 462
  of surface wave, 611
Center of
  buoyancy, 58
  pressure, 48
Centrifugal pumps, 575–583
Centroids, of areas, 635
Chezy equation, 393
Circulation, 437–442
Coefficient
  contraction, 532
  discharge, 533, 572
  of drag
    airfoil, 418, 444
    circular cylinder, 422
    cone, 431
    cube, 431
    defined, 421
    disk, 430

flat plate 421
projectile, 435
rotating cylinder, 440
rotating sphere, 441
sphere, 430, 435, 436, 441
square rod, 422
flow
orifice, 533
weir, 543
head, 572
head loss
bends, 384
pipes, 376
transitions, 384
lift, 440
power, 568
pressure, 157, 284, 291, 294, 474
resistance, in pipes, 371, 376
thrust, 567
of velocity, 533
Compressible flow
measurements
mass flow rate, 547
pressure, 545
velocity, 546
venturi meter, 548
in pipes
with friction, 494–510
Mach number distribution, 496, 507
pressure distribution, 501, 508
in variable area duct, 481
Compressible flow regimes, 468
Compressors, 579
Conjugate depth, 615
Conservation of mass, 110
Continuity, at a point, 119
Continuity equation, 110–120
Continuum, 5
Contraction coefficient, 532
Control surface, 104
Control volume, 104
Control-volume approach, 103–110
Control-volume equation, 106–109
Convective acceleration, 101
Conversion factor, 631
Couette flow, 315
Critical depth, 606, 608
Critical flow, 605–610
Critical Mach number, 436
Critical pressure ratio, 489
Critical Reynolds number

circular cylinder, 422
in pipes, 92
Cup anemometer, 524
Curved surface, hydrostatic force on, 52–55
Cylinder drag, 422

Deformation of a fluid element, 122
Density, mass, 14
function of Mach number, 472
total, 472
Differential manometer, 44
Dimensional analysis, 274–287
basic method of, 278
limitations of, 283
Dimensionless numbers, 283–287
Dimensionless parameters, 274
Dimensions, 277
Discharge rate, 96–99
Doppler effect, 467
Drag, 314, 416–437
airfoil, 418, 444
circular cylinder, 422
coefficient of, 422, 425, 430, 431
compressibility effects, 435
form, 314
induced, 445
plate, 419
roughness effects, 425
skin friction, 314
sphere, 430, 435, 436, 441
square rod, 430
strut, 430
vehicle, 447
Dynamic pressure, 156n

Eddies, 90
Elasticity, bulk modulus, 21
Electromagnetic flow meter, 538
Energy
internal, 16, 234
kinetic, 235
potential, 235
specific, 603
Energy equation, 234, 251
derivation of, 234, 238
for one-dimensional flow, 251
Energy grade line, 251–255, 385
Enthalpy, 16, 239
total, 470
Equation of state, 15

Equations
  Bernoulli, 149, 151, 156
  Chezy, 393
  continuity, 111–120
  control volume, 108
  Darcy-Weisbach, 371
  energy, 234–251
  Euler, 144
  Manning, 393
  moment-of-momentum, 206
  momentum, 182–187
Euler's equation of motion, 144
Expansion loss
  abrupt, 249
  gradual, 384
Extensive property, defined, 14, 103
External forces, 183

Falling head problem, 115–117
Floating bodies, 58
  stability of, 58–62
Flow
  in abrupt expansion, 248–249
  in boundary layer, 321
  in elbows, 382
  gradually varied, 618–625
  ideal, 92
  at inlets, 381
  irrotational, 123, 151
  laminar, 90
  measurement, 520–551
  in noncircular conduits, 395
  nonuniform, 89
  nonviscous, 151
  nozzle, exit condition, 490
  one-dimensional, 95
  in pipes
    compressible, 494–510
    incompressible, 360–390
  rate of, 96–99, 529–543
  rotational, 121
  steady, 89, 109
  subcritical, 605
  supercritical, 605
  supersonic, 468
  transonic, 468
  turbulent, 90
  two-dimensional, 95
  uniform, 89
  unsteady, 89, 109

Flow classification, 5
Flow coefficient
  orifice, 533
  venturi meter, 534
  weir, 543
Flow nozzle, 538
Flow pattern, 88, 92
  past airfoil, 94, 466
  past circular cylinder, 161
  in converging conduit, 155
  past a disk, 161
  past a plate, 123
  past a square rod, 161
  for uniform flow, 89
  for a vortex, 90
Flow work, 236
Fluid
  as a continuum, 5
  definition of, 4
  Newtonian, 20
  properties, 10–23
  viscosity of, 639, 642, 644
Fluid mechanics, 4, 7
Fluid statics, 28–62
Force
  on bends, 191–195
  body, 183
  buoyant, 55–58
  from deflected jet, 188
  elastic, 285
  external, 183
  gravity, 286
  hydrostatic, 46
  inertial, 285
  in model studies, 289, 294
  on sluice gate, 198
  surface, 183
  surface tension, 286
  viscous, 285
Free-surface model studies, 299
Friction factor, 371–376
Froude number, 284, 291, 299

Gage pressure, 33
Gas constant, 15
Gases, properties of, 641
Gas table
  for subsonic flow, 636
  for supersonic flow, 637
Generator, steam turbine, 239

Grade line
  energy, 251–255
  hydraulic, 251–255

Head coefficient, 451, 572
Head loss, definition, 244
Head loss in pipes
  coefficient, 376
  laminar, 365
  turbulent, 371
Heat flow, 234
Hot-wire anemometer, 524
Hydraulic grade line, 251–255, 385
Hydraulic jump, 613–617
Hydraulic models, 287–303
Hydraulic radius, 391
Hydrogen bubble method, 527
Hydrometer, 57
Hydrostatic force
  on curved surface, 52–55
  on plane surfaces, 46–52
Hydrostatics, 28–62

Ideal fluid, 92
Impulse turbine, 583
Induced drag, 445
Inertia, moment of, 635
Inertial force, 285
Inertial reference frame, 184
Intensive property, definition of, 14, 103
Interferometry, 550
Internal energy, 16, 234
Irrotational flow, 123

Jet deflection, 188
Jump, hydraulic, 613–618

Kinematic viscosity, 18
  of air, 642
  values for common fluids, 640, 643
  values for water, 644
Kinetic energy, 234
  correction factor, 242
Kinetic pressure, 156n, 473
Kutta condition, 442

Laminar flow, 90
  in pipes, 91, 242, 362
  over a plate, 316
  between plates, 314, 318
Laminar losses, in pipes, 365

Lapse rate, 39
Laser-Doppler anemometer, 526
Laval nozzle
  characteristics of, 484–490
  Mach number versus area ratio, 486
  mass flow rate, 487
Law of the wall, 333
Lift, 417, 437–449
Lift coefficient
  airfoil, 440–447
  definition of, 440, 443
  rotating cylinder, 440
  rotating sphere, 441
Liquid, 4
Local acceleration, 101
Losses
  contraction, 384
  entrance, 381
  expansion, 248
  hydraulic jump, 616
  laminar in pipes, 365
  transition, 384
  turbulent, pipes, 371–377
  valve, 384

Mach number, 284, 436, 466, 468
  critical, 436
  downstream of normal shock, 477
Magnus effect, 439
Manning n values, 394
Manometers, 41
  differential, 44
Mass
  kilogram, 12
  pound mass, 13
  slug, 13
Mass density, 14
Measurement
  of flow rate, 529–543
  of pressure 41–46
  of velocity, 152, 153, 520
Measurements in compressible flow, 545–551
Metacenter, 59
Metacentric height, 59
Meter, venturi, 537
Mixing length theory, 331
Model, 274
Model studies, 287–303
  without free surfaces, 291–294

ship, 302
spillway, 299
Moment
    of inertia, 635
    of momentum, 205
Momentum, angular, 205
Momentum equation, 182–187
Moody diagram, 376

Newtonian fluid, definition of, 20
Newton's second law, 143, 182
Nomenclature and dimensions, 631
Nonuniform flow, definition of, 89
Nonviscous flow, 151
Normal acceleration, 101
Normal depth, 393
Normal shock waves, 475
Nozzle flow, exit condition, 489

One-dimensional flow, definition of, 95
Open channels
    critical depth, 606
    hydraulic jump, 613
    specific energy, 603
    uniform flow in, 392
    varied flow in, 600–625
Orifice, 531

Pascal, unit of pressure, 33
Pathline, 94
Penstock, 247
Piezometer, 41
Piezometric head, 37
Pipe
    network, 389
    systems, 386
Pipes
    flow in, 360–390
    in parallel, 389
    resistance coefficient in, 371–377
    roughness in, 372–376
    turbulent flow in, 368
    velocity distribution in, 362, 369, 372, 381
    water hammer in, 199–205
Π-theorem, 277
Pitot tube, 152
Power coefficient, 568
Power law, 337, 369
Prandtl's mixing length theory, 331
Pressure

absolute, 33
center of, 48
definition of, 30
effect of separation on, 161
gage, 33
kinetic, 156n, 473
Mach number, effect of, 471
measurement of, 41–46
total, 472
vacuum, 33
water hammer, 202
Pressure coefficient, 157, 284, 291, 294
    on circular cylinder, 160, 163
    on disk, 163
    in model studies, 284, 291
Pressure distribution
    on airfoil, 417
    on automobile, 448
    on circular cylinder, 159, 163
    near curved boundaries, 155
    on a disk, 163
    on a flat plate, 420
Pressure measurements, 41–46
Pressure transducer, 45
Pressure variations
    due to acceleration, 143–148
    in atmosphere, 39–40
    with elevation, 34–40
    in flowing fluid, 142–165
Projected area, 421
Propeller characteristics, 568
Propeller theory, 564–570
Properties
    of air, 642
    of common gases, 641
    of common liquids, 643
    extensive, 14, 103
    of fluids, 10–23
    intensive, 14, 103
    of water, 644
Prototype, 274
Pump, 244, 387
Pumps
    axial flow, 571
    centrifugal, 575–583
    radial flow, 575

Radial flow pumps, 575
Reaction turbine, 587–590
Rectangular weir, 541–544
Relative roughness, 373, 377

Resistance coefficient, in pipes, 371–377
Reynolds number, 285, 291
  boundary layer, 324
  critical, circular cylinder, 425
  critical, in channels, 316
  critical, in pipes, 365
  definition of, 92
Reynolds stress, 330
Rocket motion, 195
Rotameter, 540
Rotation
  of a fluid element, 121
  of a tank of liquid, 147

Scale models, 287
Schlieren technique, 550
Separation, 123, 162
  effect on pressure, 163
Separation point, 162
Sequent depth, 615
Shaft work, 237, 243
Shasta Dam, 29
Shear stress, 17
  apparent, 330
  in boundary layers, 325
  coefficients in boundary layers, 340
  in a pipe, 360
  turbulent, in pipes, 368
Shear velocity, 329
Ship model tests, 302
Shock waves, 466, 475–480, 490
  existence of, 478
  visualization of, 549
Similitude, 287–303
  dynamic, 288–291
  geometric, 288
  at high Reynolds numbers, 296
Sluice gate, force on, 198
Sound, speed of, 462
Specific energy, 603
Specific gravity, 15
Specific heat, 16
Specific heat ratio, 465
Specific speed, 580
Specific weight, 14
Sphere drag, 430
Spillway models, 299
Stability of floating bodies, 58–62
Stagnation point, 152
Stagnation tube, 152, 520
Standard atmosphere, U.S., 39

Static tube, 521
Steady flow, definition of, 89
Stokes' law, 430
Stratosphere, 39
Streakline, 94
Streamline, definition of, 88
Streamlining, 428
Strouhal number, 427
Subcritical flow, 605
Subsonic flow gas table, 636
Supercritical flow, 605
Supersonic flow, 468
Supersonic flow gas table, 637
Supersonic wind tunnel, 486–489
Surface force, 183
Surface profiles, 619–625
Surface tension, 21, 286
Surge, 617
System, 13, 103

Tangential acceleration, 101
Temperature
  Centigrade, 12
  Fahrenheit, 13
  function of Mach number, 470
  Kelvin, 12
  Rankine, 13
  static, 471
  total, 470
Terminal velocity, 432
Thermodynamics, first law, 234
Thrust coefficient, 567
Total enthalpy, 470
Total head, 252
Transitions, in channels, 610
Transonic flow, 468
Triangular weir, 544
Troposphere, 39
Truncated nozzle, mass flow in, 492
Turbines
  gas, 591
  impulse, 583
  reaction, 587–590
  steam, 239
  wind, 591
Turbulence, in pipes, 368
Turbulent flow, 90
  in pipes, 368
Two-dimensional flow, 95, 418

Ultrasonic flow meter, 539
Uniform acceleration of a tank, 144

Uniform flow, definition of, 89
U.S. standard atmosphere, 39
Units
    English system, 13
    Système Internationale (SI), 12
Unsteady flow, 89, 109

Vane
    moving, 190
    stationary, 188
Vane anemometer, 523
Vapor bubbles, 164
Vapor pressure, 23
Varied flow, gradual, 618–625
Velocity
    average mean, 91, 98
    from the Eulerian viewpoint, 87
    gradient, 17
    from the Lagrangian viewpoint, 86
    measurement, 152, 153, 520–527
    shear, 329
    along a streamline, 87
    terminal, 432
Velocity coefficient, 533
Velocity defect law, 334
Velocity diagram for turbine vanes, 589
Velocity distribution
    in boundary layer, 325, 333, 337
    in pipes, 362, 369, 372, 381
    between plane surfaces, 318
Velocity gradient, 17
Velocity head, 156n
Vena contracta, 532
Venturi meter, 537
    flow coefficient, 534

Viscosity, 16–20
    absolute
        of air, 642
        of common fluids, 639
        of water, 644
    kinematic
        of air, 642
        of common fluids, 640
        of water, 644
Viscous effects, 521
    in turbines, 593
Viscous force, 285
Viscous sublayer, 329
Vortex flowmeter, 540
Vortex, tip, 444
Vortex shedding, 426
Vorticity, 123

Water, properties of, 644
Water hammer, 199–205
Wave, pressure, 199, 462
Wave celerity, 462, 611
Weber number, 284
Weir
    broad-crested, 608
    end contractions, 543
    rectangular, 541–544
    triangular, 544
Wind force example, 298
Wind turbine, 591
Work
    flow, 236
    shaft, 237, 243

Yaw meter, 522

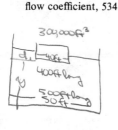

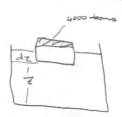

$$Z + dz - (y + d_1) \quad (h \text{ difference})$$

$$d_1 = \frac{300,000}{40} = 18.75 \text{ft} \qquad d_2 = 18.75 + \frac{(4000)(2000)}{40(400)62.4} = 26.76'$$

$$V = (50 \times 450 \times y) + [(50 \times 450) - (40 \times 400)] d_1$$

$$\forall = (50 \times 50 \times z) + [(50 \times 450) - (40 \times 400)] d_2 \qquad Z - y = -2.314 \text{ft}$$

$$h = 5.7' = -2.314 + 26.76 - 18.75$$

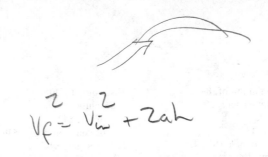

$$V_f^2 = V_{in}^2 + 2ah$$

$$P = \gamma Q h_T \, \eta$$

$$P = \frac{\gamma Q h_P}{\eta}$$

rate of change of
momentum w/in the control
volume.

Momentum

$$\Sigma F_x = \Sigma V_x \rho (\vec{V} \cdot \vec{A}) + \frac{d}{dt} \int_{cv} V \rho \, d\forall$$

Pressure Variation

$$\frac{\partial}{\partial l} (P + \gamma z) = -\rho a_l$$

Rotation

$$\frac{P}{\gamma} + z - \frac{V^2}{2g} = C$$

$$V = \omega r$$

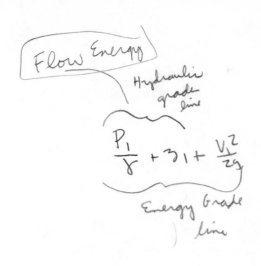

Flow Energy

Hydraulic grade line

$$\frac{P_1}{\gamma} + z_1 + \frac{V_1^2}{2g}$$

Energy Grade line

for rotation probs. the volume is constant

$$Q = V A$$

Bernuli

$$\frac{P_1}{\gamma} + z_1 + \frac{V_1^2}{2g} = \frac{P_2}{\gamma} + z_2 + \frac{V_2^2}{2g}$$

Stagnation tube

$$V = \sqrt{2gh}$$

p 152-3

Pitot tube

$$V = \sqrt{2g(h_1 - h_2)}$$

$$V_2^2 = \frac{2g}{\gamma}(P_1 - P_2) \qquad V_2 = \sqrt{\frac{2\Delta P}{\rho}}$$

$$h_1 - h_2 = \left(\frac{P_1}{\gamma} + z_1\right)\left(\frac{P_2}{\gamma} + z_2\right)$$

$$2gL + V_{in}^2 = V_f^2$$

$$\boxed{\sqrt{2gL + V_{out}^2} = \overline{V}}$$

LAMINAR
$$\Downarrow$$
$$h_f = \frac{32 \mu L V}{\gamma D^2}$$

$$V = \sqrt{2gL}$$

∗ Vel as drops

14.38
$$\Downarrow$$
$$V_B = \frac{1}{2} V_j$$

$$P = \frac{Q \gamma V_j^2 \eta}{2g}$$

$$\Sigma F = \rho Q (V_{out} - V_{inlet})$$

$$\alpha = 2 \Rightarrow \text{LAMINAR}$$
$$= 1 \Rightarrow \text{TURBULENT}$$

$$n_s = \frac{n \sqrt{Q}}{g^{3/4} h^{3/4}}$$

$$Re < 2100 \quad \text{LAMINAR}$$
$$Re > 2100 \quad \text{TURBULENT}$$

$$\Delta H = \text{pressure rise}$$

$$n_s < .23 \Rightarrow \text{radial centrifugal pump}$$

$$n_s > .60 \Rightarrow \text{axial pump}$$